Springer Proceedings in Physics

Volume 288

Indexed by Scopus

The series Springer Proceedings in Physics, founded in 1984, is devoted to timely reports of state-of-the-art developments in physics and related sciences. Typically based on material presented at conferences, workshops and similar scientific meetings, volumes published in this series will constitute a comprehensive up-to-date source of reference on a field or subfield of relevance in contemporary physics. Proposals must include the following:

- Name, place and date of the scientific meeting
- A link to the committees (local organization, international advisors etc.)
- Scientific description of the meeting
- List of invited/plenary speakers
- An estimate of the planned proceedings book parameters (number of pages/articles, requested number of bulk copies, submission deadline).

Please contact:

For Americas and Europe: Dr. Zachary Evenson; zachary.evenson@springer.com
For Asia, Australia and New Zealand: Dr. Loyola DSilva; loyola.dsilva@springer.com

Yiqian Wang · Yisheng Gao · Chaoqun Liu
Editors

Liutex and Third Generation of Vortex Identification

Workshop from Aerospace and Aeronautics World Forum 2021

Springer

Editors
Yiqian Wang
Soochow University
Suzhou, China

Yisheng Gao
Nanjing University of Aeronautics
and Astronautics
Nanjing, China

Chaoqun Liu
The University of Texas at Arlington
Arlington, TX, USA

ISSN 0930-8989 ISSN 1867-4941 (electronic)
Springer Proceedings in Physics
ISBN 978-981-19-8954-4 ISBN 978-981-19-8955-1 (eBook)
https://doi.org/10.1007/978-981-19-8955-1

This Springer imprint is published by the registered company Springer Nature Singapore Pte Ltd.
The registered company address is: 152 Beach Road, #21-01/04 Gateway East, Singapore 189721, Singapore

Preface

This volume is a collection of papers presented in the invited workshop "Liutex and Third Generation of Vortex Identification for Engineering Applications" from "Aerospace and Aeronautics World Forum 2021", Frankfurt, Germany, December 02–04, 2021 (although held online as a virtual conference). The invited workshop has 43 registered participants and 41 speakers, which represents the community effort to promote the recent developments of Liutex and third generation of vortex identification method in both theoretical side and engineering applications. Dr. Yiqian Wang and Dr. Yisheng Gao are the co-chairs of the invited workshop and Professor Chaoqun Liu is the plenary speaker of the invited workshop.

Liutex was proposed by Prof. Chaoqun Liu at the University of Texas at Arlington (UTA) in 2018 as a new physical quantity to define the local fluid rotation or vortex mathematically and systematically. Now it is established as the core of third generation of vortex identification. Professor Liu began to challenge the traditional vortex identification methods since 2014: faked "vortex breakdown" phenomenon could be observed by inappropriate value of λ_2, which led to totally different vortical structure (Liu et al., Computers and Fluids 102, 2014). This so-called threshold issue has been generally acknowledged as a major drawback of most of Eulerian local vortex identification criteria. Meanwhile, several serious problems of vorticity-based methods for vortex identification have been recognized. For example, the vorticity tube does not present the Λ-vortex at all, because the direction of vorticity lines is not always aligned with the direction of vortical structures in turbulent wall-bounded flows, especially at locations close to the wall. And vortical structures can be located in the region where the magnitude of vorticity is relatively smaller (Wang et al., Communications in Computational Physics 22(2), 2017). To alleviate the threshold issue associated with the existing Eulerian local vortex identification criteria, a new vortex identification criterion called Omega method was proposed by Prof. Liu in 2016 (Liu et al., Sci. China Phys., Mech. Astron. 59:684711, 2016), in which relative strength was used to identify the "rigidity" of the fluid motion. As a significant development before the introduction of Liutex, the Omega method is insensitive to the moderate threshold change and can capture both strong vortices and weak vortices simultaneously, and therefore it has been successfully applied in numerous engineering problems. Also, an idea of vorticity decomposition has come into being in this paper: vorticity should be decomposed into two parts, vortical vorticity contributed to rotation and non-vortical vorticity like the vorticity without rotation in laminar boundary layer flows. However, "what is vortical vorticity contributed to rotation" remains unclear.

The concept of Liutex originated in the second half of 2017 and early 2018. In the second half of 2017, the UTA team had begun thinking about a mathematical definition of vortex which represents the rigid rotation from fluid motion, especially vector form. The first attempt was the vortex vector proposed by Dr. Shuling Tian, which was obtained by a special matrix representation of velocity gradient tensor with double zero elements (the third ones in row 1 and row 2) using coordinate rotation (Tian et al., Journal of Fluid Mechanics 849, 2018). However, the implementation involved the cumbersome numerical solution of system of nonlinear equations, which was quite time-consuming. Even worse, the strict proof of the existence of such a special matrix was not yet available then. Consequently, the next priority was to prove the existence and improve the computational efficiency. Soon after, Dr. Yisheng Gao found that the real Schur decomposition could act as a foundation for both the proof of existence and an efficient implementation. In December 2017, Prof. Liu first presented the idea to extract the rigid rotation from fluid motion and the above work in a conference called "New Development and Key Issues of Vortex and Turbulence Research" held at the University of Shanghai for Science and Technology (USST), China. During the discussion, several professors suggested to provide a new name of the definition. Professor Liu accepted one suggestion to name the new definition of vortex as "Rortex". And the first public code to calculate Rortex, written by Dr. Gao, was available soon after the conference. The above-mentioned work is published on Physics of Fluids (Liu et al., Physics of Fluids 30(3), 2018), which officially marks the birth of Rortex. Rortex gives a definite answer to the question "what is vortical vorticity contributed to rotation" and accordingly provides a reasonable method for vorticity decomposition. In February 2018, Prof. Liu further proposed the definition of Rortex based on eigenvector (Gao et al., Physics of Fluids 30(8), 2018). Because the eigenvector-based definition is clear and the calculation is much faster, the real Schur decomposition is totally avoided in practice since then. So far, the early phase of the new mathematical definition of vortex came to an end. The complete work in this early phase was reported in a conference on vortex and turbulence research held in June 2018 at Zhejiang University, China. Since almost all the original ideas in this series of work were contributed by Prof. Chaoqun Liu and it was helpful to avoid unnecessary misunderstandings and arguments, in late 2018 the UTA team and many collaborators decided to rename the new vortex definition as "Liutex".

The following year witnessed the remarkable development of Liutex. First, Dr. Yiqian Wang made an important contribution to introduce the explicit formula of the Liutex vector in terms of vorticity, eigenvalues, and eigenvectors of the velocity gradient tensor in the original coordinate system (Wang et al., Journal of Hydrodynamics 31(3), 2019). The explicit formula eliminates the clumsy coordinate rotation, dramatically improving the efficiency. And it also provides physically intuitive comprehension of the Liutex vector. At the same time, the Omega method and the concept of Liutex prompted Professor Liandi Zhou, the Executive Chief Editor of Journal of Hydrodynamics, to suggest six core elements of vortex identification: (1) absolute strength, (2) relative strength, (3) local rotational axis, (4) vortex rotation axis, (5) vortex core size, and (6) vortex boundary (Liu et al., Journal of Hydrodynamics 31(2), 2019). Six core elements advocate the Liutex system and hence the Liutex system is classified as the third generation of vortex identification methods. Only the Liutex system can provide all information about six core elements while the vorticity-based first-generation and Eulerian-local-criterion-based second-generation methods all fail except for a rough estimate of the vortex boundaries by the second-generation methods. Furthermore, since Liutex has consistent and systemic definitions of scalar, vector, and tensor forms, a new velocity gradient tensor decomposition is introduced accordingly (Gao et al., Physics of Fluids 31, 2019). Another crucial breakthrough is Liutex core line by Prof. Liu (Gao et al., Journal of Hydrodynamics 31(3), 2019). Rather than the traditional representation by iso-surfaces, Liutex core line provides a unique and threshold-free line-type representation of vortical structures. And a distinct Liutex similarity law was discovered in low and medium Reynolds number turbulent boundary layers, which may promote the deeper understanding toward the physics of turbulence and promisingly lead to more reliable turbulence modeling (Xu et al., Journal of Hydrodynamics 31(6), 2019). In addition, Liutex and third generation of vortex identification were well furnished with modified Liutex-Omega method, objective Liutex method, and asymmetric Liutex theory for turbulence generation in this year. And as a main team member, Dr. Yisheng Gao develops the codes for Liutex-based methods (with some contributions from Dr. Xiangrui Dong and Dr. Jianming Liu).

The serious outbreak of the COVID-19 pandemic cannot impede the blossoming of Liutex and third generation of vortex identification. A so-called Principal Coordinate based on the velocity gradient tensor is defined. A principal decomposition, or UTA R-NR decomposition, is performed in the Principal Coordinate with a clear physical meaning, which represents the rigid rotation, stretching (compression), and shear (symmetric and antisymmetric shear). The Liutex-based UTA R-NR tensor decomposition is unique and Galilean invariant. Furthermore, a new version of principal decomposition performed in the original coordinate system is derived by Prof. Liu to replace the traditional Cauchy-Stokes decomposition, which creates the new fluid kinematics and serves as a pathway to new fluid dynamics, new vortex science, and new turbulence research. And the introduction of Liutex core tube implies a great contribution to visualize and identify vortical structures. Liutex core tube can be used to answer one of six core elements, namely, the vortex core size, and present reasonable structures than iso-surfaces. More importantly, based on the Liutex system, new governing equations for fluid dynamics, which use the vorticity tensor only, have been proposed by Prof. Liu and verified against the traditional Navier-Stokes equations. The new governing equations are identical to the N-S equations in mathematics, but have lower cost and several advantages, which may be useful for research on turbulence.

Since the introduction of Liutex in 2018, hundreds of papers on Liutex and third generation of vortex identification have been published by the community. And three professional books on Liutex and third generation of vortex identification have been published by Bentham, Elsevier, and Springer, respectively. Liutex and third generation of vortex identification have been applied by countless scientists and engineers to identify vortex; visualize vortical structures; study new physics; and, more importantly, uncover the turbulence mysterious veil. As more and more scientists apply Liutex and the third generation of vortex methods, people will reveal more and more secrets of turbulence.

This book is concerned with the recent advance of the theoretical foundation of Liutex and a variety of complex engineering applications of Liutex and third generation of vortex identification, including powertrain, energy engineering, marine engineering, aeroengine, rocket engine, fluid machinery, etc. The book has 27 chapters. Part I consists of Chaps. 1–11, which provides the theoretical foundation and applications for turbulence and transition research. Part II consists of Chaps. 12–27, which presents Liutex and third generation of vortex identification methods in engineering applications. These chapters are briefly described as follows:

Chapter 1 is entitled "Liutex and Third Generation of Vortex Definition and Identification" written by Chaoqun Liu*, Habib Ahmari, Charles Nottage, Yifei Yu, Oscar Alvarez, Vishwa Patel. Liutex is a physical quantity like velocity, vorticity, pressure, temperature, etc., describing local fluid rotation or vortex, which was ignored for centuries. Liutex was defined by the University of Arlington at Texas (UTA) team in 2018 as a vector for vortex. Its direction is the local rotation axis, and magnitude is twice local angular rotation speed. As the third generation of vortex definition and identification, Liutex has been widely applied for visualization of vortex structure to replace the first generation or vorticity, which cannot distinguish shear from rotation and the second generation such as Q, Δ, λ_2, and λ_{ci} methods, which are all scalar without rotation axis, dependent on threshold and contaminated by shear and stretching. Several new vortex identification methods have been developed, especially the Modified Liutex-Omega method, which is threshold insensitive, and the Liutex-Core-Line method, which is unique and threshold-free. According to the Liutex vector, a unique coordinate system called Principal Coordinate can be set up, and consequent Principal Decomposition of velocity gradient tensor can be made. Different from classical fluid kinematics, the Liutex-based new fluid kinematics decomposes the fluid motion to a rotational and non-rotational part (UTA R-NR decomposition). The non-rotational part can be further decomposed to stretching and shear consisting of symmetric shear and antisymmetric shear in contrast with the classical fluid kinematics, which decomposes fluid motion to deformation and vorticity, which was misunderstood as rotation. According to the constitutive relation between stress and strain, the new fluid kinematics may give significant influence to new fluid dynamics.

Chapter 2 is entitled "Galilean Variance of Streamline in Vortex/Liutex Visualization" written by Yifei Yu* and Chaoqun Liu. Intuitively, vortex exists at the position where the streamlines bend. This perception comes from the fact that people can only observe water flow, which is the streamline, by eyes. However, since streamline is a Galilean invariant quantity, this intuition may be incorrect in some situations.

Galilean variance is the feature that the issue will be different under different coordinate systems, while Galilean invariance refers to keeping the same under different coordinate systems. No bending of the streamline can be identified when the streamwise speed is far larger than the speed produced by rotation; in other words, the speed produced by rotation can be omitted. In such a situation, streamline fails to detect vortex. Therefore, a proper vortex visualization method requires Galilean invariance. Liutex is a newly invented vortex identification method that is Galilean invariant, and as a result, it can capture vortex correctly regardless of the choice of coordinate systems.

Chapter 3 is entitled "Rules of Tensor and Matrix Operation for Liutex Calculation" written by Yifei Yu*, Yinlin Dong, and Chaoqun Liu. Vectors and tensors, which are widely used in physics, have their clear definitions and rules of calculations. However, in the practical applications and the process of derivations, vectors and tensors are usually expressed in the format of matrices which do not have as clear as their original mathematical definitions. The first problem is that there is no agreement on whether to use a column or row matrix to represent a vector which also leads to the problem of how to express a second-order tensor in a matrix format. In a word, the corresponding relations between vectors, tensors, and matrices are not very clear. Another problem is how to express vectors/tensors operation in matrices operations. It is clear that the first column matrix needs to be transposed before doing matrices multiplication when it is used to represent inner products of vectors. However, when it comes to the operation that a tensor dot product is a vector, some people do not transpose the first matrix (representing the second-order tensor). Liutex is a vortex identification concept that relates many vector and tensor operations. Therefore, it is important to have clear rules for using matrices to represent vectors and tensors. To overcome these problems and keep consistency, this chapter provides rules of tensor and matrix operation for Liutex calculation. First, column matrices are used to represent vectors. Second, the matrix forms of vector/tensor operations are provided.

Chapter 4 is entitled "Liutex Core Tube for Vortex Visualization and Structure" written by Oscar Alvarez, Charles Nottage, and Chaoqun Liu*. Vortex structure visualization methods have been hampered by the preceding vortex identification methods which derived them. Popular vortex identification schemes like the Q-criterion and the λ_{ci}-Criterion only produce scalar quantities to express the vortex structure, hindering our ability to expand on these methods. Liutex, however, produces a vector quantity known as the Liutex vector. This Liutex vector provides more information about the vortex structure and can be processed to generate rich and informative vortex visualizations. With the Liutex vector, we can create the Liutex Core Line, which accurately represents the vortex structure. From the Liutex Core Line, we can generate the Liutex Core Tube. The Liutex Core Tube has many advantages over an iso-surface for vortex structure analysis. Information can be conveyed intuitively and clearly to the user using the Liutex Core Tube to visualize a vortex structure. Thus, using the Liutex Core Tube can allow for optimal analysis of any vortex structure.

Chapter 5 is entitled "Study of Vortex and Vorticity in a Laminar Flow" written by Aayush Bhattarai*, Oscar Alvarez, Shah Md. Imran Kabir, Qazu Asguqye E. Mowla, and Yifei Yu. Vortices are ubiquitous in nature. From kitchen sinks to galaxies, they can be found everywhere. Usually, the swirling motion of fluids comes to mind when one thinks of vortices. It has been extensively studied for centuries, being significantly important in various fields such as engineering, physics, chemistry, and aerospace. Still, we do not have an unambiguous and universally accepted definition of a vortex. Often vorticity is used to describe the vortex, which is accurate for rigid body rotation; however, this explanation is simply not true for the fluid flow in the boundary layer. For fluid rotation, pure shear deformation needs to be considered. In order to demonstrate that, we recreated Shapiro's experiment where he used the rigid body (vorticity meter) rotation to show that the vorticity is the same thing as the vortex. Additionally, we used dyed ink to investigate if the same results still hold in fluid rotation as they did for the vorticity meter.

Chapter 6 is entitled "POD Analysis on Losing Symmetry of Vortex Structure in the Flow Transition by Liutex Method" written by Pushpa Shrestha, Charles Nottage, and Chaoqun Liu*. The proper orthogonal decomposition (POD) is a data decomposition method to investigate and analyze complex and turbulent flow. The POD method offers the optimal lower dimensional approximation of a given dataset. In our case, POD is used with Liutex vector as an input instead of velocity vector to extract the coherent structure of late boundary layer flow transition. Mathematically, the Liutex vector field is decomposed into a sum of basis functions (spatial modes) multiplied with time coefficients (Fourier-splitting method). A singular value decomposition (SVD) algorithm is used as the POD method. Our studies show that fluid motion can be modeled/reconstructed by a few early modes as they contain a large portion of total rotational intensity. Later modes can be neglected as they do not contribute to the total rotational strength of the flow. From the reconstructed vortex structure in the flow transition, the loss of asymmetry of vortex is investigated, and it is found that the asymmetry of vortex develops at the flow transition. In fact, our study shows that the antisymmetric of the vortex starts from the middle, then the antisymmetric structure of the bottom part starts and spreads to the top level.

Chapter 7 is entitled "The Liutex Shear Interaction in Boundary Layer Transition" written by Biyu Pang, Yuan Ding, and Yiqian Wang*. The third generation of vortex identification methods based on Liutex vector is superior to previous methods in that they overcome the drawbacks of previous methods including threshold problem, shear contamination, etc. with a clear physical meaning for the Liutex vector. The direction of Liutex represents the local axis of rotation, and its magnitude is equal to twice the angular velocity of rotation. The current study focuses on the interaction between Liutex represented rotation and the residual shear part during the development of Λ vortex and hairpin vortex in boundary layer transition. The results show that shear plays an important role in the generation and dissipation of vortices and the proportion of Liutex in the whole vorticity affects the stability of a vortex. When the directions between Liutex and shear are approximately parallel, the vortex moves mainly along the flow direction and the offsets in other directions are relatively small. It is also shown that the Liutex vector can accurately extract the rigid rotation part from fluid motion and the third-generation vortex identification methods can serve as a powerful tool to study fluid dynamics.

Chapter 8 is entitled "Liutex Shear Interaction in Turbulent Channel Flow" written by Yuan Ding, Biyu Pang, and Yiqian Wang*. Vortex structures are ubiquitously present in aerodynamics and hydrodynamics. To study vortex structure, it is obligatory and of great significance to systematically define and visualize vortices. Many methods have been introduced including Q, λ_2, Δ, λ_{ci}, Ω, and Liutex vector-based methods. In this chapter, we explore the interaction between Liutex and shear on the basis of the Liutex-shear (R-S) decomposition in a turbulent channel flow with Reynolds number of 180. The results suggest that shearing may strengthen or weaken the magnitude of Liutex depending on their inclined angle. It is inferred that the interaction between Liutex and shear plays an important role in turbulence regeneration cycle.

Chapter 9 is entitled "Experimental Studies on the Evolution of Hairpin Vortex Package in the Boundary Layer of a Square Tube" written by Yuyan Li, Xiangrui Dong, Xiaoshu Cai*, Wu Zhou, and Xinran Tang. The hairpin vortex structures in the boundary layer with $Re_\theta = 159$–239 of a square tube were studied experimentally using the moving single-frame and long-exposure image (M-SFLE) method. And the Liutex vortex identification criteria are used to confirm vortices and characterize their strength in the experimental results. The flow measurement system always moves at the same or similar speed as the vortex structures in the boundary layer to track the vortex structure continuously. The experimental results show that the secondary hairpin vortex and the tertiary hairpin vortex can be induced by the primary hairpin vortex. And the extension of the vortex packet in the flow direction can lead to the occurrence of vortex merging in the near-wall turbulent boundary layer. The Q2 event plays a key role in the formation of secondary hairpin vortices and the merging behaviors. The merging process of two vortices can be divided into three stages and the variation of vortices' intensity has unique characteristics in different stages.

Chapter 10 is entitled "The Correlation between Pressure Fluctuation and Liutex Spectrum in Boundary Layer Transition" written by Vishwa Patel, Yonghua Yan, Yifei Yu, and Chaoqun Liu*. The extensive decades of research have remained unexplained about the primary source of flow frequency unsteadiness in the Boundary Layer Transition. The study aims to determine the relation between Boundary Layer low frequencies and the Liutex using the high-order LES on a large grid system. The newly developed third-generation vortex identification method—Rortex/Liutex—was adopted to investigate the motion of complex vortex structures in the boundary layer transition. The power spectrum analysis clearly shows that the pressure fluctuation (low-frequency noises) correlates highly with the Liutex frequencies.

Chapter 11 is entitled "Statistical Analysis for Liutex Growth in Flow Transition" written by Charles Nottage*, Yifei Yu, and Chaoqun Liu. In computational fluid dynamics, many researchers and textbooks accepted that vorticity is vortex. However, this is a misunderstanding of the tensors derived from the Cauchy-Stokes decomposition of the velocity gradient tensor. It was believed that the symmetric tensor A and antisymmetric tensor B (vorticity tensor) represented stretching/compression and rotation, respectively. Decomposing the vorticity tensor yields R (rotation part) and S (antisymmetric shear deformation part). Liutex, on the other hand, represents rigid rotation and the Liutex magnitude represents twice the angular speed. We analyze three flow areas in boundary layer transition: laminar, transitional, and turbulent. In laminar flow, there is no vortex structure. In transitional flow, the formation of hairpin vortex rings will begin. Finally, in turbulent flow, many vortex rings have formed. In this chapter, a DNS simulation of boundary transition is conducted, then statistical analysis is performed on the recorded results for Liutex, shear, and vorticity. The resulting values for Liutex followed the proper growth trend, starting at zero in laminar flow and steadily increasing through the transitional and turbulent flows. On the other hand, the vorticity values were much greater and remained consistent with little change throughout the flow transition periods. The analysis also revealed that the shear component negatively relates with Liutex, i.e., as Liutex increases, shear decreases. Since shear substantially impacts the vorticity value where it can be misrepresented as rotation in laminar flow, vorticity, in general, should not be considered vortex.

Chapter 12 is entitled "Three-Dimensional Vortex Structure Identification of Fluid Coupling and Analysis of Spatial-Temporal Evolution Mechanism" written by Bosen Chai*, Dong Yan, Jin Zhang, Wenjie Zuo, and Guangyi Wang. The interior of fluid coupling is full of turbulent and multi-scale vortex flow. The generation, development, and interaction of the multi-scale vortex structure dominate the overall internal flow. Accurate identification of unsteady multi-scale vortex structure characteristics is extremely important for revealing the law of flow and the mechanism of energy conversion and loss mechanism. The flow field of fluid coupling under braking condition is numerically simulated based on the advanced turbulence model. The temporal and spatial evolution characteristics of the three-dimensional vortex structure inside the pump wheel and turbine are extracted based on different vortex recognition methods, and the results are compared with the flow field visualization test. The accuracy and applicability of the extraction results for different vortex recognition methods are analyzed. The results show that: the threshold selection of Q method extraction vortex structure is blind. It is difficult to capture both strong and weak vortices at the same time. Ω method can accurately capture the weak vortex structure within a large threshold range. Ω-Liutex method is not sensitive to threshold selection, and it is the most effective in extracting vortex structure. From the perspective of the temporal and spatial evolution of the three-dimensional vortex structure, Ω-Liutex method can reveal the mechanism of energy conversion and loss best.

Chapter 13 is entitled "Numerical Simulation and Analysis of Two-Phase Flow Around Cylinder Using Pseudo-Potential Model and Liutex Method" written by Zeliang Chen, Pengxin Chen, Xiaoli Huang, Xiaoxi Zhang, Nan Gui*, Xingtuan Yang, Jiyuan Tu, and Shengyao Jiang. Two-phase systems have a wide range of applications in nuclear energy, chemical industry, petroleum, refrigeration, and other industrial processes. In this chapter, the lattice Boltzmann method that is apt for multiphase flow simulation is adopted, and the third-generation vortex identification method Liutex that is able to distinctly identify rotational vortex is utilized to analyze the vortex field. The two-phase cross flow around columns or tube bundles is widely used in industrial equipment with heat exchange. Based on the practical engineering background, this chapter presents a comparative numerical analysis on the two-phase flow around single cylinder under different Reynolds numbers and investigates the evolution of vortex field. The 2D and 3D numerical simulation results have shown the flow field, the vortex shedding pattern, as well as the drag and lift of each column. In particular, the conditions of $Re = 30$ and $Re = 120$ in 2D are calculated, respectively, corresponding to the conditions of wake vortex stability and laminar vortex street. The drag and lift forces are mainly affected by continuous phase.

Chapter 14 is entitled "Analysis of Vortices Shed by Generic Submarines Based on Liutex" written by Fenglai Huang, Liushuai Cao, and Decheng Wan*. Submarines in towed conditions shed a wealth of vortical structures from the main body and appendages, such as casing, fins, rudders, and diving planes. The interaction of these coherent structures is detrimental to its hydrodynamic performance and is a source of vibrations and emission of noise, having a negative impact on its stealth capabilities. Studies on flow control and vortex manipulation are of great importance for submarines. However, most of the current vortex identifications based on Eulerian criteria, including the Q-criterion and λ_2-criterion, fail to identify the swirl axis or orientation. To address the problem, we analyzed vortices shed by generic submarines by using an alternative eigenvector-based definition of Liutex. The platform under study is the DARPA SUBOFF submarine model and the Joubert BB2 submarine, which are now international benchmarks for underwater vehicles. This chapter is aimed at analyzing vortices shed by a submarine by using the Liutex and the traditional Q-criterion.

Chapter 15 is entitled "Vortex Identification Methods Applied to Complex Viscous Flow Field of Ship in Restricted Waters" written by Hongjie Cao, Chao Guo, and Decheng Wan*. Vortex plays a vital role in the generation and maintenance of turbulence. The flow mechanism around the ship and marine structures can be deeply analyzed through the vortex structure, which is of great significance to the study of turbulence problems. The performance of ships in restricted waters is much different from ships in open waters. The issue of the ship entering a lock is one of the most typical situations of the restricted waters. In this chapter, different generations of vortex identification methods are applied to complex viscous flow field of an 8,000t bulk carrier entering a lock. The turbulent flows are modeled by Reynolds-averaged Navier-Stokes (RANS) simulations based on the two-equation k-ω shear stress transport (SST) model. With the help of the overset grid method, the surface pressure of the ship, wave height of the free surface, and the wave field are analyzed numerically to explain the ship hydrodynamic performance. Then, based on Liutex vortex identification, the vortex structures in the viscous flow field are captured. Compared with the traditional vortex identification methods, the third-generation methods show a better ability of capturing the vortex structures and give a more accurate definition of vortex, which can be used in the study focused on the flow mechanism, especially on the problem of the ship-lock interaction. All the analysis aspects of the flow field in this work above provide a reference for the post-processing method of the complex viscous flow field of ship in restricted waters.

Chapter 16 is entitled "The Applicability of Third Vortex Identification Methods on Atmospheric Boundary Layer and Turbine Wakes" written by Shun Xu, Weiwen Zhao, and Decheng Wan*. To better understand the interaction between the complex atmospheric boundary layer (ABL) inflow and wind turbine, the third vortex identification methods including the Liutex and OmegaR methods are implemented to identity and visualize the vortex structures, and the results are compared with the Q-criterion method. Two numerical simulation stages are presented, including the precursor stage for the generation of ABL inflow and the successor stage for wind turbine subjected to it. The turbine blades are modeled based on the actuator line model (ALM), and large eddy simulations (LES) are performed for the two stages. Compared with the Q-criterion method, the Liutex method can better identify vortices in ABL inflow and turbine wakes. However, the vortices in turbine wakes are not clearly visualized by OmegaR method due to the ability of capturing weak vortices. But the small-scale vortices in turbine wakes and the large-scale vortices in the ambient atmospheric are evident based on OmegaR method. Generation and fragmentation of blade tip vortices, the interaction of vortices induced by wind turbine with the ambient atmospheric vortices, and the wake expansion effect at middle wake region are clearly visualized based on the Q-criterion and Liutex methods. The unphysical effect that the sheet-like vortex structures upstream wind turbine is wrongly presented by the Q-criterion method, whereas this error is eliminated by Liutex method.

Chapter 17 is entitled "Propeller Wake and Noise Analysis Based on the Third-Generation Vortex" written by Lianjie Yu, Weiwen Zhao, and Decheng Wan*. Vortex is one of the complex contents of fluid dynamics. The traditional first-generation vortex "Q-criterion" and second-generation "vorticity" have various problems in calculation. The third-generation vortex recognition "Rortex", which is more in line with the actual laws of physics, came into being. Based on the third-generation vortex recognition technology, this chapter conducts a detailed analysis of the propeller tail vortex on the OpenFOAM platform. Taking the DTMB4119 propeller as the research object, the wake characteristics of the propeller are studied by comparing the wake vortex considering cavitation and ignoring it. The development process is qualitatively analyzed in axial velocity, turbulent kinetic energy, turbulent viscosity, and so on. Besides, since the vortex is an important source of noise, the acoustic characteristics is analyzed by acoustic analogy. Combined with the vortex distribution, this chapter is committed to explaining the influence of the vortex on the linear and nonlinear terms of the propeller noise. The results show that the third-generation vortex recognition technology can remove the shear in the flow field and the spurious vorticity near the boundary, so that the continuous vorticity calculated by the Q-criterion becomes a separated vorticity, which is more in line with the actual laws of physics. At the same time, the cavitation affects vorticity and nonlinear sound source, which increases the overall sound pressure level.

Chapter 18 is entitled "Identification of Vortex Around High-Speed Ship Based on Liutex Method" written by Yunpeng Ma, Weiwen Zhao, and Decheng Wan*. Vortex is ubiquitous in ship navigation. The first-generation vortex identification method is based on vorticity, which is understood as twice the rotational angular velocity around the center. The iso-surface is used to explore the vortex structure of the flow field when it comes to second-generation $\boldsymbol{Q}$, Δ, and other methods. The third-generation Liutex vector defines the local rigid body rotation of the fluid, including the local rotation axis and the rotation strength. The vorticity can be decomposed into the rigid body rotation and the pure shear, and the rigid rotation part can be accurately extracted from the fluid motion. The vortex identification method overcomes the need to adjust the threshold for specific flows and can capture both strong and weak vortices at the same time. Based on the idea of Liutex, the various characteristics of vortices can be quantitatively described, such as the absolute strength of vortices, relative strength, local axis of rotation, vortex core center, vortex core size, and vortex boundary. As a prototype of a destroyer, many engineers use the DTMB model to do a lot of numerical calculations. In this chapter, the third-generation Liutex method is used to identify the vortex around the DTMB, and the vortex structure is analyzed in detail and compared with the Q-criterion and so on. The calculation results show that the Liutex method has better accuracy.

Chapter 19 is entitled "Identification of Vortical Structures of Flows Past a Surface-piercing Finite Square Cylinder with Rounded Corners" written by Songtao Chen, Weiwen Zhao, and Decheng Wan*. As one of the typical bluff bodies, square cylinders are widely applied in offshore platforms, such as semi-submersibles and tension leg platforms. In order to reduce the resistance, rounded corners are often adopted instead of the original sharp corners to passively control the flow separation. To further explore its flow mechanism, this chapter focuses on the vortical structures of flows past a vertical surface-piercing finite square cylinder with rounded corners at $Re = 2.7 \times 105$ and $Fr = 1.1$. The delayed detached-eddy simulation (DDES) is used to resolve the large-scale separation flow, and a piecewise-linear interface calculation (PLIC)-based geometric volume of fluid (VOF) method is employed to accurately capture the interface. Due to the co-existence of the free surface and the free end, the wake flow exhibits strong three-dimensional characteristics. To clearly visualize the vortical structures, the third generation of vortex identification method—Liutex—is adopted throughout the present study. The iso-surfaces of the normalized Omega-Liutex $\Omega\tilde{}_R$ represent three-dimensional structures, while the Liutex vector field and integral lines show the direction of the local rotational axis. The results show that the Liutex method and its variant could well characterize the typical vortical structures and reveal more flow details.

Chapter 20 is entitled "Comparison of Vortex Identification Methods Based on the Liutex Decomposition and Application in a Compressor Cascade" written by Weibo Zhong, Yumeng Tang, and Yangwei Liu*. Gaining a deeper insight into the complex vortex-dominated flow in aeroengine compressors is of great meaning for improving the aerodynamic performance of compressors. However, vortex regions vary depending on the vortex identification methods by which they are extracted. The Liutex-based velocity gradient decomposition provides an effective approach to quantify the rigid body rotation, pure shearing, and compression-stretching effects in fluid motion. The LT-plane constructed based on the local trace criterion provides a tool to geometrically represent the relationship of different vortex identifications. In this study, the commonly used Q-series vortex criteria and the normalized Liutex method Ω_L are analytically compared under zero threshold and non-zero threshold conditions. A typical corner separation flow in a highly loaded compressor cascade is analyzed. Vortex regions extracted by different vortex criteria with equivalent threshold values are compared. Results show that proper exclusion of pure shearing and compressing-stretching contamination helps to capture entangled vortex structures and indicate the pressure loss regions.

Chapter 21 is entitled "Research on Vortex Structure of All-Oxygen MILD Combustion Reheating Furnace Based on Third-Generation Vortex Identification Method" written by Yongfu Wu*, Shang Zhao, Hongyu Ji, Changpeng Guo, Zhenfeng Wang, and Zhongxing Liu. The whole oxygen MILD combustion technology mainly relies on the flue gas coiling and reflow effect of MILD combustion technology, so that the whole oxygen combustion can achieve MILD effect. Flue gas coiling backflow is the embodiment of gas eddy motion. There are a lot of eddy structures in coiling mixing process, so the effect of all-oxygen MILD combustion is determined by eddy structure. In this chapter, numerical simulation and experimental methods are combined to analyze the vortex structure in heating furnace with the help of Liutex method, and the threshold value 3 of vortex core is selected to draw the vortex core and vortex line. The results show that the vortex structure can directly reflect the vortex motion of gas in the furnace, and the vortex motion in the furnace is mainly concentrated in the initial section of the furnace. The size and quantity of the vortex structure in this area are stronger than that in the middle and rear of the furnace, and it is also the main combustion region of the furnace. Under certain threshold conditions, too much or too little vortex core structure in furnace will affect combustion efficiency. The combination of vortex line and streamline line can better reflect the backflow and mixing of gas in the furnace, and is more conducive to the analysis of vortex motion of gas in the furnace.

Chapter 22 is entitled "The Effect of Fluidic Pintle Nozzle on the Ablative Features" written by Zheng Yang, Kan Xie*, Xiang Li, Dongfeng Yan, Junwei Li, Ningfei Wang. Fluidic Pintle Nozzle (FPN) combines mechanical and fluidic schemes to inject secondary flow at the head of the pintle to change the mainstream gas profile. As a means of mechanical combined with fluidic regulation, this method has not completed systematic research since it was proposed, and the current research is still in the basic theoretical stage, with insufficient research on its basic laws, flow mechanism, influencing factors, ablative protection, and so on. This chapter focuses on the influence of injecting secondary flow at the head of the pintle on the ablation of the fluidic pintle nozzle of the solid rocket motor, the surface chemical reaction model under fluid-solid coupling was established, and fluent commercial software was used for numerical calculation. The basic law and influencing factors of the ablative rate with a different opening, injection position, and secondary flow temperature on the ablation rate of the nozzle were studied. The results show that the ablation rate of the nozzle is significantly different under the different openings. A low-temperature jet can effectively reduce the temperature of the pintle, reduce the ablation rate of the pintle, and has a significant protective effect on the pintle downstream of the injection hole.

Chapter 23 is entitled "Liutex Investigation of Backflow Leakage in a Shaft-Less Rim-Driven Thruster" written by Xincheng Wang, Hua Ruan, Xiaorui Bai, Chengzao Han, Huaiyu Cheng, and Bin Ji*. In the current chapter, the phenomenon of backflow leakage in a shaft-less rim-driven thruster is numerically investigated. The numerical results simulated with the *k-w* SST model are calibrated with the measured parameters. The feasibility of different vortex identification methods (vorticity method, Q-criterion, Ω method, and Liutex method) in the investigation of backflow leakage is discussed. Only the Liutex method can filter the shear layer while retaining the main vortex structure. With the assistance of the Liutex method, the complex components of backflow leakage are identified, which are gap flow, wake vortex, upstream gap vortex, and downstream gap vortex. Moreover, one phenomenon that should be especially concerned is the mixing of wake vortex and downstream gap vortex, because it can aggravate the flux of backflow leakage and raise a significant hydraulic loss.

Chapter 24 is entitled "Application of Omega Identification Method in the Ventilated Cavities Around a Surface-piercing Hydrofoil" written by Yuchang Zhi, Rundi Qiu, Renfang Huang*, and Yiwei Wang. This article summarizes a simulation campaign to study atmospheric ventilation flow patterns, forces, and stability at various speeds. Simulations are based upon cited experiments, and results show good agreement for global forces, cavity shapes, etc. The hydrofoil deceleration process includes three different flow regimes: fully ventilated flow, fully wetted flow, and partially ventilated flow. To obtain the evolution process of the vortex structure, the vortex structure is visualized using the Omega method. The results show that the evolution of the vortex structure mainly includes three types of vortex structures, i.e., a tip vortex, a Karman vortex, and an unstable vortex. The vortex structure is related to the ventilated cavity. Omega can identify weak vortices and strong vortices.

Chapter 25 is entitled "Application of Third Generation of Vortex Identification for Straight-Bladed Vertical Axis Wind Turbines" written by Guoqiang Tong, Zhenjun Xiao, Shengbing Yang, Fang Feng, Yan Li*, and Kotaro Tagawa. The Straight-Bladed Vertical Axis Wind Turbine (SB-VAWT) has a high share in the small-scale wind turbine market with the advantages of its simple structure, low noise, smart forms, as well as its suitability for off-grid operation. Currently, the performance of the SB-VAWT is lower than that of Horizontal Axis Wind Turbines (HAWTs). One of the main reasons is the complexity of the SB-VAWT flow field. Accurate identification of vortex in the flow field is critical because an effective means of improving the performance of vertical axis wind turbines is to optimize the vortex. In this study, a four-bladed SB-VAWT with rotor diameter, blade chord length, and blade height of 0.6 m, 0.125 m, and 0.5 m, respectively, is used as the object of study. The Ω method, Liutex, λ_2-criterion, and Q-criterion are used to identify the vortex of the SB-VAWT flow field, respectively, when the tip speed ratio is 1.0. The results show that the vortex structure of the SB-VAWT flow field can be obtained by all four methods. Liutex can identify more small-scale vortices, which is not available in other methods. This chapter provides a reference for the selection of vortex identification methods for SB-VAWT.

Chapter 26 is entitled "Study on Fluid-Borne Noise Around a Cylinder Based on Vortex Sound Theory with Liutex-Shear Decomposition" written by Zhiyi Yuan* and Yongxue Zhang. The flow-induced noise over a circular cylinder is widely studied as its geometry is typical and the noise is highly mattered on many occasions. The acoustic field for the flow around the cylinder is studied using delayed detached-eddy simulation and Powell vortex sound theory. Combined with Liutex-shear decomposition of vorticity, the effect of the rigid rotational and shear part of the fluid parcel on the acoustic perturbation is discussed. The results show that the numerical prediction on the acoustic field is in good agreement with the experiment. Shear often makes up a large proportion of vorticity in vortex areas compared to rigid rotational parts, and the vorticity concentration areas with a weaker shearing effect correspond to lower loss. The tonal noise is from vortex shedding, where the shear mainly contributes to noise compared to the rigid rotational part of vorticity. The possible reason is that the sound can be a kind of energy loss and the shearing effect dominates the loss.

Chapter 27 is entitled "Coherent Structures Analysis Across Cavity Interface in Cloud Cavitating Flows Using Different Vortex Identification Methods" written by Changchang Wang*, Guoyu Wang, and Biao Huang. Cavitating flows are characterized by strong compressibility inside the cavity and weak compressibility outside the cavity. The transition between these two regions with distinct acoustic features is defined as the acoustic shear layer (i.e., from cavity interface to 0.99 sonic seed in pure liquid, about $1\times10^{-7} < \alpha v < 1\times10^{-1}$) which has been preliminarily studied in the recent paper (Ocean Eng. 209(2020): 107025). The acoustic shear layer is characterized by the large sonic speed gradient which is of great acoustics importance to understand how the disturbance (i.e., shockwave) propagates between these two regions. With varying cavitation numbers and Reynold numbers, cavitating flows present different flow regimes where cloud cavitation is supposed to be the most destructive cavity regime and of great fundamental interest and engineering applications, mainly consisting of two kinds of cavity structures, namely, attached sheet cavity and shedding cloud cavity. In this work, to improve the understanding of cavity instabilities associated with wave dynamics, we examine in detail the coherent structures inside the acoustic shear layer of both the attached sheet cavity and the shedding cloud cavity, in particular, that across the cavity interface. Numerical simulation of cavitating flows around a NACA66 (mod) hydrofoil was conducted using computational fluid dynamics (CFD) tool. Vortex identification methods, including vorticity, Q-criteria, and the Liutex method, are employed to identify the flow structures within the acoustics shear layer. Results show that acoustic shear layer across the attached sheet cavity is thinner than that across the shedding cloud cavity. The acoustic shear layer consists of two regions, i.e., turbulence-dominated region and the acoustics-dominated region. Specifically, the turbulence-dominated region which is identified by the turbulent/non-turbulent interface near the cavity boundary is important for the mass, momentum, and energy transfer characteristics. The acoustics-dominated region is significant for the disturbance propagation (i.e., shockwave) between the compressible cavitation region and incompressible pure liquid region. Different vortex identification methods show different performances in terms of identifying the flow structures in these two regions. It is suggested that further work could be done to implement the acoustics characteristics into the vortex identification method to improve the identification performance in acoustics-dominated regions, i.e., low void fraction region in gas-liquid flows.

The book was finished during the pandemic period of COVID-19 which is one of the most difficult times in human history. The editors are very thankful to Stefan S. and all workshop organizers and technicians for their hard work and unprecedented courage. Dr. Wang would like to acknowledge the support from his family, especially his wife, Shang Wu, whose love, patience, and enthusiasm inspire Wang constantly, and their lovely little son who brings countless happiness to the family. Dr. Gao would like to acknowledge the support from his family. Professor Liu is grateful to his wife, Weilan Jin, for her full support by taking care of all housework. He also wants to express thanks to his daughter, Haiyan Liu, and his son, Haifeng Liu, for their support as well.

Asterisk (*) marked is the corresponding author.

Suzhou, Jiangsu, China — Yiqian Wang
Nanjing, Jiangsu, China — Yisheng Gao
Arlington, TX, USA — Chaoqun Liu

Contents

Editors and Contributors

About the Editors

Dr. Yiqian Wang is currently an Associate Professor at Soochow University, Suzhou, China. He obtained his B.E. and Ph.D. from Nanjing University of Aeronautics and Astronautics, Nanjing, China in 2010 and 2016. He had been a visiting student at the University of Texas at Arlington under the supervision of Prof. Chaoqun Liu from 2014 to 2015. After graduation, he became a Post-Doctoral Researcher at Tsinghua University for 2 years until 2019 when he joined Soochow University. He has published more than 30 journals and conference papers and has been one of the main contributors to the developments of the third generation of vortex identification methods, especially the Liutex-based system and the Omega methods. His research interests are focused on vortex identification, turbulence, and numerical methods.

Dr. Yisheng Gao is currently a Lecturer at Nanjing University of Aeronautics and Astronautics University, Nanjing, China. He received his Bachelor of Engineering degree in Aircraft Design and Engineering (2007), Master of Engineering degree in Fluid Mechanics (2009), and Ph.D. degree in Fluid Mechanics (2016) from Nanjing University of Aeronautics and Astronautics University, China. He was a Postdoctoral Fellow at the University of Texas at Arlington, Texas, USA in 2017–2019. He has published more than 20 journals and conference papers and has been one of the main contributors to the developments of the third generation of vortex identification methods. His research is concerned with computational fluid dynamics, including direct numerical simulation and discrete adjoint methods.

Dr. Chaoqun Liu received both BS (1968) and MS (1981) from Tsinghua University, Beijing, China and Ph.D. (1989) from University of Colorado at Denver, USA. He is currently the Tenured and Distinguished Professor and the Director of Center for Numerical Simulation and Modeling at University of Texas at Arlington, Arlington, Texas, USA. He has worked on high-order direct numerical simulation (DNS) and large eddy simulation (LES) for flow transition and turbulence for 32 years since 1990. As PI, he has been awarded by NASA, US Air Force, and US Navy with 50 federal research grants of over 5.7×10^6 US dollars in the United States. He has published 14 professional books, 140 journal papers, and 158 conference papers. He is the founder and main contributor of Liutex and the third generation of vortex definition and identification methods including the Omega, Liutex/Rortex, Modified Liutex-Omega, Liutex-Core-Line methods, Objective Liutex, RS vorticity decomposition, and UTA R-NR velocity gradient tensor decomposition. He is also the founder of Liutex-based new fluid kinematics.

Contributors

Habib Ahmari University of Texas at Arlington, Arlington, TX, USA

Oscar Alvarez Department of Mathematics, The University of Texas at Arlington, Arlington, TX, USA;
University of Texas at Arlington, Arlington, TX, USA

Xiaorui Bai State Key Lab of Water Resources and Hydropower Engineering Science, Wuhan University, Wuhan, China

Aayush Bhattarai University of Texas at Arlington, Arlington, TX, USA

Xiaoshu Cai Institute of Particle & Two-Phase Flow Measurement, Shanghai, China;
School of Energy and Power Engineering, University of Shanghai for Science and Technology, Shanghai, China

Hongjie Cao Computational Marine Hydrodynamic Lab (CMHL), School of Naval Architecture, Ocean and Civil Engineering, Shanghai Jiao Tong University, Shanghai, China

Liushuai Cao Computational Marine Hydrodynamics Lab (CMHL), School of Naval Architecture, Ocean and Civil Engineering, Shanghai Jiao Tong University, Shanghai, China

Bosen Chai College of Mechanical and Aerospace Engineering, Jilin University, Changchun, Jilin, China;
State Key Laboratory of Automotive Simulation and Control, Jilin University, Changchun, Jilin, China;
Sinotest Equipment Co., LTD., Changchun, Jilin, China

Pengxin Chen Institute of Nuclear and New Energy Technology, Collaborative Innovation Center of Advanced Nuclear Energy Technology, Key Laboratory of Advanced Reactor Engineering and Safety, Ministry of Education, Tsinghua University, Beijing, China

Songtao Chen Computational Marine Hydrodynamics Lab (CMHL), School of Naval Architecture, Ocean and Civil Engineering, Shanghai Jiao Tong University, Shanghai, China

Zeliang Chen Institute of Nuclear and New Energy Technology, Collaborative Innovation Center of Advanced Nuclear Energy Technology, Key Laboratory of Advanced Reactor Engineering and Safety, Ministry of Education, Tsinghua University, Beijing, China

Huaiyu Cheng State Key Lab of Water Resources and Hydropower Engineering Science, Wuhan University, Wuhan, China

Yuan Ding School of Mathematical Science, Soochow University, Suzhou, China

Xiangrui Dong Institute of Particle & Two-Phase Flow Measurement, Shanghai, China;
School of Energy and Power Engineering, University of Shanghai for Science and Technology, Shanghai, China

Yinlin Dong Department of Mathematics, University of Central Arkansas, Conway, AR, USA

Fang Feng College of Arts and Sciences, Northeast Agricultural University, Harbin, China

Nan Gui Institute of Nuclear and New Energy Technology, Collaborative Innovation Center of Advanced Nuclear Energy Technology, Key Laboratory of Advanced Reactor Engineering and Safety, Ministry of Education, Tsinghua University, Beijing, China

Changpeng Guo College of Energy and Environment, Inner Mongolia University of Science and Technology, Baotou, Inner Mongolia, China

Chao Guo Key Laboratory of Navigation Structure Construction Technology, Ministry of Transport, Nanjing Hydraulic Research Institute, Nanjing, China

Chengzao Han State Key Lab of Water Resources and Hydropower Engineering Science, Wuhan University, Wuhan, China

Biao Huang Beijing Institute of Technology, Beijing, China

Fenglai Huang Computational Marine Hydrodynamics Lab (CMHL), School of Naval Architecture, Ocean and Civil Engineering, Shanghai Jiao Tong University, Shanghai, China

Renfang Huang Key Laboratory for Mechanics in Fluid Solid Coupling Systems, Institute of Mechanics, Chinese Academy of Sciences, Beijing, China

Xiaoli Huang Institute of Nuclear and New Energy Technology, Collaborative Innovation Center of Advanced Nuclear Energy Technology, Key Laboratory of Advanced Reactor Engineering and Safety, Ministry of Education, Tsinghua University, Beijing, China

Shah Md. Imran Kabir University of Texas at Arlington, Arlington, TX, USA

Bin Ji State Key Lab of Water Resources and Hydropower Engineering Science, Wuhan University, Wuhan, China

Hongyu Ji College of Energy and Environment, Inner Mongolia University of Science and Technology, Baotou, Inner Mongolia, China

Shengyao Jiang Institute of Nuclear and New Energy Technology, Collaborative Innovation Center of Advanced Nuclear Energy Technology, Key Laboratory of Advanced Reactor Engineering and Safety, Ministry of Education, Tsinghua University, Beijing, China

Junwei Li School of Aeronautics and Astronautics, University of Electronic Science and Technology of China, Chengdu, China

Xiang Li School of Aeronautics and Astronautics, University of Electronic Science and Technology of China, Chengdu, China

Yan Li Engineering College, Northeast Agricultural University, Harbin, China

Yuyan Li Institute of Particle & Two-Phase Flow Measurement, Shanghai, China; School of Energy and Power Engineering, University of Shanghai for Science and Technology, Shanghai, China

Chaoqun Liu The University of Texas at Arlington, Arlington, TX, USA;
Department of Mathematics, The University of Texas at Arlington, Arlington, TX, USA

Yangwei Liu National Key Laboratory of Science and Technology On Aero-Engine Aero-Thermodynamics, Beihang University, Beijing, China;
School of Energy and Power Engineering, Beihang University, Beijing, China

Zhongxing Liu College of Energy and Environment, Inner Mongolia University of Science and Technology, Baotou, Inner Mongolia, China

Yunpeng Ma Computational Marine Hydrodynamic Lab (CMHL), School of Naval Architecture, Ocean and Civil Engineering, Shanghai Jiao Tong University, Shanghai, China

Qazu Asguqye E. Mowla University of Texas at Arlington, Arlington, TX, USA

Charles Nottage Department of Mathematics, The University of Texas at Arlington, Arlington, TX, USA

Biyu Pang School of Mathematical Science, Soochow University, Suzhou, China

Vishwa Patel The University of Texas at Arlington, Arlington, TX, USA

Rundi Qiu Key Laboratory for Mechanics in Fluid Solid Coupling Systems, Institute of Mechanics, Chinese Academy of Sciences, Beijing, China;
School of Future Technology, University of Chinese Academy of Sciences, Beijing, China

Hua Ruan Nation Key Laboratory On Ship Vibration and Noise, China Ship Science Research Center, Wuxi, China

Pushpa Shrestha Department of Mathematics, The University of Texas at Arlington, Arlington, TX, USA

Kotaro Tagawa Faculty of Agriculture, Tottori University, Tottori, Japan

Xinran Tang Institute of Particle & Two-Phase Flow Measurement, Shanghai, China;
School of Energy and Power Engineering, University of Shanghai for Science and Technology, Shanghai, China

Yumeng Tang National Key Laboratory of Science and Technology On Aero-Engine Aero-Thermodynamics, Beihang University, Beijing, China;
School of Energy and Power Engineering, Beihang University, Beijing, China

Guoqiang Tong Engineering College, Northeast Agricultural University, Harbin, China

Jiyuan Tu Institute of Nuclear and New Energy Technology, Collaborative Innovation Center of Advanced Nuclear Energy Technology, Key Laboratory of Advanced Reactor Engineering and Safety, Ministry of Education, Tsinghua University, Beijing, China;
School of Engineering, RMIT University, Melbourne, VIC, Australia

Decheng Wan Computational Marine Hydrodynamics Lab (CMHL), School of Naval Architecture, Ocean and Civil Engineering, Shanghai Jiao Tong University, Shanghai, China;
Ocean College, Zhejiang University, Zhoushan, China

Changchang Wang Beijing Institute of Technology, Beijing, China

Guangyi Wang College of Mechanical and Aerospace Engineering, Jilin University, Changchun, Jilin, China

Guoyu Wang Beijing Institute of Technology, Beijing, China

Ningfei Wang School of Aeronautics and Astronautics, University of Electronic Science and Technology of China, Chengdu, China

Xincheng Wang State Key Lab of Water Resources and Hydropower Engineering Science, Wuhan University, Wuhan, China

Yiqian Wang School of Mathematical Science, Soochow University, Suzhou, China

Yiwei Wang Key Laboratory for Mechanics in Fluid Solid Coupling Systems, Institute of Mechanics, Chinese Academy of Sciences, Beijing, China;
School of Future Technology, University of Chinese Academy of Sciences, Beijing, China;
School of Engineering Science, University of Chinese Academy of Sciences, Beijing, China

Zhenfeng Wang College of Energy and Environment, Inner Mongolia University of Science and Technology, Baotou, Inner Mongolia, China;
School of Metallurgy, Northeastern University, Shenyang, Liaoning, China

Yongfu Wu College of Energy and Environment, Inner Mongolia University of Science and Technology, Baotou, Inner Mongolia, China;
School of Metallurgy, Northeastern University, Shenyang, Liaoning, China

Zhenjun Xiao Engineering College, Northeast Agricultural University, Harbin, China

Kan Xie School of Aeronautics and Astronautics, University of Electronic Science and Technology of China, Chengdu, China

Shun Xu Computational Marine Hydrodynamics Lab (CMHL), School of Naval Architecture, Ocean and Civil Engineering, Shanghai Jiao Tong University, Shanghai, China

Dong Yan College of Mechanical and Aerospace Engineering, Jilin University, Changchun, Jilin, China

Dongfeng Yan Civil Aviation Flight, University of China, Guanghan, China

Yonghua Yan Jackson State University, Jackson, MS, USA

Shengbing Yang Engineering College, Northeast Agricultural University, Harbin, China

Xingtuan Yang Institute of Nuclear and New Energy Technology, Collaborative Innovation Center of Advanced Nuclear Energy Technology, Key Laboratory of Advanced Reactor Engineering and Safety, Ministry of Education, Tsinghua University, Beijing, China

Zheng Yang School of Aeronautics and Astronautics, University of Electronic Science and Technology of China, Chengdu, China

Lianjie Yu Computational Marine Hydrodynamics Lab (CMHL), School of Naval Architecture, Ocean and Civil Engineering, Shanghai Jiao Tong University, Shanghai, China

Yifei Yu Department of Mathematics, The University of Texas at Arlington, Arlington, TX, USA

Zhiyi Yuan College of Mechanical and Transportation Engineering, China University of Petroleum, Beijing, China;
Key Laboratory of Process Fluid Filtration and Separation, Beijing, China

Jin Zhang College of Mechanical and Aerospace Engineering, Jilin University, Changchun, Jilin, China

Xiaoxi Zhang Institute of Nuclear and New Energy Technology, Collaborative Innovation Center of Advanced Nuclear Energy Technology, Key Laboratory of Advanced Reactor Engineering and Safety, Ministry of Education, Tsinghua University, Beijing, China

Yongxue Zhang College of Mechanical and Transportation Engineering, China University of Petroleum, Beijing, China;
Key Laboratory of Process Fluid Filtration and Separation, Beijing, China

Shuang Zhao College of Energy and Environment, Inner Mongolia University of Science and Technology, Baotou, Inner Mongolia, China

Weiwen Zhao Computational Marine Hydrodynamics Lab (CMHL), School of Naval Architecture, Ocean and Civil Engineering, Shanghai Jiao Tong University, Shanghai, China

Yuchang Zhi School of Aeronautics and Astronautics, Sun Yat-Sen University, Guangzhou, China

Weibo Zhong National Key Laboratory of Science and Technology On Aero-Engine Aero-Thermodynamics, Beihang University, Beijing, China;
Research Institute of Aero-Engine, Beihang University, Beijing, China

Wu Zhou Institute of Particle & Two-Phase Flow Measurement, Shanghai, China;
School of Energy and Power Engineering, University of Shanghai for Science and Technology, Shanghai, China

Wenjie Zuo College of Mechanical and Aerospace Engineering, Jilin University, Changchun, Jilin, China;
State Key Laboratory of Automotive Simulation and Control, Jilin University, Changchun, Jilin, China

Part I
Theoretical Foundation and Applications for Turbulence and Transition Research

Chapter 1
Liutex and Third Generation of Vortex Definition and Identification

Chaoqun Liu, Habib Ahmari, Charles Nottage, Yifei Yu, Oscar Alvarez, and Vishwa Patel

Abstract Liutex is a physical quantity like velocity, vorticity, pressure, temperature, etc., describing local fluid rotation or vortex, which was ignored for centuries. Liutex was defined by the University of Arlington at Texas (UTA) team in 2018 as a vector for vortex. Its direction is the local rotation axis, and magnitude is twice local angular rotation speed. As the third generation of vortex definition and identification, Liutex has been widely applied for visualization of vortex structure to replace the first generation or vorticity, which cannot distinguish shear from rotation and the second generation such as Q, Δ, λ_2, and λ_{ci} methods, which are all scalar without rotation axis, dependent on threshold and contaminated by shear and stretching. Several new vortex identification methods have been developed, especially the modified Liutex-Omega method, which is threshold insensitive, and the Liutex-Core-Line method, which is unique and threshold-free. According to the Liutex vector, a unique coordinate system called Principal Coordinate can be set up, and consequent Principal Decomposition of velocity gradient tensor can be made. Different from classical fluid kinematics, the Liutex-based new fluid kinematics decomposes the fluid motion to a rotational part and a non-rotational part (*UTA R-NR* decomposition). The non-rotational part can be further decomposed to stretching and shear consisting of symmetric shear and anti-symmetric shear in contrast with the classical fluid kinematics, which decomposes fluid motion to deformation and vorticity, that was misunderstood as rotation. According to the constitutive relation between stress and strain, the new fluid kinematics may give significant influence to new fluid dynamics.

1.1 Introduction

A vortex is intuitively recognized as a rotational/swirling motion of fluids. It is ubiquitous in nature and viewed as the building blocks, muscles, and sinews of turbulent flows [1]. Quantitative understanding of vortex is essential for turbulence research

C. Liu (✉) · H. Ahmari · C. Nottage · Y. Yu · O. Alvarez · V. Patel
University of Texas at Arlington, Arlington, TX, USA
e-mail: cliu@uta.edu

Y. Wang et al. (eds.), *Liutex and Third Generation of Vortex Identification*,
Springer Proceedings in Physics 288, https://doi.org/10.1007/978-981-19-8955-1_1

and many engineering applications such as hydrodynamics, aerodynamics, thermodynamics, oceanography, meteorology, metallurgy, civil engineering, astronomy, biology, etc. In 1858, Helmholtz first defined vortex as tubes composed of so-called vortex filaments [2], which are, in fact, infinitesimal vorticity tubes. Vorticity has a rigorous mathematical definition with no clear physical meaning.

Conversely, vortex has a physical meaning but no mathematical definition until recently. Science and engineering applications have shown that the correlation between vortex and vorticity is very weak, especially in the near-wall region [3]. Also, existing vortex identification methods, which are based on eigenvalues of the velocity gradient tensor, are scalars and therefore strongly depend on the arbitrary thresholds.

Liutex is a new physical quantity introduced by Liu et al. in 2018 [4–6] that represents local fluid rotation, i.e., vortex, and is a mathematically rigorous tool for vortex characterization. The major idea of Liutex is to extract the rigid rotation part from the fluid motion to represent the vortex. Given the novelty of Liutex, it is inevitable that the vortex dynamics be re-examined to develop unique and accurate vortex identification methods independent of thresholds. This chapter presents the existing vortex definition and identification methods and their limitations, and introduces Liutex, as the third generation of vortex definition, and Liutex-based vortex Identification methods.

1.2 Three Generations of Vortex Definition and Identification

There are three generations of vortex identification methods [6], and Helmholtz's [2] definition of vortex is classified as the first. During the past four decades, several vortex identification criteria, such as Q, Δ, λ_2, and λ_{ci} methods, have been developed [7–11] and are classified as the second generation of vortex identification. They are all based on eigenvalues of the velocity gradient tensor; however, they are scalars and thus strongly dependent on the arbitrary thresholds when plotting the-iso surface to represent vortical structures. They cannot show the vortex rotation axis, which is critical for vortex structure. Furthermore, they are all contaminated by stretching (compression) and shearing. Rotational axis and uniqueness in strength are two important issues for vortex definition that cannot be solved by either the first or second vortex identification methods.

Liutex, the third generation of vortex definition and identification, was introduced by the first author Dr. Liu at the University of Texas at Arlington (UTA) [4–6]. It is defined as a vector that uses the real eigenvector of velocity gradient tensor as its direction and twice the local angular speed of the rigid rotation as its magnitude. The major idea of Liutex is to extract the rigid rotation part from fluid motion to represent vortex, which is a mathematically rigorous tool suitable for vortex characterization [12]. The location of the rotation axis is then the local maxima of Liutex (not vorticity), and the Liutex magnitude is twice the vortex angular speed, which is uniquely defined. Although similar ideas of decomposition of vorticity tensor to a pure rotation and anti-symmetric shear have been given by [13, 14], they did not find a vector definition for flow rotation or Liutex. Several vortex identification methods have been developed based on the Liutex definition [4–6, 15–39]. Examples of these methods are modified *Liutex-Omega* and *Liutex-Core-Line.* The former is threshold insensitive, and the latter is threshold-free. These methods have been shown to accurately visualize vortical structures in turbulent flows [40, 41]; however, schemes and software for automated generation and case-independent features are still open for research. The current methods still have parameters that need manual adjusting.

1.2.1 First Generation of Vortex Identification Methods (Vorticity-Based Methods)

Since Helmholtz proposed the concepts of vorticity tube and filament in 1858 [2], it has generally been believed that vortices consist of vorticity tubes, and vortex strength is measured by the magnitude of vorticity, i.e., $\nabla \times \vec{v}$ [42]. Although vorticity is widely adopted for detecting vortices, one immediate counterexample is that the average shear force generated by the no-slip wall in the laminar boundary layer is so strong that a very large amount of vorticity exists. However, no rotation motions are observed in the near-wall regions. This implies that vortex cannot be represented by vorticity. Robinson [43] found that "the association between regions of strong vorticity and actual vortices can be rather weak in the turbulent boundary layer, especially in the near-wall region." In 1990. Wang et al. [30] showed that the magnitude of vorticity could be substantially reduced along vorticity lines by entering the vortex core region near the solid wall in a flat plate boundary layer. Figure 1.1 clearly indicates that for a transitional flow over a flat plate in the near-wall region of the boundary layer, the local vorticity vector can deviate from the direction of vortical structures. Also, a vortex can appear in the region where the vorticity is smaller than the surrounding area in which the vorticity is larger than the vorticity inside the vortex. These results demonstrate that vorticity cannot be used to represent vortices, especially in the near-wall region of the boundary layers.

Vortex and vorticity are two different concepts. Vortex is a natural phenomenon, but vorticity is a mathematical definition. In fact, if a vortex cannot be ended inside the flow field, how can turbulence be generated by "vortex breakdown?" Vortex can break down, which means it is not vorticity tube. Although both vortex and vorticity are vectors, they are different vectors.

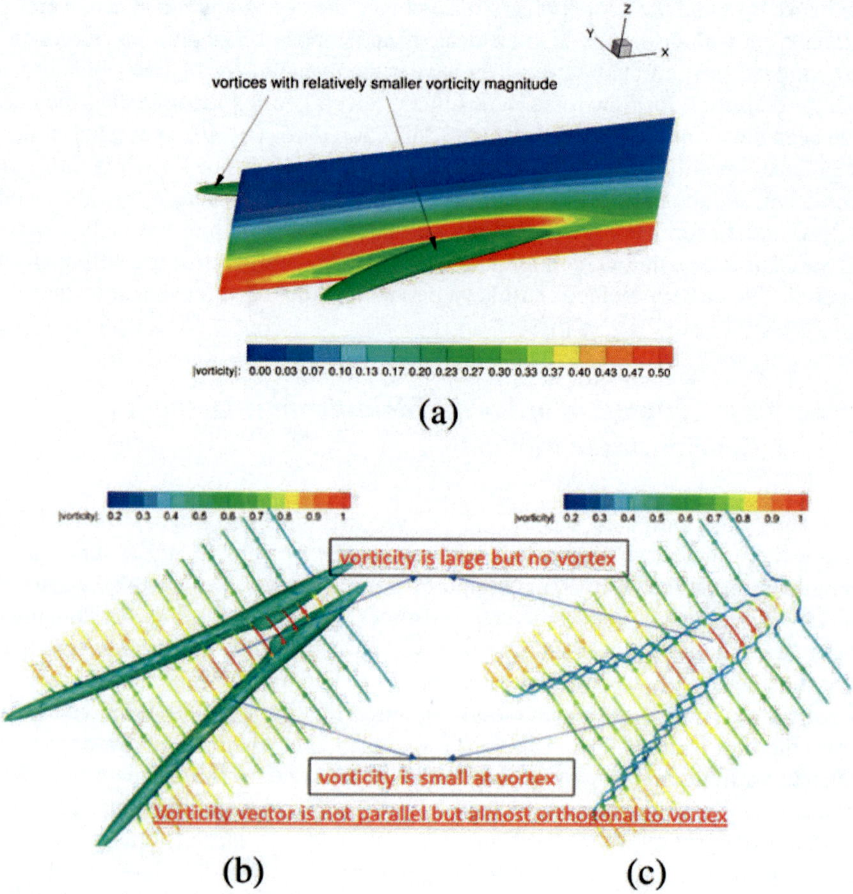

Fig. 1.1 (**a**) Vortex appears in the area where vorticity is relatively smaller, (**b**) vorticity line is not aligned with vortex, (**c**) vorticity is smaller (green) than surrounding (yellow or red)

1.2.2 Second Generation of Vortex Identification

Several vortex identification methods have been proposed during the past four decades, including the Q, λ_{ci}, and λ_2 methods. These methods are briefly explained in the following.

Q Method. The Q criterion is expressed by Eq. 1.1:

$$Q = \frac{1}{2}\left(\|\boldsymbol{B}\|_F^2 - \|\boldsymbol{A}\|_F^2\right) \tag{1.1}$$

where $\boldsymbol{A}$ and $\boldsymbol{B}$ are the symmetric and antisymmetric parts of the velocity gradient tensor, respectively, and $\| \|_F^2$ represents the Frobenius norm. Theoretically, $Q > 0$

can identify the vortex boundary, but in practice, a threshold $Q_{threshold}$ must be specified to define the regions where $Q > Q_{threshold}$. Q-criterion shows the symmetric tensor roles that balance the anti-symmetric tensor.

λ_{ci} Method. The λ_{ci} method defines the strength of vortex as the imaginary part λ_{ci} of the complex eigenvalue of the velocity gradient tensor ∇V [10].

λ_2 Method. The λ_2 method defines the strength of vortex by using the second negatively largest eigenvalue λ_2 of $A^2 + B^2$, where A and B are the symmetric and anti-symmetric parts of the velocity gradient tensor [9]. Assuming the fluid is incompressible, steady, and non-viscous, the Navier–Stokes equation is converted to $A^2 + B^2 = -\nabla(p)/\rho$, where p and ρ represent pressure and density, respectively. The rotational area is defined by the existence of two negative eigenvalues of the symmetric tensor $A^2 + B^2$ [9].

1.2.3 Limitations of Existing Vortex Identification Methods

Thresholds. Without exception, the abovementioned vortex identification methods require user-specified thresholds. Since different thresholds indicate different vortical structures, it is critical to determine an appropriate threshold. When a large threshold for Q criterion is used, *vortex breakdown* occurs in the late boundary layer transition (Fig. 1.2a); however, when a small threshold is applied, no *vortex breakdown* occurs (Fig. 1.2b), which means that an appropriate threshold is vital to these vortex identification methods. Many computational results have revealed that the threshold is case-related, empirical, sensitive, time step-related, and hard to adjust. Furthermore, it is unclear whether the specified threshold is proper or improper. There may be no single proper threshold, especially if strong and weak vortices co-exist. If the threshold is too small, weak vortices may be captured, but strong vortices could be smeared and become vague. If the threshold is too large, weak vortices will be wiped out.

Direction of Rotation. The other limitation of these vortex identification methods is that they can only provide iso-surface; no information is given about the rotation axis or vortex direction.

Strength of Vortex. A more serious question is whether the iso-surface represents the rotation strength. The answer is that they do not, because they are different from each other, though not unique; are contaminated by shear and stretching in different degrees and fail to represent the rigid rotation strength of fluid motion.

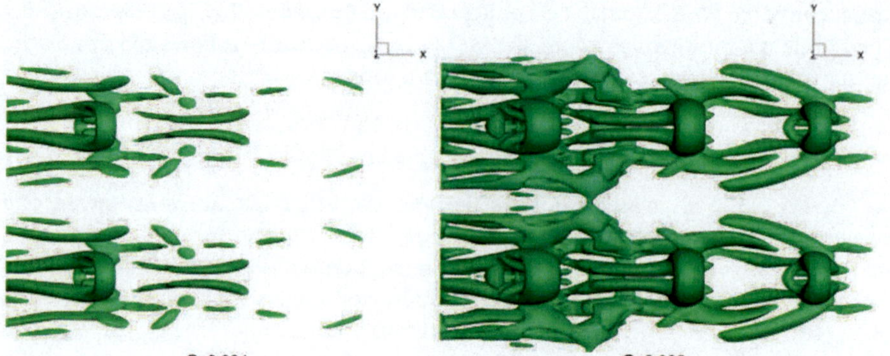

Fig. 1.2 Vortex breakdown with a large threshold of Q= 0.024 and no vortex breakdown with a small threshold of $Q = 0.002$ (same DNS dataset)

1.2.4 Mathematical Misunderstandings of the First and Second Generation of Vortex Identification Methods

The classical theory considers the vorticity vector as the rotation axis and vorticity magnitude as the strength of the vortex (angular speed) [42]. This is correct for solids but not for fluids. It can be shown mathematically that *vorticity is not a fluid rotation axis*, and the first and second generation of vortex identification methods (Q, λ_{ci}, and λ_2) cannot be used as the fluid rotation strength. These methods are contaminated by shear and have different dimensions from the fluid angular speed.

1.2.5 Contaminations of First and Second Generations of Vortex Identification Methods

Contamination of Vorticity. In a principal coordinate, vorticity $\boldsymbol{\omega}$ and its magnitude $\|\boldsymbol{\omega}\|$ can be expressed by Eqs. 1.2 and 1.3, respectively:

$$\boldsymbol{\omega} = (\eta, -\xi, R + \epsilon)^T \tag{1.2}$$

$$\|\boldsymbol{\omega}\| = \sqrt{\eta^2 + \xi^2 + (R + \epsilon)^2} \tag{1.3}$$

From Eqs. 1.2 and 1.3, it can be concluded that a vorticity vector not only includes rotation but also claims shear as part of the vortical structure. Here, ξ, η, ϵ are all shears but not rotations.

Contamination of Q Method. In a principal coordinate, the velocity gradient tensor $\nabla\vec{V}$ and Q value can be written as Eqs. 1.4 and 1.5:

$$\nabla\vec{V} = \begin{bmatrix} \lambda_{cr} & -\frac{1}{2}R & 0 \\ \frac{1}{2}R+\epsilon & \lambda_{cr} & 0 \\ \xi & \eta & \lambda_r \end{bmatrix} = \begin{bmatrix} \lambda_{cr} & \frac{1}{2}\epsilon & \frac{1}{2}\xi \\ \frac{1}{2}\epsilon & \lambda_{cr} & \frac{1}{2}\eta \\ \frac{1}{2}\xi & \frac{1}{2}\eta & \lambda_r \end{bmatrix} + \begin{bmatrix} 0 & -\frac{1}{2}R-\frac{1}{2}\epsilon & -\frac{1}{2}\xi \\ \frac{1}{2}R+\frac{1}{2}\epsilon & 0 & -\frac{1}{2}\eta \\ \frac{1}{2}\xi & \frac{1}{2}\eta & 0 \end{bmatrix} = \boldsymbol{A}_Q + \boldsymbol{B}_Q \tag{1.4}$$

$$\begin{aligned} Q &= \frac{1}{2}\left(\|\boldsymbol{B}_Q\|_F^2 - \|\boldsymbol{A}_Q\|_F^2\right) = \frac{1}{2}\left[2\left(\frac{R}{2}+\frac{\varepsilon}{2}\right)^2 + 2\left(\frac{\xi}{2}\right)^2 + 2\left(\frac{\eta}{2}\right)^2\right] \\ &\quad - \frac{1}{2}\left[2\lambda_{cr}^2 + \lambda_r^2 + 2\left(\frac{\varepsilon}{2}\right)^2 + 2\left(\frac{\xi}{2}\right)^2 + 2\left(\frac{\eta}{2}\right)^2\right] \\ &= \left(\frac{R}{2}\right)^2 + \frac{1}{2}R\cdot\varepsilon - \lambda_{cr}^2 - \frac{1}{2}\lambda_r^2 \end{aligned} \tag{1.5}$$

Therefore, the value of Q is contaminated by shear and stretching. In addition, Q contains the term of square of (R/2) indicating dimensional inconsistency with fluid rotation, as R/2 is the angular speed.

Contamination of λ_{ci} Method. In a principal coordinate, we have:

$$\frac{R}{2}\left(\frac{R}{2}+\epsilon\right) = \lambda_{ci}^2 \tag{1.6}$$

thus,

$$\lambda_{ci} = \sqrt{\frac{R}{2}\left(\frac{R}{2}+\epsilon\right)} \tag{1.7}$$

The expression of λ_{ci} includes ϵ, which is a component of the shear part and thus it is contaminated by shear.

1.2.6 Liutex

1.2.6.1 Definition of Liutex

Liutex is defined as the rigid rotation part of fluid motion [4–6]. The mathematical definition of Liutex is presented by Eq. 1.8 [32]:

$$\vec{R} = R\vec{r} \tag{1.8a}$$

$$R = (\vec{\omega} \cdot \vec{r}) - \sqrt{(\vec{\omega} \cdot \vec{r})^2 - 4\lambda_{ci}^2}, \quad \vec{\omega} \cdot \vec{r} > 0 \tag{1.8b}$$

where $\vec{R}$ and R are Liutex vector and magnitude, $\vec{r}$ is the real eigenvector of $\nabla\vec{v}$, $\vec{\omega} = \nabla \times \vec{v}$ is vorticity, and λ_{ci} is the imaginary part of the conjugate complex eigenvalues of $\nabla\vec{v}$. The condition $\vec{\omega} \cdot \vec{r} > 0$ is used to keep the definition unique and consistent when the fluid motion is pure rotation.

1.2.6.2 Vorticity Versus Vortex

The major mathematical misunderstanding of the first generation of vortex identification methods is the consideration of vorticity vector as the fluid rotation axis and vorticity magnitude as the vortex strength (angular speed). In the following, it is shown that *vorticity is not the fluid rotation axis.*

Definition 1.1 At a moment, a local fluid rotation axis is defined as a vector that can only have stretching (compression) along its length.

It is basic math that the increment of $\vec{v}$ in the direction of $d\vec{r}$ is $d\vec{v} = \nabla\vec{v}.d\vec{r}$.

Theorem 1.1 Liutex is the local fluid rotation axis.

Proof In the Liutex direction, which is the real eigenvector, $d\vec{v} = \nabla\vec{v}.\vec{r} = \lambda_r\vec{r}$.

According to Definition 1.1, Liutex is the local rotation axis as $\vec{R} = R\vec{r}$.

Theorem 1.2 Vorticity is, in general, **not** the local fluid rotation axis.

Proof

$$\begin{aligned} d\vec{v} &= \nabla\vec{v} \cdot \vec{\omega} = \boldsymbol{A} \cdot \vec{\omega} + \boldsymbol{B} \cdot \vec{\omega} = \boldsymbol{A} \cdot \left(a_1\vec{r_1} + a_2\vec{r_2} + a_3\vec{r_3}\right) + (\nabla \times \vec{v}) \times \vec{\omega} \\ &= a_1\lambda_1\vec{r_1} + a_2\lambda_2\vec{r_2} + a_3\lambda_3\vec{r_3} + 0 \neq \lambda\left(a_1\vec{r_1} + a_2\vec{r_2} + a_3\vec{r_3}\right) = \lambda\vec{\omega} \end{aligned} \tag{1.9}$$

Unless $\lambda_1 = \lambda_2 = \lambda_3 = \lambda$ or for rigid rotation where $\lambda_1 = \lambda_2 = 0$.

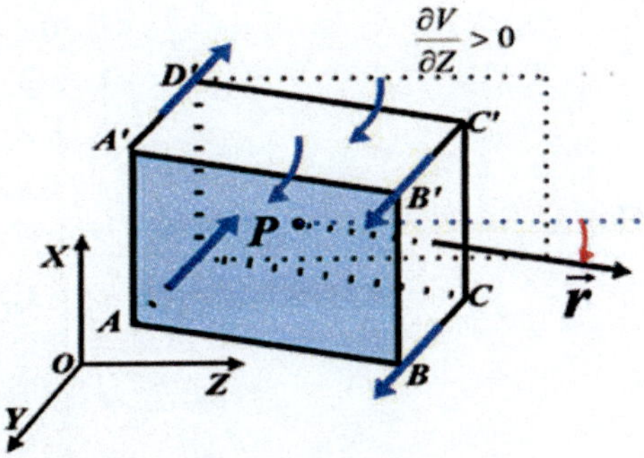

Fig. 1.3 Principal coordinates

1.2.6.3 Principal Coordinate and Principal Decomposition

Principal Coordinate. In the vortex area, the velocity gradient tensor must have one real eigenvalue and two conjugate complex eigenvalues [44]. In Fig. 1.3, the Z-axis of a principal coordinate (X, Y, Z) is aligned with the real eigenvector of $\nabla\vec{v}$ and two diagonal elements that are equal to each other. In the principal coordinate:

$$\nabla\vec{V} = \begin{bmatrix} \frac{\partial U}{\partial X} & \frac{\partial U}{\partial Y} & 0 \\ \frac{\partial V}{\partial X} & \frac{\partial V}{\partial Y} & 0 \\ \frac{\partial W}{\partial X} & \frac{\partial W}{\partial Y} & \frac{\partial W}{\partial Z} \end{bmatrix} = \begin{bmatrix} \frac{\partial U}{\partial X} & \frac{\partial U}{\partial Y} & 0 \\ \frac{\partial V}{\partial X} & \frac{\partial V}{\partial Y} & 0 \\ \frac{\partial W}{\partial X} & \frac{\partial W}{\partial Y} & \lambda_r \end{bmatrix} \tag{1.10}$$

$$\nabla\vec{V}_\theta = \begin{bmatrix} \lambda_{cr} & -\frac{1}{2}R & 0 \\ \frac{1}{2}R + \varepsilon & \lambda_{cr} & 0 \\ \xi & \eta & \lambda_r \end{bmatrix} \tag{1.11}$$

where R is the Liutex magnitude, λ_r is the real eigenvalue, λ_{cr} is the real part of the conjugated complex eigenvalues, and ξ, η, ϵ are shears.

1.2.6.4 UTA R-NR Principal Tensor Decomposition

The first author Dr. Liu has proposed a new Liutex-based tensor decomposition, i.e., *UTA R-NR tenser,* to replace the traditional Cauchy-Stokes (Helmholtz) decomposition. The *UTA R-NR* velocity gradient tensor decomposition can be written as:

$$\nabla\vec{V} = \boldsymbol{R} + \boldsymbol{NR} \tag{1.12}$$

where

$$\boldsymbol{R} = \begin{bmatrix} 0 & -R/2 & 0 \\ R/2 & 0 & 0 \\ 0 & 0 & 0 \end{bmatrix} \tag{1.13}$$

and

$$\boldsymbol{NR} = \begin{bmatrix} \lambda_{cr} & 0 & 0 \\ \epsilon & \lambda_{cr} & 0 \\ \xi & \eta & \lambda_r \end{bmatrix} \tag{1.14}$$

therefore,

$$\nabla \vec{V} = \begin{bmatrix} \lambda_{cr} & -R/2 & 0 \\ R/2 + \epsilon & \lambda_{cr} & 0 \\ \xi & \eta & \lambda_r \end{bmatrix} \tag{1.15}$$

where $\boldsymbol{R}$ stands for the rotational part of the local fluid motion, which is the tensor version of Liutex, and $\boldsymbol{NR}$ is the non-rotational part. It is clear that $\boldsymbol{NR}$ has three real eigenvalues, so $\boldsymbol{NR}$ itself implies no local rotation. The *UTA R-NR decomposition is important to vortical flow and turbulent flow.*

1.2.6.5 Vorticity RS Decomposition

Vorticity cannot be applied to represent flow rotation; otherwise, we would not need vortex identification methods like Q, λ_{ci}, λ_2, Δ. The vorticity vector must be decomposed to a rotational part (Liutex) and a non-rotational part (shear). The vorticity *RS* decomposition can be obtained from Eq. 1.16.

$$\vec{\omega} = \nabla \times \vec{V} = \vec{\boldsymbol{R}} + \vec{\boldsymbol{S}} = \begin{bmatrix} 0 \\ 0 \\ R \end{bmatrix} + \begin{bmatrix} \eta \\ -\xi \\ \epsilon \end{bmatrix} \tag{1.16}$$

where $\vec{\boldsymbol{S}} = \vec{\omega} - \vec{\boldsymbol{R}}$ can be considered as a shearing vector since the components of $\vec{\boldsymbol{S}}$ indicate the strengths of the simple shear along different axes. Note that the vorticity in the vortex legs is almost shear $\vec{\boldsymbol{S}}$, but not rotation $\vec{\boldsymbol{R}}$ (Fig. 1.4). Again, vorticity cannot represent vortex.

1.2.6.6 Liutex-Based Vortex Identification Methods

Liutex Iso-surface. Since Liutex is the rigid rotational part that is extracted from fluid motion, the Liutex vector, Liutex vector lines, Liutex tubes, and Liutex iso-surface

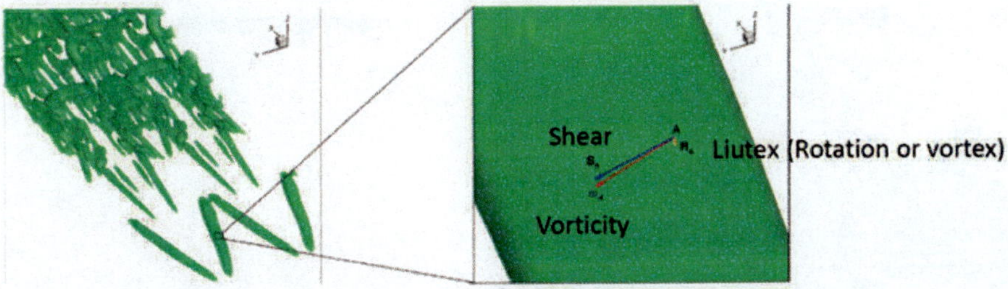

Fig. 1.4 Illustration of vorticity vector decomposition at Point A

can all be applied to display the vortex structure (Fig. 1.5). The advantage of the Liutex method is that Liutex is a vector, unlike others which are all scalar. Another benefit of Liutex is that it represents pure rotation without contamination by shears, while all other vortex identification methods are contaminated by shears. Of course, the Liutex iso-surface still needs thresholds like other scalar methods, but it is the pure rotation strength.

Omega Method. The Ω method is given by Eq. 1.17:

$$\Omega = \frac{\|\boldsymbol{B}\|_F^2}{\|\boldsymbol{A}\|_F^2 + \|\boldsymbol{B}\|_F^2 + \varepsilon} \tag{1.17}$$

where $\boldsymbol{A}$ and $\boldsymbol{B}$ are the symmetric and antisymmetric parts from Cauchy-Stokes decomposition, ε is a small positive number introduced to avoid division by zero or extremely small numbers, and $\|A\|_F$ is the Frobenius norm ε that can be determined at each time step from Eq. 1.18 [19].

$$\varepsilon = 0.001\left(\|\boldsymbol{B}\|_F^2 - \|\boldsymbol{A}\|_F^2\right)_{max} \tag{1.18}$$

The advantages of the Ω method include: (i) it is easily performed, (ii) it has a clear physical meaning, (iii) it is normalized from 0 to 1, (iv) it is insensitive to threshold adjustments by setting $\Omega = 0.52$, and (v) it can capture both strong and weak vortices simultaneously.

Modified Liutex-Omega Method. The modified Liutex-Omega method combines the ideas of both the Liutex and Omega methods, which is normalized, not contaminated, by shear and is insensitive to threshold selection. The modified Liutex-Omega method is defined by Eq. 1.19.

$$\widetilde{\Omega}_{\mathrm{R}} = \frac{\beta^2}{\beta^2 + \alpha^2 + \lambda_{cr}^2 + \frac{1}{2}\lambda_r^2 + \varepsilon} \tag{1.19}$$

where

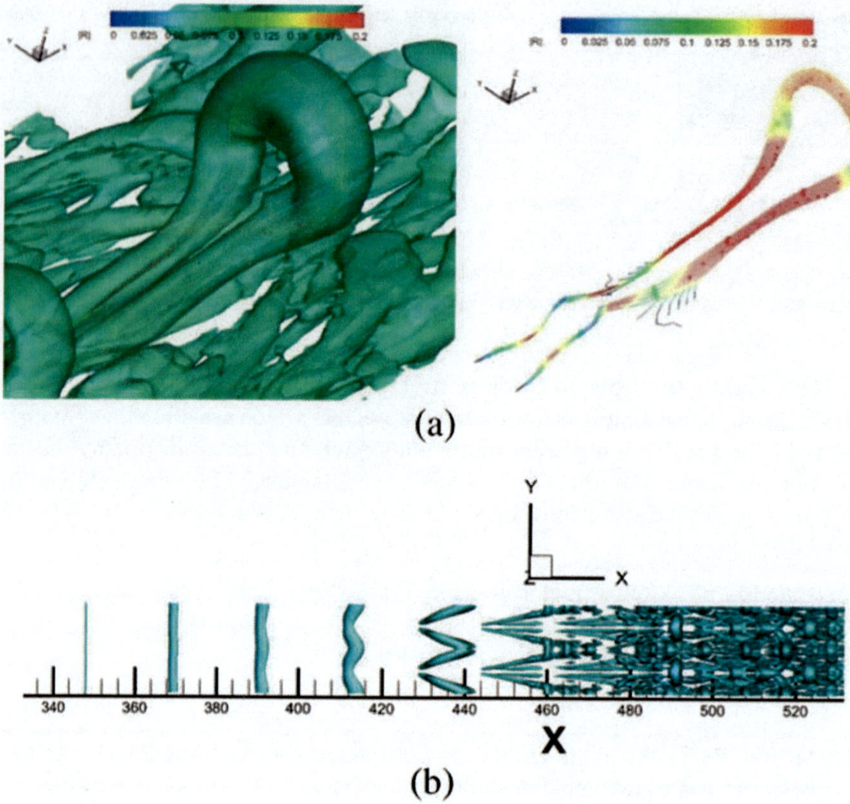

Fig. 1.5 (**a**) Liutex iso-surface and Liutex lines (color represents the rotation strength), and (**b**) Liutex iso-surface for the vortex structure in early transition ($R = 0.1$)

$$\alpha = \frac{1}{2}\sqrt{\left(\frac{\partial V}{\partial Y} - \frac{\partial U}{\partial X}\right)^2 + \left(\frac{\partial V}{\partial X} + \frac{\partial U}{\partial Y}\right)^2} \tag{1.20a}$$

$$\beta = \frac{1}{2}\left(\frac{\partial V}{\partial X} - \frac{\partial U}{\partial Y}\right) \tag{1.20b}$$

Equation 1.20 is equivalent to

$$\widetilde{\Omega}_{\mathrm{R}} = \frac{(\vec{\omega} \cdot \vec{r})^2}{2\left[(\vec{\omega} \cdot \vec{r})^2 - 2\lambda_{ci}^2 + 2\lambda_{cr}^2 + \lambda_r^2\right] + \varepsilon} \tag{1.21}$$

Liutex Core Line Method. All iso-surface methods are threshold-dependent. A Liutex core line is defined as the rotation axis of each vortex and is unique and threshold-free. The vortex core line is defined as a special Liutex line that passes through the points satisfying the condition expressed by Eq. 1.22.

$$\nabla R \times \vec{r} = 0, \quad \vec{r} \neq 0 \tag{1.22}$$

where $\vec{r}$ represents the direction of the Liutex vector. The Liutex (vortex) rotation core lines are uniquely defined without any threshold requirement (Fig. 1.6). Therefore, the Liutex core rotation axis lines with Liutex strength are derived uniquely and are believed to be the only entity capable of cleanly and unambiguously representing vortex structures. Figure 1.6c shows that it is possible to use the automatic Liutex- core-line method to show the vortex structure; still, we have a long way to go from improving the method enough that it shows the vortex structure uniquely and automatically for turbulence. Figure 1.7 provides a comparison of Q-criteria and Liutex-core-line methods and indicates that the Q-criteria is threshold-dependent and unable to capture weak vortices, in contrast to the Liutex-core-line method, which is threshold-free and able to capture both strong and weak vortices.

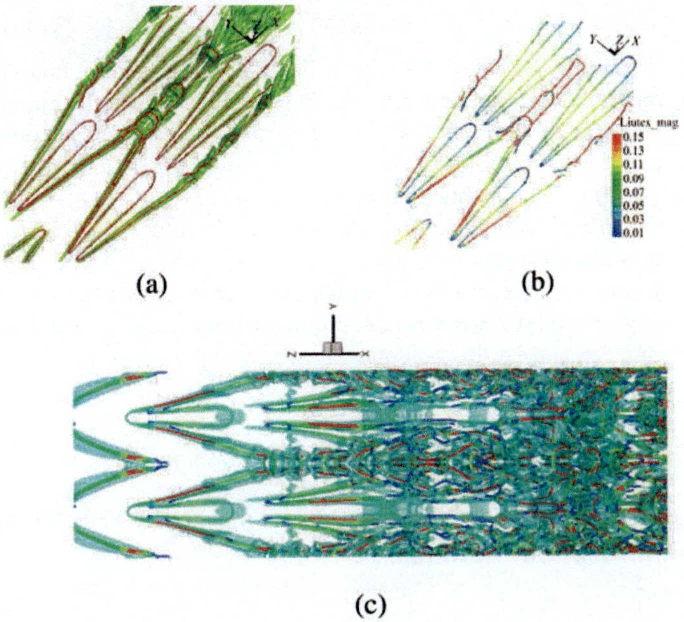

Fig. 1.6 Vortex structure in flow transition displayed by Liutex core line methods (color represents Liutex strength): **(a)** Liutex iso-surface and core lines, **(b)** Liutex core lines, **c** Liutex core lines for flow transition (preliminary results)

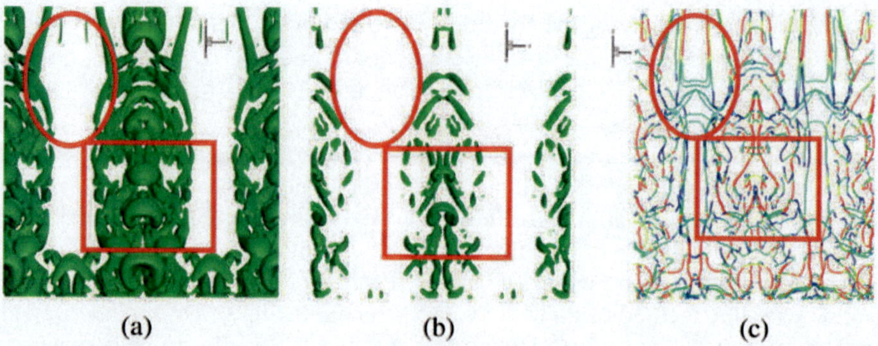

Fig. 1.7 Comparison of Q-criteria and Liutex-core-line methods (a) $Q = 0.005$, (b) $Q = 0.05$, c Liutex-core-line

1.3 Future Research Plan

Future research on Liutex and Litex-based vortex identification methods is inevitable. The following are expected to be achieved through the future research:

(a) a transformational improvement in the knowledge of fluid kinematics and fluid dynamics through a combination of mathematical approaches, computational simulations, and experimental studies, and uncover and quantify the true nature of vortex and its mathematical definition;
(b) new vortex identification methods that provide an accurate and unique vortical structure of turbulence;
(c) quantification of the evolution of vortex geometries and topologies; and
(d) determination of how the new definition of vortex can be used to quantify turbulent flows and vortex dynamics.

The future research may include the following areas.

1.3.1 Develop Unique, Accurate, and Threshold-Free Vortex Identification Methods

Vortex identification methods that are accurate and threshold-free need to be developed. Liutex is a mathematical definition for fluid rotation or vortex, but the iso-surface of Liutex is still threshold-dependent for vortex identification. The modified Liutex-Omega method is insensitive to threshold change and a nice tool for vortex identification; however, it has a small adjustable number of ε and thus is still not unique, as a threshold is still needed. The Liutex core line and tube method is the only one that provides a unique vortex structure by defining a special Liutex line that satisfies $\nabla R \times \vec{r} = 0, \vec{r} \neq 0$. However, Liutex core lines and tubes are currently located by manual methods, which is not realistic for sophisticated vortex structures in turbulent flow. Several efforts have been made, but the outcome is not yet satisfactory. The key issue is how to find the local maxima of Liutex in a 3-D flow field. The Liutex core line should pass the local Liutex maxima. For 1-D problems, the local maxima should have a first-order derivative of zero and second derivative negative. For 2-D and 3-D problems, the local maxima should have a gradient of Liutex equal to zero and the Hessian matrix, which has all negative eigenvalues.

1.3.2 Characterize Vortex Structure

The vortex structure can be studied using experimental modeling and numerical simulations to characterize the vortex structure in flow transition and turbulent flows. The mechanism of flow transition and turbulence generation and sustenance should be explored using experimental and numerical techniques.

Vortex structure is still a mystery. Even very simple questions, e.g., whether hairpin vortices exist in turbulent flows, remain unanswered. In fact, not all researchers believe that hairpin packets exist in turbulent flow. In recent studies, large numbers of hairpin packets were observed in low Reynolds number flows [44], but no hairpin vortices were observed at high Reynolds numbers [45, 46]. This disagreement exists since there is no rigorous mathematical vortex definition. Additionally, the iso-surface of existing vortex identification methods are all threshold-dependent but not unique. The new vortex identification may be used to find unique vortex structures. Flow visualization can provide hints on how large vortices are generated and how they become non-symmetric and chaotic. A qualitative comparison between the results from PIV and DNS and statistical analysis of flow field can be used to extract vortex characteristics at the range of $Re_\delta = 500$–4000. This range of friction Reynolds number covers transitional flows and low-to-high Reynolds number turbulent flows. The vortex structure and its temporal and spatial evolution can be studied by conducting experiments to capture the hairpin structure, multi-layer vortices, and vortex merging.

1.3.3 Liutex Similarity

The energy transformation paths between the large and small vortices will be investigated by DNS and experiments. The most successful theory of turbulence is the *second Kolmogorov similarity K41* [47] in the inertial subrange ($l_{EI} > l > l_{DI}$). According to K41, the energy spectrum $E_k(f)$ of turbulence must be of the form of Eq. 1.23, which is the famous −5/3 law.

$$E_k(f) = C\varepsilon_0^{\frac{2}{3}} f^{-\frac{5}{3}} \tag{1.23}$$

where f is the wavenumber and ε_0 is the turbulence energy dissipation. The similarity is based on Kolmogorov's third hypothesis, which assumes that in the inertial subrange, the turbulence energy spectrum is solely determined by energy dissipation and the wave number, independent of viscosity. This will give $E_k(f) = C\varepsilon_0^a f^b$, where the dimension of wave number f is *1/m.* The power coefficients a and b are easily obtained by dimensional analysis; however, Kolmogorov's law requires homogeneous incompressible flow with a very high Reynolds number and is for the inertial subrange. It is hard to match DNS data and experimental results in a turbulent boundary layer.

There is no similarity for vorticity and Q-criterion (Fig. 1.8a, b); however, the Liutex similarity [36] is found in a low Reynolds number turbulent boundary layer in the subrange of dissipation (Fig. 1.8c), which is important for a subgrid model in LES. The Liutex similarity, which has been reported by a number of authors, cannot be an accident because it is very meaningful for finding turbulence structure and turbulence sub-grid modeling. However, it is still unknown why the Liutex spectrum meets the −5/3 law and how the Liutex −5/3 law can be used for turbulence modeling. The exact Liutex similarity provides a powerful tool for studying turbulence structure and developing a reliable subgrid model of large eddy simulation, which is one of the goals of this research. Similarity is the foundation for subgrid modeling.

1.3.4 Correlation Analysis of Pressure Fluctuation (Noise Generation) and Liutex Spectrum by Mathematical Analysis and Experimental Modeling

A high-order large eddy simulation (LES) of a micro vortex generator (MVG) is conducted to control the flow separation induced by shock and the turbulent boundary layer interaction (SBLI) at Mach number 2.4 in a compression corner. A low frequency of pressure oscillation caused by SBLI has been one of the major hurdles of supersonic commercial aircraft design for decades. Liutex is a kinematic definition that works for both incompressible and compressible flow. We found that pressure oscillation and the Liutex spectrum are closely correlated, over 0.9 at most points (Fig. 1.9 and Table 1.1), which could help in finding the mechanism of the low-frequency noise generation and learning how to reduce or remove the noise. It cannot be an accident that the correlation is high for the turbulent boundary layer, as shown in a recent study. As the pressure fluctuation is found closely correlated with the Liutex spectrum, it is hoped that a new theory can be developed on shock waves and turbulent boundary layer interactions that will lead to new technology for controlling SBLI, which is the bottleneck of commercial aircraft design. The theory and control of SBLI require additional research.

A new technique for controlling the supersonic boundary layer flow is to distribute an array of MVGs, whose height is less than the boundary layer thickness, ahead of the region with adverse flow conditions. In contrast to the conventional vortex

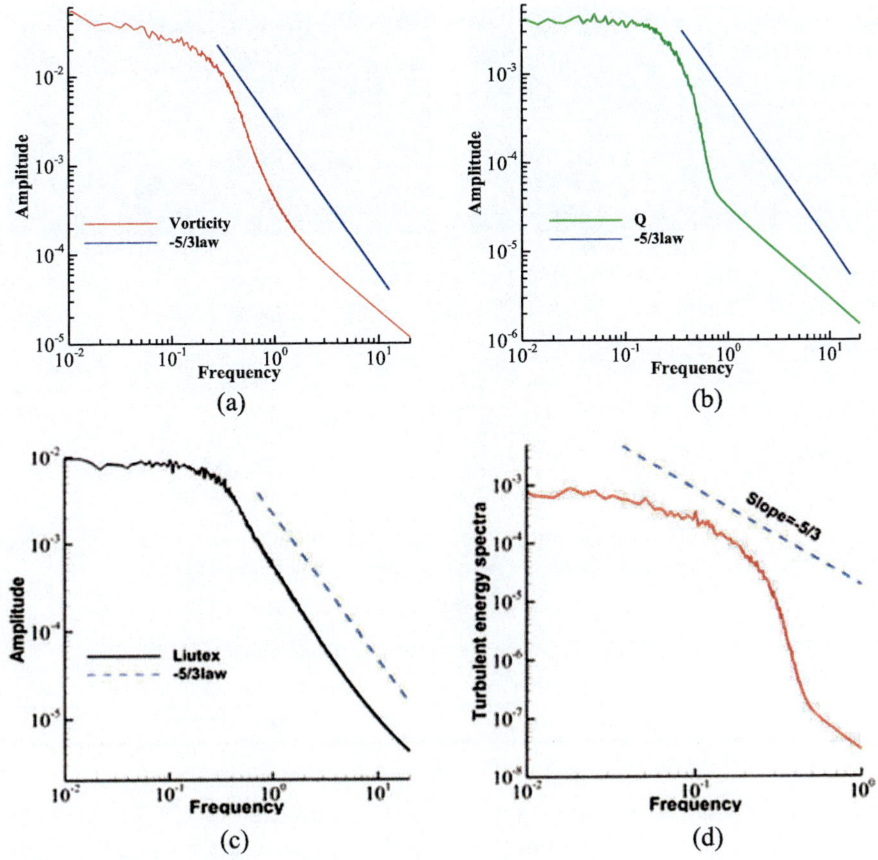

Fig. 1.8 Energy spectrum: (**a**) vorticity, (**b**) Q-criterion, (**c**) Liutex −5/3 similarity and (**d**) Kolmogorov K41 similarity

generator with a height comparable to the boundary layer thickness, an MVG's height is about 20–60% of the boundary layer thickness.

An MVG appears to reduce the size of the separation zone. It has been suggested that the boundary layer is energized by the MVGs from a system of streamwise counter-rotating vortices. Such a mechanism has recently been studied in detail in low-speed experiments. A series of ring-like vortices (Fig. 1.8b) generated behind MVG was discovered by our previous LES, and the numerical discoveries of the ring-like vortex structure were confirmed by the 3-D PIV experiment. A V-shaped separation zone was discovered on the wall boundary by the LES. The length of the separation at the two spanwise sides of the domain was estimated to be around 6–6.5 h (height of the MVG), and the length at the centerline was estimated to be around 5 h in supersonic flow with Ma = 2.5. Compared with the value of 8.2–8.4 h for the ramp-only case, the MVG significantly reduces the separation region. It was also confirmed that the interaction between ring-like vortices and a ramp shock wave is the mechanism for forming the V-shaped separation zone and the flow

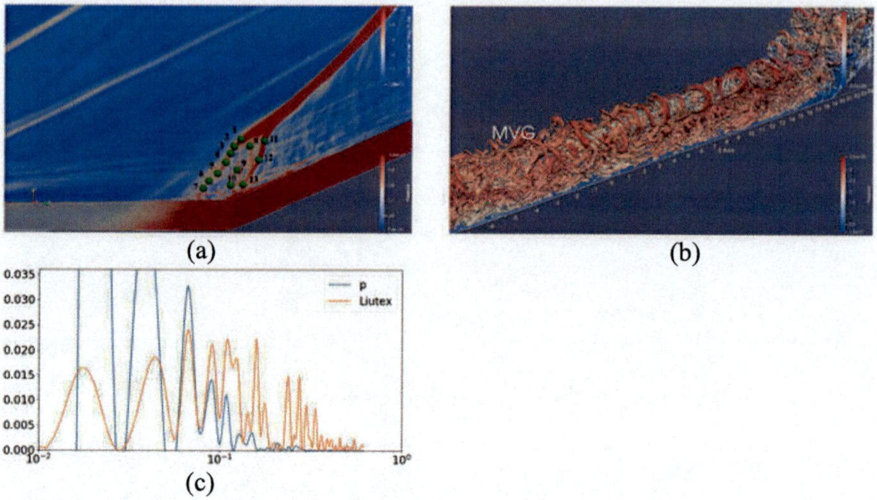

Fig. 1.9 Correlation of spectrum of pressure fluctuation and Liutex: (**a**) 13 sample points, (**b**) shock boundary layer interaction, (**c**) spectrum of pressure, and Liutex at Point 5

Table 1.1 Correlation Coefficient (r) of the Spectra at the First Seven Points in Fig. 1.9a

Point #	1	2	3	4	5	6	7
r	88.4%	86.3%	93.7%	95.5%	94.6%	97.0%	91.4%

separation reduction at the ramp corner. The vortex structure in MVG needs to be examined through numerical simulations and be compared the results with those from experiments.

1.4 Conclusions and Future Work

Based on abovementioned understanding, following conclusions can be made:

1. As the first generation of vortex identification, vorticity is a vector, but cannot be used to represent vortex. Vorticity is curl of velocity, but vortex is fluid rotation. They are two different vectors.
2. As the second generation of vortex identification criteria, Q, Δ, λ2, and λci methods are all scalars and thus strongly dependent on the arbitrary thresholds. They cannot show the vortex rotation axis, and are all contaminated by stretching (compression) and shearing.
3. Liutex is the third generation of vortex definition and identification, which is the only quantified and mathematical definition of local fluid rotation or vortex vector.

4. Liutex can give right direction and strength for group fluid rotation or natural vortices. In other words, vortex core is a special Liutex vector, Liutex tube, and concentration of Liutex lines.
5. Liutex vector, line, tube, core lines can all be correctly applied for vortex structure, but vortex structure cannot be visualized by the first and second generations since they are threshold-dependent, not unique, and shear-contaminated.
6. Modified Liutex-Omega method is insensitive to threshold change.
7. Liutex-core-line method is unique, threshold-free, accurate to visualize vortex structure including direction and strength.

Future research should integrate mathematical analysis, direct measurements, and numerical simulations to advance the knowledge and understanding of vortex structure and evolution. These advancements will improve vortex identification and quantification methods. The new vortex identification methods are accurate, unique, and threshold-free and will be used to formulate Liutex dynamics for quantified research on vortex and turbulence. The Liutex similarity will provide a powerful tool for studying turbulence structure and developing a reliable sub-grid model of LES. The results from future research will lead to quantified research of turbulence generation and sustenance and pave the way for developing new governing equations using Liutex, which may enhance turbulence calculations and modeling. These outcomes will significantly impact basic mathematics and fundamental fluid mechanics by introducing a new definition of vortex into turbulence research. New fluid kinematics can be established using the Liutex theoretical system to replace existing fluid kinematics and possibly launch new turbulence dynamics.

Acknowledgements The first author is grateful to the University of Texas at Arlington (UTA) for long-time support as a tenured faculty member. The authors also thank Texas Advanced Computing Center (TACC) for long-term support in the provision of computation hours.

Data Availability The data supporting this study's findings are available from the corresponding author upon reasonable request.

References

1. D. Küchemann, Report on the IUTAM symposium on concentrated vortex motions in fluids. J. Fluid Mech. **21**, 1–20 (1965)
2. H. Helmholtz, Über Integrale der hydrodynamischen Gleichungen, welche den Wirbelbewegungen entsprechen. Journal für die reine und angewandte Mathematik **55**, 25–55 (1858)
3. S. Robinson, S. Kline, P. Spalart, A review of quasi-coherent structures in a numerically simulated turbulent boundary layer. Tech. rep., NASA TM-102191(1989)
4. C. Liu, Y. Gao, S. Tian, X. Dong, Rortex—A new vortex vector definition and vorticity tensor and vector decompositions. Phys. Fluids **30**, 035103 (2018). https://doi.org/10.1063/1.5023001
5. Y. Gao, C. Liu, Rortex and comparison with eigenvalue-based vortex identification criteria. Phys. Fluids **30**, 085107 (2018). https://doi.org/10.1063/1.5040112

6. C. Liu, Y. Gao, X. Dong, J. Liu, Y. Zhang, X. Cai, N. Gui, Third generation of vortex identification methods: omega and Liutex/Rortex based systems. J. Hydrodyn. **31**(2), 1–19 (2019). https://doi.org/10.1007/s42241-019-0022-4
7. J. Hunt, A. Wray, P. Moin, Eddies, streams, and convergence zones in turbulent flows. Center for turbulence research proceedings of the summer program, 193 (1988)
8. M. Chong, A. Perry, B. Cantwell, A general classification of three-dimensional flow fields. Phys. Fluids A **2**, 765–777 (1990)
9. J. Jeong, F. Hussain, On the identification of a vortices. J. Fluid Mech. **285**, 69–94 (1995)
10. J. Zhou, R. Adrian, S. Balachandar, T. Kendall, Mechanisms for generating coherent packets of hairpin vortices in channel flow. J. Fluid Mech. **387**, 353–396 (1999)
11. P. Chakraborty, S. Balachandar, R.J. Adrian, On the relationships between local vortex identification schemes. J. Fluid Mech. **535**, 189–214 (2005)
12. V. Kolá˘r, J. Šístek, Consequences of the close relation between Rortex and swirling strength. Phys. Fluids **32**, 091702 (2020). https://doi.org/10.1063/5.0023732
13. V. Kolář, Vortex identification: new requirements and limitations. Int. J. Heat Fluid Flow **28**(4), 638–652 (2007)
14. Z. Li, Zhang, X., He, F., Evaluation of vortex criteria by virtue of the quadruple decomposition of velocity gradient tensor. Acta Physics Sinica, **63**(5), 054704 (2014), in Chinese
15. C. Liu, Y. Wang, Y. Yang, Z. Duan, New Omega vortex identification method. Sci. China Phys. Mech. Astron. **59**, 684711 (2016)
16. Y. Gao, J. Liu, Y. Yu, C. Liu*, A Liutex based definition and identification of vortex core center lines. J. Hydrodyn. **31**(2), 774–781 (2019a)
17. Y. Gao, Y. Yu, J. Liu, C. Liu*, Explicit expressions for Rortex tensor and velocity gradient tensor decomposition. Phys. Fluids **31**, 081704 (2019b)
18. Y. Gao, C. Liu*, Rortex based velocity gradient tensor decomposition. Phys. Fluids **31**(1), 011704 (2019c)
19. X. Dong, Y. Wang, X. Chen, Y. Zhang, C. Liu, Determination of epsilon for Omega vortex identification method. J. Hydrodyn. **30**(4), 541–548 (2018)
20. X. Dong, Y. Yan, Y. Yang, G. Dong and C. Liu*, Spectrum study on unsteadiness of shock wave -vortex ring interaction. Phys. Fluids **30**, 056101 (2018). https://doi.org/10.1063/1.5027299, with (2018b)
21. X. Dong, S. Tian, C. Liu*, Correlation analysis on volume vorticity and vortex in late boundary layer transition. Phys. Fluids **30**, 014105 (2018c)
22. X. Dong, G. Dong, C. Liu*, Study on vorticity structures in late flow transition. Phys. Fluids **30**, 104108 (2018d)
23. X. Dong, Y. Gao, C. Liu*, New normalized Rortex/vortex identification method. Phys. Fluids **31**, 011701 (2019). https://doi.org/10.1063/1.5066016
24. X. Dong, X. Cai, Y. Dong, C. Liu*, POD analysis on vortical structures in MVG wake by Liutex core line identification. J. Hydrodyn. **32**, 497–509 (2020)
25. J. Liu, Y. Gao, C. Liu*, An objective version of the Rortex vector for vortex identification. Phys. Fluids **31**(6), 065112 (2019a). https://doi.org/10.1063/1.5095624
26. J. Liu, C. Liu*, Modified normalized Rortex/vortex identification method. Phys. Fluids **31**(6), 061704 (2019b). https://doi.org/10.1063/1.5109437
27. J. Liu, Y. Gao, Y. Wang, C. Liu*, Galilean invariance of Omega vortex identification method. J. Hydrodyn. (2019c). https://doi.org/10.1007/s42241-019-0024-2
28. J. Liu, Y. Gao, Y. Wang, C. Liu*, Objective Omega vortex identification method. J. Hydrodyn. (2019d). https://doi.org/10.1007/s42241-019-0028-y
29. J. Liu, Y. Deng, Y. Gao, S. Charkrit, C. Liu*, Mathematical foundation of turbulence generation from symmetric to asymmetric Liutex. J. Hydrodyn. **31**(3), 632–636 (2019e)
30. Y. Wang, Y. Yang, G. Yang, C. Liu*, DNS study on vortex and vorticity in late boundary layer transition. Comm. Comp. Phys. **22**, 441–459 (2017)
31. Y. Wang, Y. Gao, C. Liu*, Galilean invariance of Rortex. Phys. Fluids **30**, 111701 (2018). https://doi.org/10.1063/1.5058939

32. Y. Wang, Y. Gao, J. Liu, C. Liu*, Explicit formula for the Liutex vector and physical meaning of vorticity based on the Liutex-Shear decomposition. J. Hydrodyn. (2019a). https://doi.org/10.1007/s42241-019-0032-2
33. Y. Wang, Y. Gao, C. Liu*, Letter: Galilean invariance of Rortex. Phys. Fluids **30**(11), 111701 (2019b)
34. Y. Wang, Y. Gao, H. Xu, X. Dong, J. Liu, W. Xu, M. Chen, C. Liu*, Liutex theoretical system and six core elements of vortex identification. J. Hydrodyn. **32**, 197–221 (2020)
35. W. Xu, Y. Gao, Y. Deng, J. Liu, C. Liu*, An explicit expression for the calculation of the Rortex vector. Phys. Fluids **31**, 095102 (2019a). https://doi.org/10.1063/1.5116374
36. W. Xu, Y. Wang, Y. Gao, J. Liu, H. Dou, C. Liu*, Liutex similarity in turbulent boundary layer. J. Hydrodyn. **31**(6), 1259–1262 (2019b)
37. H. Xu, X. Cai, C. Liu*, Liutex core definition and automatic identification for turbulence structures. J. Hydrodyn. **31**(5), 857–863 (2019)
38. Y. Zhang, X. Qiu, F. Chen, K. Liu, Y. Zhang, X. Dong, C. Liu*, A selected review of vortex identification methods with applications. J. Hydrodyn. **30**(5) (2018). https://doi.org/10.1007/s42241-018-0112-8
39. Y. Zhang, X. Wang, Y. Zhang, C. Liu, Comparisons and analyses of vortex identification between Omega method and Q criterion. J. Hydrodyn. **31**(2), 224–230 (2019)
40. C. Liu, H. Xu, X. Cai, Y. Gao, Liutex and Its Applications in Turbulence Research, ISBN-13: 978–0128190234, ISBN-10: 012819023X, Elsevier, Oct 2020a
41. C. Liu, Y. Gao, Liutex-based and Other Mathematical, Computational and Experimental Methods for Turbulence Structure, Vol. 2, ISSN: 2589–2711,eISSN: 2589–272X (Online), ISBN: 978–981–14–3758–8, eISBN: 978–981–14–3760–1 (Online), Bethman, April 2020b
42. C. Truesdell. The Kinematics of Vorticity. (Indiana University Publications Science Seres Nr. 14.) XVII + 232 S. Bloomington (1954). Indiana University Press
43. S. Robinson, A review of vortex structures and associated coherent motions in turbulent boundary layers, in *Structure of Turbulence and Drag Reduction*, Springer, Berlin, Heidelberg (1990)
44. X. Wu, P. Moin, Direct numerical simulation of turbulence in a nominally zero-pressure gradient flat-plate boundary layer. J. Fluid Mech. **630**, 5–41 (2009)
45. I. Marusic, B.J. McKeon, P.A. Monkewitz, H.M. Nagib, A.J. Smits, K.R. Sreenivasan6, Wall-bounded turbulent flows at high Reynolds numbers: recent advances and key issues. Phys. Fluids **22**, 065103 (2010)
46. J. Jiménez, Coherent structures in wall-bounded turbulence. J. Fluid Mech. **842**, P1 (2018). https://doi.org/10.1017/jfm.2018.144
47. AN. Kolmogorov Local structure of turbulence in an incompressible fluid at very high Reynolds numbers. Dokl. Akad. Nauk. SSSR 26: 115–18 (1941)

Chapter 2
Galilean Variance of Streamline in Vortex/Liutex Visualization

Yifei Yu and Chaoqun Liu

Abstract Intuitively, vortex exists at the position where the streamlines bend. This perception comes from the fact that people usually observe water flow, which is the streamline, by eyes. However, since streamline is a Galilean invariant quantity, this intuition may be incorrect in some situations. Galilean variance is the feature that the issue will be different under different coordinate systems, while Galilean invariance refers to keeping invariant under different inertial coordinate systems. No bending of the streamline can be identified when the streamwise speed is far larger than the speed produced by rotation; in other words, the speed produced by rotation can be omitted. In such a situation, streamline fails to detect vortex. Therefore, a proper vortex visualization method requires Galilean invariance. Liutex is a newly invented vortex identification method that is Galilean invariant, and as a result, it can capture vortex correctly regardless of the choice of coordinate systems.

2.1 Introduction

Vortices play an indispensable role in fluid dynamics research, especially turbulence. Therefore, vortex visualization has a very important research value. Influenced by the fact that vorticity represented twice angular speed for rigid bodies, scientists had first believed that vorticity is also a good indicator of fluid rotation. Using this cognitive, some successes were achieved in analyzing low-speed flow away from the boundary. At the same time, problems arose when scholars shifted their research focus to high-speed flow and flow near the boundary region. Robinson [1] found that the connection between vorticity and actual vortices can be very weak. Wang et al. [2] found vorticity strength is smaller inside the vortex region while bigger outside. The causes of this problem can be explained by R-S decomposition [3] of vorticity. R-S decomposition shows that vorticity can be decomposed into rigid rotation and shear. For rigid bodies and low-speed flow away from the boundary, the shear is small or zero if ideal, but for high-speed flow or flow near the boundary, the shear cannot

Y. Yu (✉) · C. Liu
Department of Mathematics, The University of Texas at Arlington, Arlington, TX, USA
e-mail: yifei.yu@mavs.uta.edu

Y. Wang et al. (eds.), *Liutex and Third Generation of Vortex Identification*,
Springer Proceedings in Physics 288, https://doi.org/10.1007/978-981-19-8955-1_2

be ignored. Later, some vortex visualization methods were developed to overcome this drawback of using vorticity for detecting vortices. Some popular methods of these are the Q criterion [4], λ_{ci} criterion [5] and λ_2 criterion [6]. Admittedly, using these methods can obtain more consistent results with the experimental results than vorticity, these methods also have some shortcomings.

Firstly, these methods cannot reveal what the rotation axis, important rotation information, is since they are all scalar methods. Secondly, the physical meaning of the numbers calculated from these methods is unclear. People can only know the relative rotation strength but do not know the exact angular speed. These shortcomings motivate researchers to find more proper vortex indicators. In 2018, Liu et al. [7, 8] proposed a new method called "Liutex", which is a vector. The direction of Liutex is the rotation axis, and the magnitude of Liutex represents twice rigid rotation angular speed. Many numerical and experimental results have shown that Liutex performs beyond the previous methods. In the experiments done by Guo et al. [9], Liutex is more consistent with the experimental results than other methods.

Apart from all the methods discussed above, the most human intuitive way to identify vortices is the curving of the streamline; since what people can observe in daily life is the motion of fluids which is streamline or trace in fluid concepts. Streamline is based on velocity, which is dependent on the choice of coordinates. Due to this reason, some weird results can be obtained when using streamlines to detect vortices.

People may conclude that there are no vortices within the iso-surface from Fig. 2.1a and apparent vortex from Fig. 2.1b, even though these two figures depict the same scene at the same moment. Choice of the coordinates is the only difference between these two figures. The boundary-fixed coordinate is used in Fig. 2.1a, while a uniform linear motion coordinate is used in Fig. 2.1b, and both coordinates are inertial. Therefore, different choices of coordinates may lead to different vortices detections when applying streamline. This issue can be described by Galilean invariance.

Galilean invariance is a feature in that the variables are the same under different inertial coordinates. Obviously, streamline is Galilean variant since velocity is Galilean variant. Galilean invariance should be a vital property of vortex visualization to avoid the ambiguous vortex detection results shown in Fig. 2.1.

In this paper, Galilean invariance is first introduced in Sect. 2.2. Section 2.3 explains why streamline is Galilean variant, and the proof of Liutex being Galilean invariant is shown in Sect. 2.4. In Sect. 2.5, numerical examples illustrate that streamline is Galilean variant and Liutex is Galilean invariant. Conclusions are given in Sect. 2.6.

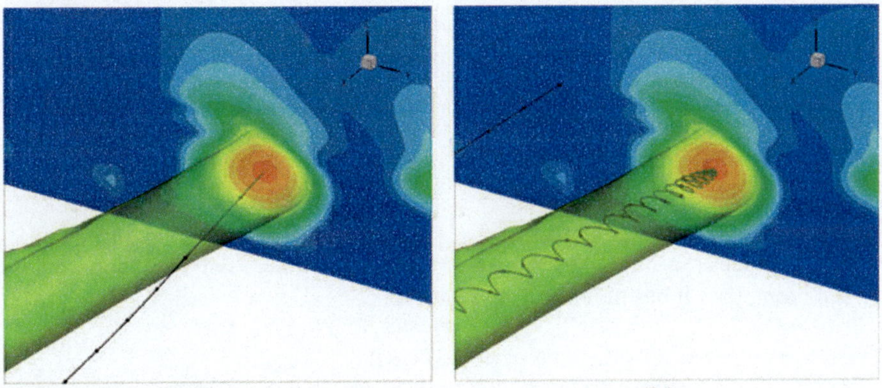

Fig. 2.1 Streamline in the **a** original frame **b** the frame moving with the u speed of the intersection point

2.2 Galilean Invariance

Galilean transformation is used to transform between two different inertial coordinates or two coordinates that only differ by a constant translation within Newtonian physics. The Galilean transformation can be uniquely defined as the decomposition of a rotation, a translation, and a uniform motion of space–time.

$$\begin{bmatrix} x' \\ y' \\ z' \end{bmatrix} = Q \begin{bmatrix} x \\ y \\ z \end{bmatrix} + \vec{c}_1 t + \vec{c}_2 = rotation + uniform\, motion + translation \tag{2.1}$$

where x, y, z are the coordinates in the original reference frame and x', y', z' are the coordinates in the new reference frame.

Galilean invariance means variables are the same in different inertial coordinates, so in other words, variables need to be the same under Galilean transformations. The Galilean variances of streamline and Liutex are checked in Sects. 2.3 and 2.4, respectively.

2.3 Galilean Variance of Streamline

Let u', v', w' be the velocity after Galilean transformation, and u, v, w be the velocity in the original coordinate. From Eq. 2.1, it can be derived that

$$\begin{bmatrix} u' \\ v' \\ w' \end{bmatrix} = \frac{d}{dt}\left(\begin{bmatrix} x' \\ y' \\ z' \end{bmatrix}\right) = \frac{d}{dt}\left(Q\begin{bmatrix} x \\ y \\ z \end{bmatrix} + \vec{c}_1 t + \vec{c}_2\right) = Q\begin{bmatrix} \frac{dx}{dt} \\ \frac{dy}{dt} \\ \frac{dz}{dt} \end{bmatrix} + \vec{c}_1$$
$$= Q\begin{bmatrix} u \\ v \\ w \end{bmatrix} + \vec{c}_1 \tag{2.2}$$

A streamline is a line whose tangent direction is the velocity direction of the chosen point, thus it has the following relation

$$d\vec{s} \times \vec{V} = 0 \tag{2.3}$$

where $\vec{s}$ represents streamline. Since the velocities in different coordinates are different, $\vec{V} \neq \vec{V}'$, their streamlines are different. Thus, streamline is not a Galilean invariant physical quantity.

2.4 Galilean Invariance of Liutex

This section gives the proof of Liutex being Galilean invariant based on Wang's paper [10]. Liutex [8] is a vector, which can be expressed as

$$\vec{R} = R\vec{r} \tag{2.4}$$

where R is the magnitude and $\vec{r}$ is the direction of Liutex. $\vec{r}$ is meanwhile the real eigenvector of the velocity gradient tensor with $\vec{r} \cdot \vec{\omega} > 0$, and R can be calculated from [11]

$$R = \vec{r} \cdot \vec{\omega} - \sqrt{(\vec{r} \cdot \vec{\omega})^2 - 4\lambda_{ci}^2} \tag{2.5}$$

where λ_{ci} is the imaginary part of the complex conjugate eigenvalues.

To prove Liutex is Galilean invariant, it needs to show that its direction and magnitude are Galilean invariant.

Let $\nabla\vec{v}'$ and $\nabla\vec{v}$ be the velocity gradient tensor in the new and original reference frames, respectively. Let Q be the invertible rotation matrix of these two reference frames. Then $\nabla\vec{v}'$ and $\nabla\vec{v}$ have the following relation

$$\nabla\vec{v}' = \begin{bmatrix} \frac{\partial u'}{\partial x'} & \frac{\partial u'}{\partial y'} & \frac{\partial u'}{\partial z'} \\ \frac{\partial v'}{\partial x'} & \frac{\partial v'}{\partial y'} & \frac{\partial v'}{\partial z'} \\ \frac{\partial w'}{\partial x'} & \frac{\partial w'}{\partial y'} & \frac{\partial w'}{\partial z'} \end{bmatrix} = Q\nabla\vec{v}Q^{-1} = Q\begin{bmatrix} \frac{\partial u}{\partial x} & \frac{\partial u}{\partial y} & \frac{\partial u}{\partial z} \\ \frac{\partial v}{\partial x} & \frac{\partial v}{\partial y} & \frac{\partial v}{\partial z} \\ \frac{\partial w}{\partial x} & \frac{\partial w}{\partial y} & \frac{\partial w}{\partial z} \end{bmatrix} Q^{-1} \tag{2.6}$$

Suppose $\vec{r}$ is the Liutex direction in the original xyz coordinate, in other words, $\vec{r}$ is the eigenvector of $\nabla\vec{v}$.

$$\nabla\vec{v}\cdot\vec{r}=\lambda_r\vec{r} \tag{2.7}$$

Do the following manipulation, it has

$$Q\nabla\vec{v}\cdot\vec{r}=Q\nabla\vec{v}\left(Q^{-1}Q\right)\cdot\vec{r}=\left(Q\nabla\vec{v}Q^{-1}\right)Q\cdot\vec{r}=\nabla\vec{v}'Q\cdot\vec{r} \tag{2.8}$$

Equation 2.7 both sides multiplied by Q from the left

$$Q\nabla\vec{v}\cdot\vec{r}=\lambda_r Q\vec{r} \tag{2.9}$$

Combine Eqs. 2.8 and 2.9, it can be obtained

$$\nabla\vec{v}'\cdot(Q\vec{r})=\lambda_r(Q\vec{r}) \tag{2.10}$$

From this, it can be seen that $Q\vec{r}$ is the real eigenvector of $\nabla\vec{v}'$.

Suppose $\vec{r}=\left[x_1\ y_1\ z_1\right]^T-\left[x_0\ y_0\ z_0\right]^T$ and $\vec{r}'=\left[x_1'\ y_1'\ z_1'\right]^T-\left[x_0'\ y_0'\ z_0'\right]^T$ where $\vec{r}'$ is the corresponding vector of $\vec{r}$ after Galilean transformation.

$$\begin{bmatrix}x_0'\\y_0'\\z_0'\end{bmatrix}=Q\begin{bmatrix}x_0\\y_0\\z_0\end{bmatrix}+\vec{c}_1t+\vec{c}_2 \tag{2.11}$$

$$\begin{bmatrix}x_1'\\y_1'\\z_1'\end{bmatrix}=Q\begin{bmatrix}x_1\\y_1\\z_1\end{bmatrix}+\vec{c}_1t+\vec{c}_2 \tag{2.12}$$

$$\vec{r}'=\begin{bmatrix}x_1'\\y_1'\\z_1'\end{bmatrix}-\begin{bmatrix}x_0'\\y_0'\\z_0'\end{bmatrix}=Q\left\{\begin{bmatrix}x_1\\y_1\\z_1\end{bmatrix}-\begin{bmatrix}x_0\\y_0\\z_0\end{bmatrix}\right\}=Q\vec{r} \tag{2.13}$$

So, $\vec{r}'$ is the real eigenvector of $\nabla\vec{v}'$.

In the expression of R (Eq. 2.5), it contains $\vec{\omega}$, $\vec{r}$ and λ_{ci}. It has been proved that $\vec{r}$ is Galilean invariant, and it is known to us that $\vec{\omega}$ and λ_{ci} are both Galilean invariant physical quantities. So, R is Galilean invariant as well.

It has been proved that both magnitude and direction of Liutex are Galilean invariant. Thus, Liutex is a Galilean invariant physical quantity.

2.5 Numerical Example

In a DNS research of boundary layer transition, the grid level is 1920 × 128 × 241, representing the total amount in streamwise(x), spanwise(y), and wall-normal(z) directions. The first interval length in the normal direction at the origin is 0.43 in wall units ($Z^+ = 0.43$). The flow parameters are listed in Table 2.1, including Mach number, Reynolds number, and others. In this case, $x_{in} = 300.79\delta_{in}$ represents the distance between the leading edge and the inlet, Lx, Ly, and $\mathrm{Lz_{in}}$ are the lengths of the computational domain and T_w is the wall temperature.

The streamline passing through a selected point is shown in Fig. 2.2. The iso-surface is the iso-surface of Liutex, and the black line is the streamline. In this figure, the streamline looks like a straight line by eye observation which implies no rotation exists according to people's intuition.

However, if discussed in a new frame, the conclusion of rotation existence based on the curve of streamlines can be different. Establish a new frame such that the frame moves with the streamwise speed of the selected point (intersection of the plane and black line). The streamline in the new frame is shown in Fig. 2.3.

In the new figure, the streamline looks like a spiral. From these two figures, it can be seen clearly that streamlines are not Galilean invariant.

To test the Galilean invariance of Liutex, a rotation matrix Q is selected. Let $\gamma_1 = 36°$, $\gamma_2 = 69°$, $\gamma_3 = 78°$, then

Table 2.1 Flow parameters

M_∞	Re	x_{in}	Lx	Ly	$\mathrm{Lz_{in}}$	T_w	T_∞
0.5	1000	$300.79\delta_{in}$	$798.03\delta_{in}$	$22\delta_{in}$	$40\delta_{in}$	273.15 K	273.15 K

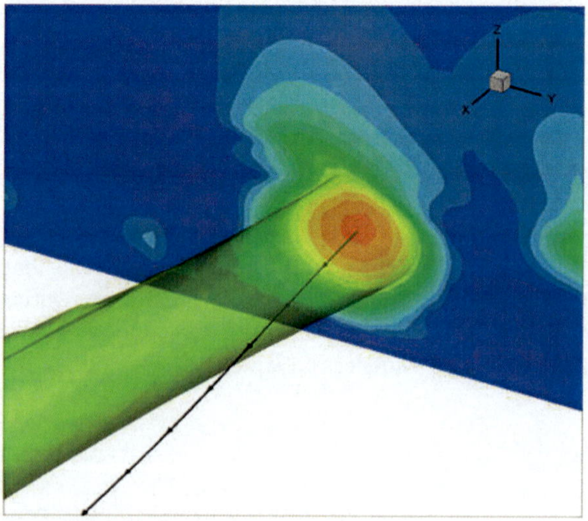

Fig. 2.2 Streamline in original frame

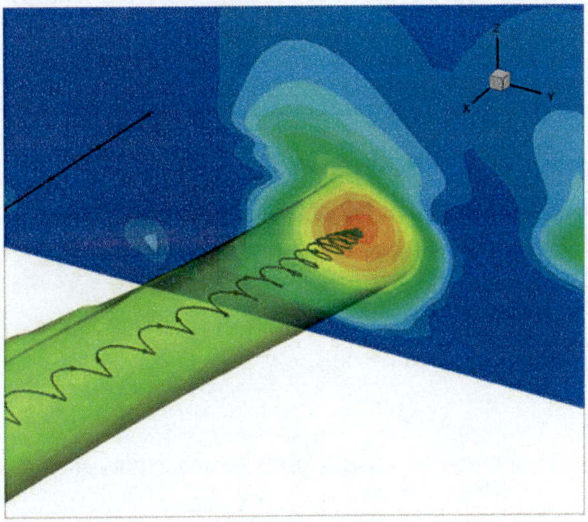

Fig. 2.3 Streamline in a frame moving with u speed of the intersection point

$$Q = Q_x Q_y Q_z = \begin{bmatrix} 1 & 0 & 0 \\ 0 & cos\gamma_1 & sin\gamma_1 \\ 0 & -sin\gamma_1 & cos\gamma_1 \end{bmatrix} \begin{bmatrix} cos\gamma_2 & 0 & -sin\gamma_2 \\ 0 & 1 & 0 \\ sin\gamma_2 & 0 & cos\gamma_2 \end{bmatrix} \begin{bmatrix} cos\gamma_3 & sin\gamma_3 & 0 \\ -sin\gamma_3 & cos\gamma_3 & 0 \\ 0 & 0 & 1 \end{bmatrix} = \begin{bmatrix} 0.2686 & 1.2637 & -0.5752 \\ -0.6644 & 0.7156 & 0.2157 \\ 0.7437 & 0.6042 & 0.2862 \end{bmatrix} \tag{2.14}$$

The iso-surface of Liutex in the original frame and the frame after Galilean transformation are shown in Figs. 2.4 and 2.5, respectively.

It can be clearly seen that it keeps the same after Galilean invariant.

Fig. 2.4 Liutex iso-surface in the original frame

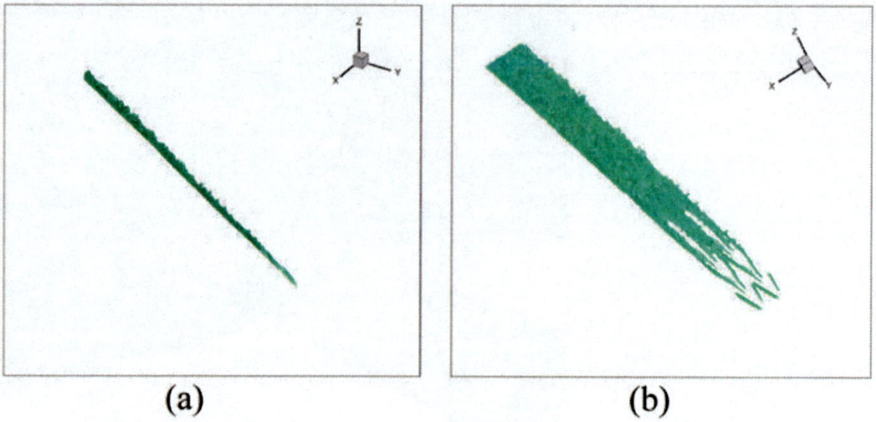

Fig. 2.5 Liutex iso-surface in the frame after Galilean transformation

2.6 Conclusion

Liutex is Galilean invariant, while streamline is not. Being Galilean invariant is an important requirement of vortex identification methods. It is not proper to use the curve of the streamlines to detect vortices because different frames lead to different results. Liutex keeps the same after Galilean transformation, so Liutex is more reliable than streamlines.

Acknowledgements The author would like to thank the Department of Mathematics of the University of Texas at Arlington, where the UTA Team is housed and where the birthplace of Liutex is. The authors are grateful to Texas Advanced Computational Center (TACC) for providing computation hours.

References

1. S.K. Robinson, Coherent motions in the turbulent boundary-layer. Annu. Rev. Fluid Mech. **23**, 601–639 (1991)
2. Y. Wang et al., DNS study on vortex and vorticity in late boundary layer transition. Commun. Comput. Phys. **22**(2), 441–459 (2017)
3. Y. Gao et al., Explicit expressions for Rortex tensor and velocity gradient tensor decomposition. Phys. Fluids **31**(8) (2019)
4. J.C. Hunt, A.A. Wray, P. Moin, *Eddies, streams, and convergence zones in turbulent flows.* Studying Turbulence Using Numerical Simulation Databases, 2. Proceedings of the 1988 Summer Program (1988)
5. J. Zhou et al., Mechanisms for generating coherent packets of hairpin vortices in channel flow. J. Fluid Mech. **387**, 353–396 (1999)
6. J. Jeong, F. Hussain, On the identification of a vortex. J. Fluid Mech. **285**, 69–94 (1995)
7. S. Tian et al., Definitions of vortex vector and vortex. J. Fluid Mech. **849**, 312–339 (2018)

8. C. Liu et al., Rortex—A new vortex vector definition and vorticity tensor and vector decompositions. Phys. Fluids **30**(3) (2018)
9. Y.-A. Guo et al., Experimental study on dynamic mechanism of vortex evolution in a turbulent boundary layer of low Reynolds number. J. Hydrodyn. **32**(5), 807–819 (2020)
10. Y.W.Y.G.C. Liu, *Letter: Galilean invariance of Rortex.* Phys. Fluids **30**(11) (2018)
11. Y.-Q. Wang et al., Explicit formula for the Liutex vector and physical meaning of vorticity based on the Liutex-Shear decomposition. J. Hydrodyn. **31**(3), 464–474 (2019)

Chapter 3
Rules of Tensor and Matrix Operation for Liutex Calculation

Yifei Yu, Yinlin Dong, and Chaoqun Liu

Abstract Vectors and tensors, which are widely used in physics, have their clear definitions and rules of calculations. However, in the practical applications and the process of derivations, vectors and tensors are usually expressed in the format of matrices which do not have as clear as their original mathematical definitions. The first problem is that there is no agreement on whether to use a column or row matrix to represent a vector which also leads to the problem of how to express a second-order tensor in a matrix format. In a word, the corresponding relations between vectors, tensors, and matrices are not very clear. Another problem is how to express vectors/tensors operation in matrices operations. It is clear that the first column matrix needs to be transposed before doing matrices multiplication when it is used to represent inner products of vectors. However, when it comes to the operation that a tensor dot product is a vector, some people do not transpose the first matrix (representing the second-order tensor). Liutex is a vortex identification concept that relates many vector and tensor operations. Therefore, it is important to have clear rules for using matrices to represent vectors and tensors. To overcome these problems and keep consistency, this paper provides rules of tensor and matrix operation for Liutex calculation. First, column matrices are used to represent vectors. Second, the matrix forms of vector/tensor operations are provided.

3.1 Introduction

Tensor is of significant importance in math as it can depict some important physical features, such as the velocity gradient tensor and stress tensor. The original mathematical definition of tensor is based on the summation of unit dyads with coefficients, resulting in the abstract and difficulty of tensor operation. In practical use

Y. Yu (✉) · C. Liu
Department of Mathematics, The University of Texas at Arlington, Arlington, TX, USA
e-mail: yifei.yu@mavs.uta.edu

Y. Dong
Department of Mathematics, University of Central Arkansas, Conway, AR, USA

Y. Wang et al. (eds.), *Liutex and Third Generation of Vortex Identification*,
Springer Proceedings in Physics 288, https://doi.org/10.1007/978-981-19-8955-1_3

of tensor, scientists pretend to express tensors in the format of matrices for those tensors whose orders are two so that people can see the components of the tensor more clearly. Meanwhile, the operation of tensors is equivalent to the operation of matrices. However, there is no uniform agreement on the corresponding relationship of vector/tensor and matrix, and tensor operations and matrix operations. One point is using whether array matrix or column matrix to represent a vector. Admittedly, it does not raise much confusion for expressing a vector by a matrix as people can recognize a vector easily whenever it is expressed by an array matrix or a column matrix; a disagreement occurs when expressing a tensor. Taking velocity gradient tensor, a second-order tensor, as an example, there are two widely used matrix forms of velocity gradient tensor.

One is $\begin{bmatrix} \frac{\partial u}{\partial x} & \frac{\partial u}{\partial y} & \frac{\partial u}{\partial z} \\ \frac{\partial v}{\partial x} & \frac{\partial v}{\partial y} & \frac{\partial v}{\partial z} \\ \frac{\partial w}{\partial x} & \frac{\partial w}{\partial y} & \frac{\partial w}{\partial z} \end{bmatrix}$, while the other one is $\begin{bmatrix} \frac{\partial u}{\partial x} & \frac{\partial v}{\partial x} & \frac{\partial w}{\partial x} \\ \frac{\partial u}{\partial y} & \frac{\partial v}{\partial y} & \frac{\partial w}{\partial y} \\ \frac{\partial u}{\partial z} & \frac{\partial v}{\partial z} & \frac{\partial w}{\partial z} \end{bmatrix}$.

Many researchers write velocity gradient tensor matrices by using arrays as vectors [1, 2], while some others use columns as vectors [3]. For vector/tensor operations, it is obvious that the inner product $\vec{u} \cdot \vec{v}$ is $\begin{bmatrix} u_1 & u_2 & u_3 \end{bmatrix} \begin{bmatrix} v_1 & v_2 & v_3 \end{bmatrix}^T$ if vectors are expressed by array matrices, or $\begin{bmatrix} u_1 \\ u_2 \\ u_3 \end{bmatrix}^T \begin{bmatrix} v_1 \\ v_2 \\ v_3 \end{bmatrix}$ if vectors are expressed by column matrices. Nevertheless, when it comes to the inner product that involving tensors, scientists may face some discrepancy. People hold an idea that matrix multiplication can represent vector/tensor inner product, however, because of the disagreement of how to express a tensor by a matrix, people can get two different results which are transpose to each other. Take velocity gradient tensor $grad\vec{v}$ and a vector $\vec{u} = \begin{bmatrix} u_1 & u_2 & u_3 \end{bmatrix}^T$ as an example.

One result is $\begin{bmatrix} \frac{\partial u}{\partial x} & \frac{\partial u}{\partial y} & \frac{\partial u}{\partial z} \\ \frac{\partial v}{\partial x} & \frac{\partial v}{\partial y} & \frac{\partial v}{\partial z} \\ \frac{\partial w}{\partial x} & \frac{\partial w}{\partial y} & \frac{\partial w}{\partial z} \end{bmatrix} \begin{bmatrix} u_1 \\ u_2 \\ u_3 \end{bmatrix}$ and the other result is $\begin{bmatrix} \frac{\partial u}{\partial x} & \frac{\partial v}{\partial x} & \frac{\partial w}{\partial x} \\ \frac{\partial u}{\partial y} & \frac{\partial v}{\partial y} & \frac{\partial w}{\partial y} \\ \frac{\partial u}{\partial z} & \frac{\partial v}{\partial z} & \frac{\partial w}{\partial z} \end{bmatrix} \begin{bmatrix} u_1 \\ u_2 \\ u_3 \end{bmatrix}$.

Note that $grad\vec{v} \cdot \vec{u}$. can only lead to one result, therefore one of the above matrix operations does not proper reflect the result of inner product $grad\vec{v} \cdot \vec{u}$. If people do not take care of this problem, some errors may occur in the research.

Many tensors are involved in the derivation of fluid dynamics, such as velocity gradient tensor and stress tensor [4–6]. For simplicity, scientists pretend to use matrices to represent tensors in their practical derivations. To make sure the matrices and matrix operations are correctly used to represent vector/tensor and their operations, a set of agreements is proposed in this paper to reduce the chaos of using matrices. This set of agreements includes the way to express vector/tensor by matrices, the matrix operations corresponding with the tensor operations. It can help scientists ensure that matrices and matrix operations are correctly used in the derivations.

Vortex is another important element for fluid dynamics. People used to believe vorticity is an indicator of vortex [7–10]; however, many experiments and simulations have shown it is just an approximation where the shear is small. Robinson has pointed out that the correlation between vorticity and vortex strength is weak, especially in the near boundary region [11]. Liutex is an emerging concept proposed by Liu et al. [12, 13], representing rigid body rotation of fluid. The magnitude of Liutex reflects twice angular speed and the direction of Liutex is the swirling axis. Many numerical and experimental results have shown that Liutex performs beyond vorticity and the other vortex identification methods [14, 15].

This paper is organized as follows. The rules for using matrices are provided in Sect. 3.2, after which the proper matrix form of velocity gradient tensor is discussed in Sect. 3.3. In Sect. 3.4, two calculation examples based on Liutex from [6] are reviewed, and the conclusion is given in Sect. 3.5.

3.2 Express Tensor and Its Operation by Matrix

Column matrices are used to represent vectors.

Definition 3.1 The matrix form of a vector $\vec{u} = u_i\vec{e}_i$ is

$$U = \begin{bmatrix} u_1 \\ u_2 \\ \vdots \\ u_n \end{bmatrix} \tag{3.1}$$

where Einstein summation convention is used in the expressions.

Since a tensor can be expressed by the outer product of two vectors, we define the matrix form of the outer product before that of a tensor.

Definition 3.2 Suppose $\vec{u} = u_i\vec{e}_i$ and $\vec{v} = v_i\vec{e}_i$. The matrix form of the outer product $\vec{u} \otimes \vec{v}$ is

$$\vec{u} \otimes \vec{v} = \begin{bmatrix} u_1 \\ u_2 \\ \vdots \\ u_n \end{bmatrix} \begin{bmatrix} v_1 \\ v_2 \\ \vdots \\ v_n \end{bmatrix}^T = \begin{bmatrix} u_1v_1 & u_1v_2 & \cdots & u_1v_n \\ u_2v_1 & u_2v_2 & \cdots & u_2v_n \\ \vdots & \vdots & \ddots & \vdots \\ u_nv_1 & u_nv_2 & \cdots & u_nv_n \end{bmatrix} \tag{3.2}$$

Note that $\vec{u} \otimes \vec{v} = \sum_{i=1}^{n} \sum_{j=1}^{n} u_i v_j \vec{e}_i \vec{e}_j$ which corresponds with the matrix above.

Definition 3.3 The matrix form of a second-order tensor $A_{ij} = \vec{u} \otimes \vec{v} = \sum_{i=1}^{n} \sum_{j=1}^{n} u_i v_j \vec{e}_i \vec{e}_j$ is

$$A = \begin{bmatrix} u_1v_1 & u_1v_2 & \cdots & u_1v_n \\ u_2v_1 & u_2v_2 & \cdots & u_2v_n \\ \vdots & \vdots & \ddots & \vdots \\ u_nv_1 & u_nv_2 & \cdots & u_nv_n \end{bmatrix} \tag{3.3}$$

The matrix form of vector/tensor operations can be found with the above definitions. Let $\vec{u} = u_i\vec{e}_i$, $\vec{v} = v_i\vec{e}_i$, $A_{ij} = a_{ij}\vec{e}_i\vec{e}_j$ and $B_{ij} = b_{ij}\vec{e}_i\vec{e}_j$, where Einstein notation is used in the expressions. U, V, A, and B are the matrix forms of $\vec{u}$, $\vec{v}$, A_{ij} and B_{ij} according to Definitions 3.1 and 3.3.

$$\vec{u} \cdot \vec{v} = (u_1, u_2, \ldots, u_n) \cdot (v_1, v_2, \ldots, v_n) = \sum_{i=1}^{n} u_i v_i = U^T V \tag{3.4}$$

$$\begin{aligned} \vec{u} \cdot A_{ij} &= u_i\vec{e}_i \cdot a_{jk}\vec{e}_j\vec{e}_k = u_i a_{jk}\delta_{ij}\vec{e}_k = u_i a_{ik}\vec{e}_k = u_i a_{ij}\vec{e}_j \\ &= \begin{bmatrix} u_1a_{11} + u_2a_{21} + \cdots + u_na_{n1} \\ u_1a_{12} + u_2a_{22} + \cdots + u_na_{n2} \\ \vdots \\ u_1a_{1n} + u_2a_{2n} + \cdots + u_na_{nn} \end{bmatrix} \\ &= \begin{bmatrix} a_{11} & a_{21} & \cdots & a_{n1} \\ a_{12} & a_{22} & \cdots & a_{n2} \\ \vdots & \vdots & \ddots & \vdots \\ a_{1n} & a_{2n} & \cdots & a_{nn} \end{bmatrix} \begin{bmatrix} u_1 \\ u_2 \\ \vdots \\ u_n \end{bmatrix} = A^T U = \left(U^T A\right)^T \end{aligned} \tag{3.5}$$

$$\begin{aligned} A_{ij} \cdot \vec{u} &= a_{ij}\vec{e}_i\vec{e}_j \cdot u_k\vec{e}_k = a_{ij}u_k\delta_{jk}\vec{e}_i = a_{ij}u_j\vec{e}_i \\ &= \begin{bmatrix} a_{11}u_1 + a_{12}u_2 + \cdots + a_{1n}u_n \\ a_{21}u_1 + a_{22}u_2 + \cdots + a_{2n}u_n \\ \vdots \\ a_{n1}u_1 + a_{n2}u_2 + \cdots + a_{nn}u_n \end{bmatrix} \\ &= \begin{bmatrix} a_{11} & a_{12} & \cdots & a_{1n} \\ a_{21} & a_{22} & \cdots & a_{2n} \\ \vdots & \vdots & \ddots & \vdots \\ a_{n1} & a_{n2} & \cdots & a_{nn} \end{bmatrix} \begin{bmatrix} u_1 \\ u_2 \\ \vdots \\ u_n \end{bmatrix} = AU \end{aligned} \tag{3.6}$$

$$A_{ij} \cdot B_{ij} = a_{ij}\vec{e}_i\vec{e}_j \cdot b_{mn}\vec{e}_m\vec{e}_n = a_{ij}b_{mn}\delta_{jm}\vec{e}_i\vec{e}_n = a_{ij}b_{jn}\vec{e}_i\vec{e}_n$$
$$= a_{ij}b_{jk}\vec{e}_i\vec{e}_k$$
$$= \begin{bmatrix} a_{11}b_{11} + a_{12}b_{21} + \cdots + a_{1n}b_{n1} & a_{11}b_{12} + a_{12}b_{22} + \cdots + a_{1n}b_{n2} & \cdots & a_{11}b_{1n} + a_{12}b_{2n} + \cdots + a_{1n}b_{nn} \\ a_{21}b_{11} + a_{22}b_{21} + \cdots + a_{2n}b_{n1} & a_{21}b_{12} + a_{22}b_{22} + \cdots + a_{2n}b_{n2} & \cdots & a_{21}b_{1n} + a_{22}b_{2n} + \cdots + a_{2n}b_{nn} \\ \vdots & \vdots & \ddots & \vdots \\ a_{n1}b_{11} + a_{n2}b_{21} + \cdots + a_{nn}b_{n1} & a_{n1}b_{12} + a_{n2}b_{22} + \cdots + a_{nn}b_{n2} & \cdots & a_{n1}b_{1n} + a_{n2}b_{2n} + \cdots + a_{nn}b_{nn} \end{bmatrix}$$
$$= \begin{bmatrix} a_{11} & a_{12} & \cdots & a_{1n} \\ a_{21} & a_{22} & \cdots & a_{2n} \\ \vdots & \vdots & \ddots & \vdots \\ a_{n1} & a_{n2} & \cdots & a_{nn} \end{bmatrix} \begin{bmatrix} b_{11} & b_{12} & \cdots & b_{1n} \\ b_{21} & b_{22} & \cdots & b_{2n} \\ \vdots & \vdots & \ddots & \vdots \\ b_{n1} & b_{n2} & \cdots & b_{nn} \end{bmatrix} = AB \tag{3.7}$$

To summarize, the matrix forms of vector/tensor operations obey the following rules,

$$\vec{u} \cdot \vec{v} = U^T V \tag{3.8}$$

$$\vec{u} \cdot A_{ij} = \left(U^T A\right)^T = A^T U \tag{3.9}$$

$$A_{ij} \cdot \vec{u} = AU \tag{3.10}$$

$$A_{ij} \cdot B_{ij} = AB \tag{3.11}$$

3.3 Matrix Form of Velocity Gradient Tensor

People are used to using ∇f to refer to the gradient of the function f, and, obeying this habit, $\nabla\vec{v}$ is widely used to refer to the velocity gradient tensor, where $\vec{v}$. is the velocity function. ∇f is equivalent to the gradient of f when f is a scalar function, but things could be complicated for vectors, so in the following part of this section, a question is answered: which should be the proper symbol/definition of velocity gradient tensor. As said in the introduction, there are two ways which are transposed to each other to represent the velocity gradient tensor; therefore, in this section $grad\vec{v}$ is used to refer to the velocity gradient to avoid confusion. According to Definition 3.2, $\nabla\overrightarrow{v}$ corresponds with the matrix

$$\nabla\vec{v} = \nabla \otimes \vec{v} = \begin{bmatrix} \frac{\partial}{\partial x} \\ \frac{\partial}{\partial y} \\ \frac{\partial}{\partial z} \end{bmatrix} \begin{bmatrix} u \\ v \\ w \end{bmatrix}^T = \begin{bmatrix} \frac{\partial}{\partial x} \\ \frac{\partial}{\partial y} \\ \frac{\partial}{\partial z} \end{bmatrix} \begin{bmatrix} u & v & w \end{bmatrix} = \begin{bmatrix} \frac{\partial u}{\partial x} & \frac{\partial v}{\partial x} & \frac{\partial w}{\partial x} \\ \frac{\partial u}{\partial y} & \frac{\partial v}{\partial y} & \frac{\partial w}{\partial y} \\ \frac{\partial u}{\partial z} & \frac{\partial v}{\partial z} & \frac{\partial w}{\partial z} \end{bmatrix} \tag{3.12}$$

While in some other papers, $grad\overrightarrow{v}$ is represented by $\left(\nabla\overrightarrow{v}\right)^T$ [16, 17].

Velocity gradient tensor $grad\vec{v}$ does have its physical meaning that the increment of $\vec{v}$ ($d\vec{v}$) along some direction $d\vec{r} = (dx, dy, dz)$ can be evaluated by $grad\vec{v}$ inner product $d\vec{r}$, i.e.

$$d\vec{v} = grad\vec{v} \cdot d\vec{r} \tag{3.13}$$

where $d\vec{v} = \begin{bmatrix} \frac{\partial u}{\partial x}dx + \frac{\partial u}{\partial y}dy + \frac{\partial u}{\partial z}dz \\ \frac{\partial v}{\partial x}dx + \frac{\partial v}{\partial y}dy + \frac{\partial v}{\partial z}dz \\ \frac{\partial w}{\partial x}dx + \frac{\partial w}{\partial y}dy + \frac{\partial w}{\partial z}dz \end{bmatrix}$.

Therefore, by checking whether Eq. (3.13) is satisfied, we can determine which one is more appropriate.

$$\nabla\vec{v} \cdot d\vec{r} = \begin{bmatrix} \frac{\partial u}{\partial x} & \frac{\partial v}{\partial x} & \frac{\partial w}{\partial x} \\ \frac{\partial u}{\partial y} & \frac{\partial v}{\partial y} & \frac{\partial w}{\partial y} \\ \frac{\partial u}{\partial z} & \frac{\partial v}{\partial z} & \frac{\partial w}{\partial z} \end{bmatrix} \begin{bmatrix} dx \\ dy \\ dz \end{bmatrix} = \begin{bmatrix} \frac{\partial u}{\partial x}dx + \frac{\partial v}{\partial x}dy + \frac{\partial w}{\partial x}dz \\ \frac{\partial u}{\partial y}dx + \frac{\partial v}{\partial y}dy + \frac{\partial w}{\partial y}dz \\ \frac{\partial u}{\partial z}dx + \frac{\partial v}{\partial z}dy + \frac{\partial w}{\partial z}dz \end{bmatrix} \neq d\vec{v} \tag{3.14}$$

$$(\nabla\vec{v})^T \cdot d\vec{r} = \begin{bmatrix} \frac{\partial u}{\partial x} & \frac{\partial u}{\partial y} & \frac{\partial u}{\partial z} \\ \frac{\partial v}{\partial x} & \frac{\partial v}{\partial y} & \frac{\partial v}{\partial z} \\ \frac{\partial w}{\partial x} & \frac{\partial w}{\partial y} & \frac{\partial w}{\partial z} \end{bmatrix} \begin{bmatrix} dx \\ dy \\ dz \end{bmatrix} = \begin{bmatrix} \frac{\partial u}{\partial x}dx + \frac{\partial u}{\partial y}dy + \frac{\partial u}{\partial z}dz \\ \frac{\partial v}{\partial x}dx + \frac{\partial v}{\partial y}dy + \frac{\partial v}{\partial z}dz \\ \frac{\partial w}{\partial x}dx + \frac{\partial w}{\partial y}dy + \frac{\partial w}{\partial z}dz \end{bmatrix} = d\vec{v} \tag{3.15}$$

Therefore, in fact $(\nabla\vec{v})^T$ is a more accurate presentation of $grad\vec{v}$.

Some discussions about velocity gradient tensor can be found in the following.

We are not saying it is wrong to use $\nabla\vec{v}$ to represent $grad\vec{v}$. We just state $\nabla\vec{v}$ is not $grad\vec{v}$ under Eq. (3.13). If it is defined $d\vec{v} = d\vec{r} \cdot grad\vec{v}$, i.e., $d\vec{r}$ inner products $grad\vec{v}$, then $\nabla\vec{v}$ becomes the correct way to refer to velocity gradient tensor. As long as the definitions are consistent, it is a correct system.

These are not the only matrix forms to represent vector/tensor or their operations. As long as all the definitions are adjusted correspondingly, an array matrix can be used to represent a vector.

3.4 Some Calculation Examples Using Matrix Expressions

Two derivation examples using the rules of tensor and matrix operations based on Liutex from [6] are reviewed in this section.

A. Explanation of the source of stress

In Ref. [6], it shows that it is a misunderstanding that the stress is produced by symmetric strain only in fluid dynamics. The classic NS equation is

$$\frac{\partial(\rho\vec{v})}{\partial t} + \nabla \cdot (\rho\vec{v}\vec{v}) = \rho\vec{f} - \nabla p - \frac{2}{3}\nabla[\mu(\nabla \cdot \vec{v})] + \nabla \cdot \left[\mu\left(\nabla\vec{v} + (\nabla\vec{v})^T\right)\right] \tag{3.16}$$

which indicates that the symmetric tensor $\nabla\vec{v} + (\nabla\vec{v})^T$ makes contributions to the stress while the anti-symmetric tensor $\nabla\vec{v} - (\nabla\vec{v})^T$ does not. However, this is a misunderstanding.

The $\nabla\vec{v}$ can be decomposed into a symmetric part and an anti-symmetric part by Cauchy-Stokes decomposition.

$$\nabla\vec{v} = \frac{1}{2}\left[\nabla\vec{v} + (\nabla\vec{v})^T\right] + \frac{1}{2}\left[\nabla\vec{v} - (\nabla\vec{v})^T\right] \tag{3.17}$$

Suppose the viscosity coefficient is μ, then the stress tensor can be written by

$$F = \frac{1}{2}\mu\left[\nabla\vec{v} + (\nabla\vec{v})^T\right] + \frac{1}{2}\mu\left[\nabla\vec{v} - (\nabla\vec{v})^T\right] \tag{3.18}$$

And,

$$\nabla \cdot F = \nabla \cdot \left\{\frac{1}{2}\mu\left[\nabla\vec{v} + (\nabla\vec{v})^T\right] + \frac{1}{2}\mu\left[\nabla\vec{v} - (\nabla\vec{v})^T\right]\right\} \tag{3.19}$$

For incompressible flow, we have $\nabla \cdot (\nabla\vec{v})^T = \vec{0}$ because

$$\nabla \cdot (\nabla\vec{v})^T = \left(\begin{bmatrix} \frac{\partial}{\partial x} & \frac{\partial}{\partial y} & \frac{\partial}{\partial z} \end{bmatrix}\begin{bmatrix} \frac{\partial u}{\partial x} & \frac{\partial u}{\partial y} & \frac{\partial u}{\partial z} \\ \frac{\partial v}{\partial x} & \frac{\partial v}{\partial y} & \frac{\partial v}{\partial z} \\ \frac{\partial w}{\partial x} & \frac{\partial w}{\partial y} & \frac{\partial w}{\partial z} \end{bmatrix}\right)^T = \begin{bmatrix} \frac{\partial^2 u}{\partial x^2} + \frac{\partial^2 v}{\partial y \partial x} + \frac{\partial^2 w}{\partial z \partial x} \\ \frac{\partial^2 u}{\partial x \partial y} + \frac{\partial^2 v}{\partial y^2} + \frac{\partial^2 w}{\partial z \partial y} \\ \frac{\partial^2 u}{\partial x \partial z} + \frac{\partial^2 v}{\partial y \partial z} + \frac{\partial^2 w}{\partial z^2} \end{bmatrix} = \begin{bmatrix} \frac{\partial}{\partial x}\left(\frac{\partial u}{\partial x} + \frac{\partial v}{\partial y} + \frac{\partial w}{\partial z}\right) \\ \frac{\partial}{\partial y}\left(\frac{\partial u}{\partial x} + \frac{\partial v}{\partial y} + \frac{\partial w}{\partial z}\right) \\ \frac{\partial}{\partial z}\left(\frac{\partial u}{\partial x} + \frac{\partial v}{\partial y} + \frac{\partial w}{\partial z}\right) \end{bmatrix} \tag{3.20}$$

Noted that for incompressible flow, $\frac{\partial u}{\partial x} + \frac{\partial v}{\partial y} + \frac{\partial w}{\partial z} = 0$.

$$\nabla \cdot (\nabla\vec{v})^T = \begin{bmatrix} \frac{\partial}{\partial x}\left(\frac{\partial u}{\partial x} + \frac{\partial v}{\partial y} + \frac{\partial w}{\partial z}\right) \\ \frac{\partial}{\partial y}\left(\frac{\partial u}{\partial x} + \frac{\partial v}{\partial y} + \frac{\partial w}{\partial z}\right) \\ \frac{\partial}{\partial z}\left(\frac{\partial u}{\partial x} + \frac{\partial v}{\partial y} + \frac{\partial w}{\partial z}\right) \end{bmatrix} = \begin{bmatrix} \frac{\partial}{\partial x} \cdot 0 \\ \frac{\partial}{\partial y} \cdot 0 \\ \frac{\partial}{\partial z} \cdot 0 \end{bmatrix} = \vec{0} \tag{3.21}$$

Thus,

$$\nabla \cdot \left\{\mu\left[\nabla\vec{v} + (\nabla\vec{v})^T\right]\right\} = \nabla \cdot \left\{\mu\left[\nabla\vec{v} - (\nabla\vec{v})^T\right]\right\} \tag{3.22}$$

from which we can see the symmetric tensor $\nabla\vec{v} + (\nabla\vec{v})^T$ and the anti-symmetric tensor $\nabla\vec{v} - (\nabla\vec{v})^T$ have the same contribution to the stress.

B. **New NS equation**

A new NS equation is derived in Ref. [6]

$$\nabla\cdot(\nabla\vec{v})^T=\begin{bmatrix}\frac{\partial}{\partial x}\left(\frac{\partial u}{\partial x}+\frac{\partial v}{\partial y}+\frac{\partial w}{\partial z}\right)\\ \frac{\partial}{\partial y}\left(\frac{\partial u}{\partial x}+\frac{\partial v}{\partial y}+\frac{\partial w}{\partial z}\right)\\ \frac{\partial}{\partial z}\left(\frac{\partial u}{\partial x}+\frac{\partial v}{\partial y}+\frac{\partial w}{\partial z}\right)\end{bmatrix}=\nabla(\nabla\cdot\vec{v}) \tag{3.23}$$

So, Eq. (3.16) can be transformed to

$$\frac{\partial(\rho\vec{v})}{\partial t}+\nabla\cdot(\rho\vec{v}\vec{v})=\rho\vec{f}-\nabla p+\frac{4}{3}\nabla[\mu(\nabla\cdot\vec{v})]+\nabla\cdot\left[\mu\left(\nabla\vec{v}-(\nabla\vec{v})^T\right)\right] \tag{3.24}$$

This new NS equation has several benefits. At first, it reduces the number of unknown variables because $\nabla\vec{v}+(\nabla\vec{v})^T$ in the classic NS equation has six unknown variables while $\nabla\vec{v}-(\nabla\vec{v})^T$ only has three unknown variables. Secondly, the vorticity tensor is Galilean invariant which means the results keep the same under different coordinates.

3.5 Conclusion

For a long time, using matrices to represent vector/tensor and their operations has been very confusing. In this paper, a rule is provided to eliminate this confusion. Based on this system, it is explained the reason for the statement proposed by Liu in Ref. [6] that it is not only the symmetric tensor $\nabla\vec{v}+(\nabla\vec{v})^T$ has contributions to stress in NS equation but the vorticity tensor $\nabla\vec{v}-(\nabla\vec{v})^T$ does the same contribution as $\nabla\vec{v}+(\nabla\vec{v})^T$. The derivation of a new NS equation, provided by Liu [6], is also reviewed.

References

1. W. Sheng, A revisit of Navier-Stokes equation. Eur. J. Mech. B. Fluids **80**, 60–71 (2020)
2. V. Kolář, Vortex identification: new requirements and limitations. Int. J. Heat Fluid Flow **28**(4), 638–652 (2007)
3. L. Zhen, Z. Xi-Wen, H. Feng, Evaluation of vortex criteria by virtue of the quadruple decomposition of velocity gradient tensor. Acta Physica Sinica **63**(5) (2014)
4. C. Liu, Letter: Galilean invariance of Rortex. Phys. Fluids **30**(11) (2018)
5. C. Liu, Rortex based velocity gradient tensor decomposition. Phys. Fluids **31**(1) (2019)
6. C. Liu, Z. Liu, New governing equations for fluid dynamics. AIP Adv. **11**(11) (2021)
7. C.K. Batchelor, G. Batchelor, *An Introduction to Fluid Dynamics*. Cambridge University Press (2000)
8. J.-Z. Wu, H.-Y. Ma, M.-D. Zhou, *Vorticity and vortex dynamics*. Springer Science & Business Media 11 (2007)

9. P.G. Saffman, *Vortex Dynamics*. Cambridge University Press 1995
10. A.J. Maja, A.L. Bertozzi, A. Ogawa, Vorticity and incompressible flow. Cambridge texts in applied mathematics. Appl. Mech. Rev. **55**(4), B77–B78 (2002)
11. S.K. Robinson, A review of vortex structures and associated coherent motions in turbulent boundary layers. Struct. Turbulence Drag Reduct., p. 23–50 (1990)
12. C. Liu et al., Rortex—a new vortex vector definition and vorticity tensor and vector decompositions. Physi. Fluids **30**(3) (2018)
13. C. Liu et al., Third generation of vortex identification methods: omega and Liutex/Rortex based systems. J. Hydrodyn. **31**(2), 205–223 (2019)
14. Y.-N. Zhang et al., A selected review of vortex identification methods with applications. J. Hydrodyn. **30**(5), 767–779 (2018)
15. Y.-A. Guo et al., Experimental study on dynamic mechanism of vortex evolution in a turbulent boundary layer of low Reynolds number. J. Hydrodyn. **32**(5), 807–819 (2020)
16. Y. Yu et al., Investigation of correlation between vorticity, Q, λci, $\lambda 2$, Δ and Liutex. Comput. Fluids, 225 (2021)
17. Y. Yu et al., Principal coordinates and principal velocity gradient tensor decomposition. J. Hydrodyn. **32**(3), 441–453 (2020)

Chapter 4
Liutex Core Tube for Vortex Visualization and Structure

Oscar Alvarez, Charles Nottage, and Chaoqun Liu

Abstract Vortex structure visualization methods have been hampered by the preceding vortex identification methods which derived them. Popular vortex identification schemes like the Q-Criterion and the lambda_ci-Criterion only produce scalar quantities to express the vortex structure, hindering our ability to expand on these methods. Liutex, however, produces a vector quantity known as the Liutex vector. This Liutex vector provides more information about the vortex structure and can be processed to generate rich and informative vortex visualizations. With the Liutex vector, we can create the Liutex Core Line, which accurately represents the vortex structure. From the Liutex Core Line, we can generate the Liutex Core Tube. The Liutex Core Tube has many advantages over an iso-surface for vortex structure analysis. Information can be conveyed intuitively and clearly to the user using the Liutex Core Tube to visualize a vortex structure. Thus, using the Liutex Core Tube can allow for optimal analysis of any vortex structure.

4.1 Introduction

Vortex visualization tools for turbulent flow analysis have come a long way since the use of rheoscopic fluids. This book contains a lot of information about Liutex and the many advantages of Liutex compared to other vortex identification methods. In this chapter, we talk about the Liutex Core Tube, how it is made, and compare it to the Liutex iso-surface as we explore it.

The Liutex Core Tube is a Liutex vector-based vortex visualization tool [1–3]. Since it is Liutex-based, it benefits from the advantages of the Liutex method [4, 5]. The Liutex core tube is not weighed down by the disadvantages of the vortex identification methods of previous generations [6, 8–11]. One of the main disadvantages of the previous generations is that they are contaminated by shear [7]. They cannot express pure fluid rotation and thus cannot represent a vortex. A good example of this

O. Alvarez · C. Nottage · C. Liu (✉)
Department of Mathematics, The University of Texas at Arlington, Arlington, TX 76019, USA
e-mail: cliu@uta.edu

Y. Wang et al. (eds.), *Liutex and Third Generation of Vortex Identification*,
Springer Proceedings in Physics 288, https://doi.org/10.1007/978-981-19-8955-1_4

is the turbulent boundary layer near the wall, where high shear exists but little rotation. Only Liutex can accurately represent the local rotational structure of a fluid—a vortex.

One main advantage of Liutex is the fact that Liutex is a vector-based method. Liutex being vector-based is what makes the Liutex core tube possible. Without the Liutex vector to create Liutex Lines, we could not create the Liutex core tube. As well as Liutex lines, we also use the Liutex iso-surface to generate the Liutex core tube structure. It is to say that without Liutex, the Liutex core tube technique could not exist.

4.2 Data

All the visuals and graphics used for the figures in this chapter were generated by the same snapshot of Direct Numerical Simulation (DNS) data of the early transition boundary layer of a flat plate being dragged (Fig. 4.1). The iso-surface shown in green is the Liutex magnitude at a value of 0.07. The software program we use to visualize the data is Tecplot 360. We used Tecplot 360 to render the graphics for all the figures.

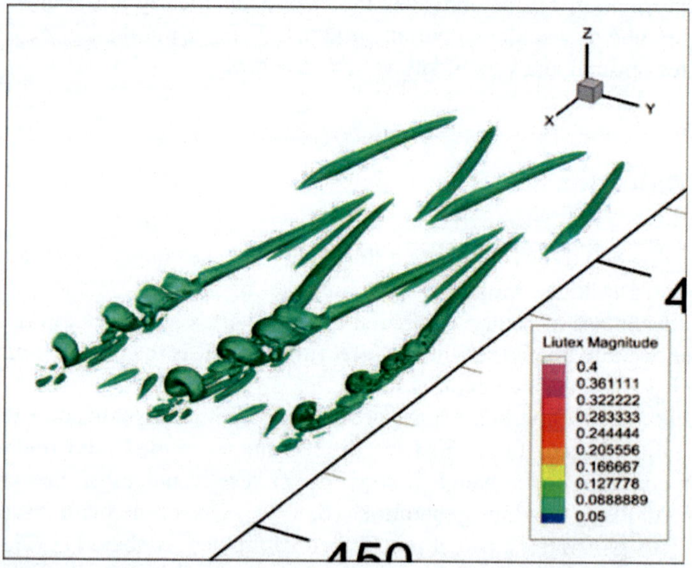

Fig. 4.1 DNS data with Liutex Magnitude iso-surface at a value of 0.07 in the early transition boundary layer of a flat plate

4.3 Liutex Core Tube

4.3.1 Constructing the Liutex Core Tube

In order to begin to understand the structure produced by the Liutex core tube, we must first know how the Liutex core tube is created.

First, we must select a region of interest. This is the region where we will direct our focus and the place where we begin to construct the Liutex core tube. The region of interest for this chapter will be at the frontmost center hairpin structure, which can be seen in Fig. 4.3.

We start by creating a slice that intersects our region of interest. In our case, the region of interest is a hairpin vortex structure. We place the slice, essentially cutting the hairpin structure into two pieces (Figs. 4.2 and 4.3). We can now work on either side of the slice.

We now must draw the Liutex Magnitude Gradient lines along the hairpin structure. To do this, we set the stream trace tool to use the components of the Liutex magnitude gradient vector ∇R where.

$$\nabla R = \begin{bmatrix} \frac{dR}{dx} \\ \frac{dR}{dy} \\ \frac{dR}{dz} \end{bmatrix}. \tag{4.1}$$

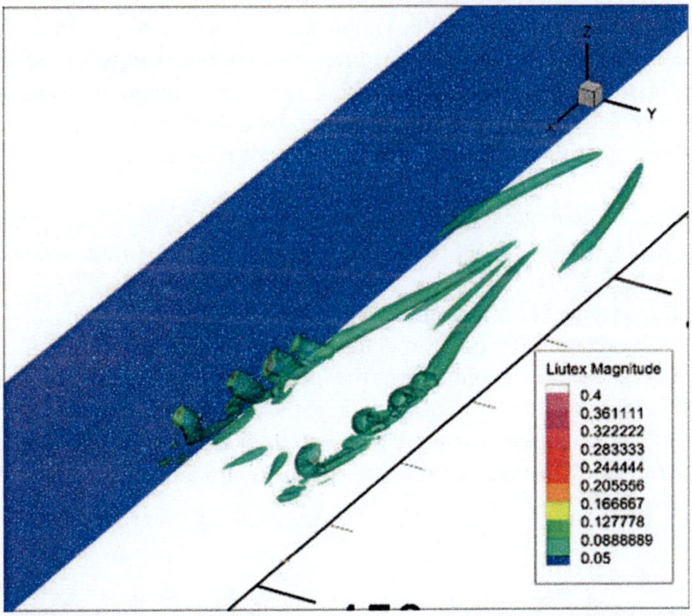

Fig. 4.2 Vertical slice taken in the y-direction

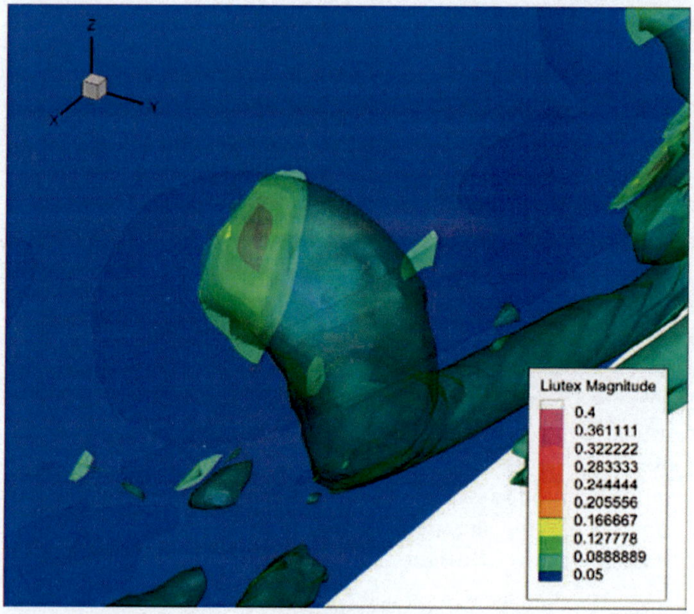

Fig. 4.3 Region of interest; The center front hairpin vortex structure

Using the tool, we create Liutex magnitude gradient lines on and along the iso-surface of the hairpin. We can see the results in Fig. 4.4.

Looking at the contour of the slice (representing the Liutex magnitude), you can see that the Liutex magnitude gradient lines concentrate themselves at the center of the hairpin structure. The location where the Liutex magnitude gradient lines concentrate intersecting the slice can be said to be a local maximum. In Fig. 4.5, it is clearly shown that the Liutex magnitude gradient lines are converging to a local maximum point on the slice.

The Liutex Line generated at the local maximum point location is called the Liutex Core Line [3, 12], but that is out of the scope of this chapter, so we will continue to focus on the Liutex Core Tube.

The local max point is important because we will use the Liutex magnitude at the x, y, and z location of the point to determine the strength/size of the Liutex core tube. We record the Liutex magnitude at the local max point, R_{max}. To determine the strength/size of the Liutex Core Tube, we select an arbitrary percentage s between 0 and 100%. We have chosen a percentage $s = 0.9$, or a strength/size of 90%. Using Eq. 4.2, we can find the value of the iso-surface that needs to be drawn using the Liutex magnitude.

$$R_{tube} = s \cdot R_{max} \tag{4.2}$$

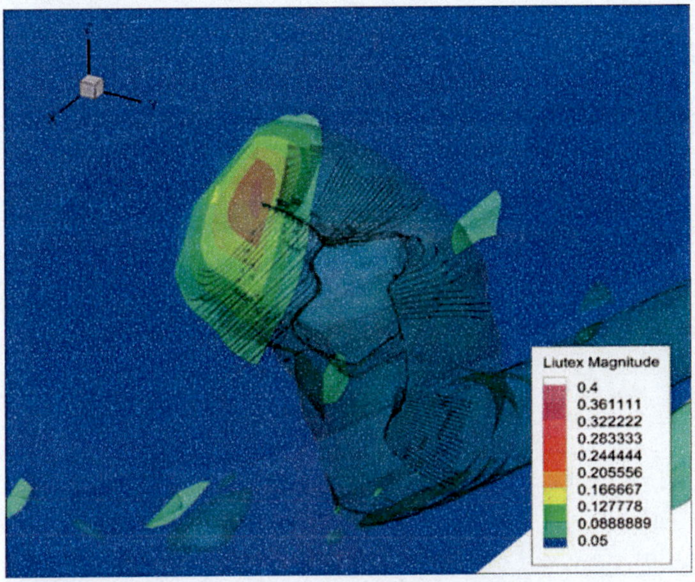

Fig. 4.4 Liutex Magnitude Gradient Lines leading and converging to the local max point

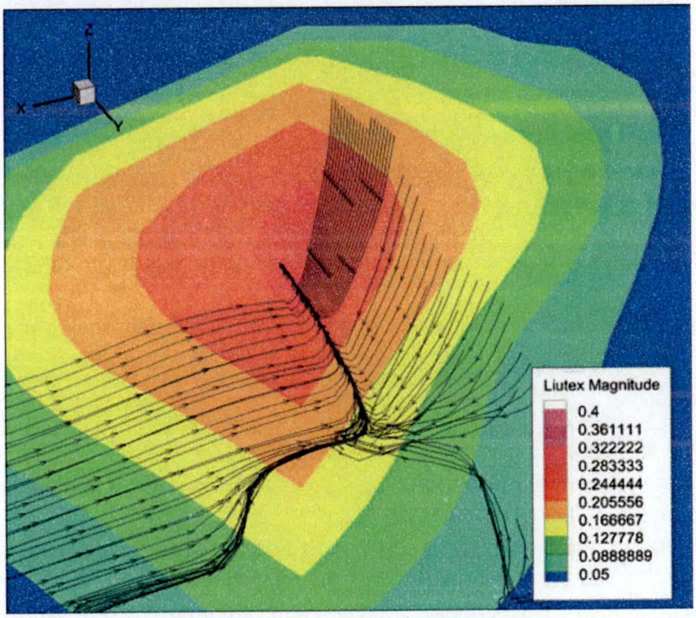

Fig. 4.5 A closer and slightly rotated view of Fig. 4.4. You can see the Liutex magnitude gradient lines converging to the core of the vortex structure. The local max point is where the concentration of Liutex magnitude gradient lines converge and intersect the slice

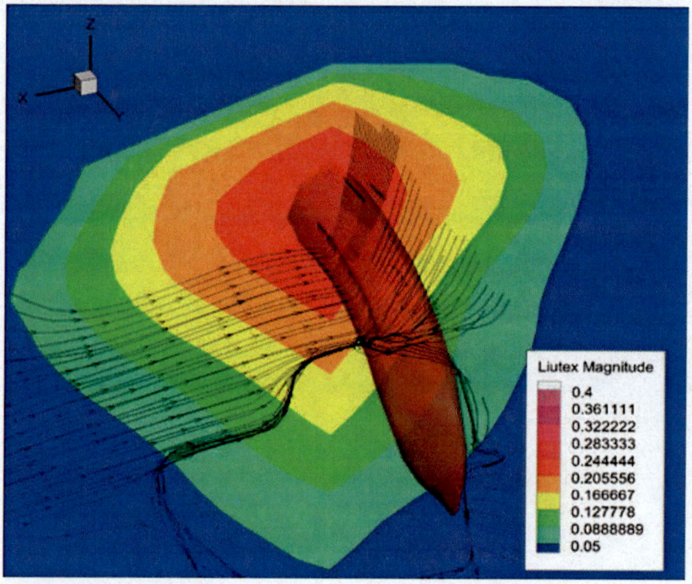

Fig. 4.6 Iso-surface at 90% of the local max value where the Liutex magnitude gradient lines intersect the slice

where R_{tube} is the relative strength/size of the Liutex Core tube.

We now draw the iso-surface at the value of R_{tube} using the Liutex magnitude (Fig. 4.6).

For more complex and intricate vortex structures, we recommend higher percentage values, $s > 90\%$. This prevents the generation of unstable Liutex lines further away from the core center. The closer you remain to the Liutex core center (local max point), the more stable the Liutex lines will be relative to the Liutex core center.

This threshold s is arbitrary, and thus the Liutex core tube does not provide a unique vortex structure. Nevertheless, the structure produced by this method is rich in the amount of information that is provided within only one instance.

Figure 4.7 shows the grid associated with the slice and the iso-surface we just created. The software we use (Tecplot 360) interpolates the data so we can more finely select our points in 3-D space (x, y, and z) rather than relying on the grid index points (i, j, and k).

The iso-surface and the slice will serve as a guide to where we place the following Liutex lines. We will create Liutex lines at the intersection of the Liutex magnitude iso-surface and the slice. To do this, we change the stream trace tool vector components to the Liutex direction vector's components. Once that is done, we trace the tool at the intersection of the iso-surface and the slice, creating Liutex lines all around the vortex structure created by the iso-surface (Fig. 4.8).

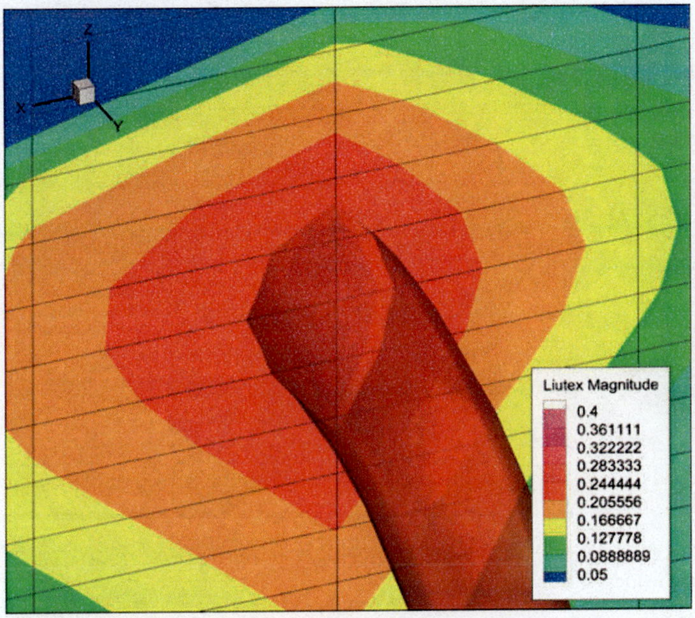

Fig. 4.7 Grid showing the nodes on the vertical slice intersecting the Liutex magnitude iso-surface at 90% of the local max value

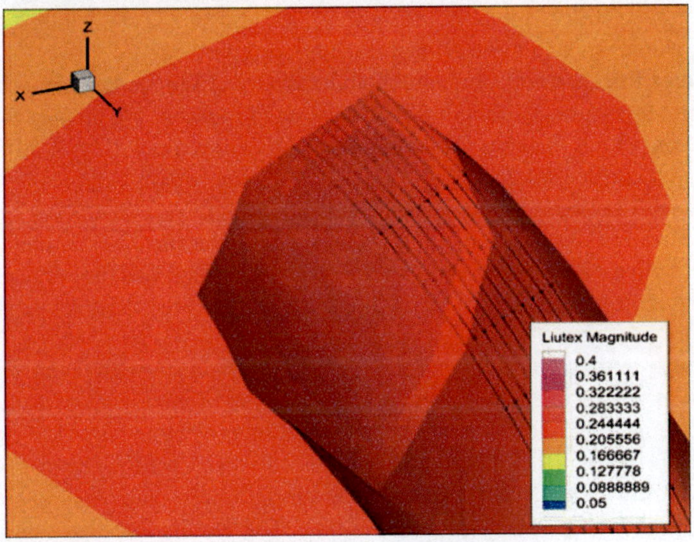

Fig. 4.8 Creating the Liutex Lines (stream traces) using the Liutex direction vector components. These are created at the intersection between the vertical slice and iso-surface at 90% value at the local max point

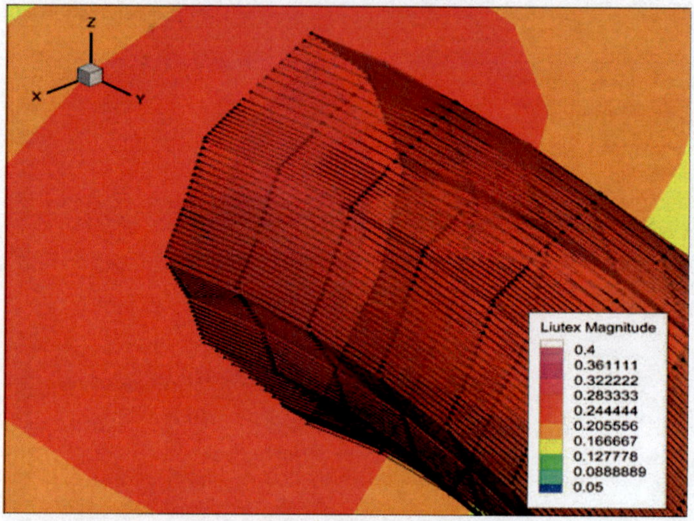

Fig. 4.9 Liutex Lines covering the entire intersection between the vertical slice and iso-surface at 90% value at the local max point

We continue to create Liutex lines until we have done so around the entire intersection of the slice and iso-surface (Fig. 4.9). The density of Liutex lines at the intersection is arbitrary. However, the more Liutex lines made at the intersection, the higher the resolution and more refined the rendering will be.

The Liutex lines are not complete, though, as we need to apply a contour to them so they can represent the strength of the local rigid rotation. We change their color to multicolor, which finalizes the creation of the Liutex lines (Fig. 4.10). Here in Fig. 4.10, you can see the grid of the slice and how the software allows the creation of Liutex lines between nodes within the cells.

The finalization of placing the Liutex lines also represents the completion of the Liutex Core Tube. We can now remove the slice and iso-surface to reveal the Liutex core tube structure. We zoom out to see the entire Liutex core tube that was generated (Fig. 4.11).

It can be seen that the Liutex core tube was able to describe the vortex structure around the region of interest. As we get closer to the region of interest, the Liutex core tube seems to be more "put together"; the Liutex vectors stay closer together. We can see the recognizable hairpin structure as we stay near the region of interest.

While introducing the Liutex Core Tube during a presentation, it was noted by a user that this technique is "grid-dependent". The results that this tool provides depend on where you decide to place your region of interest. For the work done in this chapter, we selected a region that was simple as to illustrate some of the tool's pros and cons. A notable pro is that it can better describe a local region rather than using iso-surface in some situations. A con is that the core tube structure can diverge

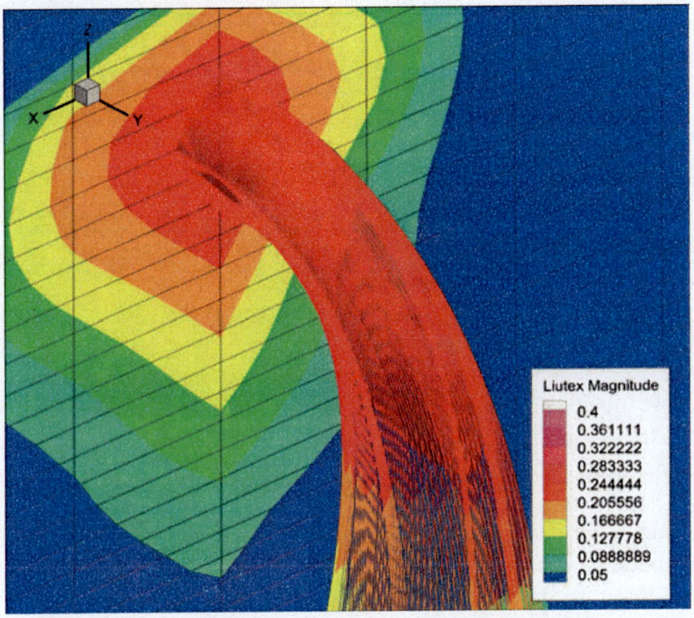

Fig. 4.10 Applying a contour defined by the Liutex Magnitude, we complete the illustration of the Liutex Lines and the Liutex Core Tube

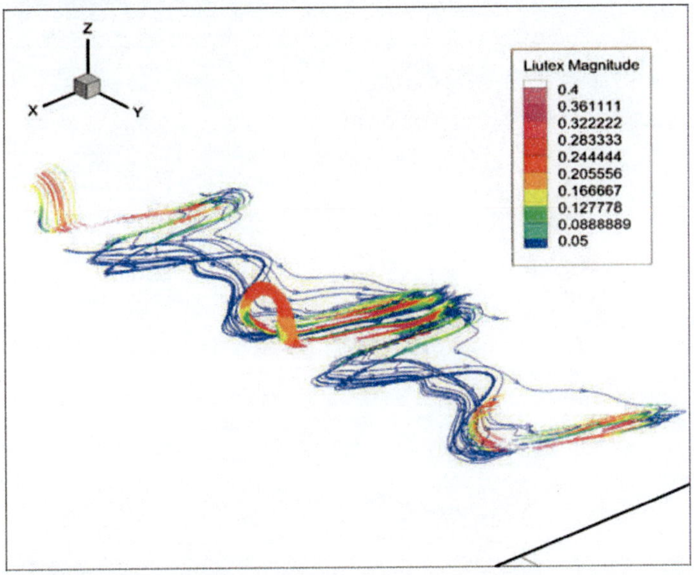

Fig. 4.11 Generated Liutex Core Tube structure

the further it travels away from where it was initialized. This con can be mitigated by increasing the percentage value s closer to 100%.

4.3.2 Liutex Core Tube Versus Iso-Surface

We can combine the iso-surface and the Liutex core tube in the same frame to find differences between the two methods (Fig. 4.12).

Notice how the Liutex core tube fills in the gaps that the Liutex iso-surface does not represent. The iso-surface represented here is the same 90% strength/size R_{tube} value that we used to generate the Liutex core tube. Another thing you can see is that the Liutex core tube can represent multiple Liutex magnitude values in one instance. If we were to try to represent all of the iso-surface values simultaneously, the graphic would become messy and difficult to analyze. The Liutex core tube is comprised of Liutex lines, so it has all of the same benefits as the vector-based Liutex method. Since the Liutex core tube is vector-based, it can simultaneously represent multiple Liutex magnitude values.

One thing that must be stated is that the Liutex Core Tube is not an iso-surface. Although it does appear to have a surface, it is comprised of Liutex lines and is, therefore, a vector-based vortex visualization technique. Figure 4.13 shows and compares the iso-surface and Liutex core tube structures. We can see that the Liutex core

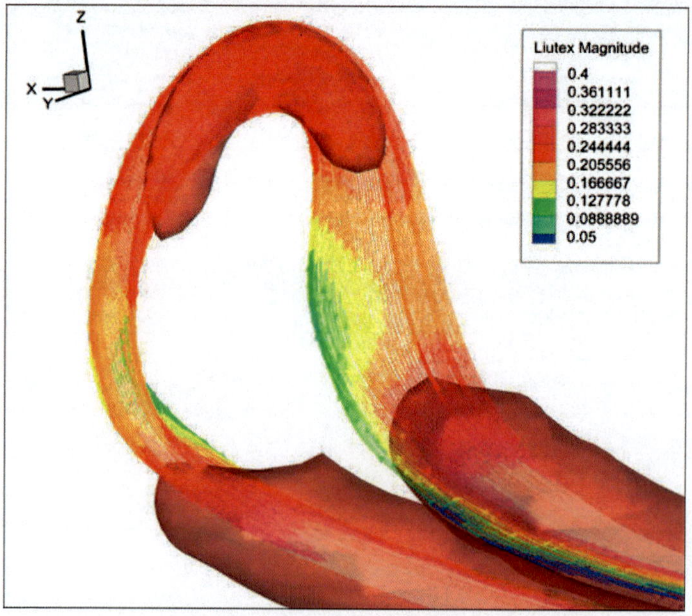

Fig. 4.12 Liutex Core Tube hairpin structure with iso-surface

tube does not cover every vortex structure that the iso-surface generates globally. However, the core tube remains almost local to the region of interest. The Liutex core tube also connects the separated parts of the iso-surface and shows the vortex structure between them. Of course, the representation of the vortex structure is better intact near the region of interest (Fig. 4.12).

Figures 4.14 and 4.15 express the clear distinction between the Liutex Core Tube and the Liutex iso-surface representations of the hairpin vortex structure. This distinction is notable as the figure created using the Liutex core tube is perceptibly more defined. In the case where data is limited to a single region of interest, the Liutex core tube could be used to extract more visual information from that particular region.

The way that the Liutex magnitude changes along the Liutex core tube is a significant feature of this tool. The results of the Liutex core tube are irreproducible if attempting only to use iso-surfaces. Iso-surfaces never intersect, so you cannot find when they change as you move along the structure. You cannot use the stream traces of velocity to create the Liutex core tube as they are not Galilean invariant and do not represent the strength of local rigid rotation of the fluid. It is thanks to the vector-based Liutex method that this technique is possible.

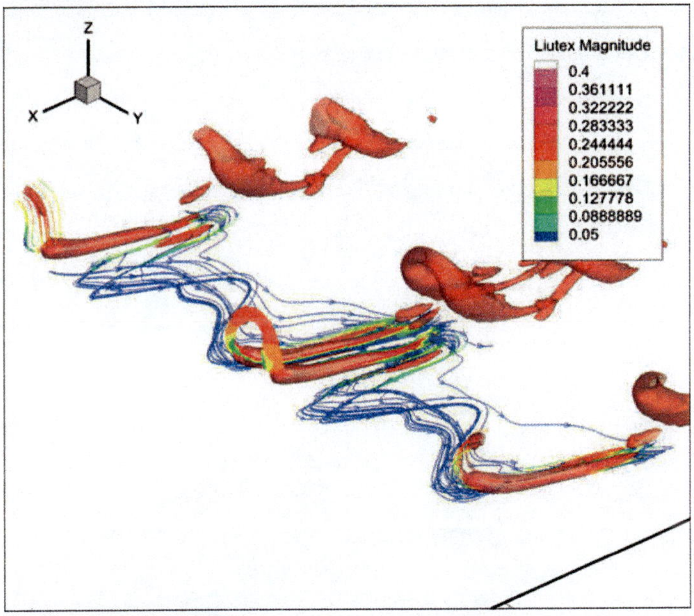

Fig. 4.13 Liutex magnitude iso-surface compared to the Liutex core tube at 90% strength of the local max point

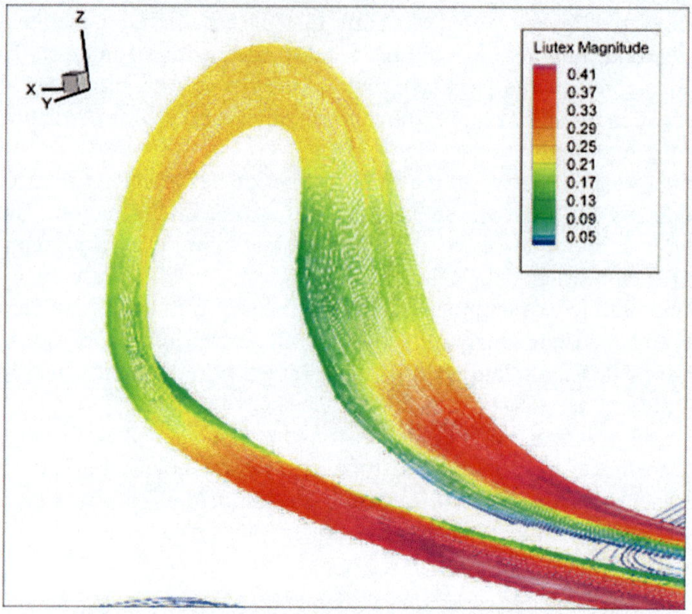

Fig. 4.14 Liutex Core Tube structure at the region of focus

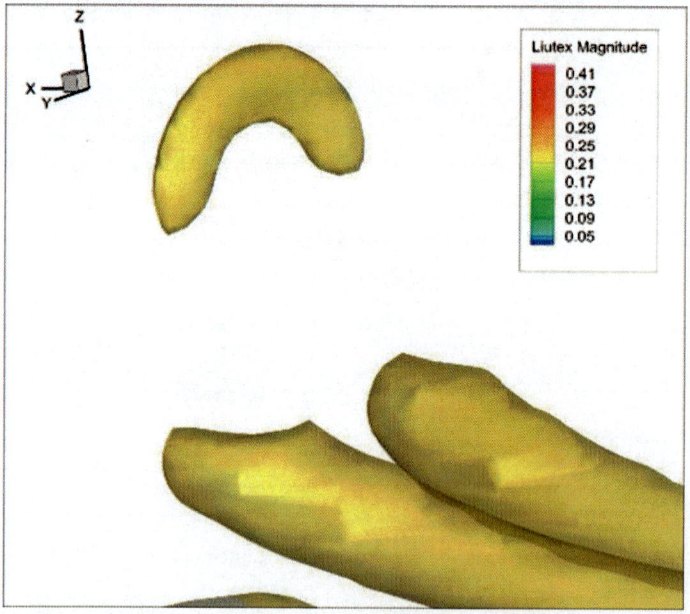

Fig. 4.15 Liutex Magnitude iso-surface structure at the region of focus

4.4 Conclusion

The Liutex Core Tube is a Liutex-based vortex visualization tool that can be used to create more friendly and intuitive graphics of turbulent structures. Although the Liutex core tube structure is not a unique vortex structure, the density of information that can be extracted and illustrated in only one instance is compelling. When strategically placed, the Liutex Core Tube is an efficient and effective way of visualizing vortex structures and can help illustrate certain scenarios better than other/older methods.

References

1. O. Alvarez, Y. Yu, C. Liu, Liutex-Core-Tubes for Vortex Structure. Paper presented at AIAA SCITECH 2022 Forum (2022). https://doi.org/10.2514/6.2022-0480
2. O. Alvarez et al., Visualizing Liutex core using Liutex lines and tubes, in *Liutex and Third Generation of Vortex Definition and Identification.* Springer Nature Switzerland, pp. 169–182 (2021). https://doi.org/10.1007/978-3-030-70217-5_2
3. O. Alvarez et al., Liutex Core Line for Vortex Structure in Turbulence. Paper presented at ASME 2021 Fluid Engineering Division Summer Meeting, (2021). https://doi.org/10.1115/FEDSM2021-66012
4. C. Liu et al., Rortex—A new vortex vector definition and vorticity tensor and vector decompositions. Phys. Fluids (2018)
5. Y. Gao, C. Liu, Rortex and comparison with eigenvalue-based vortex identification criteria. Phys. Fluids (2018)
6. C. Liu et al., Third generation of vortex identification methods: omega and Liutex/Rortex based systems. J. Hydrodyn. (2019)
7. P. Shrestha et al., Stretching and shearing contamination analysis for Liutex and other vortex identification methods. Adv. Aerodyn. **3**(1), 8 (2021). https://doi.org/10.1186/s42774-020-00060-9
8. Y. Yu et al., Incorrectness of the second-generation vortex identification method and introduction to Liutex, in *Liutex and Third Generation of Vortex Definition and Identification.* Springer Nature Switzerland, pp. 37–56 (2021). https://doi.org/10.1007/978-3-030-70217-5_2
9. Y. Yu et al., Correlation analysis among vorticity, Q method and Liutex. J. Hydrodyn. **32**(2) (2020). https://doi.org/10.1007/s42241-020-0069-2
10. Y. Yu et al., Investigation of correlation between vorticity, Q, λ_ci, λ_2. Δ and Liutex. Comput. Fluids **225**(55), 104977 (2021). https://doi.org/10.1016/j.compfluid.2021.104977
11. Y. Yu, O. Alvarez, V. Patel, C. Liu, Introduction of Liutex and Third Generation of Vortex Identification Methods. Paper presented at AIAA SCITECH 2022 Forum (2022). https://doi.org/10.2514/6.2022-2012
12. Y. Gao et al., A Liutex based definition and identification of vortex core center lines. J. Hydrodyn. **31**, 445–454 (2019). https://doi.org/10.1007/s42241-019-0048-7

Chapter 5
Study of Vortex and Vorticity in a Laminar Flow

Aayush Bhattarai, Oscar Alvarez, Shah Md. Imran Kabir, Qazu Asguqye E. Mowla, and Yifei Yu

Abstract Vortices are ubiquitous in nature. From kitchen sinks to galaxies, they can be found everywhere. Usually, the swirling motion of fluids comes to mind when one thinks of vortices. It has been extensively studied for centuries, being significantly important in various fields such as engineering, physics, chemistry, and aerospace. Still, we do not have an unambiguous and universally accepted definition of a vortex. Often vorticity is used to describe the vortex, which is accurate for rigid body rotation; however, this explanation is simply not true for the fluid flow in the boundary layer. For fluid rotation, pure shear deformation needs to be considered. In order to demonstrate that, we recreated Shapiro's experiment where he used the rigid body (vorticity meter) rotation to show that the vorticity is the same thing as the vortex. Additionally, we used dyed ink to investigate if the same results still hold in fluid rotation as they did for the vorticity meter.

5.1 Introduction

Many vortex identification methods have been established to scrutinize vortical structure in a fluid flow. In 1858, Helmholtz put forward the idea of a vorticity tube/filament to represent the vortex in the fluid flow [1]. Mathematically, it was found that the magnitude of vorticity is twice the angular speed of rotation, and the direction of vorticity is the swirling axis for a solid body. With this result, many scientists were convinced with the explanation of vortex through the concentration of vorticity and other vorticity methods [1, 2]. This result holds for a rigid body but not for fluids. However, vorticity tubes cannot constitute vortices in the turbulent viscous flow as rotation strength is minimal near the wall and shear stress is dominant. In 1991, Robinson expressed that the relation between actual vortices and firm vorticity can be relatively weak [3]. People began to be skeptical about the vorticity-based methods, classified as the first-generation vortex identification methods.

A. Bhattarai (✉) · O. Alvarez · S. Md. Imran Kabir · Q. A. E. Mowla · Y. Yu
University of Texas at Arlington, Arlington, TX, USA
e-mail: aayush.bhattarai@mavs.uta.edu

Y. Wang et al. (eds.), *Liutex and Third Generation of Vortex Identification*,
Springer Proceedings in Physics 288, https://doi.org/10.1007/978-981-19-8955-1_5

To overcome these inadequacies, new identification methods, such as Q, Δ, λ_2, λ_{ci}, and Ω were introduced by different experts in this field. These methods were able to better judge the presence of local rotational motion but came with limitations. Hunt et al. presented the Q-criterion method to visualize vortical structure more efficiently [4]. However, it is threshold-sensitive to express the area $Q > Qthreshold$ as a vortex. With the help of critical point theory, the Δ-criterion was proposed by Perry and Chong, which could depict vortical structure much better [5]. Unfortunately, to visualize the iso-surface plotting effectively, we need to choose the proper threshold as this method is threshold-sensitive, too.

To deal with the downsides of the Q- and Δ-criterion, Jeong and Hussain presented the λ_2 method [6]. Still, this method is threshold-sensitive and only works well for a steady inviscid flow. The λ_{ci} vortex identification method, similar to that of the Δ-criterion, was introduced by Zhou et al. [7]. This was an improvisation of the Δ-criterion method, which visualizes the vortex structure using the imaginary part of the complex eigenvalues of the velocity gradient tensor. The limitation of this method was that it was based on the concept of the arbitrary threshold. These methods to alleviate the deficiencies of vorticity-based vortex identification are categorized as the second-generation method. The major drawback was the user-specified threshold, and the different thresholds would show distinct vortex structures. In addition, these second-generation vortex identification methods were contaminated by shears to some degree.

The way of science is to keep pushing the boundary until the model is strictly accurate; Liu et al. published the new vortex identification method named Ω-method [8]. One of the significant advantages of this method is that it is not sensitive to the moderate threshold change. Still, all the mentioned identification methods are scalar, while the fluid rotation has magnitude and direction. In 2018, Liu gave us the Liutex/Rortex method, considered one of the most significant breakthroughs in modern fluid mechanics [9]. This method represents the vortex as a vector and can give a local direction and strength of the fluid rotation. Shrestha et al. applied all three generations (Liutex based) vortex identification methods to Direct Numerical Simulation (DNS) data to observe the vortex structure in the flow transition. They reported that the Modified Liutex-Omega method is not affected by threshold change and can accurately show the iso-surface of the vortex structure [10]. Finally, Liutex-based methods provided the mathematical definition for the vortex.

5.2 Vortex Identification Methods

Numerous vortex identification methods have been introduced to explain vortex structures in the past few decades. According to Liu [11], we can classify these vortex identification methods into three generations, starting from the first to the third. The first is the vorticity-based method, while the second is an eigenvalue-based method. Similarly, third-generation methods are Liutex-based methods. In this chapter, we will briefly discuss all these identification methods.

5.2.1 *Vorticity-Based Method*

The first-generation vortex identification method consists of vorticity lines, vorticity tubes, and vorticity filaments. As proposed by Helmholtz in 1858 [1], it represents the idea of vortices containing vorticity tubes; the magnitude of the vorticity gives its strength. Here, the mathematical definition of vorticity is a curl of velocity, i.e.,

$$\text{Vorticity} = \text{Curl}\ \boldsymbol{v} = \nabla \times \boldsymbol{v} = \begin{vmatrix} \boldsymbol{i} & \boldsymbol{j} & \boldsymbol{k} \\ \frac{\partial}{\partial \mathrm{x}} & \frac{\partial}{\partial \mathrm{y}} & \frac{\partial}{\partial \mathrm{z}} \\ \mathrm{u} & \mathrm{v} & \mathrm{w} \end{vmatrix} \tag{5.1}$$

$$\text{Vorticity} = \boldsymbol{i}\left(\frac{\partial \mathrm{w}}{\partial \mathrm{y}} - \frac{\partial \mathrm{v}}{\partial \mathrm{z}}\right) - \boldsymbol{j}\left(\frac{\partial \mathrm{w}}{\partial \mathrm{x}} - \frac{\partial \mathrm{u}}{\partial \mathrm{z}}\right) + \boldsymbol{k}\left(\frac{\partial \mathrm{v}}{\partial \mathrm{x}} - \frac{\partial \mathrm{u}}{\partial \mathrm{y}}\right) \tag{5.2}$$

Helmholtz presented three theorems in fluid mechanics that explain the three-dimensional motion of fluid particles in the surrounding area of vortex filaments:

1. Helmholtz's first theorem: The strength of a vortex filament is constant along its length.
2. Helmholtz's second theorem: A vortex filament cannot end in a fluid; it must extend to the boundaries of the fluid or form a closed path.
3. Helmholtz's third theorem: In the absence of external rotational forces, an initially irrotational fluid remains irrotational [12].

5.2.2 *Second-Generation Vortex Identification Method*

Second-generation vortex identification methods are eigenvalue-based methods. These are also based on closed or spiraling streamlines. These methods were developed to undertake the limitations of the vorticity-based method. However, they come with their drawbacks as well. The identification methods that fall under these categories are the Q criterion, Δ criterion, λ_{ci} criterion, and λ_2 criterion. We will briefly discuss each of them below.

5.2.2.1 Q Criterion

Given by Hunt et al. [4], it is one of the most widely used methods to visualize vortex structure. Q is defined as the residual of the vorticity tensor norm squared subtracted from the strain-rate tensor norm squared. Mathematically,

$$\mathrm{Q} = \frac{1}{2}\left(\|B\|_F^2 - \|A\|_F^2\right)$$

where A, B are the symmetric and antisymmetric parts of the velocity gradient tensor and $\|*\|_F^2$ represents the Frobenius norm.

$$A = \frac{1}{2}\left(\nabla\vec{v} + \nabla\vec{v}^T\right) = \begin{bmatrix} \frac{\partial u}{\partial x} & \frac{1}{2}\left(\frac{\partial u}{\partial y} + \frac{\partial v}{\partial x}\right) & \frac{1}{2}\left(\frac{\partial u}{\partial z} + \frac{\partial w}{\partial x}\right) \\ \frac{1}{2}\left(\frac{\partial v}{\partial x} + \frac{\partial u}{\partial y}\right) & \frac{\partial v}{\partial y} & \frac{1}{2}\left(\frac{\partial v}{\partial z} + \frac{\partial w}{\partial y}\right) \\ \frac{1}{2}\left(\frac{\partial w}{\partial x} + \frac{\partial u}{\partial z}\right) & \frac{1}{2}\left(\frac{\partial w}{\partial y} + \frac{\partial v}{\partial z}\right) & \frac{\partial w}{\partial z} \end{bmatrix} \tag{5.3}$$

$$B = \frac{1}{2}\left(\nabla\vec{v} - \nabla\vec{v}^T\right) = \begin{bmatrix} 0 & \frac{1}{2}\left(\frac{\partial u}{\partial y} - \frac{\partial v}{\partial x}\right) & \frac{1}{2}\left(\frac{\partial u}{\partial z} - \frac{\partial w}{\partial x}\right) \\ \frac{1}{2}\left(\frac{\partial v}{\partial x} - \frac{\partial u}{\partial y}\right) & 0 & \frac{1}{2}\left(\frac{\partial v}{\partial z} - \frac{\partial w}{\partial y}\right) \\ \frac{1}{2}\left(\frac{\partial w}{\partial x} - \frac{\partial u}{\partial z}\right) & \frac{1}{2}\left(\frac{\partial w}{\partial y} - \frac{\partial v}{\partial z}\right) & 0 \end{bmatrix} \tag{5.4}$$

The region with $Q > 0$ can be thought of as a vortex.

5.2.2.2 Criterion

This method finds the vortex in the region where the velocity gradient tensor has complex eigenvalues. The characteristics equation where λ_1, λ_2, and λ_3 are the eigenvalues of the 3×3 matrix of velocity gradient tensor can be written as

$$\lambda^3 + I_1\lambda^2 + I_2\lambda + I_3 = 0 \tag{5.5}$$

where I_1, I_2, and I_3 are the first, second, and third invariants of the characteristic equation. Mathematically they are given as

$$I_1 = -(\lambda_1 + \lambda_2 + \lambda_3) = -tr(\nabla\vec{v}) \tag{5.6}$$

$$I_2 = \lambda_1\lambda_2 + \lambda_2\lambda_3 + \lambda_3\lambda_1 = -\frac{1}{2}\left[\mathrm{tr}\left(\nabla\vec{v}^2\right) - \mathrm{tr}(\nabla\vec{v})^2\right] \tag{5.7}$$

$$I_3 = -\lambda_1\lambda_2\lambda_3 = -\det(\nabla\vec{v}) \tag{5.8}$$

where *tr* represents the trace of a matrix.

This method is a scalar method and is sensitive to the selection of iso-surface thresholds.

5.2.2.3 λ_{ci} Criterion

This method is a further exploration of the Δ criterion. When zero threshold is applied, this method gives the same result as that of the Δ criterion. Here, the vortex structure is visualized using the imaginary part of complex eigenvalues of the velocity gradient tensor. The tensor formation of a velocity gradient tensor can be written as

$$\nabla\vec{v} = \left[\vec{v}_r\vec{v}_{cr}\vec{v}_{ci}\right]\begin{bmatrix}\lambda_r & 0 & 0\\ 0 & \lambda_{cr} & \lambda_{ci}\\ 0 & -\lambda_{ci} & \lambda_{cr}\end{bmatrix}\left[\vec{v}_r\vec{v}_{cr}\vec{v}_{ci}\right]^{-1} \tag{5.9}$$

where λ_r is the real eigenvalue, $\vec{v}_r$ is eigenvector, $\lambda_{cr} \pm i\lambda_{ci}$ are complex eigenvalues with corresponding eigenvectors $\vec{v}_{cr} \pm i\vec{v}_{ci}$.

5.2.2.4 λ_2 Criterion

This vortex identification is based on the cyclostrophic balance. This balance happens when centrifugal forces and horizontal pressure gradients push each other equally in the opposite direction. This is when we have minimal pressure on the axis of rotation. In this method, the calculation is done in a vortical region when we have minimum pressure on the axis of rotation, as mentioned above. Pressure representation in a Hessian matrix, the symmetric part S of the incompressible Navier–Stokes equation gradient can be written as:

$$S = A^2 + B^2 = -\frac{\nabla(\nabla p)}{\rho} \tag{5.10}$$

where p represents pressure.

5.2.3 *Third-Generation Vortex Identification Method*

Starting with the Liutex method, we have the Liutex-Omega method (Ω_L), Modified Liutex-Omega method ($\tilde{\Omega}_L$), and the Liutex Core Lines method, all of which fall in third-generation vortex identification methods. The significant difference between the third and other generation methods is that Liutex represents vortex as a vector while the second generation represents vector as scalar and eigenvalue related. This gives the vortex both magnitude and the direction of the rotation. Here, I will only be discussing the Liutex method.

5.2.3.1 Liutex Method

As stated above, the prominent feature of the Liutex method is that it represents the vortex as a vector. Liutex is defined as

$$\vec{R} = R\vec{r} \tag{5.11}$$

where R is the magnitude of Liutex and $\vec{r}$ is the Liutex direction.

Mathematically, Liutex represents a rigid rotation of fluids. Following the Ref. [13], the formula for the magnitude and direction of Liutex can be given as

$$R = \vec{\omega} \cdot \vec{r} - \sqrt{(\vec{\omega} \cdot \vec{r})^2 - 4\lambda_{ci}^2} \tag{5.12}$$

where $\vec{r}$ is the real eigenvector of the velocity gradient tensor $(\nabla \vec{v})$ and λ_{ci} is the imaginary part of the conjugate complex eigenvalues of $\nabla\vec{v}$.

5.3 Vortex Versus Vorticity

With the completion of the discussion of different generations of vortex identification methods, we have reached the part where we will discuss the difference between the vortex and vorticity and the limitation of first-generation vorticity-based vortex identification methods.

Let, $\vec{v} = (u, v, w)$ be velocity, $\vec{S_a} = (S_x, S_y, S_z)$ be angular speed and $\vec{r} = (x, y, z)$ is a location vector. Here velocity can be written as,

$$\vec{v} = \vec{S_a} \times \vec{r} = (S_x, S_y, S_z) \times (x, y, z) = (S_y z - S_z y, S_z x - S_x z, S_x y - S_y x), \tag{5.13}$$

since vorticity is a velocity curl.

Mathematically,

$$\nabla \times \vec{v} = \left(\frac{\partial}{\partial x}, \frac{\partial}{\partial y}, \frac{\partial}{\partial z}\right) \times (S_y z - S_z x, S_z x - S_x z, S_x y - S_y x) = (2S_x, 2S_y, 2S_z) = 2\vec{S_a} \tag{5.14}$$

This implies that vorticity is twice that of the angular speed. This way, vorticity was associated with vortex, which is valid for rigid body rotation since shear is zero to negligible in this case. Hower, for fluid rotation, we need to consider pure shear deformation.

In simple words, vorticity can be written as, Vorticity = Vortex + Shear.

In a rigid body with negligible shear, Vorticity = Vortex.

However, shear needs to be considered for a fluid body, so Vorticity ≠ Vortex.

Therefore, the first-generation vortex identification method (vorticity-based) is contaminated by shear, especially in the boundary layer flow transition [14]. In order to express this, we recreated and modified Shapiro's experiment, where he showed vorticity as the rotation axis and vorticity as the strength of the vortex.

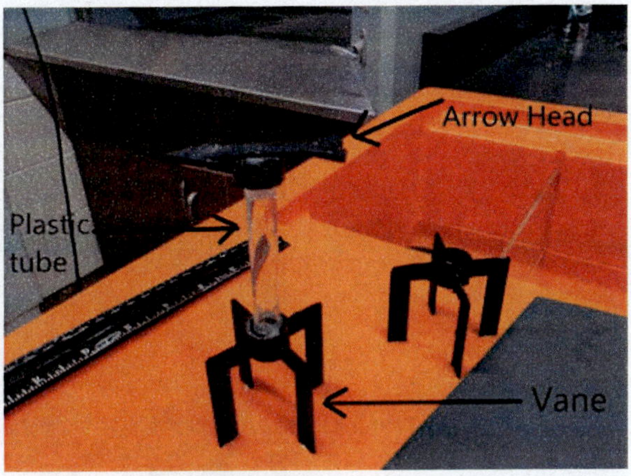

Fig. 5.1 Vorticity meter

5.3.1 Vorticity Meter and Long Channel Device

Two primary apparatus used in this experiment were the vorticity meter and the long channel device. To observe the vorticity in a fluid flow, we designed, and 3-D printed the vorticity meter similar to the one used by Shapiro in one of his experiments (Fig. 5.1).

The vanes are designed so that they are at right angles and can act as paddle wheels. Vanes are attached to a plastic tube which helps the vorticity meter float vertically. We can also change the water level inside the tube to make sure it remains afloat and does not "drown". Finally, the arrowhead gives us the direction of rotation as the vanes turn about their axis.

Another apparatus is a long channel device that consists of an open channel of rectangular cross-section supported at each end by frames. One of the frames is adjustable so that the slope of the channel can be varied. The channel walls are made from clear acrylic plastic to achieve complete visibility of the flow characteristics. This Flow Channel provides a low-cost experiment with accuracy comparable to larger-scale channel investigations. The nominal dimensions of the P6245 channel are $55 \times 175 \times 2500$ mm (W × H × L) (Fig. 5.2).

5.3.2 Reynolds Number and Laminar Flow

To determine if the flow is laminar or turbulent, we need to calculate the Reynolds number of the flow. The parameters required to achieve that are velocity and hydraulic radius of the flow and the kinematic viscosity. This viscosity is temperature-dependent, which can be determined by checking lab temperature. As the flume's width is constant, we must determine the velocity and flow depth to calculate the Reynolds number. We used a regular ruler to determine the flow depth (the ruler

Fig. 5.2 Cussons P6245 flow channels

was attached to the flume), and velocity was determined by observing how much water the flume was discharging at a particular time. The unit we used was 'liters per minute'.

Here is the complete procedure for laminar flow calculation:

(1) To determine if the flow is laminar or turbulent, we need to calculate the Reynolds number of the flow. The equation for the Reynolds number is,

$$Re = \frac{VR}{\nu} \tag{5.15}$$

where,

V = Average velocity of the flow (m/s),

R = Hydraulic radius of the flow (m), and

ν = kinematic viscosity $\left(m^2/s\right)$.

(2) Velocity can be determined by observing how much water the flume was discharging in a particular time divided by the flow area.

$$V = \frac{Q}{A} \tag{5.16}$$

where,

Q = Discharge of the flow (m^3/s)

A = Area of the flow section (m^2)

(3) Discharge of the flow can be determined by measuring the volume of water at a particular time.

$$Q = \frac{V}{t} \tag{5.17}$$

where,

V = collected volume of water (*in liter*)

t = time for the water collection (*s*)

(4) Hydraulic Radius can be defined by the area divided by the wetted perimeter of the channel.

$$R = \frac{A}{P} \tag{5.18}$$

where, P is the wetted perimeter of the channel.

The wetted perimeter can be defined as $P = W + 2 \times D$,

where,

W = Width of the channel *(m)*

D = Depth of the water *(m)*

If we measure all the parameters and insert them in Eq. 5.1, we will find the *Re* of the flow. For an open channel, if the $Re < 500$, the flow keeps laminar.

5.3.3 Modified Shapiro's Experiment

With all the essential things described, we are ready to dive into our experiment. First, we recreated the same experiment Shapiro did to show that vortex and vorticity are the same through rigid body rotation. The experimental setup was a long channel device with a laminar flow and vorticity meter on it. As expected, the vorticity meter rotated as it moved in the direction of the fluid flow (Fig. 5.3).

After that, we removed the vorticity meter and used dyed ink instead. We observed that the dyed ink just moved in a straight line without any form of rotation (Fig. 5.4).

The vorticity meter used in Shapiro's and our first experiment was an instrument half-submerged in water to detect the vortices formed in the water. However, the tool itself is rigid; hence it will only rotate for a rigid body where shear is negligible or zero. However, the shear cannot be ignored in a real fluid. This means the fluid does not rotate, but the meter must rotate as it is a rigid body.

The major takeaway from this experiment is that even though there is no vortex in a streamline flow, the vorticity meter still rotated while the ink just moved in a straight line. This means vorticity is not a proper way to represent vortex.

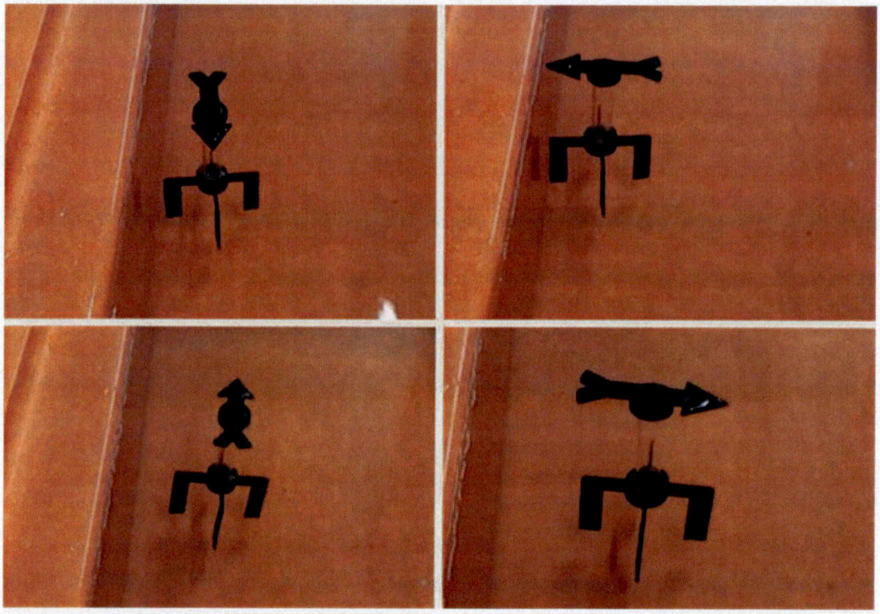

Fig. 5.3 Vorticity meter rotation in a laminar flow

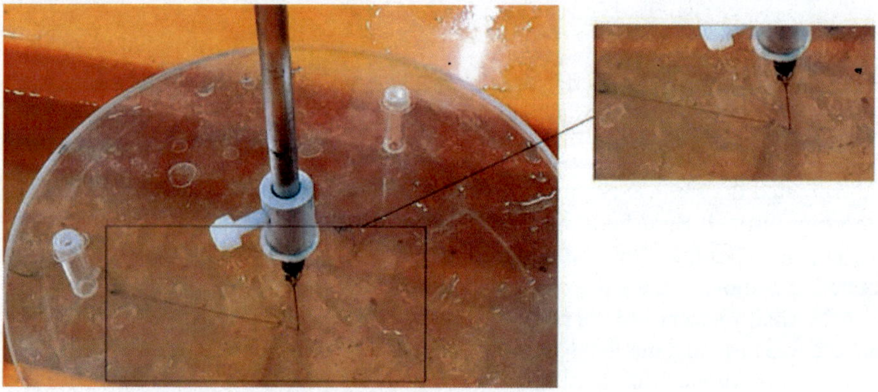

Fig. 5.4 Dyed ink injector

5.3.4 Mathematical Analysis

Rewriting the derivation that shows vorticity is twice the angular speed.

Let, $\vec{v} = (u, v, w)$ be velocity, $\overrightarrow{S_a} = \left(S_x, S_y, S_z\right)$ be angular speed and $\vec{r} = (x, y, z)$ is a location vector. Here velocity can be written as,

$$\vec{v} = \overrightarrow{S_a} \times \vec{r} = \left(S_x, S_y, S_z\right) \times (x, y, z) = \left(S_y z - S_z y, S_z x - S_x z, S_x y - S_y x\right) \tag{5.19}$$

since vorticity is a velocity curl. Mathematically,

$$\nabla \times \vec{v} = \left(\frac{\partial}{\partial x}, \frac{\partial}{\partial y}, \frac{\partial}{\partial z}\right) \times \left(S_y z - S_z x, S_z x - S_x z, S_x y - S_y x\right) = \left(2S_x, 2S_y, 2S_z\right) = 2\overrightarrow{S_a} \tag{5.20}$$

Here, the derivation works for a rigid body, but it won't for the fluid as the derivation does not take decomposition into account. We will explore this from a different aspect. Let us describe angular speed and vorticity through the change of velocity, i.e., $\frac{\partial u}{\partial x}, \frac{\partial u}{\partial y}, \frac{\partial u}{\partial z}, \frac{\partial v}{\partial x}, \frac{\partial v}{\partial y}, \frac{\partial v}{\partial z}, \frac{\partial w}{\partial x}, \frac{\partial w}{\partial y}, \frac{\partial w}{\partial z}$.

$$\nabla \times \vec{v} = \left(\frac{\partial}{\partial x}, \frac{\partial}{\partial y}, \frac{\partial}{\partial z}\right) \times (u, v, w) = \left(\frac{\partial w}{\partial y} - \frac{\partial v}{\partial z}, \frac{\partial u}{\partial z} - \frac{\partial w}{\partial x}, \frac{\partial v}{\partial x} - \frac{\partial u}{\partial y}\right) \tag{5.21}$$

A velocity gradient tensor is a tensor made up of all derivatives of the velocity.

$$\nabla\vec{v} = \begin{bmatrix} \frac{\partial u}{\partial x} & \frac{\partial u}{\partial y} & \frac{\partial u}{\partial z} \\ \frac{\partial v}{\partial x} & \frac{\partial v}{\partial y} & \frac{\partial v}{\partial z} \\ \frac{\partial w}{\partial x} & \frac{\partial w}{\partial y} & \frac{\partial w}{\partial z} \end{bmatrix} \tag{5.22}$$

We will get,

$$\nabla \times \vec{v} = (\nabla\vec{v})_{32} - (\nabla\vec{v})_{23}, (\nabla\vec{v})_{13} - (\nabla\vec{v})_{31}, (\nabla\vec{v})_{21} - (\nabla\vec{v})_{12} \tag{5.23}$$

$$\text{or, } \nabla \times \vec{v} = \left(\frac{\partial w}{\partial y} - \frac{\partial v}{\partial z}\right) - \left(\frac{\partial u}{\partial z} - \frac{\partial w}{\partial x}\right) - \left(\frac{\partial v}{\partial x} - \frac{\partial u}{\partial y}\right) \tag{5.24}$$

Suppose the angular speed for the x-axis and y-axis are zero, and there is only angular speed along the z-axis. In that case, the first two terms will be equal to zero, and we will have

$$\nabla \times \vec{v} = (\nabla\vec{v})_{21} - (\nabla\vec{v})_{12} = \frac{\partial v}{\partial x} - \frac{\partial u}{\partial y} \tag{5.25}$$

There can be two possibilities here.

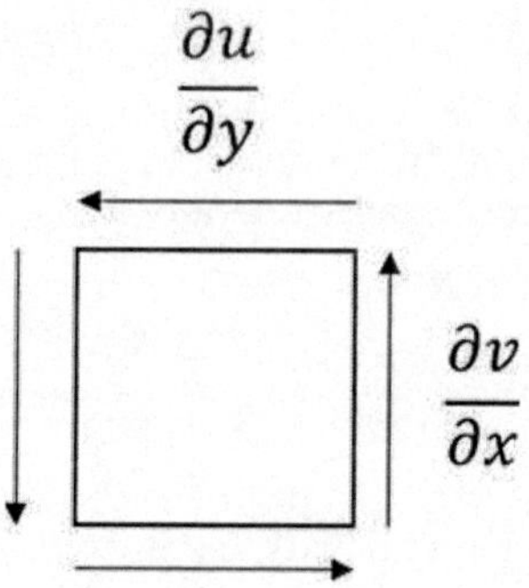

Here, the object is rotating in an anti-clockwise direction as both $\frac{\partial v}{\partial x}$ *and* $\frac{\partial u}{\partial y}$ are in an anti-clockwise direction.

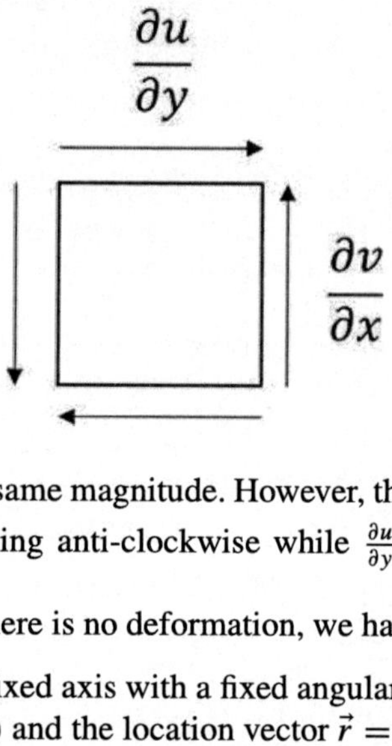

Next, $\frac{\partial v}{\partial x}$ and $\frac{\partial u}{\partial y}$ have the same magnitude. However, their direction is in the opposite direction as $\frac{\partial v}{\partial x}$ is moving anti-clockwise while $\frac{\partial u}{\partial y}$ is moving in a clockwise direction.

For a rigid body, since there is no deformation, we have $\left|\frac{\partial u}{\partial y}\right| = \left|\frac{\partial v}{\partial x}\right|$. Suppose an object is rotating around a fixed axis with a fixed angular speed $\overrightarrow{S_a} = (0, 0, S_z)$. Let the velocity be $\vec{v} = (u, v, 0)$ and the location vector $\vec{r} = (x, y, 0)$.

So,

$$\vec{v} = \overrightarrow{S_a} \times \vec{r} = (0, 0, S_z) \times (x, y, 0) = (-S_z y, S_z x, 0) \tag{5.26}$$

$$\frac{\partial u}{\partial y} = \frac{\partial - S_z y}{\partial y} = -S_z \tag{5.27}$$

$$\frac{\partial v}{\partial x} = \frac{\partial S_z x}{\partial x} = S_z \tag{5.28}$$

With this, we can see that the result that implies vorticity is rotation means $\left|\frac{\partial u}{\partial y}\right| = \left|\frac{\partial v}{\partial x}\right|$ i.e., the absolute value of change in the x-component (u) of velocity with respect to y and change in the y-component (v) of velocity with respect to x are equal. The matrices to represent rigid body rotation at different axes are as follows.

Rotation matrix (rigid rotation) around the z-axis:

$$\begin{bmatrix} 0 & -a & 0 \\ a & 0 & 0 \\ 0 & 0 & 0 \end{bmatrix} \tag{5.29}$$

Rotation matrix (rigid rotation) around the y-axis:

$$\begin{bmatrix} 0 & 0 & a \\ 0 & 0 & 0 \\ -a & 0 & 0 \end{bmatrix} \tag{5.30}$$

Rotation matrix (rigid rotation) around the x-axis:

$$\begin{bmatrix} 0 & 0 & 0 \\ 0 & 0 & -a \\ 0 & a & 0 \end{bmatrix} \tag{5.31}$$

However, in most cases for fluid, $\frac{\partial u}{\partial y} \neq -\frac{\partial v}{\partial x}$. Also, when $\frac{\partial u}{\partial y} \neq -\frac{\partial v}{\partial x}$, there will be a problem in decomposing the velocity gradient tensor to a rotational matrix. The velocity gradient tensor does not only have a rotation but is coupled with rotation, shear, and stretching. Hence, the classical theorem is to use Cauchy-Stokes decomposition to decompose the velocity gradient tensor. We have,

$$\nabla\vec{v} = \begin{bmatrix} \frac{\partial u}{\partial x} & \frac{\partial u}{\partial y} & \frac{\partial u}{\partial z} \\ \frac{\partial v}{\partial x} & \frac{\partial v}{\partial y} & \frac{\partial v}{\partial z} \\ \frac{\partial w}{\partial x} & \frac{\partial w}{\partial y} & \frac{\partial w}{\partial z} \end{bmatrix}$$

From Cauchy-Stokes decomposition,

$$\nabla\vec{v} = A + B \tag{5.32}$$

where,

$$A = \frac{1}{2}\left(\nabla\vec{v} + \nabla\vec{v}^T\right) \tag{5.33}$$

$$B = \frac{1}{2}\left(\nabla\vec{v} - \nabla\vec{v}^T\right) \tag{5.34}$$

Now,

$$A = \begin{bmatrix} \frac{\partial u}{\partial x} & \frac{1}{2}\left(\frac{\partial u}{\partial y} + \frac{\partial v}{\partial x}\right) & \frac{1}{2}\left(\frac{\partial w}{\partial x} + \frac{\partial u}{\partial z}\right) \\ \frac{1}{2}\left(\frac{\partial u}{\partial y} + \frac{\partial v}{\partial x}\right) & \frac{\partial v}{\partial y} & \frac{1}{2}\left(\frac{\partial v}{\partial z} + \frac{\partial w}{\partial y}\right) \\ \frac{1}{2}\left(\frac{\partial w}{\partial x} + \frac{\partial u}{\partial z}\right) & \frac{1}{2}\left(\frac{\partial v}{\partial z} + \frac{\partial w}{\partial y}\right) & \frac{\partial w}{\partial z} \end{bmatrix} \tag{5.35}$$

$$B = \begin{bmatrix} 0 & -\frac{1}{2}\left(\frac{\partial v}{\partial x} - \frac{\partial u}{\partial y}\right) & \frac{1}{2}\left(\frac{\partial u}{\partial z} - \frac{\partial w}{\partial x}\right) \\ \frac{1}{2}\left(\frac{\partial v}{\partial x} - \frac{\partial u}{\partial y}\right) & 0 & -\frac{1}{2}\left(\frac{\partial w}{\partial y} - \frac{\partial v}{\partial z}\right) \\ -\frac{1}{2}\left(\frac{\partial u}{\partial z} - \frac{\partial w}{\partial x}\right) & \frac{1}{2}\left(\frac{\partial w}{\partial y} - \frac{\partial v}{\partial z}\right) & 0 \end{bmatrix} \tag{5.36}$$

Here, the classical theorem thinks A and B represent a deformation and rotation matrix, respectively. Also, B is the matrix representing vorticity, which does not consider deformation.

5.3.5 *Vorticity and Liutex*

A local fluid rotation axis is defined as a vector that can only have stretching (compression) along its length. One of the fundamental properties is that the rotational axis cannot be stretched or compressed or deformed or rotate itself in any other direction than along its length. Also, the change in velocity of the rotational axis can only be in its rotational axis direction.

All rotations must follow this property of the rotational axis. We can write the increment of $\vec{v}$ in the direction of $d\vec{r}$ is $d\vec{v} = \nabla\vec{v} \cdot d\vec{r}$. By following the property of the rotational axis, it must satisfy that $d\vec{v} = \nabla\vec{v} \cdot d\vec{r} = \alpha d\vec{r}$ along the rotation axis, which indicates $d\vec{r}$ is the real eigenvector of $\nabla\vec{v}$.

First, let us analyze if the vorticity satisfies this fundamental concept of the rotational axis. We have,

$$d\vec{v} = \nabla\vec{v} \cdot \vec{\omega} \tag{5.37}$$

where $\vec{\omega}$ is vorticity. From Cauchy-Stokes decomposition,

$$d\vec{v} = (A + B) \cdot \vec{\omega} \tag{5.38}$$

$$d\vec{v} = A \cdot \vec{\omega} + B \cdot \vec{\omega} \tag{5.39}$$

Assuming $\vec{\omega} = a_1\overrightarrow{r_1} + a_2\overrightarrow{r_2} + a_3\overrightarrow{r_3}$ where $\overrightarrow{r_1}, \overrightarrow{r_2}$ *and* $\overrightarrow{r_3}$ and B is a rotation (vorticity) matrix; we have,

$$d\vec{v} = A \cdot \left(a_1\overrightarrow{r_1} + a_2\overrightarrow{r_2} + a_3\overrightarrow{r_3}\right) + (\nabla \times \vec{v}) \times \vec{\omega} \tag{5.40}$$

Now, A $\overrightarrow{r_1} = \lambda \overrightarrow{r_1}$,

$$d\vec{v} = a_1\lambda_1\overrightarrow{r_1} + a_2\lambda_2\overrightarrow{r_2} + a_3\lambda_3\overrightarrow{r_3} + 0 \tag{5.41}$$

Unless $\lambda_1 = \lambda_2 = \lambda_3 = \lambda$

$$d\vec{v} = a_1\lambda_1\overrightarrow{r_1} + a_2\lambda_2\overrightarrow{r_2} + a_3\lambda_3\overrightarrow{r_3} + 0 \neq \lambda\left(a_1\overrightarrow{r_1} + a_2\overrightarrow{r_2} + a_3\overrightarrow{r_3}\right) = \lambda\vec{\omega} \tag{5.42}$$

From here, we can see that, in general, Vorticity is not a local fluid rotational axis.

Now, let us compare if Liutex satisfies this condition. Based on the definition of Liutex (Sect. 2.3.1), Liutex is defined as

$$\vec{R} = R\vec{r}$$

where R is the magnitude of Liutex and $\vec{r}$ is the Liutex direction. Here, R is given as

$$R = \vec{\omega} \cdot \vec{r} - \sqrt{(\vec{\omega} \cdot \vec{r})^2 - 4\lambda_{ci}^2}$$

where $\vec{r}$ is the real eigenvector of the velocity gradient tensor $(\overrightarrow{\nabla v})$ and λ_{ci} is the imaginary part of the conjugate complex eigenvalues of $\nabla\vec{v}$. Since the Liutex direction $(\overrightarrow{r})$ is an eigenvector of the velocity gradient tensor, it automatically satisfies this condition.

We investigate the magnitude calculation from vorticity and Liutex in Couette flow by finding the flaw in vorticity to represent rotation direction (Fig. 5.5).

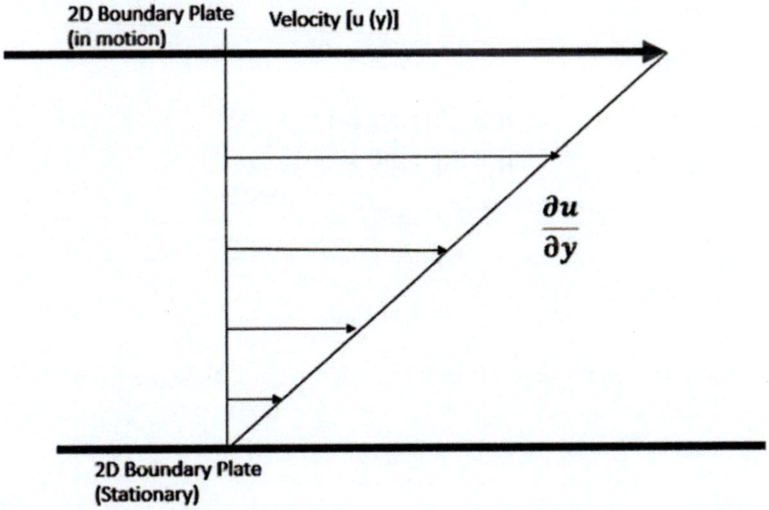

Fig. 5.5 Couette flow

Some of the assumptions made in Couette Flow are:

- Couette flow is constant pressure flow ($\frac{\partial p}{\partial y}$; the variation of pressure in the y-direction is zero).
- The only nonzero velocity component is u (v = w = 0), as the flow is consistent and directed in the x-direction.

Here, the velocity gradient tensor takes the given form.

$$\nabla\vec{v} = \begin{bmatrix} \frac{\partial u}{\partial x} & \frac{\partial u}{\partial y} \\ \frac{\partial v}{\partial x} & \frac{\partial v}{\partial y} \end{bmatrix} = \begin{bmatrix} 0 & a \\ 0 & 0 \end{bmatrix}; \text{ where a is some magnitude}$$

Now, calculate the magnitude of vorticity in a Couette flow.

$$\text{Vorticity} = \nabla \times \vec{v} = \begin{bmatrix} \mathbf{i} & \mathbf{j} & \mathbf{k} \\ \frac{\partial}{\partial x} & \frac{\partial}{\partial y} & \frac{\partial}{\partial z} \\ u & v & w \end{bmatrix} = (0,\ 0,\ -\mathrm{a})$$

Even though there is no rotation in the Couette flow, the vorticity still gives us some magnitude value. This is the same as our experiment result, where the vorticity meter rotated even when there was no rotation in the laminar (streamline) flow.

With this, we will calculate the magnitude of Liutex in a Couette flow. The magnitude of Liutex is given as follows:

$$R = \vec{\omega} \cdot \vec{r} - \sqrt{(\vec{\omega} \cdot \vec{r})^2 - 4\lambda_{ci}^2}$$

Calculating the eigenvalues of Liutex from the characteristic equation:

$$|A - \lambda I| = 0$$

$$\left| \begin{bmatrix} 0 & a \\ 0 & 0 \end{bmatrix} - \begin{bmatrix} \lambda & 0 \\ 0 & \lambda \end{bmatrix} \right| = 0$$

$$\left| \begin{bmatrix} \lambda & -a \\ 0 & \lambda \end{bmatrix} \right| = 0$$

$$\Rightarrow \lambda = 0, 0$$

which are both real eigenvalues. Hence, λ_{ci} goes to zero. Now, we have,

$$R = \vec{\omega} \cdot \vec{r} - \sqrt{(\vec{\omega} \cdot \vec{r})^2} = 0$$

Here, we have the magnitude of Liutex as zero, which is accurate for the Couette flow as there is no rotation in the system.

We can conclude that Liutex is more reasonable than vorticity in investigating the vortex's magnitude and rotation.

5.4 Conclusion and Discussion

The two major misunderstandings regarding vortex come from the derivation of vorticity as an equivalent of a vortex and the use of a solid body to show fluid rotation. Shapiro's experiment used rigid body rotation to show that vorticity is the same as the vortex. This is not essentially wrong for a rigid body rotation as it lacks shear. Vorticity can act as a rotation when there is negligible to zero shear, so the vorticity becomes rotation for the rigid body. However, fluid does not have the same properties as a rigid body. In fluid rotation, a pure shear deformation needs to be considered.

Proper understanding of fundamental concepts such as vortex and vorticity in fluid mechanics can help us to expand our research extensively. This can also aid turbulence research as vortices constitute a significant component of turbulent flow. Turbulence is a fluid motion characterized by chaotic pressure and flow velocity changes. By detecting the vortex precisely, we will learn about the turbulence in the fluid flow properly. These results can be used in engineering designs to tackle the pressure and flow velocity changes in fluid motion.

In conclusion, vortex and vorticity are two different things in a fluid motion. Vorticity cannot represent vortex in a fluid flow as a pure shear deformation is present in the fluid, especially near the wall. Hence, vorticity does not always imply rotation.

References

1. H. Helmholtz, On the integrals of the hydrodynamic equations corresponding to vortex motions. Journal für die reine und angewandte Mathematik **55**, 22–25 (1858). ((in German))
2. P. Saffman, *Vortices dynamics* (Cambridge University Press, Cambridge, UK, 1992)
3. S.K. Robinson, Coherent motion in the turbulent boundary layer. Ann. Rev. Fluid Mech. **23**, 601–639 (1991)
4. J.C.R. Hunt, A.A. Wray, P. Moin, Eddies, stream, and convergence zones in turbulent flows. Center for Turbulent Research Report CTR-S88, pp. 193–208 (1988)
5. M.S. Chong, A.E. Perry, A general classification of three-dimensional flow fields. Phys. Fluids A **2**(5), 765–777 (1990)
6. J. Jeong, F. Hussain, On the identification of a vortex. J. Fluid Mech. **285**, 69–94 (1995)
7. J. Zhou, R. Adrian, S. Balachandar, T.M. Kendall, Mechanisms for generating coherent packets of hairpin vortices in channel flow. J. Fluid Mech. **387**, 353–396 (1999)
8. C. Liu, Y. Wang, Y. Yang et al., New omega vortex identification method. Sci. China Phys. Mech. Astron. **59**(8), 684711 (2016)
9. C. Liu, Y. Gao, S. Tian, X. Dong, Rortex- a new vortex vector definition and vorticity tensor and vector decompositions. Phys. Fluids **30**, 035103 (2018). Cliff, N. R. (1988)
10. P. Shrestha, A. Bhattarai, C. Liu, Application of Liutex and some second-generation vortex identification methods to direct numerical simulation data of a transitional boundary layer, in

C. Liu, Y. Wang (eds.), Liutex and Third Generation of Vortex Definition and Identification. Springer, Cham (2021). https://doi.org/10.1007/978-3-030-70217-5_19
11. C. Liu, Y. Gao, X. Dong, Y. Wang, J. Liu, Y. Zhang, X. Cai, N. Gui, Third generation of vortex identification methods: omega and Liutex/Rortex based systems. J. Hydrodyn. **31**(2), 205–223 (2019)
12. Helmholtz's theorems [EB/OL]. https://en.wikipedia.org/wiki/Helmholtz%27s_theorems
13. W. Xu, Y. Gao, Y. Deng J. Liu, C. Liu, An explicit expression for the calculation of the Rortex vector. Phys. Fluids **31**(9), 095102.
14. P. Shrestha, C. Nottage, Y. Yu et al., Stretching and shearing contamination analysis for Liutex and other vortex identification methods. Adv. Aerodyn. **3**, 8 (2021). https://doi.org/10.1186/s42774-020-00060-9

Chapter 6
POD Analysis on Losing Symmetry of Vortex Structure in the Flow Transition by Liutex Method

Pushpa Shrestha, Charles Nottage, and Chaoqun Liu

Abstract The proper orthogonal decomposition (POD) is a data decomposition method to investigate and analyze complex turbulent flow. The POD method offers the optimal low-dimensional approximation of a given data set. In our case, POD is used with Liutex vector as an input instead of velocity vector to extract the coherent structure of late boundary layer flow transition. Mathematically, the Liutex vector field is decomposed into a sum of basis functions (spatial modes) multiplied by time coefficients (Fourier-splitting method). A singular value decomposition (SVD) algorithm is used to perform the POD method. Our studies show that fluid motion can be modeled/reconstructed by a few leading modes as they contain a large portion of total rotational intensity. Trailing modes can be neglected as they do not contribute to the total rotational strength of the flow. From the reconstructed vortex structure in the flow transition, the loss of symmetry of vortex is investigated, and it is found that the asymmetry of vortex develops at the flow transition. In fact, our study shows that the antisymmetric of the vortex starts from the middle, and then the antisymmetric structure of the bottom part starts and spreads to the top level.

6.1 Introduction

The proper orthogonal decomposition (POD) method is a highly applied data analysis and modeling method in fluid mechanics. The POD method allows us to reconstruct a flow with the first few most energetic modes, keeping the data structure intact. Sometimes, the fluid motion is not easily visible in raw data; in this case, reconstructed POD modes can model the fluid motion accurately and efficiently. This method essentially offers an orthogonal basis to represent a given set of data where optimal low-dimensional approximations for the given data set are calculated [1, 2]. The bases are also known as POD modes. These POD modes best represent the data. The leading modes represent most of the rotational intensity of incompressible flow, whereas high-ordered POD modes represent very few portions of total rotational

P. Shrestha · C. Nottage · C. Liu (✉)
Department of Mathematics, The University of Texas at Arlington, Arlington, TX 76019, USA
e-mail: cliu@uta.edu

Y. Wang et al. (eds.), *Liutex and Third Generation of Vortex Identification*,
Springer Proceedings in Physics 288, https://doi.org/10.1007/978-981-19-8955-1_6

strength. We choose the Liutex vector as an input vector instead of a velocity vector [3]. Most POD analysis has been done with the velocity vector as an input. However, in this study, we use the Liutex vector as an input as Liutex represents the local rigid rotation part of fluid motion without shear and stretching/compression contamination [4]. The velocity modes are related to the kinetic energy content, whereas the Liutex mode represents the rotation intensity. Since the Liutex vector represents local rigid rotation of fluids without any shear or stretching contamination, the Liutex vector is applied instead of the velocity vector as an input for POD analysis.

A vortex is known as the rotational fluid movement. Over the past three decades, several vortex identification methods such as Q, Δ, λ_2, andλ_{ci} have been suggested and used in direct numerical simulation (DNS) data to analyze and visualize the vortex structure numerically in the transitional boundary layer [5–8]. These methods are based on scalar quantities and need a proper threshold to capture the vortex boundary. According to Liu et al. [9], vorticity-based methods are classified as the first generation (1G) of vortex identification methods, eigenvalue -based methods such as Q, Δ, λ_2, andλ_{ci} are regarded as the second generation (2G) of vortex identification methods, and the Liutex method [10, 11], Liutex-Core–Line method, and other Liutex-based methods [12, 13] are regarded as the third-generation (3G) of vortex identification methods. The third generation of vortex identification methods is considered the best among the prevalent methods as it can present accurate vortex boundaries along with the direction [14, 15]. Since then, the novel Liutex has been used abundantly in the literature by many fellow researchers and scientists [16–18].

The POD method is one of the most broadly applied modal decomposition and dimensionality reduction techniques to analyze vortex structure. There are two versions of the POD method. Initially, the POD method was proposed by Lumley [1] in 1967 to explore the turbulent flow. In 1987 Sirovich [2] introduced the other version of POD known as snapshot POD. Both versions of POD are equivalent to the singular value decomposition (SVD) method. So SVD method is used in this paper for POD analysis of late flow transition. In the POD method with Liutex input vector, the orthogonal modes are ranked according to their rotational intensity. The first mode contains the largest rotational intensity of the flow and gradually decreases as we go on to the next mode. Due to limited computer memory, POD modes with high intensity are used to optimize the computation. Many researchers have used the POD method to study the flow structure. POD was used in research [19, 20] to analyze the flow structures in various cases and scenarios. Dong et al. applied POD analysis on vortical structures in MVG wake by Liutex core line identification [21]. The POD method has also been applied to flow transition in the boundary layer. Gunes used the POD method to reconstruct a transitional boundary layer with and without control [19]. Yang et al. studied the POD analyses on vortex structure in the late transition [20].

In this paper, POD is applied to study and analyze the conversion of the symmetric vortex to asymmetry in the flow transition with a Liutex input vector to reconstruct the data. The modified Liutex-Omega vortex identification method is applied to DNS data to capture the vortex structure of flow transition with iso-surfaces of $\Omega_L = 0.52$ and $\varepsilon = 0.001(b-a)_{max}$. Due to the limited computer memory capacity, snapshot POD is used here.

6.2 Numerical Setup

First, a snapshot matrix A is taken from DNS data of the flat plate boundary layer with 100 snapshots in time between $t = 20.505T$ to $t = 21.00T$, where T is the period of Tollmein-Schlichting wave, to study orthogonal basis functions (POD modes) of the coherent vortex structures in the transitional flow. Then, we have chosen the proper subzone in the late boundary layer transition defined by the parameters given in Table 6.1 to study the POD of flow structure (Fig. 6.1).

The table gives the starting and ending points of index i, j, k along x, y and z directions.

The snapshot matrix A between the timesteps $t = 20.5T$ to $t = 21.0T$ is given by,

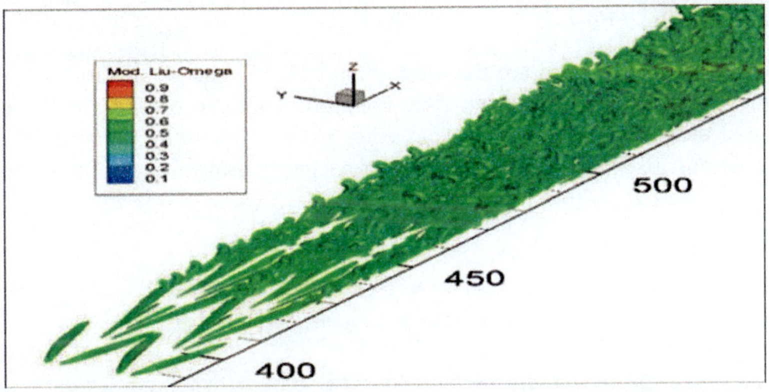

Fig. 6.1 Vortex structure of transitional boundary layer by modified Liutex-Omega method

Table 6.1 Parameters of subzone

Grid direction	Starting index	Ending index
i	500	580
j	1	128
k	1	200

$$A = \begin{pmatrix} L_{x\,500,1,1}^{(j)} \\ \vdots \\ L_{x\,580,128,200}^{(j)} \\ L_{y\,500,1,1}^{(j)} \\ \vdots \\ L_{y\,580,128,200}^{(j)} \\ L_{z\,500,1,1}^{(j)} \\ \vdots \\ L_{z\,580,128,200}^{(j)} \end{pmatrix} \text{for}\, j = 1, \ldots, 100,$$

where ${L_x}^{(j)}$, ${L_y}^{(j)}$ and ${L_z}^{(j)}$ are Liutex vectors in x, y, z directions in the flow fields at several time steps $t = (20.50 + 0.005j)T$ where j is from 1 to 100.

6.3 Proper Orthogonal Decomposition (POD)

The POD modes, also known as the orthogonal basis, are ranked by fluctuating rotational intensity, where leading modes have the dominant rotational intensity, and trailing modes have weak or no rotational intensity.

Definition 6.1 POD modes are the orthogonal basis for the given data set. In fluid mechanics, they are a set of deterministic spatial functions received from the decomposition of the random vector field representing the turbulent fluid motion. Each of these functions, also known as POD modes, can capture some portion of the rotational strength of the flow.

Let $u(\text{x, y, z, t})$ denote the vector field in the flow with fluctuating velocity. Then,

$$u(\text{x, y, z, t}) = U(\text{x, y, z}) - U'(\text{x, y, z}), \tag{6.1}$$

where $U(\text{x, y, z})$ is the velocity vector, and $U'(\text{x, y, z})$ is the temporal mean velocity vector (assumed to be stationary). Then, the POD method decomposes the random vector field $u(\text{x, y, z, t})$ into a sum of orthogonal basis functions/POD modes $\Phi_k(x, y, z)$ multiplied by random time coefficients $\text{a}_k(t)$, i.e.,

$$u(\text{x, y, z, t}) = \sum_{k=1}^{\infty} \text{a}_k(t)\Phi_k(x, y, z) \tag{6.2}$$

Here, Φ_k is the matrix of eigenvectors of the covariance matrix $\frac{1}{m-1}U^T U$, where m is rows of U.

In matrix form, it can be written as:

$$A = \Phi Q \tag{6.3}$$

where the matrix Φ contains the spatial modes $\Phi_k(x, y, z)$ and Q contains the temporal coefficients $a_k(t)$.

6.4 POD Analysis Inflow Transition by Modified Liutex Omega Method

We have used SVD to perform the POD method, and the singular values are ordered in ascending order. Figure 6.2 shows the contribution of each mode to total fluid rotational intensity. The amount of the rotational intensity possessed by higher modes gradually decreases, and ultimately, they converge to zero.

The following figures represent the structure of the first six POD modes. The other modes have similar vortex structures, so we have not included them (Fig. 6.3).

Mode 1, also known as mean flow, has the dominant streamwise vortex stricture. Mode 2, 3, and 4 also have the streamwise vortex structure but are less intense than mode 1. Spanwise characteristic dominates the vortex structure in higher modes as higher modes show more fluctuation distributions of vortex structures. In other words, the streamwise vortex structure is dominant in the leading modes, whereas the spanwise characteristic is dominant in the trailing modes. This nature of POD modes can be seen through the interior vortex structure, shown in Fig. 6.4.

Since the leading modes with streamwise structures possess significantly higher rotational strength than the other trailing modes, the original vortex structures can be reconstructed by a few leading modes. So, we can model the original flow by reconstructing the data of the first few POD modes that have major contribution to

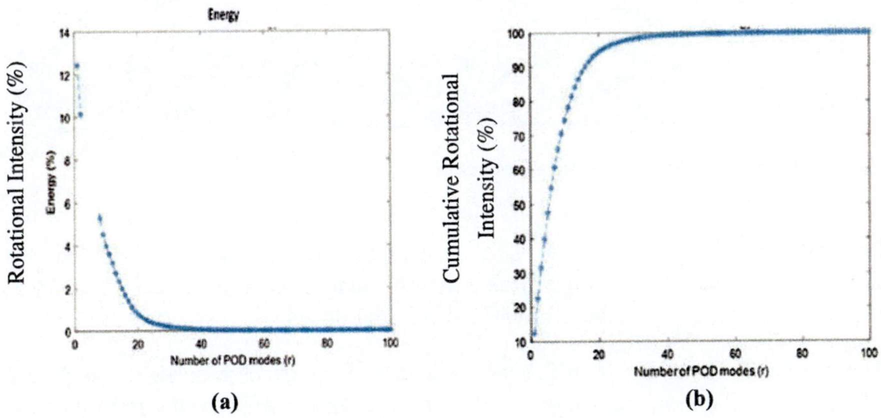

Fig. 6.2 **a** Rotational intensity at various POD modes. **b** Cumulative rotational strength of POD modes

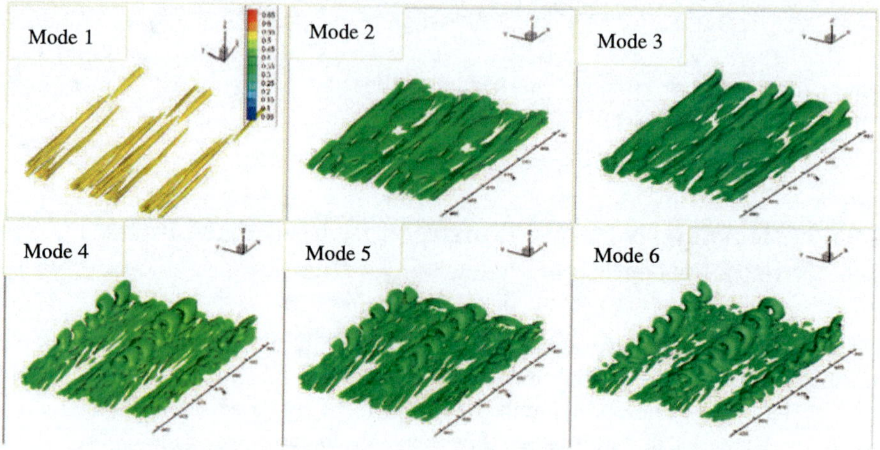

Fig. 6.3 Vortex structures of the first 10 modes with iso-surfaces of $\widetilde{\Omega}_L = 0.52$

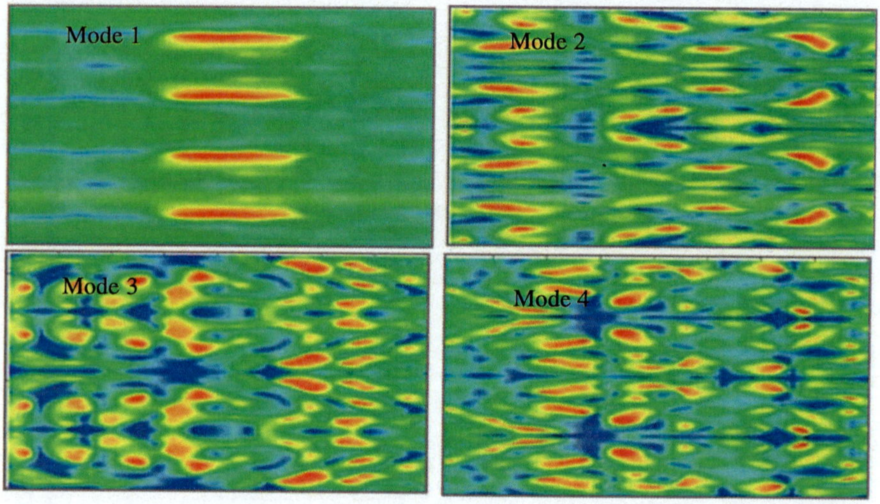

Fig. 6.4 Interior structure of some POD modes by XY-cross section

the fluid rotation. This will reduce the size of the original data, keeping most of the features of data (fluctuating vortex strength) intact. This can be seen in Fig. 6.5, where the first five POD modes model the original fluid flow.

Definition 6.2 The coefficients that we get when the eigenvector matrix Φ_k in Eq. 6.2 are scaled by the singular values of the original matrix are known as POD time coefficients. The POD time coefficients $a_k(t)$ function as a weight factor for each mode.

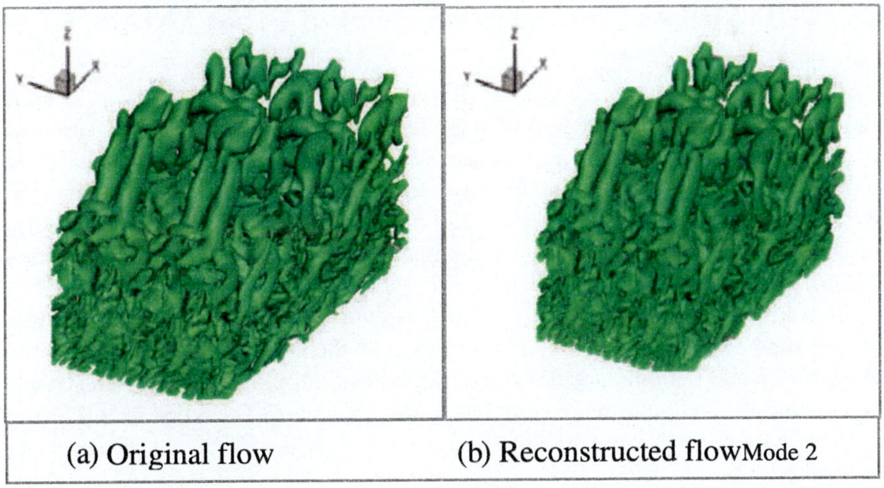

Fig. 6.5 Vortex structures of original and reconstructed flow by first five modes with of $\widetilde{\Omega}_L = 0.52$

The time coefficients are representative of flow dynamics. Each column of matrix Φ_k represents the time evolution of the respective mode. The first column gives the time coefficient of the mean flow (1st mode). Similarly, the second column of the eigenvector matrix Φ_k gives the time coefficient of the second mode, and so on. Then, these modes are scaled by the corresponding singular values. The following graphs show the POD time coefficients of the first five modes. The remaining time coefficients show a similar fluctuating structure, so they are omitted (Fig. 6.6).

These graphs demonstrate the dynamics (fluctuations) of fluid motion. The mean flow has the least fluctuation, while the higher modes have more significant fluctuations. Some similar fluctuations can be seen in the trailing modes.

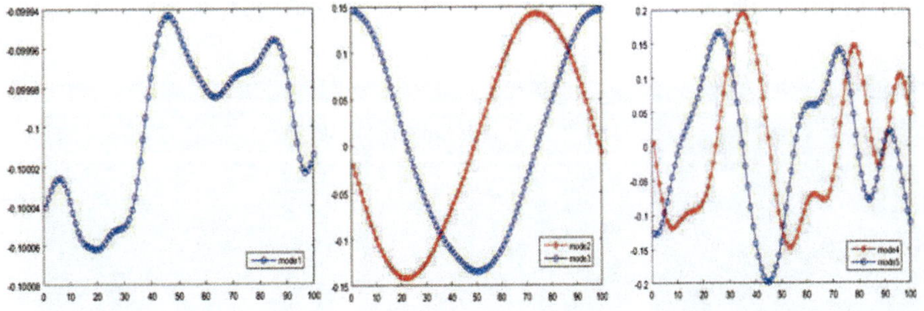

Fig. 6.6 POD time coefficients of mode 1 to mode 5, where the x-axis represents time steps, and the y-axis represents time coefficients

6.5 POD Analysis on Losing Symmetry of Vortex Structure

The vortex structure in the early stage is symmetric. However, it starts to lose the symmetry in the transition stage, and at the late transition stage, the vortex structure is entirely chaotic, making it asymmetric. We have probed the zone where vortex symmetry is being lost and examined the areas from where it starts to lose symmetry. The following figure shows that the top part of the vortex structure is symmetric near X = 470, but the bottom is already antisymmetric at the same time step and position (Fig. 6.7).

From Fig. 6.8a, it can be seen that at X = 470, the vortex structure at the top, middle, and bottom are all symmetric. In Fig. 6.8b the middle part is antisymmetric while the top and bottom are still symmetric. However, in Fig. 6.8c, we can see that the top, middle, and bottom are completely antisymmetric. The index for grid-level along the z-axis for bottom, middle and top are k = 0–1, k = 1–4 and k = 4+ respectively.

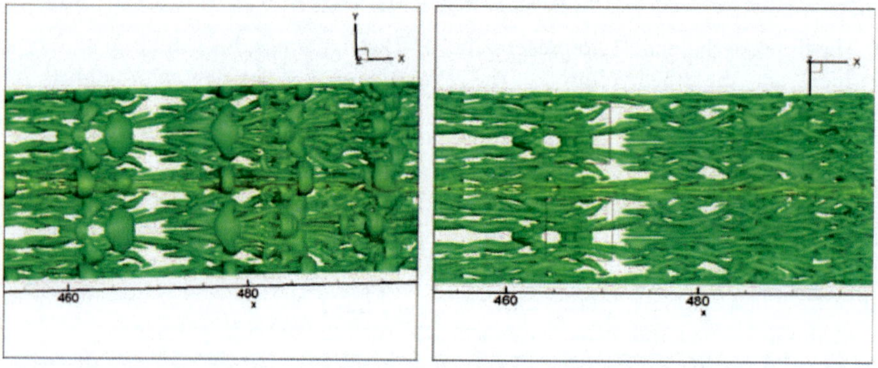

Fig. 6.7 Top and bottom views of the vortex structure at the same position

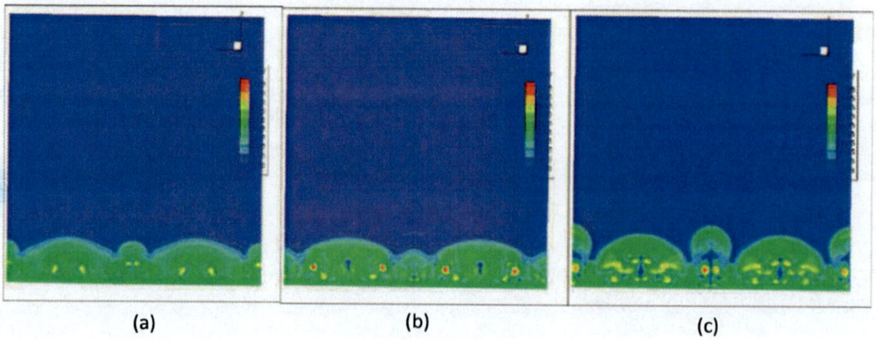

Fig. 6.8 YZ-slice of flow structure at X = 470, X = 475, and X = 480

Definition 6.3 The function $f(x, y, z)$ is symmetric about plane $y = 0$ axis along the spanwise direction in the domain $-\pi \leq y \leq \pi$ if $f(x, -y, z) = f(x, y, z)$ for $\forall x, y, z \epsilon Z^+$.

Definition 6.4 Let $L_{i,j,k}$ be the Liutex magnitude where i, j, k are indexes for grid points along the x, y, and z directions, respectively. Let the domain of grid points be $i \in [i_1, i_2]$, $j \in [2, 128]$, and $k \in [1, 200]$. The YZ plane at y = 0 (i.e. $j = 64$) is the axis of symmetry. $L_{j,k}$ and $L_{128-j,k}$ are the Liutex magnitudes for $j = 2, 3, \ldots, 64$ in the YZ-plane on the left and right of the axis of symmetry, respectively.

Two points, x(:,j, k) and x(:, 128-j, k), at any $i \in [i_1, i_2]$ are symmetric about the plane $y = 0$ at $z = k$ for $k = 1, 2, \ldots, 200$ if $|L_{j,k} - L_{130-j,k}| < 0.001$; otherwise, these two points are not symmetric about the plane y = 0.

Let $m(d_i)$ and $n(d_i)$ are the number of pairs of points that are symmetric and antisymmetric, respectively. The antisymmetric index is defined by

$$\alpha_i = \left(\frac{n(d_i)}{m(d_i) + n(d_i)}\right). \tag{6.4}$$

For $i \in [i_1, i_2]$, the contour of Liutex magnitude in the YZ-plane is symmetric about the plane $y = 0$ if $\alpha_i < 0.01$. Otherwise, the contour of Liutex magnitude in the YZ-plane is antisymmetric about $y = 0$. So, the vortex structure is symmetric about $y = 0$ if $\alpha_i < 1$, $\forall i \in [i_1, i_2]$. Otherwise, the vortex structure is antisymmetric about $y = 0$.

From the above definition, if $\alpha_i \geq 1$, the vortex structure is antisymmetric. The higher the value of α_i, the more antisymmetric the vortex structure is. If $\alpha_i < 1$, symmetric vortex exists (Figs. 6.9 and 6.10).

The figure is the YZ-slice contour of the reconstruction of the vortex structure by the first five modes at timestep t = 20.00 T, where T is the period of T-S wave.

This figure shows that in reconstructed data, the bottom part has crossed the antisymmetric index threshold (i.e., 0.01) at around X = 468. Then the middle and top part crosses that limit at X = 470 and X = 475, respectively, indicating that the antisymmetric vortex structure starts from the middle, then spreads to the bottom, and then to the top of the boundary layer vortex structure in transitional flow.

6.6 Conclusion

POD can be used to reduce the dimension of the large data, keeping most of the features intact. This will reduce the cost and time of computation. So, this technique has been so effectively applied in the literature. From the POD analysis on losing symmetry of vortex structure in the boundary layer transition, we can conclude the followings:

Fig. 6.9 YZ-slice of reconstructed flow by the first five POD modes at X = 470

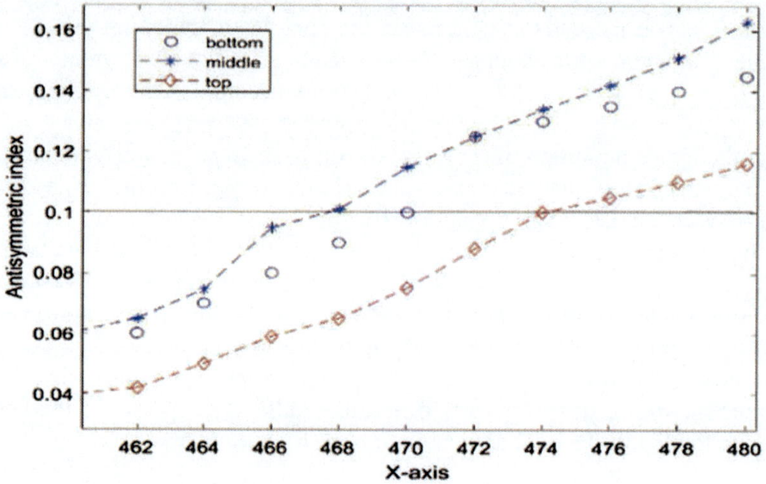

Fig. 6.10 The antisymmetric index of the top, bottom, and middle part of the reconstructed vortex structure by the first five POD modes

1. Mode 1, known as the mean flow, has most of the rotational intensity of the flow. Leading modes have dominant rotational strength, while trailing modes contribute less to fluid rotation.
2. The first three/four modes have streamwise characteristic, but when we take higher modes, the spanwise characteristic of the flow becomes dominant.

3. The antisymmetry of the vortex structure starts from the middle of the boundary layer, and then the antisymmetric structure of the bottom part starts and spreads to the top level. In the late transition stage of the boundary layer, the vortex structure is entirely asymmetric.

References

1. J.L. Lumley, The structure of inhomogeneous turbulent flows. Atmospheric Turbulence and Radio Wave Propagation, pp.166–178 (1967)
2. L. Sirovich, Turbulence and the dynamics of coherent structures. Part I: coherent structures. Q. Appl. Math. **45**(3), 561–571 (1987)
3. S. Charkrit, P. Shrestha, C. Liu, Liutex core line and POD analysis on hairpin vortex formation in natural flow transition. J. Hydrodyn. **32**, 1109–1121 (2020). https://doi.org/10.1007/s42241-020-0079-0
4. P. Shrestha, C. Nottage, Y. Yu et al., Stretching and shearing contamination analysis for Liutex and other vortex identification methods. Adv. Aerodyn. **3**, 8 (2021). https://doi.org/10.1186/s42774-020-00060-9
5. M.S. Chong, A.E. Perry, A general classification of three-dimensional flow fields. Phys. Fluids A **2**(5), 765–777 (1990)
6. J. Jeong, F. Hussain, On the identification of a vortex. J. Fluid Mech. **285**, 69–94 (1995)
7. J.C.R. Hunt, A.A. Wray, P. Moin, Eddies, stream, and convergence zones in turbulent flows. Center for Turbulent Research Report CTR-S88, 193–208 (1988)
8. J. Zhou, R. Adrian, S. Balachandar, T.M. Kendall, Mechanisms for generating coherent packets of hairpin vortices in channel flow. J. Fluid Mech. **387**, 353–396 (1999)
9. C. Liu, Y. Gao, X. Dong, Y. Wang, J. Liu, Y. Zhang, X. Cai, N. Gui, Third generation of vortex identification methods: omega and Liutex/Rortex based systems. J. Hydrodyn. **31**(2), 205–223 (2019)
10. C. Liu, Y. Gao, S. Tian, X. Dong, Rortex-a new vortex vector definition and vorticity tensor and vector decompositions. Phys. Fluids **30**, 035103 (2018)
11. Y. Gao, C. Liu, Rortex and comparison with eigenvalue-based identification criteria. Phys. Fluid **30**, 085107 (2018)
12. Y. Gao, J. Liu, Y. Yu, C. Liu, A Liutex based definition and identification of vortex core center lines. J. Hydrodyn. **31**(3) (2019)
13. O. Alvarez, Y. Yu, P. Shrestha, D. Almutairi, C. Liu, Visualizing liutex core using liutex lines and tubes, in C. Liu, Y. Wang (eds.), *Liutex and Third Generation of Vortex Definition and Identification*. Springer, Cham (2021). https://doi.org/10.1007/978-3-030-70217-5_10
14. Y. Yu, P. Shrestha, O. Alvarez, C. Nottage, C. Liu, Investigation of correlation between vorticity, Q, λci, λ2, Δ and Liutex. Comput. Fluids **225**, 104977 (2021), ISSN 0045-7930.https://doi.org/10.1016/j.compfluid.2021.104977
15. Y. Yu, P. Shrestha, O. Alvarez, C. Nottage, C. Liu. Correlation analysis among vorticity, Q method and Liutex. J. Hydrodyn. **32** (2020).https://doi.org/10.1007/s42241-020-0069-2
16. P. Shrestha, A. Bhattarai, C. Liu, Application of liutex and some second-generation vortex identification methods to direct numerical simulation data of a transitional boundary layer, in: C. Liu, Y. Wang (eds.), *Liutex and Third Generation of Vortex Definition and Identification*. Springer, Cham (2021). https://doi.org/10.1007/978-3-030-70217-5_19
17. Y. Yu, P. Shrestha, O. Alvarez, C. Nottage, C. Liu, Incorrectness of the second-generation vortex identification method and introduction to liutex, in C. Liu , Y. Wang (eds.), *Liutex and Third Generation of Vortex Definition and Identification*. Springer, Cham (2021). https://doi.org/10.1007/978-3-030-70217-5_2

18. C. Nottage, Y. Yu, P. Shrestha, C. Liu, Dimensional and theoretical analysis of second-generation vortex identification methods, iIn C. Liu, Y. Wang (eds.), *Liutex and Third Generation of Vortex Definition and Identification.* Springer, Cham (2021). https://doi.org/10.1007/978-3-030-70217-5_3
19. H. Gunes, Proper orthogonal decomposition reconstruction of a transitional boundary layer with and without control. Phys. Fluids **16** (2004), Article ID 2763
20. Y. Yang, S. Tian, C. Liu, POD Analyses on Vortex Structure in Late-stage Transition," AIAA paper 2018–0821, January 2018
21. X. Dong, X. Cai, Y. Dong, C. Liu, POD analysis on vortical structures in MVG wake by Liutex core line identification. J. Hydrodyn. (2020)

Chapter 7
The Liutex Shear Interaction in Boundary Layer Transition

Check for updates

Biyu Pang, Yuan Ding, and Yiqian Wang

Abstract The third generation of vortex identification methods based on Liutex vector are superior to previous methods in that they overcome the drawbacks of previous methods including threshold problem, shear contamination, etc. with a clear physical meaning for the Liutex vector. The direction of Liutex represents the local axis of rotation, and its magnitude is equal to twice the angular velocity of rotation. The current study focuses on the interaction between Liutex represented rotation and the residual shear part during the development of Λ vortex and hairpin vortex in boundary layer transition. The results show that shear plays an important role in the generation and dissipation of vortices and the proportion of Liutex in the whole vorticity affects the stability of a vortex. When the directions between Liutex and shear is approximately parallel, the vortex moves mainly along the flow direction and the offsets in other directions are relatively small. It is also shown that the Liutex vector can accurately extract the rigid rotation part from fluid motion and the third-generation vortex identification methods can serve as a powerful tool to study fluid dynamics.

7.1 Introduction

In both nature and industrial applications, vortex structure is everywhere, especially in turbulence. It plays an important role in the study of the generation and maintenance of turbulence. Therefore, a fully understanding and a mathematical definition of vortex with systematical methods to accurately identify the vortex structure are of great significance in understanding the flow mechanism, turbulent structures [1, 2]. So far, there is still no widely accepted vortex definition and identification method. The vorticity-based vortex identification methods, classified as the first generation here, confuse the concepts of vortex and vorticity, and regards the magnitude of vorticity as the intensity of local rotational motion. However, Robinson [3] found that there was a low correlation between the high vorticity region and the actual

B. Pang · Y. Ding · Y. Wang (✉)
School of Mathematical Science, Soochow University, Suzhou 215006, China
e-mail: yiqian@suda.edu.cn

Y. Wang et al. (eds.), *Liutex and Third Generation of Vortex Identification*,
Springer Proceedings in Physics 288, https://doi.org/10.1007/978-981-19-8955-1_7

vortex structure. Therefore, several vortex identification methods based on Cauchy-Stokes decomposition were introduced, here categorized as the second-generation methods. These methods improve the visualization of vortices, but still suffer from some issues. Firstly, the physical meanings of the methods are not clear and different from each other. Secondly, as iso-surface based methods, the problem of threshold selection is introduced. Thirdly, the information of velocity gradient tensor $\nabla \boldsymbol{V}$ is not fully utilized. These problems of traditional vortex identification methods need to be dealt with [4]. The first part of this paper revisits the vortex identification methods, mainly the new introduced third generation of vortex identification methods based on Liutex vector. The second part describes the shear contamination analysis under the Liutex-shear decomposition. In the third part, taking the boundary layer transition as an example, the interaction between Liutex and shear is discussed. The fourth part summarizes the conclusions of current study.

7.2 Vortex Identification Method

To deal with the issues of first and second-generation vortex identification methods, the team led by Professor Chaoqun Liu of the University of Texas at Arlington carried out various research works and proposed vortex identification methods, including Ω vortex identification method, Liutex vector, Ω-Liutex method and so on [4–6]. Liu et al. [5] proposed the Ω vortex identification method to overcome the problem of threshold selection, and further proposed the Liutex vector that can accurately identify the vortex structure. This definition decomposes the vorticity $\boldsymbol{\omega}$ into a rotating part $\boldsymbol{R}$ and a residual pure shear part $\boldsymbol{\omega} - \boldsymbol{R}$. Liutex is a vector whose direction represents the local axis of rotation and whose magnitude represents twice the rotational angular velocity of rigid motion. It not only solves the problem of threshold selection, but also answers the problem of six elements of vortex definition. After that, Dong et al. [7] proposed a new normalized Ω_R vortex identification method.

7.2.1 *Vortex, Vorticity, Velocity Gradient Tensor*

Broadly speaking, vortex refers to the rotational motion of fluid, and vorticity is defined as the curl of velocity vector (i.e., $\boldsymbol{\omega} = \nabla \times \boldsymbol{V}$). Its physical meaning is not clear. Robinson [3] pointed out that near the wall of the turbulent boundary layer, the correlation between the vorticity concentration area and the actual vortices is rather low. Wang et al. [8] found in the boundary layer transition the Λ vortex has smaller vorticity at the vortex core than that in the surroundings. This shows that vortex and vorticity are different, and vorticity cannot represent the rotational motion of fluid. The pattern of local fluid motion can be derived from the velocity gradient tensor $\nabla \boldsymbol{V}$ [9], and most Eulerian vortex identification methods are dependent on the velocity gradient tensor.

7.2.2 Cauchy-Stokes Decomposition

The first and major second-generation methods are generally based on Cauchy-Stokes decomposition, including vorticity-based vortex identification method, Q method, λ_2 method and Ω method. The most fundamental problem of these methods is that Cauchy-Stokes decomposition cannot represent the decomposition of fluid motion. Generally, Cauchy-Stokes decomposition is understood as twice the angular velocity of rigid rotation of fluid element around its center. However, in practical application, there is a contradiction between the fluid rotation motion based on Cauchy-Stokes decomposition and vorticity. For example, in the two-dimensional simple shear layer, there is a fluid rotation part according to Cauchy-Stokes decomposition, but in fact, there is no rotational motion in the flow field [9].

7.2.3 Velocity Gradient Tensor

The other approach is based on the analysis of the velocity gradient tensor ∇V itself, including Δ methods and λ_{ci} method. Both of them define vortex as the region where ∇V has two complex conjugate eigenvalues. They are scalar vortex identification methods, which ignore the information such as the rotation axis of fluid rotation. Q, λ_2, Δ and λ_{ci} methods are fuzzy to represent the rotation intensity of vortex structure, and they are different from each other. In fact, these methods contain varying degrees of shear contamination [9, 10].

7.2.4 Liutex Vector

Liu et al. [6] pointed out that the vorticity should be decomposed into a rotating part and a non-rotating part, and proposed the Liutex vector which can accurately capture the vortex structure. Liutex is a vector that can accurately describe the local rigid rotational motion of the fluid. The definition of Liutex vector can be summarized as follows.

7.2.4.1 Direction of Liutex

In the initial xyz coordinate system, the velocity gradient tensor ∇V and its eigenvalues are calculated. When ∇V has a real eigenvalue λ_r and two complex conjugate eigenvalues $\lambda_{cr} \pm i\lambda_{ci}$, there would only tension or compression in the direction of the real eigenvector $\boldsymbol{v}_r$ corresponding to the real eigenvalue λ_r, and there is no rotational motion in this direction. The rotational motion can only occur in the plane perpendicular to the real eigenvector $\boldsymbol{v}_r$, that is, $\boldsymbol{v}_r$ is the rotation axis of the local fluid element, The complex conjugate eigenvalues $\lambda_{cr} \pm i\lambda_{ci}$ indicate that the instantaneous streamline presents a circular or spiral shape, that is, there is a vortex structure. The unit direction vector of Liutex vector is $\boldsymbol{r} = \boldsymbol{v}_r$ and we require $\langle \boldsymbol{\omega} \cdot \boldsymbol{v}_r \rangle > 0$.

7.2.4.2 Strength of Liutex

Firstly, **Q** rotation is used to rotate the initial xyz coordinate system to $x_Q y_Q z_Q$, so that the rotated z_Q is in the same direction as the rotation axis $\boldsymbol{r}$, and the velocity gradient tensor $\nabla \boldsymbol{V}_Q$ after rotation becomes

$$\nabla \boldsymbol{V}_Q = \begin{bmatrix} \frac{\partial u_Q}{\partial x_Q} & \frac{\partial u_Q}{\partial y_Q} & 0 \\ \frac{\partial v_Q}{\partial x_Q} & \frac{\partial v_Q}{\partial y_Q} & 0 \\ \frac{\partial w_Q}{\partial x_Q} & \frac{\partial w_Q}{\partial y_Q} & \frac{\partial w_Q}{\partial z_Q} \end{bmatrix} \tag{7.1}$$

(u_Q, v_Q, w_Q) is the velocity component in $x_Q y_Q z_Q$ coordinate system after **Q** rotation. Here, the formula of Rodrigues is used to solve $\boldsymbol{Q}$. $\boldsymbol{Q}$ is the coordinate rotation matrix. See [4] for specific expressions.

A second rotation (**P** rotation) is used to rotate the reference frame around the z_Q-axis and the corresponding velocity gradient tensor $\nabla \boldsymbol{V}_P$ can be written as

$$\nabla \boldsymbol{V}_P = \boldsymbol{P} \nabla \boldsymbol{V}_Q \boldsymbol{P}^T \tag{7.2}$$

with

$$\nabla \boldsymbol{V}_\theta = \boldsymbol{P}_r \nabla \boldsymbol{V}_Q \boldsymbol{P}_r^T = \begin{bmatrix} \left.\frac{\partial u}{\partial x}\right|_\theta & \left.\frac{\partial u}{\partial y}\right|_\theta & 0 \\ \left.\frac{\partial v}{\partial x}\right|_\theta & \left.\frac{\partial v}{\partial y}\right|_\theta & 0 \\ \left.\frac{\partial w}{\partial x}\right|_\theta & \left.\frac{\partial w}{\partial y}\right|_\theta & \left.\frac{\partial w}{\partial z}\right|_\theta \end{bmatrix} \tag{7.3}$$

where the rotation matrix $\boldsymbol{P}_r$ is given by

$$\boldsymbol{P}_r = \begin{bmatrix} cos\theta & sin\theta & 0 \\ -sin\theta & cos\theta & 0 \\ 0 & 0 & 1 \end{bmatrix} \tag{7.4}$$

And α and β are defined as [10]

$$\alpha = \frac{1}{2}\sqrt{\left(\frac{\partial v_Q}{\partial y_Q} - \frac{\partial u_Q}{\partial x_Q}\right)^2 + \left(\frac{\partial v_Q}{\partial x_Q} + \frac{\partial u_Q}{\partial y_Q}\right)^2} \tag{7.5}$$

$$\beta = \frac{1}{2}\left(\frac{\partial v_Q}{\partial x_Q} - \frac{\partial u_Q}{\partial y_Q}\right) \tag{7.6}$$

Then, according to Liu et al. [11] the Liutex magnitude/rotational strength is defined as

$$R = \begin{cases} 2(\beta - \alpha), & \alpha^2 - \beta^2 < 0 \\ 0, & \alpha^2 - \beta^2 \geq 0 \end{cases} \tag{7.7}$$

and here we assume $\beta > 0$ (if $\beta < 0$, we can first rotate the local axis to the opposite direction of v_r to make it positive).

7.2.4.3 Explicit Formula for the Liutex Vector

Liutex is a vector defined as $\boldsymbol{R} = R\boldsymbol{r}$, where R is the magnitude of Liutex, and $\boldsymbol{r}$ is the direction of Liutex. $\boldsymbol{r}$ is the normalized real eigenvector of the velocity gradient tensor such that $\boldsymbol{\omega} \cdot \boldsymbol{r} > 0$. From Wang et al. [11] study on the explicit formula for the Liutex vector and physical meaning of vorticity based on the Liutex-Shear decomposition, an explicit formula of the Liutex vector can be derived as

$$\boldsymbol{R} = \left(\boldsymbol{\omega} \cdot \boldsymbol{r} - \sqrt{(\boldsymbol{\omega} \cdot \boldsymbol{r})^2 - 4\lambda_{ci}^2} \right) \boldsymbol{r} \tag{7.8}$$

7.2.4.4 Advantage of Liutex

As the third-generation vortex identification methods, Liutex-based vortex identification system outperforms the previous two generations, as mainly reflected by:

- Liutex vector is an accurate representation of the physical amount of vortex with a clear physical meaning.
- The vortex identification method represented by Liutex solves the problem of threshold selection for the second-generation of vortex identification methods.
- The Liutex method is able to represent quantitatively the six elements of vortex.

7.3 Shear Contamination

According to the concept of Liutex vector, the vorticity is decomposed into rotating part and non-rotating part. The non-rotating part is pure shear. Mathematically, Cauchy-Stokes decomposition is correct, but the physical meaning is not clear and depends on the choice of coordinates. In different coordinate systems, the forms of rotation matrix, tension matrix, shear matrix and deformation matrix generated by

Table 7.1 Contamination by stretching and shearing on different criteria

Methods	Q	Δ	λ_2	λ_{ci}	Liutex
Contamination by stretching	Yes	Yes	Yes	No	No
Contamination by shearing	Yes	Yes	Yes	Yes	No

Source Copy of Table 1, Page 18, Stretching and shearing contamination analysis for Liutex and other vortex identification methods by Pushpa Shrestha et al. [10]

velocity gradient tensor decomposition will also be different, which makes the analysis of pollution uncertain. Shrestha et al. [10] studied the pollution matrix in the principal coordinate system and obtained the theoretical pollution analysis results in the principal coordinate system (see Table 7.1).

7.3.1 Principal Decomposition

Because of the uniqueness of the principal coordinates, we selected the principal coordinates for analysis, i.e., by coordinate rotation so that the new z axis points in the direction of the local axis of rotation and so that the x and y axis are stretched or compressed the same in the plane of rotation.

Definition 1. The Principal Decomposition is the decomposition of velocity gradient tensor under the Principal Coordinates [10] i.e.,

$$\nabla V = \begin{bmatrix} \lambda_{cr} & -\frac{R}{2} & 0 \\ \frac{R}{2}+\varepsilon & \lambda_{cr} & 0 \\ \xi & \eta & \lambda_r \end{bmatrix} = \begin{bmatrix} 0 & -\frac{R}{2} & 0 \\ \frac{R}{2} & 0 & 0 \\ 0 & 0 & 0 \end{bmatrix} + \begin{bmatrix} 0 & 0 & 0 \\ \varepsilon & 0 & 0 \\ \xi & \eta & 0 \end{bmatrix} + \begin{bmatrix} \lambda_{cr} & 0 & 0 \\ 0 & \lambda_{cr} & 0 \\ 0 & 0 & \lambda_r \end{bmatrix}$$

$$= \boldsymbol{R} + \boldsymbol{S} + \boldsymbol{C}$$

Here $\boldsymbol{R}$ represents the rotation part, $\boldsymbol{S}$ represents the shear part, and $\boldsymbol{C}$ represents the stretching and compression part in three axis directions. ε, λ_{cr} and λ_r are components of either the shear part or stretching part, or both [12].

Shrestha et al. [10] studied the influence of shear or tension on different vortex identification methods through theoretical contamination analysis, and the results are shown in Table 7.1.

7.3.2 Shear in Boundary Layer Transition

Boundary layer transition is a transition process from laminar flow to turbulence, during which Λ vortex structure, hairpin vortex and annular vortex are generated. With the emergence of high frequency annular vortex in the outer layer, the flow near the wall of the boundary layer first appears disorder. The research shows that in the initial stage of flow disorder, the flow disorder is affected by the instability of local high shear layer. Therefore, it is necessary to study the effect of Liutex shear interaction in boundary layer transition [13–17]. In the transition phase, the inflow boundary is disturbed by T-S waves, and the unstable two-dimensional T-S waves develop along the flow direction, and there is only high shear near the wall without rotational motion. As the flow develops downstream, the 2-D T-S waves begin to show spanwise changes from 2-D disturbance to 3-D disturbance, and accompanied by the emergence of spikes and decay. The schematic of flow transition on a flat plan can be described in the Book of Boundary Layer Theory by Schlichting and Gersten [18]. Next, taking the boundary layer transition as an example, we calculate the cosine values of the angle between Liutex and shear by making the inner product of Liutex vector and shear vector. Specific vortices are selected as the research object, and the cosine value change of Λ vortex moving along the flow direction at a certain time, the development and change of hairpin vortex with time are analyzed. Take the position where the inclined angle is approximately 0 as vortex core lines, the trend for percentage of Liutex over vorticity, and the Liutex shear interaction in the process of boundary layer transition are detailed.

7.4 Numerical Study and Observation

In order to study the transition process of flat plate boundary layer, the numerical data satisfying the characteristics of small-scale vortex and chaos are selected to represent the fully developed turbulence. Meyer et al. [19] believe that the chaos is that "the inclined high shear layer between the Λ vortex legs shows an increased phase jitter from its tip to the wall region." The visualization results in the transition process are displayed by carefully observing and analyzing DNS data [20–25].

7.4.1 Case Setup

The grid level is 1280 × 256 × 241, representing the number of grids in streamwise (x), spanwise (y), and wall normal (z) directions. The grid is stretched in the normal direction and uniform in the streamwise and spanwise directions. Here, $x_{in} = 300.792\delta_{in}$ represents the distance between leading edge and inlet, $Lx = 672.021\delta_{in}$, $Ly = 22\delta_{in}$, $Lz_{in} = 40\delta_{in}$ are the lengths of the computational domain in x-, y-, and z-directions, respectively, δ_{in} represents inflow displacement thickness [16].

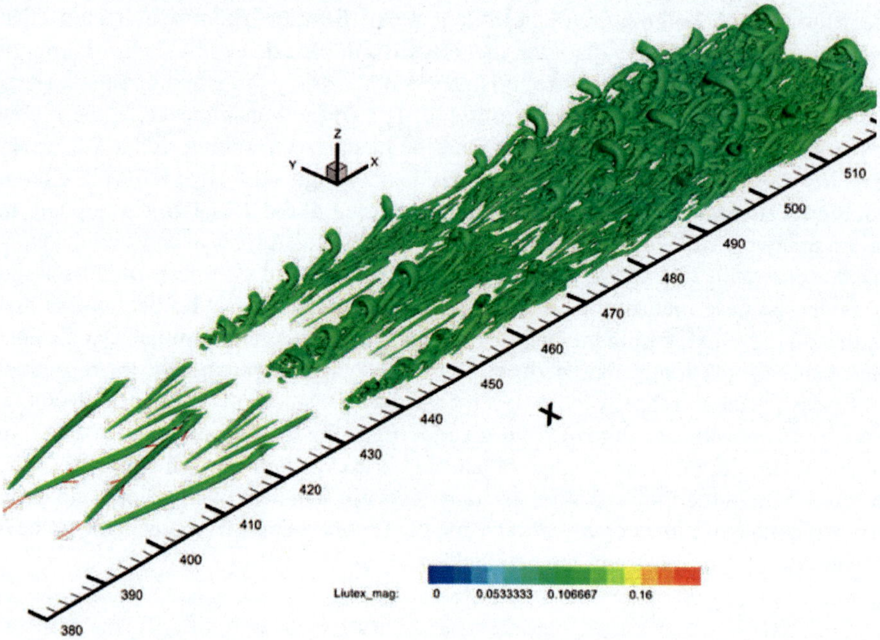

Fig. 7.1 Multiple ring generation and vortex structures

7.4.2 Visualization of Simulation Results

The DNS code "DNSUTA" has been carefully verified by NASA Langley and UTA researchers to ensure that the DNS results are correct and reliable. The following are the visualization results and analysis of the simulation [16].

7.4.2.1 Vortex Structures and Flow Transition

Figure 7.1 shows the visualization of flow transition by an iso-surface of Liutex magnitude equal to 0.1. It includes Λ vortex, hairpin vortex and annular vortex. We can see hairpin structures are pretty clear. In particular, the Liutex lines can accurately describe the structure of vortex. In order to study the role of Liutex and shear in the formation and development of small-scale vortices, we need to pay attention to the location and change information of Liutex and shear.

7.4.2.2 Liutex and Shear Interaction

In the boundary layer transition, two slices $x = 408.98\delta_{in}$ and $y = 10.99\delta_{in}$ are selected, and the point with the largest Liutex magnitude is taken as the seed points of the Liutex respectively. The $cos\theta$ values of Liutex and shear at the vortex core of Λ vortex and hairpin vortex are calculated and compared.

The structure of Λ vortex and hairpin vortex is shown in Fig. 7.2. Through numerical simulation, we get that the integral average value of $cos\theta$ is 0.723 on the vortex core line in Fig. 7.2a, and the integral average value of $cos\theta$ is 0.962 on the vortex core line in Fig. 7.2b.

Next, we take Λ vortex as the research object and extract the Liutex core line data representing the vortex core, as listed in Table 7.2. The variation of $cos\theta$ values on Liutex core line with flow direction is studied when $t = 5.5T$ (T is the period of T-S wave), as shown in the Fig. 7.3. Variation of shear magnitude value on the Liutex core line with flow direction is shown in Fig. 7.3.

It can be seen that the $cos\theta$ value of Λ vortex leg begins to decrease in the middle and rear section, indicating that the inclined angle between Liutex and shear increases, and the shear value in the front section is greater than that in the middle and rear section (see Fig. 7.3). The shear magnitude value in the rear section of Λ vortex leg increases, while the cosine value decreases. Combined with the structural evolution of vortex in Fig. 7.4, it can be seen that the effect of shear in the rear section of Λ vortex leg affects the shape of vortex. An indication is that if the angle between Liutex and shear vectors in the vortex core is approximately zero, then this part of the vortex is more likely to be stable in the sense that the shape would basically remain the same. On the contrary, the vortex core where the angle is large tend to deform.

7.4.2.3 Comparison Between Different Times

In order to quantitatively study the change of the angle between Liutex and shear with different times and understand the interaction of Liutex and shear, we select five points on the Λ vortex core, and track the value of $cos\theta$ and shear magnitude at the same position but different time. The values are shown in Table 7.3. In addition, for the hairpin vortex, we select the maximum point of Liutex magnitude on the plane perpendicular to the y-axis as shown in Fig. 7.2b, and record the development of Liutex and shear on this particular point. It is found that on this point Liutex is approximately parallel to the shear. The time development of Liutex and shear magnitudes is listed in Table 7.5.

Λ vortex

We select five positions, which are $x_1 = 403.397\delta_{in}$, $x_2 = 406.396\delta_{in}$, $x_3 = 409.326\delta_{in}$, $x_5 = 412.349\delta_{in}$ and $x_6 = 415.397\delta_{in}$.

It can be seen that the structure of hairpin vortex is gradually generated at the rear end of Λ vortex leg in Fig. 7.4b. In Fig. 7.4c, the Λ vortex leg dissipates. Combined with Fig. 7.5, we find that the $cos\theta$ value of the front end (1cos, 2cos) of the leg is decreasing, the $cos\theta$ value of the middle and rear end (3cos, 4cos, 5cos) of the leg is increasing, and the angle between Liutex and shear is gradually decreasing.

It can be seen from the Fig. 7.6 that the closer to the rear end of the Λ vortex leg, the smaller the change value of shear with time. According to Fig. 7.4a, we find that

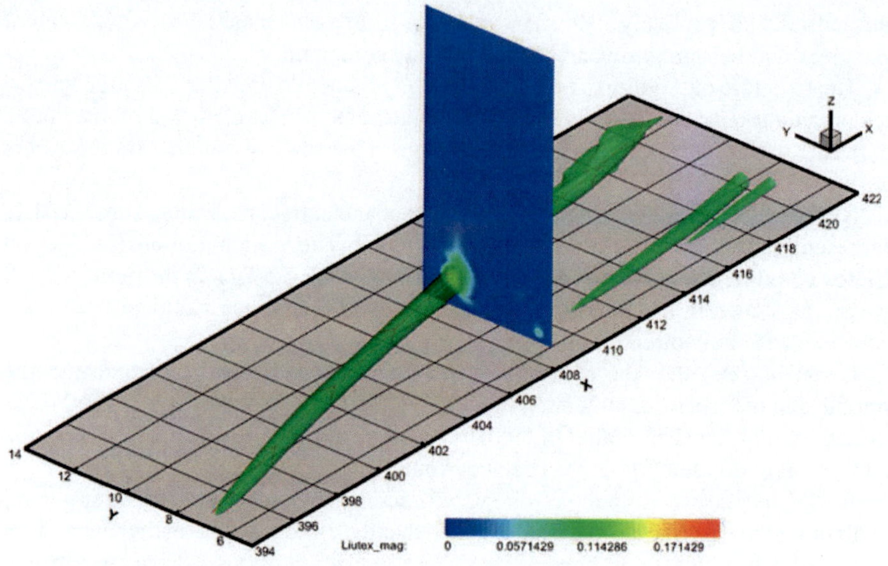

(a) vortex and slice at the position $X = 408.98$

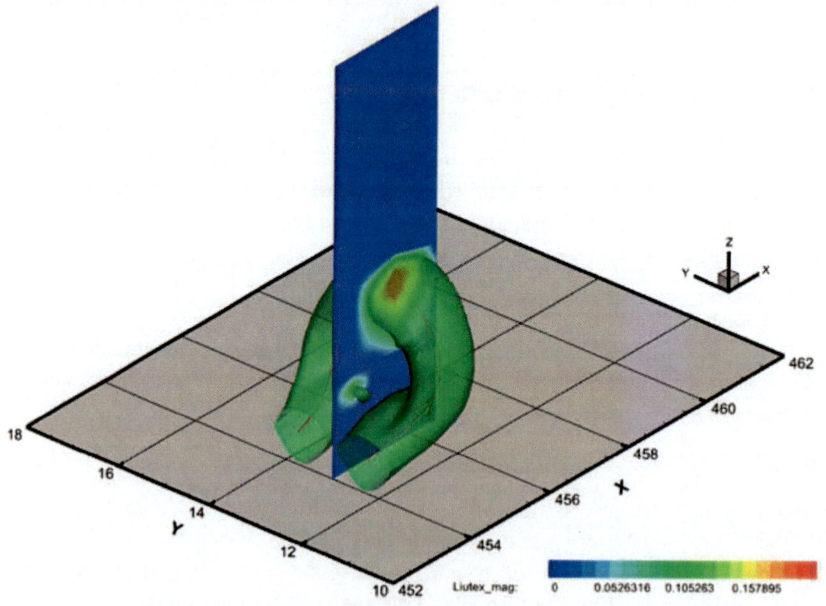

(b) Hairpin vortex and slice at the position $Y = 10.99$

Fig. 7.2 Λ vortex, hairpin vortex and Liutex lines ($t = 5.5T$)

Table 7.2 Distribution of Liutex, vorticity, shear magnitudes on the Λ vortex core ($t = 5.5T$)

X	Liutex_mag	Vorticity	Shear_mag	$cos\theta$
394.325	0.089	0.414	0.345	0.718
395.321	0.129	0.448	0.334	0.839
396.338	0.154	0.465	0.323	0.884
397.355	0.161	0.457	0.308	0.894
398.457	0.160	0.431	0.281	0.901
399.476	0.153	0.399	0.255	0.905
400.326	0.146	0.372	0.235	0.910
401.349	0.137	0.345	0.215	0.915
402.373	0.132	0.323	0.197	0.929
403.397	0.127	0.309	0.187	0.931
404.338	0.127	0.300	0.177	0.937
405.366	0.129	0.295	0.172	0.927
406.396	0.133	0.296	0.170	0.907
407.342	0.138	0.300	0.172	0.873
408.376	0.150	0.310	0.175	0.830
409.326	0.165	0.322	0.177	0.784
410.362	0.193	0.342	0.177	0.713
411.398	0.240	0.372	0.172	0.620
412.349	0.300	0.425	0.177	0.553
413.394	0.333	0.472	0.187	0.615
414.356	0.342	0.480	0.184	0.627
415.397	0.379	0.460	0.158	0.359
416.421	0.321	0.448	0.233	0.290
417.345	0.161	0.402	0.333	0.220

the shear value at the front end of the leg is relatively large and the evolution of Λ vortex structure of phase transition can be clearly seen (Figs. 7.4b and 7.4c).

Hairpin vortex

We select a hairpin vortex in the boundary layer transition process, extract the Liutex line data when $t = 5.5T$ (see in Table 7.4), and plot the value of $cos\theta$ varies with Y coordinate, as shown in the Fig. 7.7. It can be seen from the figure that the hairpin vortex is relatively stable, and the $cos\theta$ values on the vortex ring is between 0.93 and 1, indicating that the inclined angle between Liutex and shear at the vortex is quite small, and the angle at the highest part of the hairpin vortex is 0, with the highest point as the axis, and the $cos\theta$ values on both sides are very symmetrical (see Fig. 7.7).

It can be seen from Fig. 7.8 that the shear magnitude value and vorticity have the same change trend. The Liutex magnitude value, shear magnitude value, vorticity and $cos\theta$ value have good symmetry with respect to the section $Y = 10.99\delta_{in}$. The high

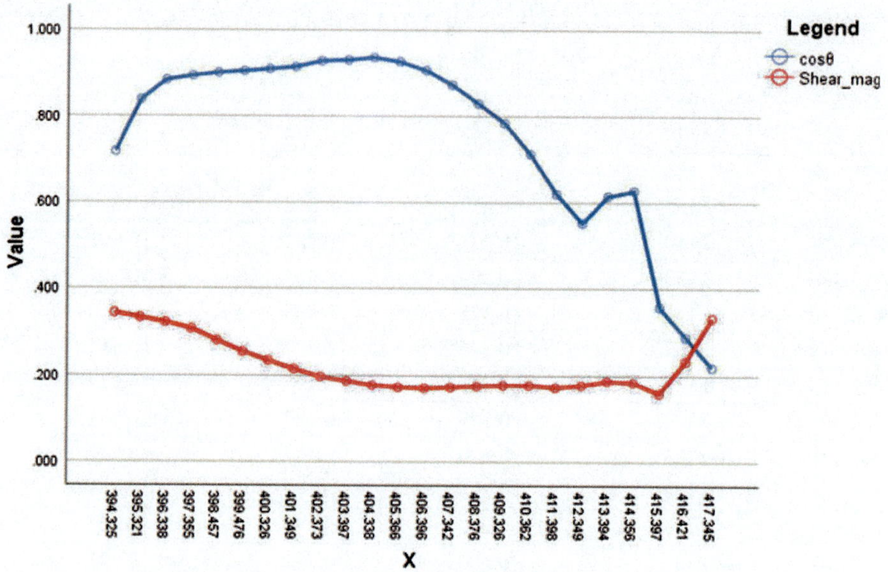

Fig. 7.3 The distribution of $cos\theta$ and shear magnitude on the Liutex core line of Λ vortex ($t = 5.5T$)

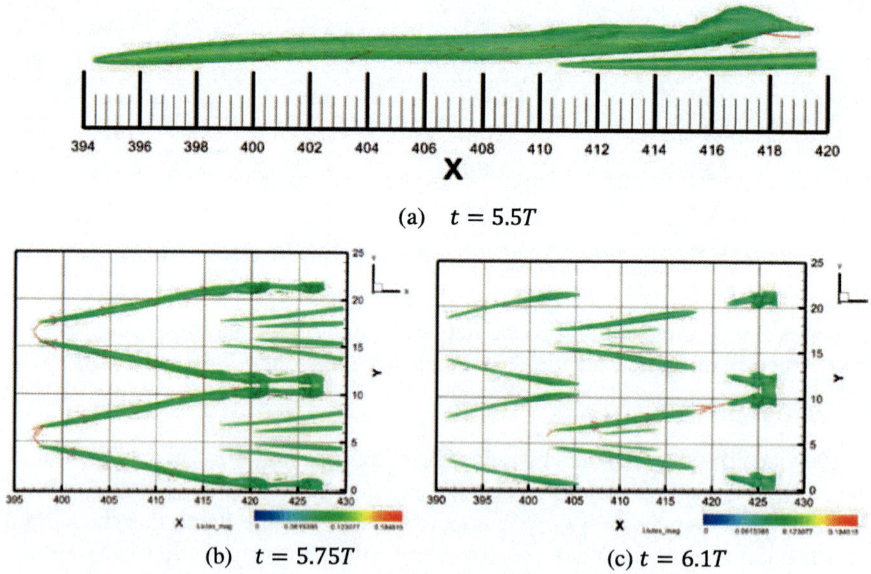

Fig. 7.4 The evolution of Λ vortex structure of transition (where T is the period of T-S wave)

Table 7.3 The value of $cos\theta$ and shear magnitude development at selected points

t	$T1 = 5.5T$	$T2 = 5.6T$	$T3 = 5.75T$	$T4 = 6.0T$	$T5 = 6.1T$
$cos\theta_1$	0.931	0.908	0.724	0.821	0.432
$cos\theta_2$	0.907	0.915	0.740	0.834	0.816
$cos\theta_3$	0.784	0.881	0.819	0.882	0.886
$cos\theta_4$	0.553	0.805	0.853	0.928	0.937
$cos\theta_5$	0.359	0.406	0.871	0.966	0.979
Shear_mag1	0.187	0.213	0.286	0.210	0.213
Shear_mag2	0.170	0.183	0.227	0.249	0.240
Shear_mag3	0.177	0.172	0.186	0.214	0.232
Shear_mag4	0.177	0.164	0.166	0.164	0.193
Shear_mag5	0.158	0.127	0.155	0.153	0.147

$cos\theta_{\#}$ represents the $cos\theta$ value of the #-th point, Shear_mag # represents the shear magnitude value of the #-th point. (# = 1, 2, 3, 4, 5)

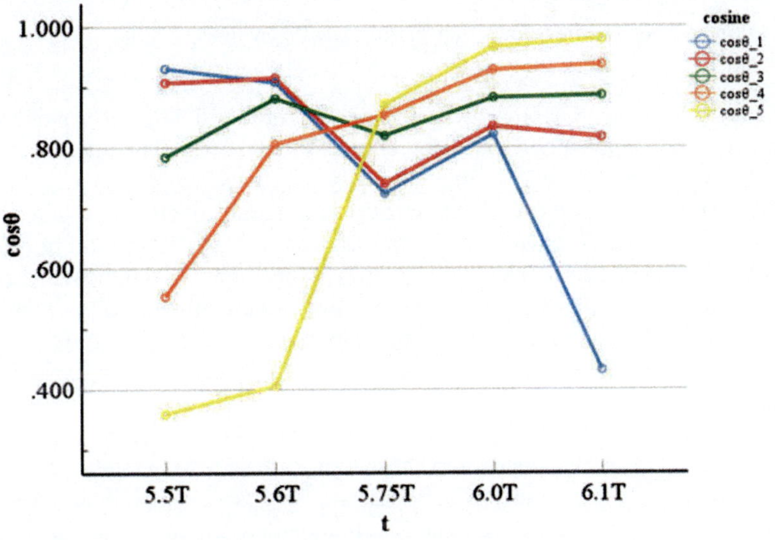

Fig. 7.5 The value of $cos\theta$ varies with five different times

shear value is mainly concentrated near the wall. The farther away from the wall, the smaller the shear magnitude value, and the Liutex magnitude value increases accordingly, but on the whole, the change range of Liutex magnitude value is far less than the shear value.

The above Fig. 7.9 shows the evolution of hairpin vortex structure of transition over time. Before $t = 6.1T$, the shape of hairpin vortex is relatively stable. When $t = 6.1T$ to $t = 6.3T$, the legs of hairpin vortex become thinner and fold inward,

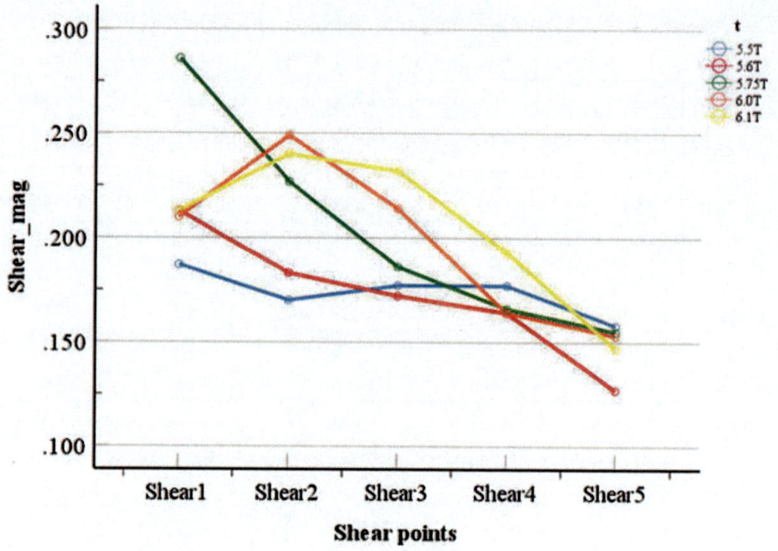

Fig. 7.6 The value of shear magnitude varies with five different times

and the integral mean value of $cos\theta$ decreases. After $t = 6.3T$, the legs of hairpin vortex begin to disappear and the shape becomes ring vortex. The integral average value of $cos\theta$ on the Liutex core line gradually increases, and the shear magnitude value at the maximum point of Liutex magnitude gradually decreases to 0 over time, while the size of Liutex magnitude is almost unchanged. As shown in Fig. 7.2b, at the different time, the $cos\theta$ values of the maximum Liutex magnitude at the slice at the position $Y = 10.99\delta_{in}$ are 1. According to the mathematical definition of vortex in the third-generation vortex identification method, the proportion of Liutex magnitude over vorticity can be analyzed since for the considered points they are in the same direction.

The proportion of Liutex magnitude in the hairpin vortex increases with time as shown in Fig. 7.10. When Liutex and shear vectors are approximately parallel, the vortex moves greatly along the flow direction and the offset in other directions is very small. Combined with Fig. 7.9, we can find that the hairpin vortex and the induced annular vortex are relatively stable.

7.5 Conclusion

According to the analysis of the specific vortices in the boundary layer transition, the following conclusions can be summarized.

Table 7.4 Distribution of Liutex, vorticity, shear magnitudes on a hairpin vortex core ($t = 5.5T$)

Y	Liutex_mag	Vorticity	Shear_mag	$cos\theta$
9.933	0.205	0.656	0.452	0.995
9.938	0.203	0.665	0.463	0.995
9.948	0.202	0.673	0.472	0.995
9.962	0.202	0.680	0.479	0.992
9.982	0.202	0.675	0.475	0.984
10.006	0.204	0.666	0.466	0.975
10.034	0.206	0.661	0.460	0.967
10.066	0.209	0.655	0.453	0.957
10.102	0.211	0.638	0.435	0.947
10.142	0.214	0.628	0.423	0.940
10.185	0.215	0.604	0.398	0.932
10.232	0.217	0.590	0.383	0.930
10.281	0.218	0.570	0.361	0.930
10.334	0.218	0.550	0.341	0.932
10.391	0.219	0.535	0.324	0.939
10.451	0.220	0.521	0.307	0.949
10.515	0.221	0.509	0.294	0.960
10.584	0.222	0.500	0.282	0.971
10.657	0.224	0.495	0.273	0.981
10.735	0.225	0.489	0.266	0.988
10.817	0.226	0.489	0.263	0.994
10.903	0.228	0.490	0.262	0.998
10.990	0.228	0.490	0.262	1.000
11.078	0.228	0.489	0.262	0.999
11.164	0.227	0.488	0.262	0.996
11.247	0.225	0.489	0.265	0.990
11.326	0.224	0.494	0.272	0.983
11.400	0.223	0.499	0.280	0.983
11.470	0.221	0.505	0.289	0.962
11.535	0.220	0.520	0.305	0.952
11.596	0.219	0.533	0.321	0.941
11.653	0.218	0.546	0.336	0.934
11.707	0.218	0.565	0.357	0.930
11.757	0.217	0.584	0.377	0.929
11.804	0.216	0.602	0.396	0.932
11.848	0.214	0.623	0.418	0.938

(continued)

Table 7.4 (continued)

Y	Liutex_mag	Vorticity	Shear_mag	$cos\theta$
11.889	0.211	0.633	0.430	0.945
11.926	0.209	0.650	0.447	0.955
11.959	0.207	0.662	0.460	0.965
11.988	0.204	0.666	0.465	0.973
12.013	0.203	0.673	0.473	0.982
12.033	0.202	0.682	0.481	0.991
12.049	0.201	0.674	0.473	0.994
12.060	0.202	0.666	0.465	0.996

Table 7.5 Liutex core line data at hairpin vortex core with times

t	Shear_mag	Liutex_mag	Vorticity	Percentage (%)	Integral average of $cos\theta$
5.5T	0.262	0.229	0.49	46.73	0.962
5.55T	0.225	0.208	0.433	48.04	0.946
5.6T	0.183	0.218	0.4	54.50	0.962
5.65T	0.172	0.207	0.379	54.62	0.961
5.7T	0.136	0.184	0.32	57.50	0.936
5.75T	0.141	0.186	0.324	57.41	0.941
5.8T	0.129	0.175	0.304	57.57	0.941
5.85T	0.14	0.159	0.299	53.18	0.939
5.9T	0.099	0.163	0.262	62.21	0.883
5.95T	0.099	0.163	0.263	61.98	0.85
6T	0.112	0.159	0.271	58.67	0.789
6.05T	0.113	0.153	0.265	57.74	0.747
6.1T	0.111	0.155	0.266	58.27	0.724
6.15T	0.104	0.159	0.263	60.46	0.680
6.2T	0.058	0.165	0.223	73.99	0.564
6.25T	0.057	0.174	0.231	75.32	0.428
6.3T	0.062	0.18	0.241	74.69	0.486
6.35T	0.069	0.183	0.252	72.62	0.643
6.4T	0.064	0.189	0.253	74.70	0.745
6.45T	0.057	0.192	0.249	77.11	0.750

1. Liutex can well capture vortices in boundary layer transition with its direction as the local rotational axis and magnitude as twice the angular velocity speed of local rotational motion.
2. Different from other vortex identification methods, Liutex is not contaminated by shear.

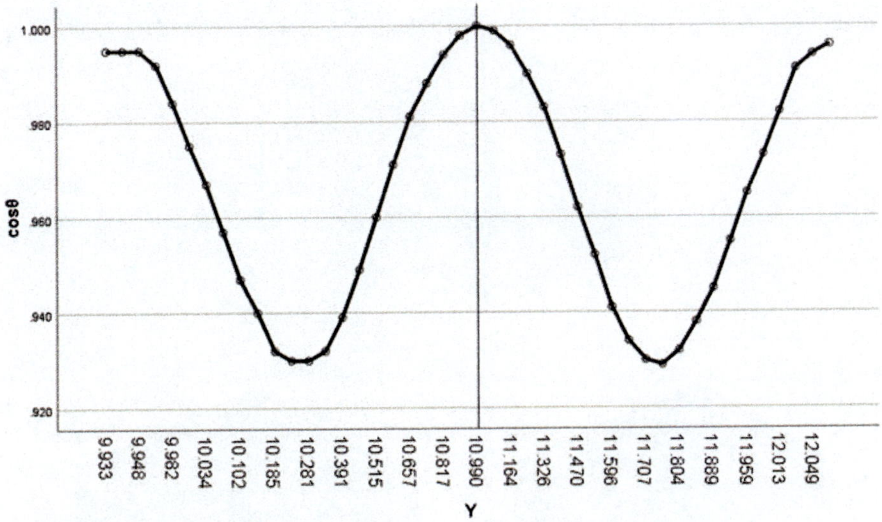

Fig. 7.7 The $cos\theta$ distribution on the Liutex core line of hairpin vortex ($t = 5.5T$)

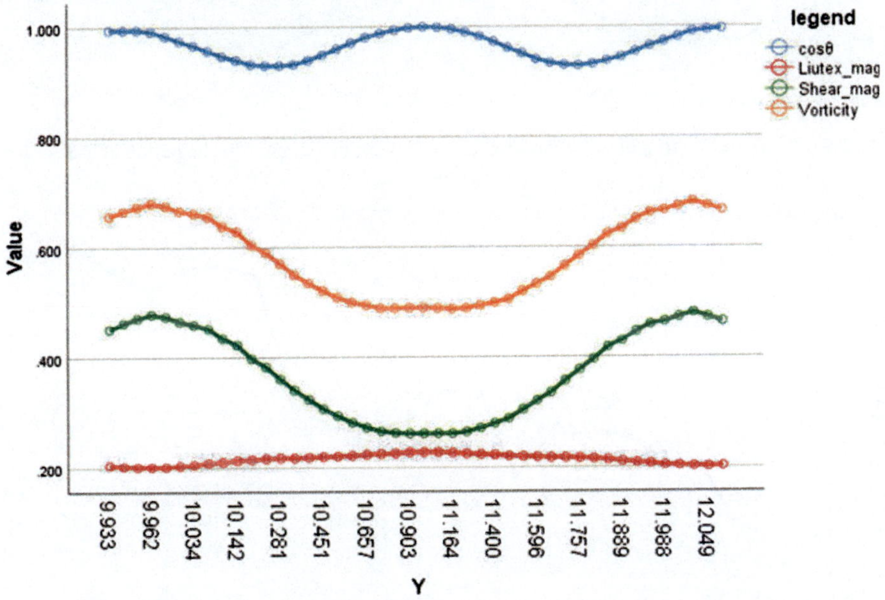

Fig. 7.8 The distribution of Liutex, vorticity, shear magnitudes on a hairpin vortex core ($t = 5.5T$)

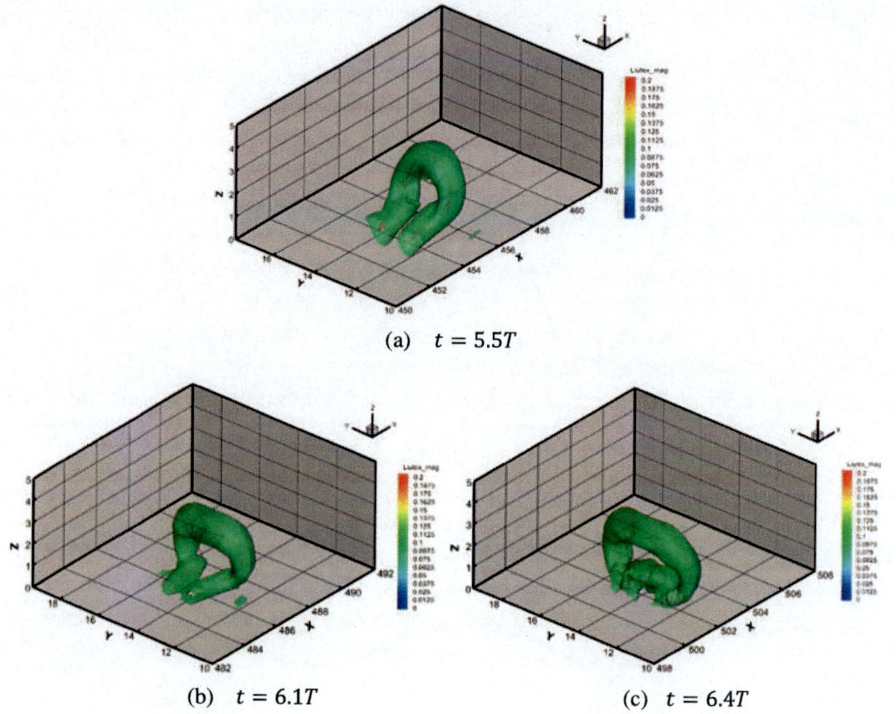

Fig. 7.9 The evolution of hairpin vortex structure of transition (where T is the period of T-S wave)

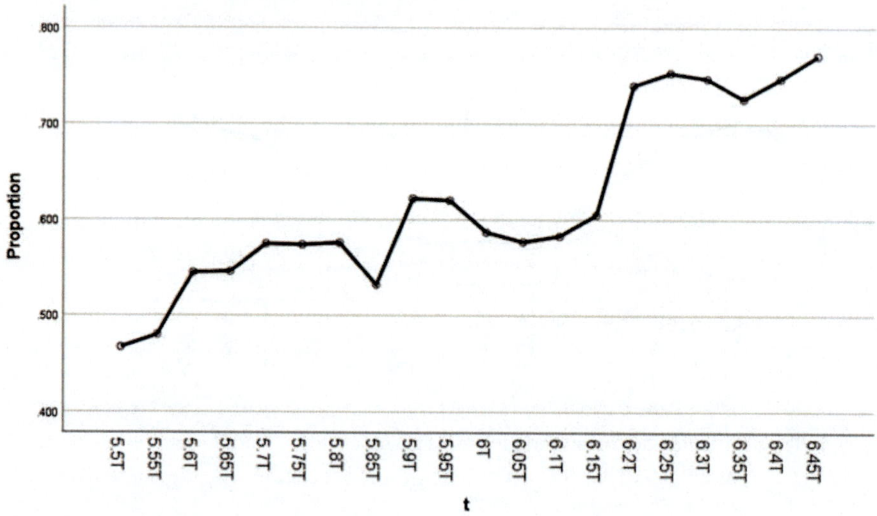

Fig. 7.10 The proportion of Liutex_mag/Vorticity varies with times

3. In the process of boundary layer transition, from the data, the inclined angle between Liutex and shear ranges from 0 to $\pi/2$. High shear is mainly concentrated before vortex generation and vortex dissipation, indicating that shear plays an important role in vortex formation and development.
4. Because of the Liutex shear interaction, Λ vortex gradually develops into hairpin vortex and some parts of it dissipates with time. At the same time, the hairpin vortex will generally maintain a stable state because the inclined angle between Liutex and shear is approximately zero, and then slowly become ring vortex. When the inclined angle between Liutex and shear is approximately parallel, the vortex mainly moves along the flow direction and the offset in other directions is very small.

References

1. C. Liu, Y. Yan, P. Lu, Physics of turbulence generation and sustenance in a boundary layer. J. Comput. Fluids. 353–384 (2014)
2. I.M. Wallace, Highlights from 50 years of turbulent boundary layer research. J. Turbul. 1–70 (2013)
3. S.K. Roblnson, Coherent motion in the turbulent boundary layer. J. Annu. Rev. Fluid. Mech. 601–639 (1991)
4. C. Liu, Y. Gao, X. Dong, et al., Third generation of vortex identification methods: Omega and Liutex/Rortex based systems. J. Hydrodyn. 205–223 (2019)
5. C. Liu, Y. Wang, Y. Yang, et al., New omega vortex identification metho. J. Sci. China. Phys. Mech. **59**(8) (2016)
6. C. Liu, Y. Gao, S. Tian, et al., Rortex-a new vortex vector definition and vorticity tensor and vector decompositions. J. Phys. Fluids. 30 (2018)
7. X. Dong, Y. Gao, C. Liu, New normalized Rortex/Vortex identification method. J. Phys. Fluids. 31 (2019)
8. Y. Wang, Y. Yang, G. Yang, et al., DNS study on vortex and vorticity in late boundary layer transition. J. Commu. Comput. Phys. 441–459 (2017)
9. C. Liu, Liutex-third generation of vortex definition and identification methods. J. Aerodyn. 413–431 (2020)
10. P. Shrestha, C. Nottage et al., Stretching and shearing contamination analysis for Liutex and other vortex identification methods. J. Aia. (2021). https://doi.org/10.1186/s42774-020-00060-9
11. Y. Wang, Y. Gao, J. Liu, et al., Explicit formula for the Liutex vector and physical meaning of vorticity based on the Liutex-Shear decomposition. J. Hydrodyn. 464–474 (2019)
12. Y. Yu, P. Shrestha, O. Alvarez, et al., Investigation of correlation between vorticity, Q, λci, λ2, Δ and Liutex. J. Comput. Fluids. 225 (2021)
13. X. Liu, L. Chen, C. Liu, Study of mechanism of ring-like vortex formation in late flow transition. J. AIAA 2010–1456 (2010)
14. P. Lu, C. Liu, Numerical study on mechanism of small vortex generation in boundary layer transition. J. AIAA 2011–2287 (2011)
15. F. Ducrous, P. Comte, M. Lesieur, Large-eddy simulation of transition to turbulence in a boundary layer developing spatially over a flat plate. J. Fluid. Mech. 1–36 (1996)
16. P. Lu, C. Liu, DNS study on mechanism of small length scale generation in late boundary layer transition. J. Phys. D 11–24 (2012)
17. P. Lu, M. Thapa, C. Liu, Numerical investigation on chaos in late boundary layer transition to turbulence. J. Comput. Fluids. 68–76 (2014).

18. H. Schlichting, K. Gersten, *Boundary Layer Theory*, 8th revised edn. (Springer, New York, 2000)
19. S. Bake, D. Meyer, U. Rist, Turbulence mechanism in Klebanoff transition: a quantitative comparison of experiment and direct numerical simulation. J. Fluid. Mech. 217–243 (2002)
20. L. Kleiser, T.A. Zang, Numerical simulation of transition in wall-bounded shear flows. J. Annu. Rev. Fluid. Mech. 495–537 (1991)
21. Y.S. Kachanov, in *Recent Results in Laminar-Turbulent Transition.* On a universal mechanism of turbulence production in wall shear flows, vol. 86 (Springer, Berlin, 2003), pp. 1–12
22. V.I. Borodulin, V.R. Gaponenko, Y.S. Kachanov, et al., Late-stage transition boundary-layer structure: direct numerical simulation and experiment. J. Theor. Comput. Fluid Dyn. 317–337 (2002)
23. W. Xu, Y. Wang, Y. Gao, et al., Liutex similarity in turbulent boundary layer. J. Hydrodyn. 1259–1262 (2019)
24. H. Guo, V.I. Borodulin, Y.S. Kachanov, C. Pan, J.J. Wang, X.Q. Lian, S.F. Wang, Nature of sweep and ejection events in transitional and turbulent boundary layers. J. Turbul. 1–51 (2010)
25. J. Cousteix, in *turbulence et couche limite*. Vol. 2, ed. by Cepadues, (Toulouse, Septembre, 1989), p. 627

Chapter 8
Liutex Shear Interaction in Turbulent Channel Flow

Yuan Ding, Biyu Pang, and Yiqian Wang

Abstract Vortex structures are ubiquitously present in aerodynamics and hydrodynamics. To study vortex structure, it is obligatory and of great significance to systematically define and visualize vortices. Many methods have been introduced including Q, λ_2, Δ, λ_{ci}, Ω and Liutex vector-based methods. In this paper, we explore the interaction between Liutex and shear on the basis of the Liutex-shear (**R**-**S**) decomposition in a turbulent channel flow with Reynolds number of 180. The results suggest that shearing may strengthen or weaken the magnitude of Liutex depending on their inclined angle. It is inferred that the interaction between Liutex and shear plays an important role in turbulence regeneration cycle.

8.1 Introduction

In nature and industrial applications, vortex structures are ubiquitous, especially in turbulent flow. It plays an important role in studying the generation and maintenance of turbulence. Therefore, comprehensive understanding and mathematical definition of eddy currents, and accurate identification of eddy current structures are of great significance for understanding the flow mechanism and turbulent structure of eddy currents [1, 2].

The first part of this paper briefly revisits the first-generation vortex identification method based on vorticity, the second-generation vortex identification method based on eigenvalues of velocity gradient tensor, and the third-generation vortex identification system based on the Liutex vector, with the decomposition of vorticity into a rotational part represented by Liutex and a pure shear part. The second part discusses the interaction between shear and Liutex using channel turbulence as an example. The third part summarizes the findings of current research.

Y. Ding · B. Pang · Y. Wang (✉)
School of Mathematical Science, Soochow University, Suzhou 215006, China
e-mail: yiqian@suda.edu.cn

Y. Wang et al. (eds.), *Liutex and Third Generation of Vortex Identification*,
Springer Proceedings in Physics 288, https://doi.org/10.1007/978-981-19-8955-1_8

8.2 Vortex Identification Methods

8.2.1 Brief Description of the First- and Second-Generation Vortex Identification Methods

Vortical structures are commonly seen in various flows, especially in turbulence, where numerous vortices present in different scales and intensity. The concept of vortex filament was proposed by Helmholtz [3] in 1858. Thereafter, many textbooks in fluid mechanics define vortex as vorticity tube, and view the magnitude of vorticity as the intensity of vortex (rotation strength). Here we classify the vorticity-based vortex identification methods as the first generation.

However, even though the vortex definition of Helmholtz [3] still appears in the textbook of fluid mechanics as a classical theory, people have realized that vorticity and vortex are two different concepts. Both are vectors, but vary in direction and magnitude, thus the equal sign cannot be drawn. Specific to the turbulent motion, Robinson [4] discovered that vorticity can be very strong in region where actual vortices are very weak. Wang et al. [5] found in the direct numerical simulation results that not only the vorticity and the direction of the vortex are completely different, but also there is a region with strong rotation but small vorticity, and the vorticity can be larger while there is no rotation. The evidence clearly states that vorticity cannot represent vortex.

In order to deal with the shortcomings of the first-generation vortex identification methods, many scalar vortex identification methods, including Q, λ_2, Δ, λ_{ci}, Ω and so on, have been proposed. Although their theoretical bases are different, they are basically determined by the eigenvalues of the local velocity gradient tensor $\nabla \boldsymbol{v}$, so Liu et al. [6] summarized them as the second-generation vortex identification methods.

However, the second-generation vortex identification methods have some obvious defects: (1) The vortex is a physical phenomenon in which the fluid rotates around an axis, so it should be a vector with magnitude and direction, but the second-generation vortex identification methods are scalars with only magnitude but no direction, so it is difficult to locate the axis of rotation. (2) Because the second-generation criteria are scalar, and generally iso-surfaces of a selected threshold are used to visualize the vortices. Thus, the identification is dependent on an artificially given threshold, and different thresholds could lead to different visualized vortex structures. (3) In addition, the second-generation vortex identification methods are contaminated by tension and compression shear to varying degrees. Therefore, the second-generation vortex identification methods can not accurately identify and capture the vortex structure.

8.2.2 *The Third Generation Vortex Identification Method*

8.2.2.1 Liutex Vortex Definition

To overcome the problems of the previous two generation vortex identification methods, a new physical quantity-Liutex vector is introduced by Liu et al. [7–10], the direction of which represent the local rotational axis while the magnitude equals twice the angular velocity of local rotational motion. According to the Liutex framework, the fluid motion can be decomposed into a rigid body rotational part and a completely unrotated part (including tension, compression, shearing and deformation).

The Liutex vector can be defined as R = Rr, where R is the magnitude of Liutex and r is the local rotation axis. Wang et al. [11] proposed an explicit expression of R, i.e.

$$R = Rr = \left(\omega \cdot r - \sqrt{(\omega \cdot r)^2 - 4\lambda_{ci}^2}\right) r$$

where ω is the vorticity vector, r is the local rotation axis, and λ_{ci} is the imaginary part of ∇v' s complex eigenvalues. In order to let r be unique, it is required that $\omega \cdot r > 0$. Liutex has been proved to be unique and Galilean invariant [6].

8.2.2.2 R-S Decomposition

Based on the definition of Liutex vector R, vorticity ω can be further decomposed into rigid body rotation part R and a shear part $\omega - R$,

$$\omega = R + (\omega - R) = R + S$$

where R and S are the rotational and shear part of the vorticity vector respectively.

8.2.2.3 Advantages of the Third Generation of Vortex Identification Methods

In contrast to the previous two generations of vortex identification methods, the third-generation of vortex identification methods have several advantages:

- Liutex is not only a vortex identification method, itself is a physical quantity with direction and magnitude.
- The Liutex vector gives a systematically and uniquely method to extract the rigid rotating part of the fluid motion and solves the problem of the second generation of vortex identification methods of being shear contaminated.
- The Liutex vortex identification method solves the problem of threshold selection of the second-generation vortex identification methods.

- Liutex answered the six major elements of the definition of vortex, i.e., the absolute intensity, relative intensity of the vortex, the local rotational axis, position of vortex core, magnitude of vortex core, and the boundaries of the vortex.

8.3 Observation and Analysis

For the simulation of turbulent channel flow at $Re_{\tau} = 180$, the grid size is chosen to be $256 \times 129 \times 128$, representing the grid number along the flow direction (X), wall-normal direction (Y) and spanwise direction (Z) [12–15].

8.3.1 The Nature of Vorticity in Channel Turbulence Flow

Figure 8.1 shows that in turbulent channel flow, the shear is mainly concentrated near the upper and lower wall. As assumed by many, the shear concentration near the wall acts as a bank for the turbulent fluctuations and is critical for the turbulence regeneration cycle.

Figure 8.2 demonstrates that the Liutex core line can accurately capture vortex structures.

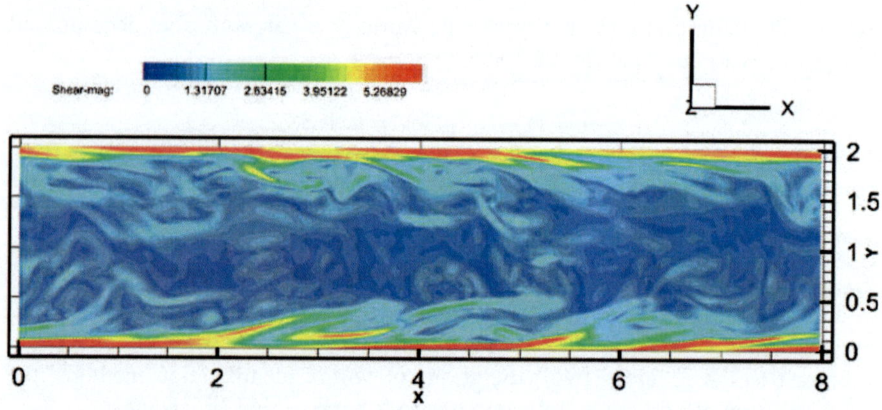

Fig. 8.1 Shear distribution at $z = 2$

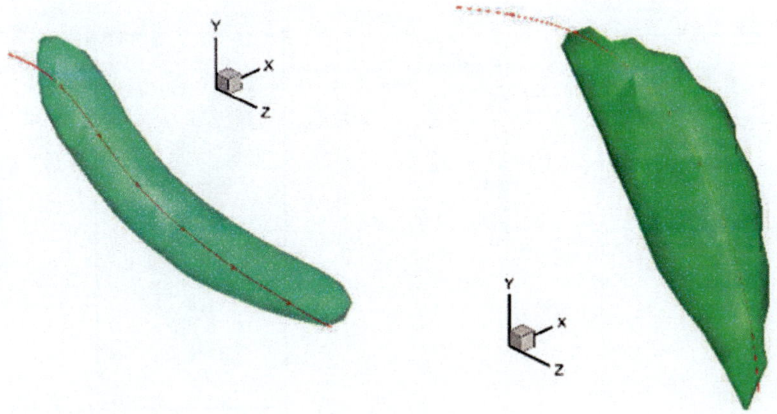

Fig. 8.2 Vortex 2 and Liutex line

8.3.2 *Correlation of Liutex and Shear at Several Different Transient Moment*

In the current study, we will mainly discuss three vortices, namely those that are strengthen slowly over time, those that dissipate gradually, and those that remain in a relatively steady state in the flow field, and hereafter we will refer to these three vortices by using Vortex 1, Vortex 2, and Vortex 3.

8.3.2.1 Vortex 1 (Vortex Structure with Generation)

To quantitatively study the time-dependent change in the angle between Liutex and the shear vector and to understand the interaction of Liutex and the shear, several points were selected at the vortex core of Vortex 1, and the values of $\cos\theta$, shear magnitude at different times of the same location were tracked. Contrasting the five moments $t = 260$, $t = 261$, $t = 262$, $t = 263$, $t = 264$, the evolution of the Vortex 1 is shown in Fig. 8.3, and the collated data are shown in Fig. 8.4.

From Fig. 8.3, we observe that Vortex 1, which increases with time, flows mainly along the direction X with little shift in position in the normal and spanwise directions. At the same time the shape was changed and changed from a striped vortex to a meniscus vortex.

Each line in Fig. 8.4 represents, at a given moment, the distribution of the $\cos\theta$ values of this vortex along the positive Y axis. We can see that at the given five times, along the positive Y axis, the $\cos\theta$ values almost always decrease slowly first and then increase gently. At the same time, with the advancement of time, when the time t increases, the $\cos\theta$ values of the front section of the vortex gradually decrease, and the $\cos\theta$ values of the rear section first decrease and then increase.

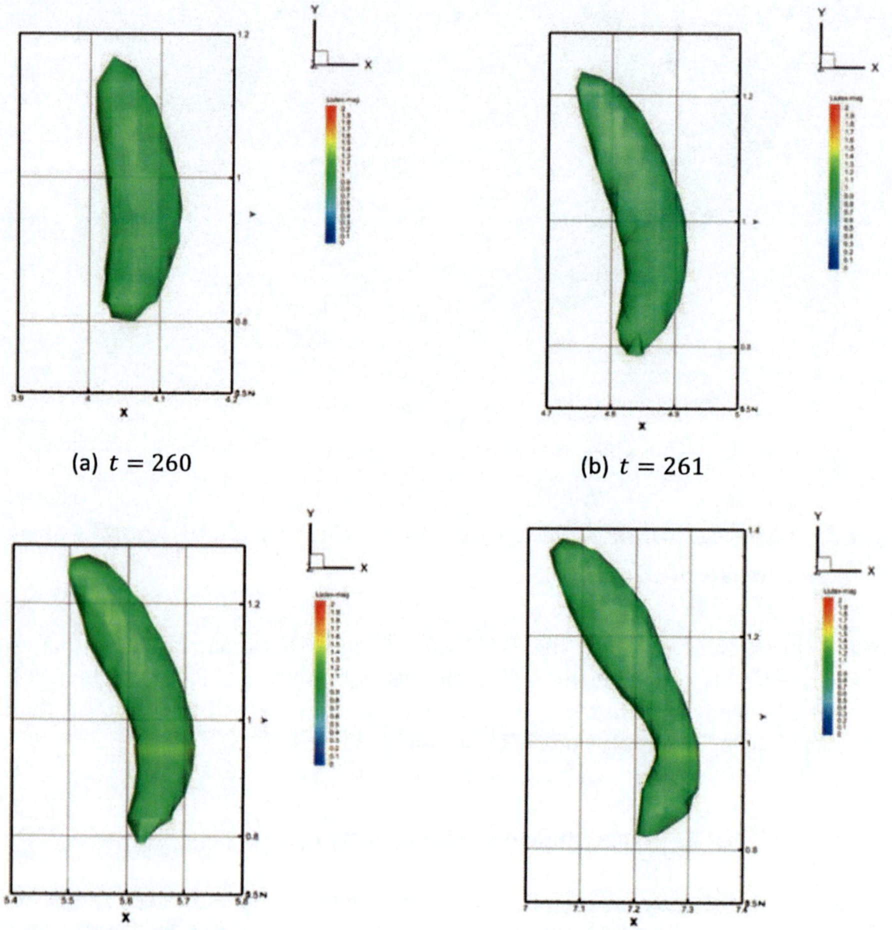

(a) $t = 260$ (b) $t = 261$

(c) $t = 262$ (d) $t = 264$

Fig. 8.3 The evolution of Vortex 1 structure (Iso-surfaces of $R = 1.1$)

Combined with Fig. 8.3, we found that in the interval of $y = 0.95$ to $y = 1.15$, the values of $\cos\theta$ at each moment have significant concaves. It is precisely because the angle between the Liutex and the shear vector changes significantly that the vortex begins to change in shape.

Pay special attention to the point of $\cos\theta = 0$, when $t = 262$. At this instant, the angle between the Liutex and the shear vector is almost vertical. The value of shear magnitude at this point changes the direction of the Liutex (the rotating part) to some extent, causing the shape of the vortex to curve from that moment on.

It can be seen that the change of the angle between the Liutex and the shear vectors can change the shape of the vortex to a certain extent.

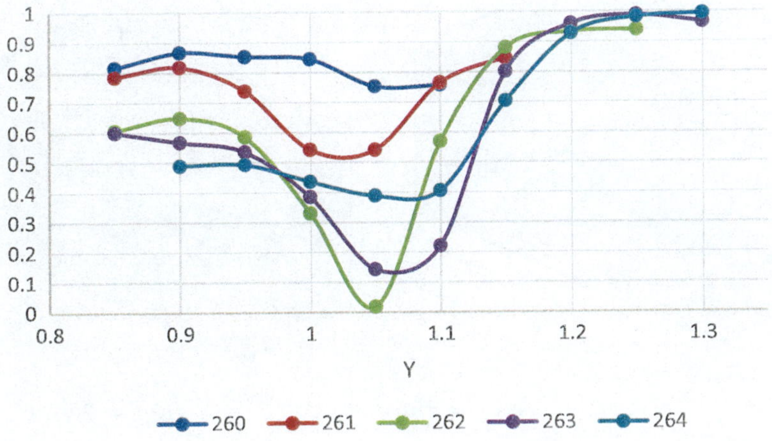

Fig. 8.4 The value of $\cos\theta$ varies with Y coordinate

8.3.2.2 Vortex 2 (Vortex Structure with Dissipation)

In order to understand the interaction between Liutex and shear, select several points at the vortex core of the 2 Vortices, and track the values of $\cos\theta$, shear magnitude, and Liutex magnitude at the same position at different times. Comparing the five moments of $t = 256$, $t = 256.5$, $t = 257$, $t = 257.5$, $t = 257.7$, the evolution of the Vortex 2 is shown in Fig. 8.5. Organizing the data and draw it as Fig. 8.6.

Through Fig. 8.6, we can observe that the vortex is essentially parallel to the flow direction X and that the wall-normal Y and spanwise Z positions are largely unchanged with flow along the flow direction X. At the same time, it is found that with the increase of time t, the middle and rear sections of this vortex begin to dissipate slowly.

It can be seen from Fig. 8.6a that the values of Liutex magnitude at each moment increase first and then decrease along the flow direction X. Comparing the five lines in the graph, it is found that the growth and decline rates at different times are almost the same.

From Fig. 8.6b, c, it can be found that the shear value and $\cos\theta$ value also have similar conclusions as the Liutex value. On the front and middle part of this vortex, the increasing and decreasing rates of shear values at different times are roughly the same. The growth rate of the $\cos\theta$ value of the front part of the vortex is also about the same.

Combined with Fig. 8.5, as time t increases, the front segment of this vortex remains in a steady state, while the middle and rear segments of the vortex begin to slowly dissipate.

Note that at $t = 257$, in the interval from $x = 3.75$ to $x = 3.95$, the shear and Liutex values are slowly decreasing, but the $\cos\theta$ values are stable. Similar findings were also found for the corresponding $t = 257.5$ and $t = 257.7$. One phenomenon

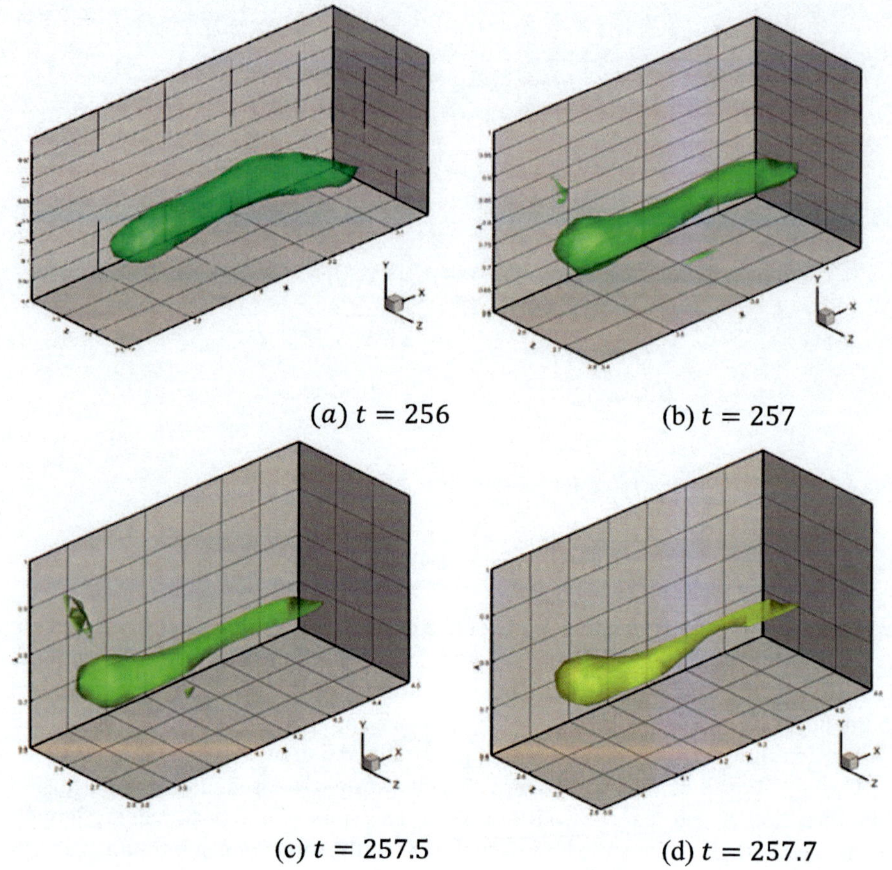

Fig. 8.5 The evolution of this vortex structure (Iso-surface of $R = 1.2$)

is that at the above three times, the $\cos\theta$ values remain roughly unchanged, that is, when $\cos\theta \approx 0.9$, as the shear and Liutex values decrease at the same time, the vorticity also decreases, which shows that the middle and rear sections of the vortex start slowly dissipation. This is similar to the conclusion of Vortex 1 in the previous article. It also shows that the interaction between Liutex and shear may have a certain influence on the dissipation of vorticity.

The above observations of Vortex 1 and Vortex 2 also verify the rationality of **R**-**S** decomposition from another aspect.

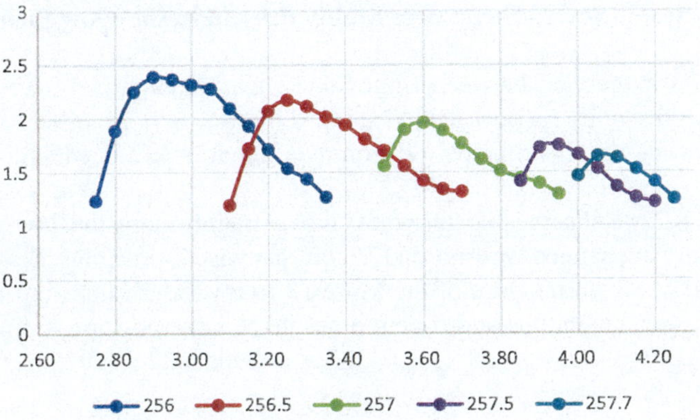

(a) The values of Liutex magnitude vary with X coordinate

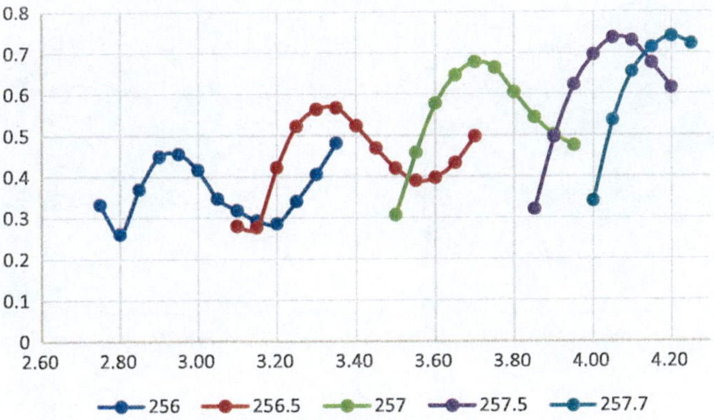

(b) The values of shear magnitude vary with X coordinate

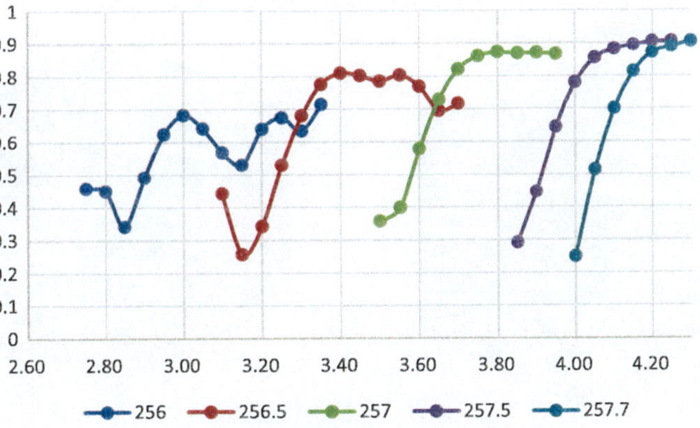

(c) The values of $\cos\theta$ vary with X coordinate

Fig. 8.6 The value of Liutex, shear magnitude and $\cos\theta$ development at selected points

8.3.2.3 Vortex 3 (Vortex Structures Stably Present in the Flow Field)

To explore the interaction between Liutex and shear quantitatively, we selected a vortex stably present in the flow field visualized by $R = 0.9$. We observe the development of vorticity in this time period from $t = 286$ to $t = 291$,which is shown in Fig. 8.7.

From Fig. 8.7, we observe that this vortex moves mainly along the flow direction X with essentially unchanged wall-normal Y and spanwise Z positions. Several points were selected at the vortex core of the Vortex 3 to track the values of $\cos\theta$, shear magnitude, Liutex magnitude at different times in the same position. Comparing the five moments of $t = 286$, $t = 287.5$, $t = 289$, $t = 290$, and $t = 291$, thus we obtain Fig. 8.8.

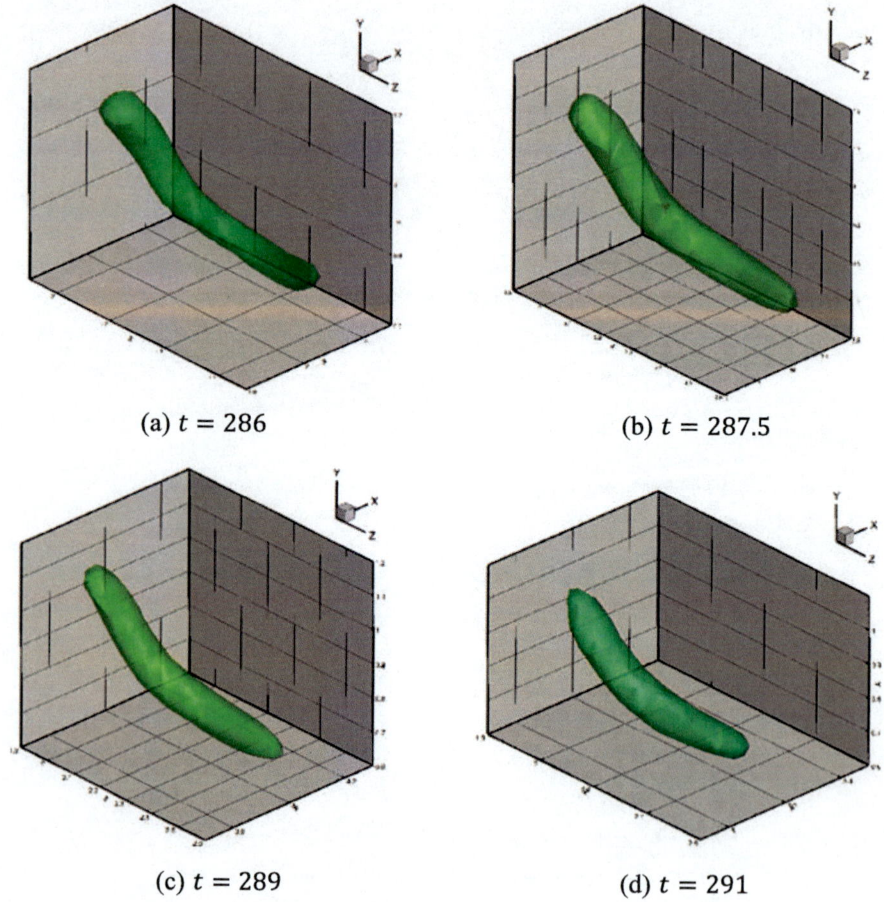

Fig. 8.7 The evolution of the selected vortex structure (Iso-surface of $R = 0.9$)

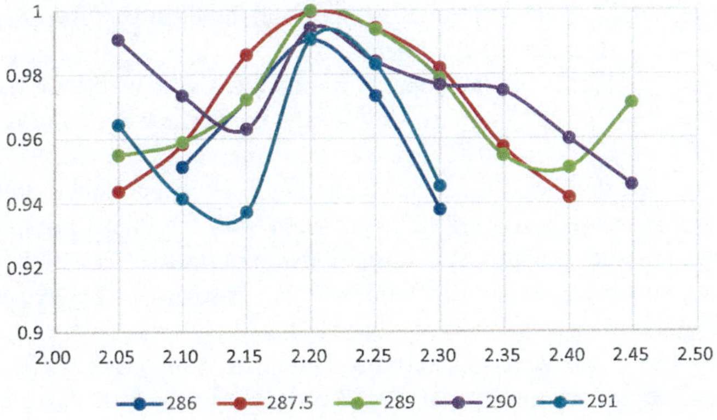

(a) The values of θ vary with Z coordinate

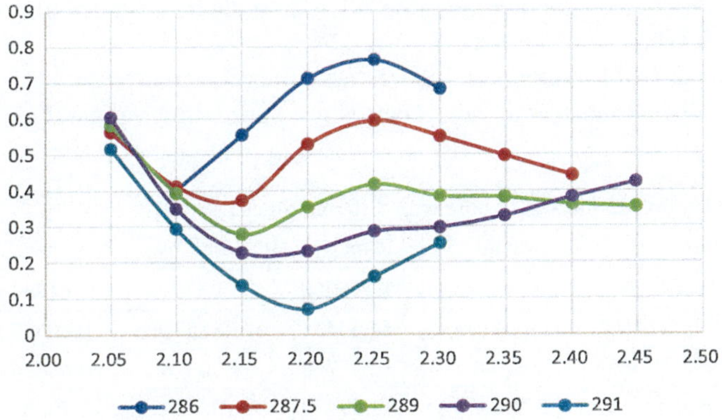

(b) The values of shear magnitude vary with Z coordinate

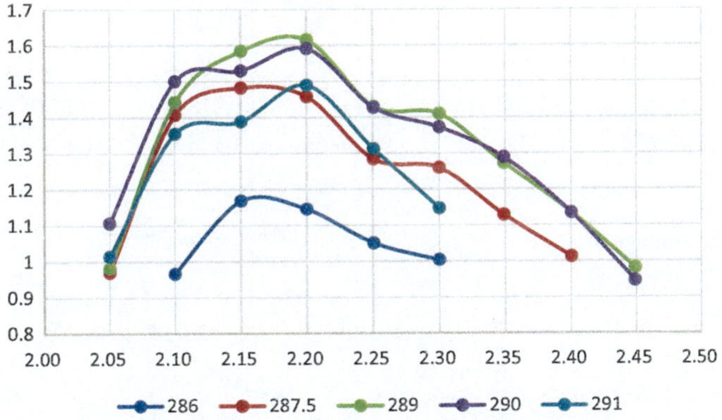

(c) The values of Liutex magnitude vary with Z coordinate

Fig. 8.8 The value of Liutex, shear magnitude and cosθ development at selected points

We observe that the angle between the Liutex and the shear vectors remain roughly between 0° and 20° through the flow field.

Referring to Fig. 8.7, along the positive Z-axis, the Liutex values of the front and middle section of the vortex increase slowly, and the shear values almost decreases gradually; the Liutex values of the rear section of the vortex decrease and the shear values increase. At the same time, since the angle between Liutex and the shear vectors are stable between 0° and 20°, combined with the R-S decomposition, the Liutex and shear values are almost in a relationship of trade-offs, which leads to the shape of this vortex in the entire flow field, which is basically no change and a stable state is maintained.

Note that when $z = 2.2$, $\cos\theta$ is almost equal to 1, which means that the Liutex and the shear vectors are almost parallel. According to the **R-S** decomposition, it makes sense to study the ratio of Liutex and vorticity when the Liutex and shear vectors are parallel. So, we extracted the data in the interval from $z = 2.2$, from $t = 286$ to $t = 290.5$, as shown in Table 8.1. The scale diagram for drawing Liutex and vorticity is shown in Fig. 8.9.

The proportion of Liutex magnitude in the Vortex 3 increases with time as shown in Fig. 8.9. When Liutex and shear vectors are approximately parallel, the vortex

Table 8.1 Liutex line data at vortex core with different times ($z = 2.2$)

t	$\cos\theta$	S-mag	R-mag	Vorticity	R-mag/Vorticity
286	0.99084501	0.711346	1.145050	1.850897	0.618646175
286.5	0.993325869	0.634879	1.255475	1.885733	0.665775485
287	0.997872565	0.623451	1.345768	1.96678	0.684249873
287.5	0.999585416	0.528455	1.457450	1.984041	0.734586575
288	0.999873993	0.435954	1.550731	1.984747	0.781324575
288.5	0.999837379	0.354144	1.615610	1.967586	0.821112461
289	0.999835688	0.354193	1.625498	1.967522	0.826165113
289.5	0.99852363	0.250944	1.634375	1.882443	0.868220105
290	0.994376453	0.231346	1.591201	1.818680	0.874920652
290.5	0.984576591	0.222311	1.529666	1.746104	0.876045325

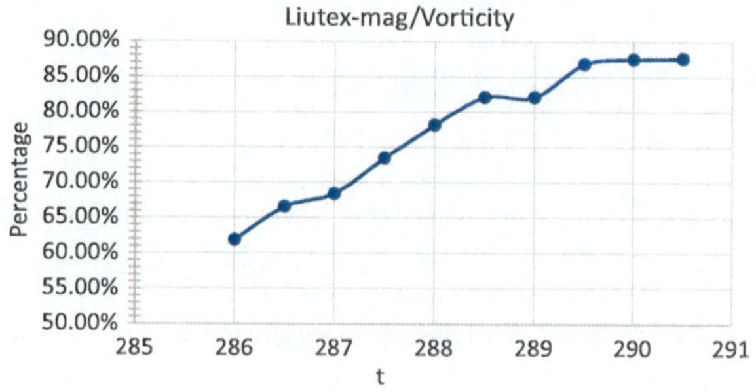

Fig. 8.9 The percentage of R-mag/vorticity varies with t

moves greatly along the flow direction and the offset in other directions is very small. Combined with Fig. 8.7, we can find that the Vortex 3 are relatively stable. Furthermore, the interaction between Liutex and shear may also have some implications for keeping the vortex stable.

8.4 Conclusions and Analysis

In summary, this paper research from the point of view of correlation analysis in turbulent channel flow, some conclusions can be drawn as follows:

1. Liutex can well capture vortices in turbulent channel flow with its direction as the local rotational axis and magnitude as twice the angular velocity speed of local rotational motion.
2. For the channel turbulence, the shear is mainly concentrated near the upper and lower wall surfaces, and the closer the wall shear value is, the greater the value.
3. The **R-S** decomposition based on the Liutex vector is reasonable, and Liutex is not contaminated by shear.
4. The change between the angle between the Liutex and the shear vectors in the Vortex 1 change the shape of the vortex, that is, when the angle between the shear and the Liutex vector is $\frac{\pi}{2}$, the shear changes the magnitude of the vortex rotation axis to a certain extent, and direction. For the Vortex 2 and Vortex 3, when the angle between Liutex and the shear vector remains stable, the interaction between the two may have an effect on the steady flow of the vortices or on the partial dissipation.

References

1. C. Liu, Y. Yan, P. Lu, Physics of turbulence generation and sustenance in a boundary layer. J. Comput. Fluids. 353–384 (2014)
2. I.M. Wallace, Highlights from 50 years of turbulent boundary layer research. J. Turbul. 1–70 (2013)
3. H. Helmholtz. Über integrale der hydrodynamischen gleichungen, welche den wirbelbewegungen entsprechen. Journal für die reine und angewandte Mathematic **55**, 25–55 (1858)
4. S.K. Roblnson, Coherent motion in the turbulent boundary layer. J. Annu. Rev. Fluid. Mech. 601–639 (1991)
5. Y. Wang, Y. Yang, G. Yang, et al., DNS study on vortex and vorticity in late boundary layer transition. J. Commun. Comput. Phys. 441–459 (2017)
6. C. Liu, Y. Wang, Y. Yang, et al., New omega vortex identification metho. J. Sci. China. Phys. Mech. **59**(8) (2016)
7. C. Liu, Y. Gao, X. Dong, et al., Third generation of vortex identification methods: Omega and Liutex/Rortex based systems. J. Hydrodyn. 205–223 (2019).
8. C. Liu, Y. Gao, S. Tian, et al., Rortex-a new vortex vector definition and vorticity tensor and vector decompositions. J. Phys. Fluids. 30 (2018)

9. X. Dong, Y. Gao, C. Liu, New normalized Rortex/Vortex identification method. J. Phys. Fluids. 31 (2019)
10. C. Liu, Liutex-third generation of vortex definition and identification methods. J. Aerodyn. 413–431 (2020)
11. Y. Wang, Y. Gao, J. Liu, et al., Explicit formula for the Liutex vector and physical meaning of vorticity based on the Liutex-Shear decomposition. J. Hydrodyn. 464–474 (2019)
12. P. Shrestha, C. Nottage et al., Stretching and shearing contamination analysis for Liutex and other vortex identification methods. J. Aia. (2021). https://doi.org/10.1186/s42774-020-00060-9
13. Y. Yu, P. Shrestha, O. Alvarez, et al., Investigation of correlation between vorticity, Q, λci, λ2, Δ and Liutex. J. Comput. Fluids. 225 (2021)
14. S. Laizet, E. Lamballais, High-order compact schemes for incompressible flows: a simple and efficient method with quasi-spectral accuracy. J. Comput. Phys. **228**, 5989–6015 (2008)
15. S. Laizet, N. Li, Incompact 3d: a powerful tool to tackle turbulence problems with up to O (105) computational cores. Int. J. Numer. Meth. Fluids **67**, 1735–1757 (2011)

Chapter 9
Experimental Studies on the Evolution of Hairpin Vortex Package in the Boundary Layer of a Square Tube

Yuyan Li, Xiangrui Dong, Xiaoshu Cai, Wu Zhou, and Xinran Tang

Abstract The hairpin vortex structures in the boundary layer with $Re_\theta = 159–239$ of a square tube were studied experimentally using the moving single- frame and long-exposure image (M-SFLE) method. And the Liutex vortex identification criteria are used to confirm vortices and characterize their strength in the experimental results. The flow measurement system always moves at the same or similar speed as vortices in the boundary layer to track vortex structures continuously. The experimental results show that the secondary hairpin vortex and the tertiary hairpin vortex can be induced by the primary hairpin vortex. And the extension of the vortex packet in the flow direction can lead to the occurrence of vortex merging in the near-wall turbulent boundary layer. The Q2 event plays a key role in the formation of secondary hairpin vortices and the merging behaviors. The merging process of two vortices can be divided into three stages and the strength of the vortex structures has unique characteristics in different stages.

9.1 Introduction

Wall turbulent is one of the most common flow situations in the field of human production and life. The quasi-order structures are considered to be the most important structures for revealing the law of wall turbulence. The famous hydrogen bubble visualization flow experiment carried out by Kline et al. [1] in 1967 found that there are organized flow vortex structures in the near-wall area and low-velocity streaks in the viscous bottom layer, and they observed that the streaks may oscillate, lift, break, and decline, which is an important progress in the study of pseudo-ordered structures.

Y. Li (✉) · X. Dong · X. Cai · W. Zhou · X. Tang
Institute of Particle & Two-Phase Flow Measurement, Shanghai 200093, China
e-mail: liyuyan469@163.com

School of Energy and Power Engineering, University of Shanghai for Science and Technology, Shanghai 200093, China

Y. Wang et al. (eds.), *Liutex and Third Generation of Vortex Identification*,
Springer Proceedings in Physics 288, https://doi.org/10.1007/978-981-19-8955-1_9

Kim et al. [2] visualized the low-velocity flat plate turbulent boundary layer in 1971 using hydrogen-bubble and also hot-wire measurements, proved that the generation of turbulent kinetic energy is closely related to the turbulent burst. At present, most of the important achievements on the hairpin vortices and the hairpin packages are based on two-dimensional experiments. Several typical coherent structures have been found in turbulent boundary layers, including high and low-speedstreaks (LSSs), bursts, streamwise vortices, hairpin vortex, large-scale motion and very large-scale motion (VLSMs), Λ-vortices and so on. More and more turbulent flow visualization experiments point to that the generation and development of hairpin vortices have important effects on other coherent structures. The predecessor of the hairpin vortex is the Ω-shaped vortex structure, which is proposed by Theodorsen [3] to explain the formation of Λ-shaped vortex. Head and Bandyopadhyay [4] first provided flow visualization photos of hairpin vortex structures through smoke experiments, and found that the crack without smoke presents an angle of 40°–50° degrees with the ground. The development of PIV technology and high-precision numerical simulation technology has greatly promoted the study of quasi-order structures in turbulent boundary layers. Zhou et al. [5, 6] discovered that a single hairpin vortex moves downstream may eventually develops into a hairpin vortex packet through numerical simulation. The simulation results show that the hairpin vortex packet can be automatically formed by the self-replication without additional external force. Adrian et al. [7] found the hairpin vortex packets imprints in the turbulent boundary layer through two-dimensional PIV experiments, and proposed an ideal model of the hairpin vortex packet, in which the characteristics of the old and new packets were clarified. The theoretical model of Marusic [8] pointed out that the hairpin vortex packet is the core structure of the wall turbulent layer, and the model can explain many flow characteristics in the logarithmic region. The 2-D imprints of the hairpin vortices are not enough for people to clearly understand the essence of the hairpin vortex structures, so it is imperative to study vortices in boundary layer is with 3D technology. Ganapathisubramani et al. [9] conducted a dual-plane PIV experiment using three cameras, and they obtained the most probable inclination angle of the hairpin vortex from the measured data. Elsinga et al. [10] and Dennis et al. [11] used three-dimensional PIV technology to extract the hairpin vortex envelope structures in the zero pressure gradient boundary layer, which provided new clues for revealing the structural characteristics of the three-dimensional vortex envelope. However, 3D flow visualization measurement systems are usually very complex and expensive, and the system has strict requirements on the density of tracer particles, image resolution and light source intensity. In addition, the current flow visualization researches are very insufficient on the dynamic evolution of the flow structures. Cai et al. [12] proposed a 2-D M-SFLE method, which can well study the dynamic evolution of the vortex structures in the turbulent boundary layer, and the measurement system is simple and easy to implement. After that, the Moving-PIV (M-PIV) method was also proposed by Cai's group, and the dynamic evolution process of a single hairpin vortex was quantitatively measured at an inclination angle of 53°. The 2-D information of the turbulent boundary layer has not been fully mined, and there are still many possibilities in this experimental field.

In this paper, some of the latest experimental results of the study of the hairpin vortex structures in the rectangular pipeline boundary layer based on the MSFLE method are mainly introduced, in order to provide some new 2-D experimental evidence for further understanding of the function and role of the hairpin structures in the turbulent boundary layer.

9.2 Measurement Setup

9.2.1 Measurement System

The experimental measurement system is composed of a circulating water tunnel and an image acquisition system, as shown in Fig. 9.1. The circulating water tunnel includes a constant pressure water tank, a rectangular measuring pipe section, a water storage tank, a flow regulating valve, a pump and a rotameter. Taking pure water as the measurement object, and the flow rate in the pipeline is adjusted by the flow regulating valve. The constant pressure water tank is composed of an inner barrel and an outer barrel, and the water flow of the inner barrel keeps overflowing. A flow stabilizer is set downstream to ensure the stability of the flow velocity at the entrance of the horizontal measurement section. The size of the horizontal rectangular tube is 80 mm × 80 mm × 2200 mm, which is made of transparent plexiglass material. And this section of the pipeline is used to take visualization images of fluid flow. A flat plate with a thickness of 4 mm and a length of 1500 mm is placed at the bottom of the pipe. The front edge of the flat plate is designed to be elliptical, and a metal mixing line is set at a distance of 50 mm from the front edge of the flat plate, in order to strengthen the disturbance and promote the formation of fully developed turbulence. The image acquisition system includes a XIMEA industrial black and white camera, a semiconductor continuous laser, a servo motor guide rail, a host computer and related control software. The resolution of the camera is 1280 × 1024, the pixel size is 4.8 um, and the target size is 6.1 mm × 4.9 mm. The wavelength of the laser is 450 nm and the power is 4 W. A cylindrical lens is used to form a laser sheet light with a thickness of about 1 mm to illuminate the flow field as shown in Fig. 9.2. And the laser sheet light is arranged on the normal centerline of the pipe along the flow-normal direction of the pipeline to reduce the influence of the boundary layers on the two side walls. The camera and the laser are fixed on the guide rail, and they move synchronously and uniformly downstream along the guide rail, so that the vortex structures of different speeds in the boundary layer can be tracked, also the evolution of the vortices over time.

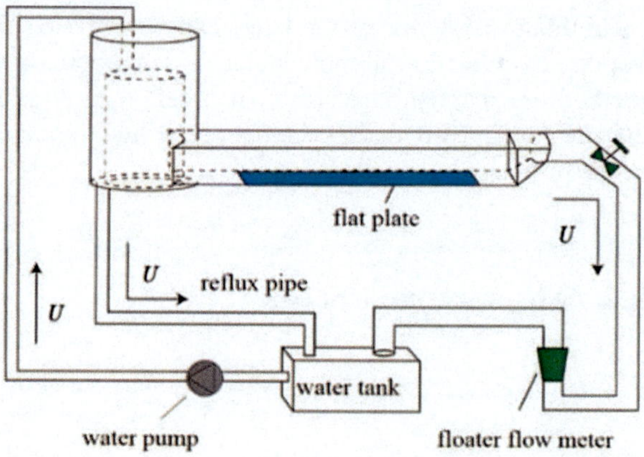

Fig. 9.1 Schematic diagram of the experimental measurement system

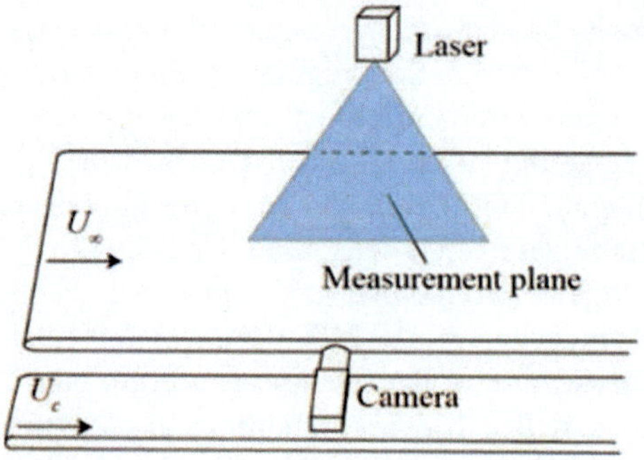

Fig. 9.2 Schematic diagram of measurement method

9.2.2 Experimental Flow Parameters

The flow rate of related experiments was set to 1.4 m^3/h, and the corresponding flow velocity is approximately $U_\infty = 82\,\text{mm/s}$. A large number of experimental data in rectangular pipes show that more hairpin vortices can be easily tracked under this flow condition. Hollow glass microbeads were used as tracer particles, with the average diameter of 15um and density of 0.8 g/ml. The measured position of the transparent pipe is the area between x = 400 mm–1200 mm from the front edge of the plate, and the measurement area is located in the rectangular range of 47 mm × 38 mm above the plate at the bottom. The boundary layer thickness $\delta 1$ and $\delta 2$ at the initial position x = 400 mm and the end position x = 1200 mm are 20 mm and 30 mm. The Reynolds numbers $Re_{\theta 1}$ and $Re_{\theta 2}$ based on the thickness of momentum loss calculated according to the formula $Re_\theta = \theta U_\infty / \nu$ are 159 and 239, where ν is the fluid kinematic viscosity coefficient. Setting the movement speed of the measurement system to $U_c \approx 0.9U_\infty$, so the camera could follow and record the vortices for a long time under above Reynolds condition.

9.2.3 MSFLE Measurements and Image Processing

In this paper, the M-SFLE experimental measurement method was used to qualitatively and quantitatively analyze the hairpin vortices and the hairpin packets in the turbulent boundary layer of rectangular pipes. Many experimental and numerical simulation results have confirmed that there are a large number of hairpin vortex structures in the wall turbulent boundary layer. However, many traditional flow visualization methods only measure the flow at a single point, and it is difficult to study the development and evolution of the hairpin vortex structures in the time dimension. M-SFLE combines motion shooting technology with the single-frame and long-exposure measurement method (SFLE) to track the hairpin vortices and other quasi-ordered structures in the temporal and spatial dimensions. The basic idea of the SFLE method to measure the particle velocity is to set a longer exposure time for the camera, so long trajectories of the particle can be obtained in one frame of image due to motion blur. And the average velocity of individual particles can be calculated. The velocity of the tracer particles can be obtained by the following formula:

$$v = \frac{s}{m\Delta t}. \tag{9.1}$$

where v is the moving speed of the tracer particle, s is the length of the trajectory in the image, m is the magnification of the camera lens, and Δt is the exposure time. The direction of movement of the particles is judged according to the correlation of two adjacent frames of images.

In the experiment using the SFLE method, the camera position was fixed, so several particle trajectories with the same direction and different lengths were obtained in the resulting image, as shown in Fig. 9.3a. However, the vortex structures in the boundary layer cannot be directly resolved from these images. The difference between the SFLE method and the MSFLE method can be seen in Fig. 9.3. The M-SFLE method makes the entire imaging system move at a uniform speed along the flow direction. The speed of the camera is similar to vortex structures in the turbulent boundary layer. Several long-exposure images were taken continuously, so the vortices are visualized. And the evolution of the vortex structures during successive shots can be studied through a complete set of experimental images. As shown in Fig. 9.3b, the two-dimensional spanwise vortices were visualized by the measurement system under the M-SFLE method. The upstream vortex on the left in the image is located near the wall, and the other larger vortex is farther away from the wall. According to the correlation between adjacent images, it can be known that both vortices are clockwise.

Through the measurement results of the M-SFLE method, the two-dimensional section of the vortices and the nearby flow structures can be directly observed. The interaction between the pseudo-order structures in the wall turbulent boundary layer and their evolution process could be further studied.

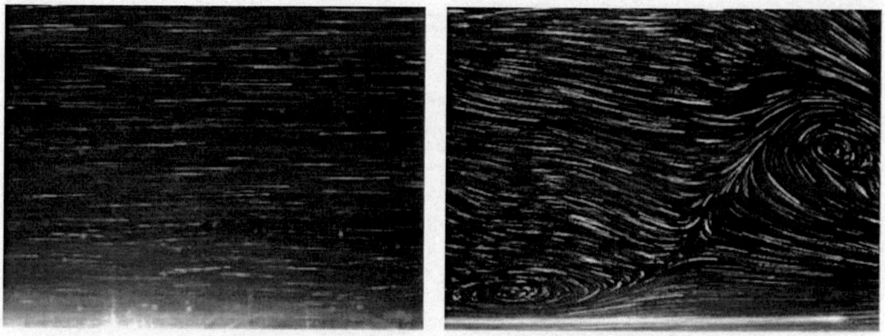

(a) Original image from SFLE (b) Original image from M-SFLE

Fig. 9.3 Comparison of SFLE and M-SFLE methods

Hairpin vortex structures could be quantitatively investigated by applying image processing procedures to their visualized images. The image processing is summarized into three procedures, the identification and extraction of the trajectory, the calculation of the velocity, and the interpolation of the velocity field.

There are many particle trajectories in an experimental result image, and these trajectories can be processed into single-pixel width lines by the skeleton-extracted trace extraction method. Several procedures are required before extracting the skeletons of the trajectories, including denoising and sharpening, adaptive thresholding, erosion and dilation, and removing short objects. The process and results of an original experimental picture using skeleton extraction method can be seen in the Fig. 9.4.

The length of the trails needs to be calculated, after obtaining single-pixel width traces by extracting the traces' skeleton. At the pixel level, particle trajectories in the M-SFLE experiments are curves rather than straight lines, so the length of the trajectories cannot be obtained simply by calculating the distance between the start

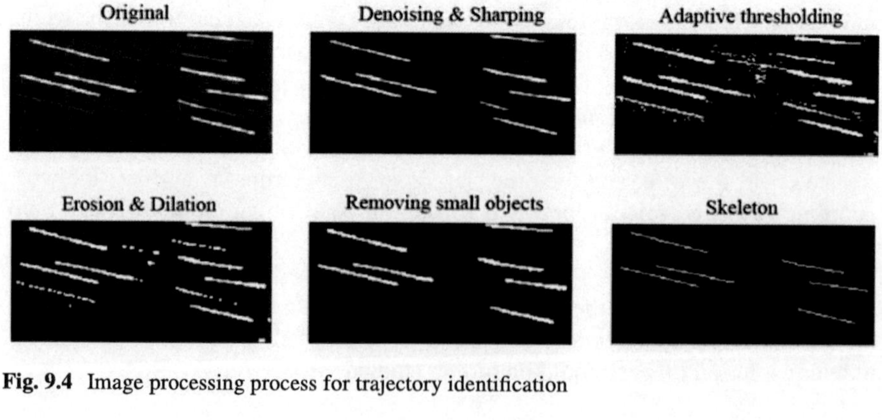

Fig. 9.4 Image processing process for trajectory identification

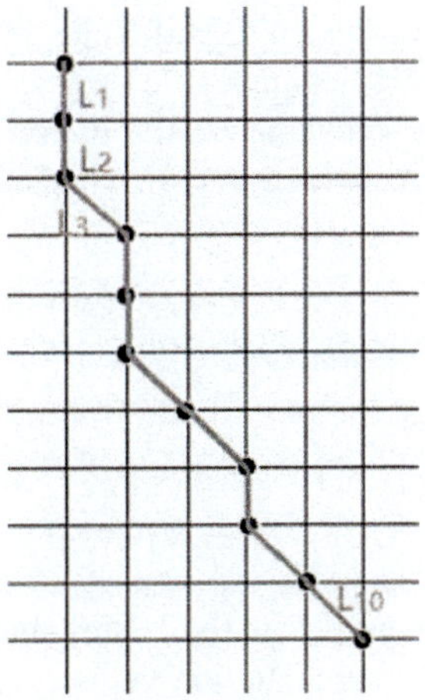

Fig. 9.5 Schematic diagram of the line integration of the trajectories

point and end point of the trajectory which works well in SFLE experiments. In this paper, the line integral method was used to calculate the length of the curved trajectories. The schematic diagram of the line integration of the trajectories is shown in the Fig. 9.5. The total length of a trajectory is obtained by superimposing the distance between each two adjacent pixels. Since the direction of the velocity of the trajectory is ambiguous, the angular velocity discrimination method was used to determine the direction of the velocity in this paper. First, the moving direction of particle trajectories was judged by the correlation of adjacent images. Then a rotation center was defined artificially, and the motion directions of the tracer particles around the center are all subject to the angular velocity direction. The above method works well when there are fewer flow structures in the experimental image, but it has limitations when dealing with complex flow situations. Therefore, more effective speed ambiguity judgment methods need to be further developed. In addition, the velocity field was reconstructed by Radial Basis Function (RBF) interpolation [13].

9.2.4 Characterization of the Vortex

How to define a vortex mathematically has always been one of the most important topics in turbulent dynamics. Several criteria based on velocity gradient tensors have been proposed by researchers for the identification of vortex structures, such as Δ, λ_{ci}, Q, λ_2, and Ω criterion [6, 14–17]. Those methods show their own advantages in vortex identification in some fluid researches, but they all have some limitations in terms of applicability or threshold setting. Liu et al. [17–19] pointed out that vorticity cannot accurately describe the rotational motion of fluid, and they proposed a new generation of vortex identification methods, which uses the Liutex vector to describe a vortex. The vorticity ω can be decomposed into two parts $\boldsymbol{R}$ and $\boldsymbol{S}$, The vorticity can be decomposed into two parts, where $\boldsymbol{R}$ describes the rotational characteristics of the fluid, and S describes the shear characteristics of the fluid. For the velocity gradient tensor ∇V, when there are a pair of complex eigenvalues $\lambda_{cr} \pm \lambda_{ci}$ and a real eigenvalue λ_r, the direction of the liutex vector coincides with the direction of the real eigenvectors. The Liutex vector is mainly described by the following formulas.

$$\boldsymbol{R} = R\boldsymbol{r} \tag{9.2}$$

$$\nabla \boldsymbol{V} \cdot \boldsymbol{r} = \lambda_r \boldsymbol{r} \tag{9.3}$$

$$R = \boldsymbol{\omega} \cdot \boldsymbol{r} - \sqrt{(\boldsymbol{\omega} \cdot \boldsymbol{r})^2 - 4\lambda_{ci}^2} \tag{9.4}$$

where R is the Liutex magnitude, and $\boldsymbol{r}$ is the normalized direction of $\boldsymbol{R}$. In order to calculate the velocity gradient tensors and use Liutex vector to characterize the vortex structures, the third-dimensional information z = 0, w = 1 was inserted into the experimental results after image processing in this paper.

9.3 Results and Discussion

9.3.1 Generation of Secondary Vortices

Many flow characteristics of wall turbulence are thought to be related to hairpin vortices and hairpin packets. The hairpin vortex packet model proposed by Adrian et al. [7] is the most widely accepted conceptual model of wall turbulence. In the model, low speed streaks are induced by the vortex legs of all hairpin vortices in the packet. And the ejection or 'Q2' event is co-induced by the head and legs of a hairpin vortex. They also discovered the imprint of the hairpin vortex packets through PIV experiments. Cai et al. [12] visualized a relatively complete hairpin vortex structure with relatively symmetrical legs through a MSFLE experiment, and the legs of the vortex deformed and twisted as it moved downstream. Numerous studies have shown that hairpin vortices can be created in several ways, and a new hairpin vortex induced by the initial hairpin vortex is one of the most important ways. Through numerical simulation, Smith et al. [20] found that the secondary vortices can be induced by the preexisting primary hairpin vortices, and pointed out that the pressure gradient is important for this process. Zhou et al. [6] found that a new secondary vortex can be formed not only upstream of the primary hairpin vortex, but also can be formed downstream. This mechanism that the new vortex structures generate by self-replication of the initial vortices is considered to be an important way to maintain turbulent flow.

In this paper, the complete generation process of a secondary hairpin vortex was captured through M-SFLE experimental method, as shown in Fig. 9.6. The structures were measured on the pipe's flow-normal plane, and the positive x-axis represents the direction of fluid flow. The flow parameters of this group of experiments are described in Sect. 9.2.2. The exposure time of the camera is 200 ms, and the magnification of the lens is 0.13X. Four moments are selected to describe the formation process of a secondary hairpin vortex, as shown in Fig. 9.6a–d. The time t of the evolution process was converted into dimensionless variable by $t^* = tU_\infty/\delta_0$. The abscissa X and ordinate Y represent the position of the actual field of view in a single frame picture. As the value of t or t^* increases, the position of the field of view is closer to the downstream of the pipe.

As shown in Fig. 9.6a, when $t^* = 75.5$, there were two clockwise spanwise vortices in the field of view. The larger-sized vortex named P was located closer to the downstream and farther from the wall, and the vortex A was located upstream of the pipe and closer to the wall. Upstream of the vortex P, there was a long sloping shear line extending from the vortex head of P to the wall in the picture, and the inclination angle between the shear layer and the wall is approximately 25°. There was a Q2 event below the shear layer. This is due to the co-induction of the legs and heads of the hairpin vortex P from a three-dimensional point of view, which leads to the uplift of the low speed fluid near the wall. The sweep or Q4 event encounters the

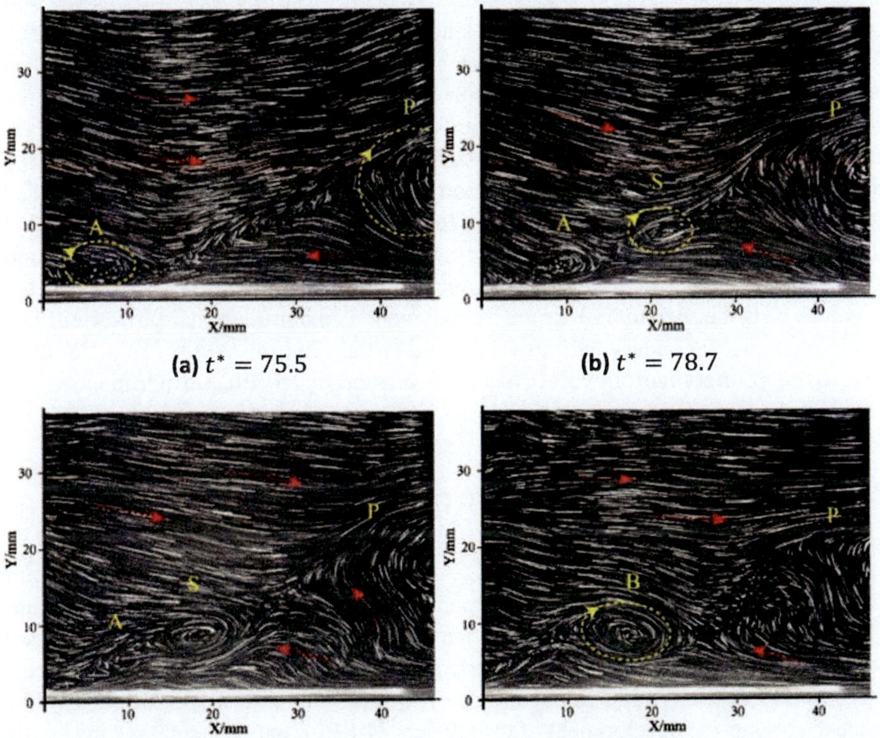

(a) $t^* = 75.5$ (b) $t^* = 78.7$

(c) $t^* = 83.7$ (d) $t^* = 88.6$

Fig. 9.6 Generation of a secondary hairpin vortex by M-SFLE method

Q2 event, and promotes the formation of the inclined shear layer. when $t^* = 78.7$, a new spanwise vortex S was generated on the shear layer under the induction of the primary hairpin vortex P, and S is the spanwise head of the secondary vortex, which is shown in Fig. 9.6b. The newly formed secondary hairpin vortex S also rotates clockwise. Comparing the X coordinates of the vortex centers in Fig. 9.6a and b, it can be seen that the moving speed of the primary vortex had slowed down, but the height of the vortex center from the wall did not change significantly. when $t^* = 75.5$, the spanwise head of the newly formed secondary hairpin vortex S was gradually separated from the shear layer away from the primary hairpin vortex P, and its velocity slowed down as shown in Fig. 9.6c. Since S and A were close enough, the mutual induction between them made the two vortices move closer to each other and they finally merged. A new spanwise vortex B was formed due to the merger when $t^* = 88.6$. The new vortex caused by the merger still rotates clockwise as shown in Fig. 9.6d. The interaction between vortex P and B became weaker, since P gradually moving away from B and moving downstream at a faster speed.

The generation of new hairpin vortices due to the induction of primary hairpin vortices was also found in the DNS results as shown in Fig. 9.7. As the primary hairpin vortex moves downstream, the legs and the head gradually lift upward, and the induction between the two legs is strengthened. A strong local shear layer will be formed due to the encounter of the rising low-velocity fluid and the high-velocity fluid. The connection points of the legs and neck of the primary hairpin vortex will be bent upward and bonded, then the new spanwise vortex can be generated on the shear layer and gradually fall off from the primary body to become a complete secondary hairpin vortex. If the primary and secondary hairpin vortices are strong enough, they can continue to induce the third generation of hairpin vortices. Zhou et al. [6] and Li et al. [21] have reported the formation of tertiary hairpin vortices through numerical simulation studies. The generation process of a tertiary hairpin vortex was also found in the M-SFLE experiment. As shown in the Fig. 9.8, T is the tertiary hairpin vortex structure, S′ is the secondary hairpin vortex, and P′ is the primary hairpin vortex. The mechanism that secondary hairpin vortex and tertiary hairpin vortex can be induced by a strong primary hairpin vortex has been verified by M-SFLE experiments.

9.3.2 *Merging of Hairpin Vortices*

Through numerical simulation studies, Adrian et al. [22] found that the hairpin vortices will inevitably grow along the spanwise direction as they move downstream. This means that different packets will inevitably collide, and the result of contact between different structures would be merged, or destroyed due to opposite forces. Mergers and destructions will establish new quasi-structural orders for wall turbulence to support the development of turbulence. M-SFLE experimental results in this paper show that the extension of the vortex packet in the flow direction can lead to the occurrence of vortex merging in the near-wall turbulent boundary layer.

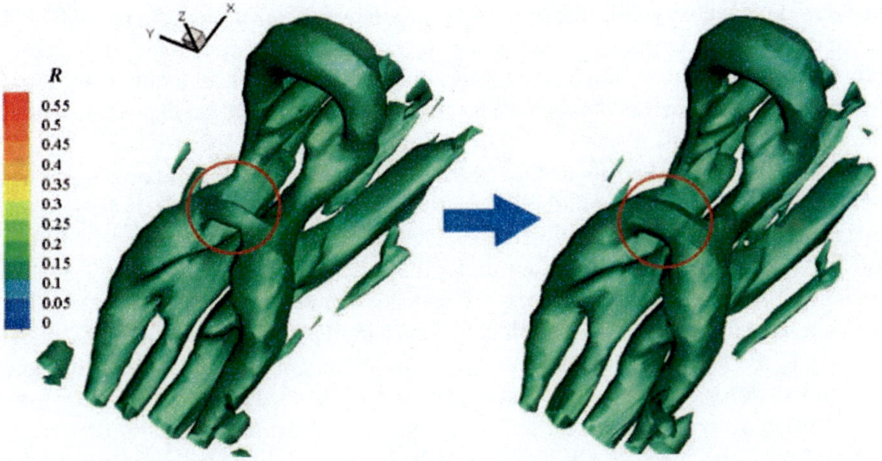

Fig. 9.7 Generation of a secondary hairpin vortex in DNS, |R|=0.12 (Date resources come from Liu et al. [18])

Fig. 9.8 Generation of a tertiary hairpin vortex based on M-SFLE

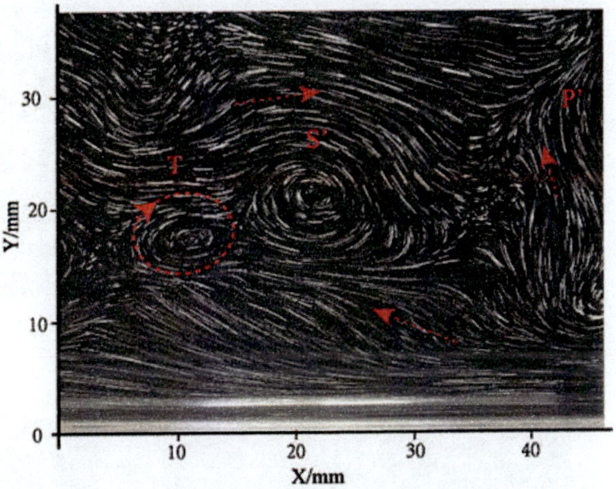

As shown in Fig. 9.6c, two vortices at different positions in the flow direction merged and formed a new vortex. The downstream vortex is farther from the wall, while the upstream vortex is lower in height. Observing the entire merging process, it can be found that before the merging, the downstream vortex S experienced a brief and obvious deceleration motion, which caused the two vortex structures to move closer and promoted their mutual induction. Finally, the A vortex which has a smaller size was rolled into the outer layer of the vortex S, and formed a new vortex B. In the vortex merging experiments of Guo et al. [23], the initial vortices have the same size and intensity are considered to be necessary conditions for vortex merging. However, the experimental results in this paper show that two original hairpin vortices of different sizes may still merge in wall turbulence. Considering the resolution, a set of images with larger sized vortex structures was used to study the vortex merger problem in the turbulent boundary layer, as shown in Fig. 9.9. The flow parameters of this group of experiments are described in Sect. 9.2.2, and the exposure time of the camera is 200 ms.

As shown in Fig. 9.9a, when $t^* = 14.0$, E and F were clockwise-rotating spanwise vortices. As can be seen from the figure, the size of F was significantly larger than E, and vortex E was located upstream of vortex F. A local shear layer had been already formed due to the Q2 event under the head of vortex F. Then, the speed of vortex F suddenly slowed down, and the distance between E and F became smaller, which resulted in a strong induction between E and F, as shown in the Fig. 9.9b. At $t^* = 22.4$, the fluid on the out layer of E and F was entangled, which pushed E and F to move closer to each other. It can also be seen from the figure that the vortex E was lifted away from the wall due to the ejection of the low-speed fluid under F. When $t^* = 25.6$, The vortex E was gradually decomposed and wrapped around the outer layer of vortex F, which is shown in Fig. 9.9c. Finally, a new spanwise vortex

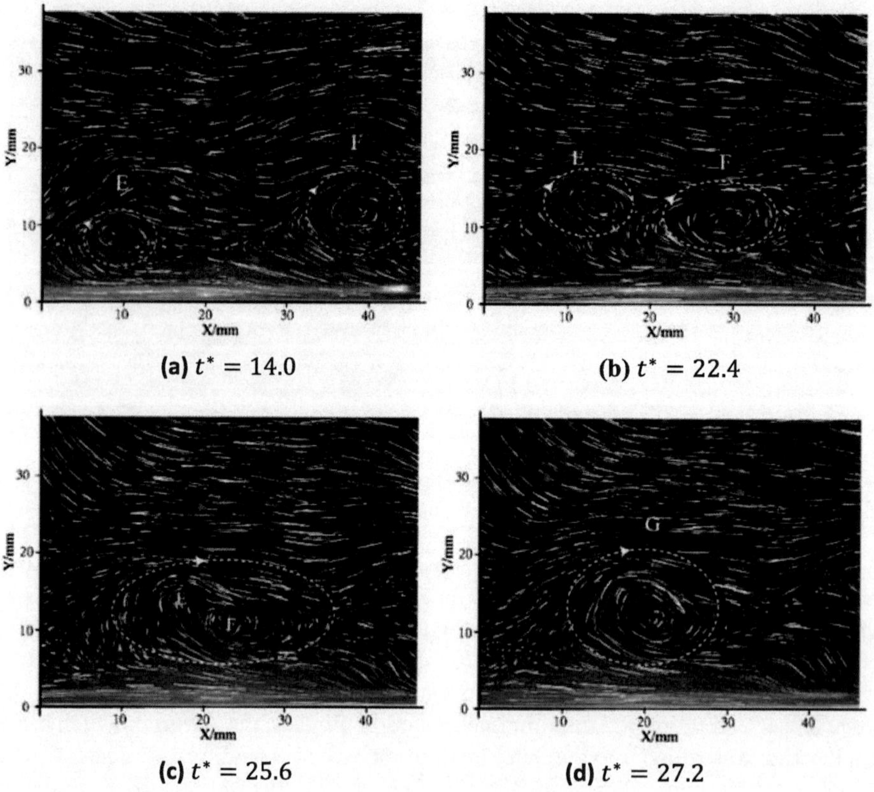

(a) $t^* = 14.0$ (b) $t^* = 22.4$

(c) $t^* = 25.6$ (d) $t^* = 27.2$

Fig. 9.9 The merger process of two spanwise clockwise vortices

G was formed as shown in the Fig. 9.9d. An inclined shear layer located downstream of vortex G was clearly visualized, which was induced by the downstream hairpin vortex structure.

Image processing was performed on this experimental results of vortex merging, and the vortex structures are characterized by the Liutex vector and vorticity respectively. Figure 9.10 shows the results of the quantitative analysis of the above vortex merging experimental data based on M-SFLE method. The figures on the left show the $|\omega|$ contour with the streamline distribution, and figures on the right show the R contour and the streamline distribution. It can be seen from the figures that there would be large value of $|\omega|$ at the region with strong shear, while Liutex could accurately characterize the rotational motion of the fluid.

For the further study of the vortex merging mechanism, R_{Int} is proposed to characterize the strength of the vortex. R_{Int} is obtained by integrating each point in the vortex region where R > 0. R_{Int} can be calculated by the formula

$$R_{\mathrm{Int}} = \int_{\Delta s} R ds \tag{9.5}$$

where ΔS represents the region of $R > 0$. R_{Int} was calculated for each vortex during the merging process to study the variation of vortex strength with time and the interaction between the two vortices. As can be seen from Fig. 9.9a, vortex E and vortex F begin to gradually approach each other and induce each other after $t^* = 14.0$. As is shown in Fig. 9.11, The vortex merging process can be divided into three stages. Between $t^* = 14.0$ and $t^* = 22.4$, the intensity of E increases, while the intensity of F decreases, and the total intensity of the two vortices has not changed almost. At this stage, the weak vortex will absorb the energy of the strong vortex, and the distance between the two vortices will continue to decrease. Between $t^* = 22.4$ and $t^* = 27.2$, The changing trend of the intensity values of the two vortices was consistent, and the total intensity of the two vortices was significantly reduced in Fig. 9.11. At this stage, the vortices are mainly affected by the nearby fluid, and the specific variation of their strength depends on the nearby fluid environment. The final stage is that the new vortex structure is fully formed and continues to develop downstream. As is shown in Fig. 9.11, the new vortex G was completely formed, and its strength was relatively stable in a short time.

And it can be seen from Figs. 9.9 and 9.11 that the two vortex structures that merged were located in the same packet, cause the two vortices shared the injection event of the packet. The Q2 event induced by the vortex located in the downstream position is important driving force for the merger in the turbulent boundary layer.

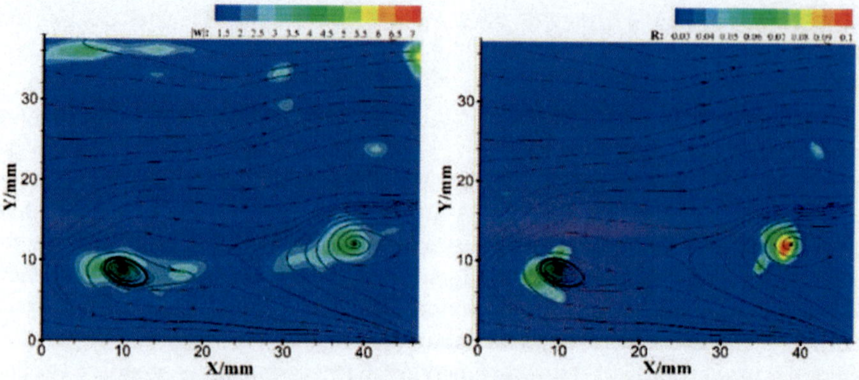

(a) $t^{*} = 14.0$

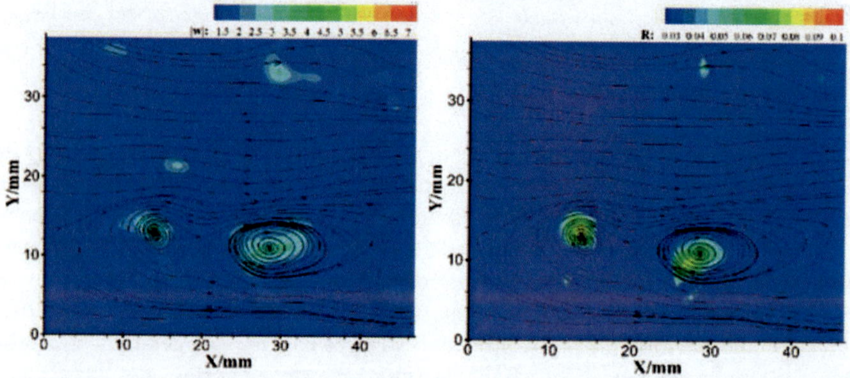

(b) $t^{*} = 22.4$

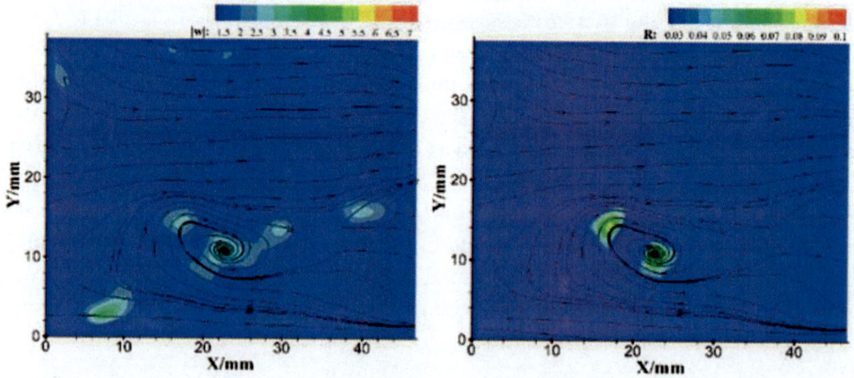

(c) $t^{*} = 25.6$

Fig. 9.10 (Color online) Process of merger of two vortices and evolution, vorticity magnitude $|\omega|$ (left) and Liutex magnitude R (right)

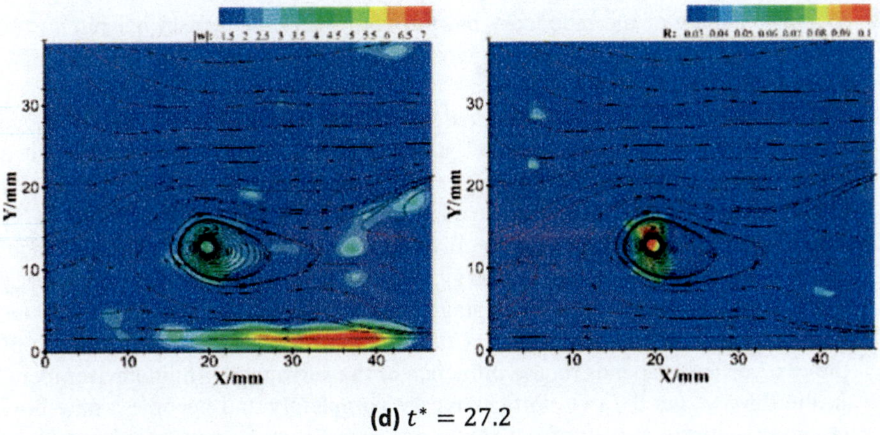

Fig. 9.10 (continued)

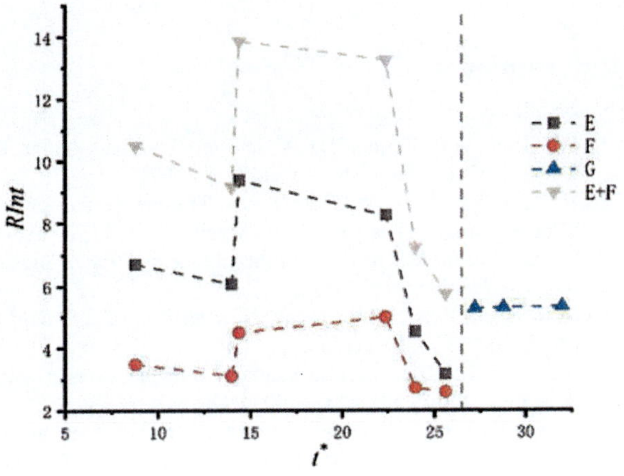

Fig. 9.11 (Color online) R_{Int} distribution of vortices for merging experiments

9.4 Conclusions

In this paper, the mechanism of the generation of secondary hairpin vortices induced by primary hairpin vortices are studied experimentally based on M-SFLE method, and the phenomenon of vortex merging in hairpin vortex packet in the turbulent boundary layer with $Re_\theta = 159$–239 was found. Some conclusions are obtained by analyzing the experimental results qualitatively and quantitatively.

1. The M-SFLE measurement method could visualize vortex structures and other flow structures well in the turbulent boundary layer and capture the evolution of the vortex structure.

2. The mechanism of the secondary hairpin vortex and the tertiary hairpin vortex induced by the primary hairpin vortex has been verified in the M-SFLE experiments in this paper.
3. The extension of the vortex packet in the flow direction can lead to the occurrence of vortex merging in the near-wall turbulent boundary layer. The Q2 event induced by the vortex located in the downstream position is important driving force for the merger in the turbulent boundary layer.
4. The merger of two vortices can be divided into three stages. In the first stage, the two vortices will exchange energy, and the weak vortex absorbs the energy of the strong vortex. In the second stage, the weak vortex gradually disintegrates and wraps around the outer layer of the strong vortex, and the total strength of the two vortices depends on the influence of the surrounding fluid environment. In the third stage, the two vortices merge completely and become a new flow structure, and the energy of the newly generated vortex structure will increase over a period of time.

References

1. S.J. Kline, W.C. Reynolds, F.A. Schraub, P.W. Runstadler, The structure of turbulent boundary layers. J. Fluid Mech. **30**(4), 741–773 (1967)
2. H. Kim, S.J. Kline, W.C. Reynolds, The production of turbulence near a smooth wall in a turbulent boundary layer. J. Fluid Mech. **50**(1), 133–160 (1971)
3. T. Theodorsen, Mechanisms of turbulence, in *Proceedings of the 2nd Midwestern Conference on Fluid Mechanics* (1952)
4. M.R. Head, P. Bandyopadhyay, New aspects of turbulent boundary-layer structure. J. Fluid Mech. **107**, 297–338 (1981)
5. J. Zhou, R.J. Adrian, S. Balachandar, Autogeneration of near-wall vortical structures in channel flow. Phys. Fluids **8**(1), 288–290 (1996)
6. J. Zhou, R.J. Adrian, S. Balachandar, T. Kendall, Mechanisms for generating coherent packets of hairpin vortices in channel flow. J. Fluid Mech. **387**, 353–396 (1999)
7. R.J. Adrian, C.D. Meinhart, C.D. Tomkins, Vortex organization in the outer region of the turbulent boundary layer. J. Fluid Mech. **422**, 1–54 (2000)
8. I. Marusic, On the role of large-scale structures in wall turbulence. Phys. Fluids **13**(3), 735–743 (2001)
9. B. Ganapathisubramani, E.K. Longmire, I. Marusic, S. Pothos, Dual-plane PIV technique to determine the complete velocity gradient tensor in a turbulent boundary layer. Exp. Fluids **39**(2), 222–231 (2005)
10. G.E. Elsinga, R.J. Adrian, B.W. Van Oudheusden, F. Scarano, Three-dimensional vortex organization in a high -Reynolds-number supersonic turbulent boundary layer. J. Fluid Mech. **644**, 35–60 (2010)
11. D.J. Dennis, T.B. Nickels, Experimental measurement of large-scale three-dimensional structures in a turbulent boundary layer. Part 1. Vortex packets. J. Fluid Mech. **673**, 180–217 (2011)
12. X.R. Tang, X.R. Dong, X.S. Cai, W. Zhou, Liutex identification on hairpin vortex structures in a channel based on msfle and moving-PIV. J. Hydrodyn. **33**(6), 1119–1128 (2021)
13. W. Fan, Z. Wu, C. Xiaoshu, Image processing algorithm for particle trajectory image and reconstruction study on flow field. J. Exp. Fluid Mech. **33**(4), 100–107 (2019)

14. A.E. Perry, M.S. Chong, A description of eddying motions and flow patterns using critical-point concepts. Annu. Rev. Fluid Mech. **19**(1), 125–155 (1987)
15. J.C. Hunt, A.A. Wray, P. Moin, Eddies, streams, and convergence zones in turbulent flows, in *Studying Turbulence Using Numerical Simulation Databases, 2. Proceedings of the 1988 Summer Program* (1988)
16. J. Jeong, F. Hussain, On the identification of a vortex. J. Fluid Mech. **285**, 69–94 (1995)
17. C. Liu, Y. Wang, Y. Yang, Z. Duan, New omega vortex identification method. Sci. China Phys. Mech. Astron. **59**(8), 1–9 (2016)
18. C. Liu, Y. Gao, S. Tian, X. Dong, Rortex—a new vortex vector definition and vorticity tensor and vector decompositions. Phys. Fluids **30**(3), 035103 (2018)
19. Y.S. Gao, J.M. Liu, Y.F. Yu, C. Liu, A Liutex based definition and identification of vortex core center lines. J. Hydrodyn. **31**(3), 445–454 (2019)
20. C.R. Smith, J.D.A. Walker, A.H. Haidari, U. Sobrun, On the dynamics of near-wall turbulence. Philos. Trans. R. Soc. Lond. Ser. A: Phys. Eng. Sci. **336**(1641), 131–175 (1991)
21. H. Li, D. Wang, H. Xu, Hairpin vortex formation mechanisms based on LXC-Liutex core line method, in *Liutex and Third Generation of Vortex Definition and Identification* (Springer, Cham, 2021), pp. 201–214
22. R.J. Adrian, S. Balachandar, Z.C. Lin, Spanwise growth of vortex structure in wall turbulence. KSME Int. J. **15**(12), 1741–1749 (2001)
23. G.U.O. Yanang, D.O.N.G. Xiangrui, C.A.I. Xiaoshu, Z.H.O.U. Wu, Experimental studies on vortices merging based on MSFLE and Liutex. 空气动力学学报 **38**(3), 432–440 (2020)

Chapter 10
The Correlation Between Pressure Fluctuation and Liutex Spectrum in Boundary Layer Transition

Vishwa Patel, Yonghua Yan, Yifei Yu, and Chaoqun Liu

Abstract The extensive decades of research have remained unexplained about the primary source of flow frequency unsteadiness in the Boundary Layer Transition. The study aims to determine the relation between Boundary Layer low frequencies and the Liutex using the high order LES on a large grid system. The newly developed Third Generation vortex identification method—Rortex/Liutex, was adopted to investigate the motion of complex vortex structures in the boundary layer transition. The power spectrum analysis clearly shows that the pressure fluctuation (low-frequency noises) correlates highly with the Liutex frequencies.

Nomenclature

LES	Large Eddy Simulation
Ma	Mach Number
Re	Reynolds number based on the momentum thickness
δ	Boundary Layer thickness
x, y, z	Spanwise, normal, and streamwise coordinate axes
u, v, w	Spanwise, normal, and streamwise velocity

10.1 Introduction

Supersonic aircraft are subject to high levels of pressure fluctuations. These intense fluctuations cause vibration leading to structural problems, total pressure loss-making flow unstable and distorted, engine unstart, drag rise, and high wall heating. The

V. Patel · Y. Yu · C. Liu (✉)
The University of Texas at Arlington, Arlington, TX 76019, USA
e-mail: cliu@uta.edu

Y. Yan
Jackson State University, Jackson, MS 39217, USA

Y. Wang et al. (eds.), *Liutex and Third Generation of Vortex Identification*,
Springer Proceedings in Physics 288, https://doi.org/10.1007/978-981-19-8955-1_10

noise generated by the airflow passing over a lifting surface is a major issue for a wide range of engineering applications. Some researchers have focused on the low-frequency unsteadiness of the SWBLI, which is still a challenging problem in the research. There are primarily two thoughts on the driving factor of low-frequency unsteadiness: (1) Upstream incoming flow variants or (2) downstream instability. Transition and SBLI have always been complicated problems [1]. SBLI topic is that which has seen extensive study over the past decades.

On the one hand, the shock acts as a diffusion causing a premature boundary-layer transition that otherwise would take place further downstream of the shock origin point. On the other hand, when a transition occurs in the interaction domain, it extremely affects the event by promoting momentum exchanges through the development of instabilities reverting into turbulence. Even though having its practical significance, the interaction between transition and shock waves is a delicate question that is far from fully illustrated.

Over the last decade and a half, some important progress has been made in applying direct numerical simulations in computation aeroacoustics. Reference [2] used DNS to investigate the noise sources in a low-Reynolds-number turbulent jet at Mach 0.9, and one of the first simulations of wall-bounded flows followed, with [3] considering sound radiation in turbulent channels. Then using both LES in order to increase the chord-Reynolds number to practical values [4, 5] or DNS [6] simulated turbulent flow over a trailing edge and direct noise simulations of full airfoil configurations have followed. DNS is the preferred tool for such fundamental studies due to the absence of any physical modeling. Conducting compressible DNS allows for a perfectly represented hydrodynamic phenomena such as turbulence and transition to turbulence; simultaneously, it gives access to the acoustic field.

Mainly, in computation, the contribution aims at outlining the challenges in conducting DNS with the application of noise. The higher consistency multi-block structured curvilinear Compressible Navier-Stocks are developed for exploiting high-performance computing systems. Solving the unsteady compressible Navier-Stocks equations directly to study aerodynamic noise brings several computational challenges. Some of those challenges are pointed out by [7], who stated that the extent of the acoustic field is considerably larger than other similar flow fields due to the greater wavelength of acoustic waves compared to flow structures. This study performs correlation analysis with pressure fluctuation and Liutex spectra, choosing the desired domain where flow is well developed.

10.2 Numerical Algorithm and Case Description

10.2.1 Governing Equations

To govern the flow field, a non-dimensional compressible Navier–Stokes system can be written in the following conservative form:

$$\frac{\partial Q}{\partial t}+\frac{\partial A}{\partial x}+\frac{\partial B}{\partial y}+\frac{\partial C}{2z}=\frac{1}{Re}\left(\frac{\partial A_v}{\partial x}+\frac{\partial B_v}{\partial y}+\frac{\partial C_v}{\partial z}\right) \tag{10.1}$$

where the vector of conserved quantities Q, inviscid flux vector A, B, and C and viscous flux vector $A_v B_v,\ and\ C_v$ are

$$Q=\begin{pmatrix}\rho\\ \rho u\\ \rho v\\ \rho w\\ \mathrm{e}\end{pmatrix}, A=\begin{pmatrix}\rho u\\ \rho u^2+p\\ \rho uv\\ \rho uw\\ (\mathrm{e}+p)u\end{pmatrix}, B=\begin{pmatrix}\rho v\\ \rho vu\\ \rho v^2+p\\ \rho vw\\ (\mathrm{e}+p)v\end{pmatrix}, C=\begin{pmatrix}\rho w\\ \rho wu\\ \rho wv\\ \rho w^2+p\\ (\mathrm{e}+p)w\end{pmatrix} \tag{10.2}$$

$$A_v=\begin{pmatrix}0\\ \sigma_{xx}\\ \sigma_{xy}\\ \sigma_{xz}\\ u\sigma_{xx}+v\sigma_{xy}+w\sigma_{xz}+q_x\end{pmatrix}, B_v=\begin{pmatrix}0\\ \sigma_{xy}\\ \sigma_{yy}\\ \sigma_{yz}\\ u\sigma_{xy}+v\sigma_{yy}+w\sigma_{yz}+q_y\end{pmatrix}$$

$$C_v=\begin{pmatrix}0\\ \sigma_{xz}\\ \sigma_{yz}\\ \sigma_{zz}\\ u\sigma_{xz}+v\sigma_{yz}+w\sigma_{zz}+q_z\end{pmatrix} \tag{10.3}$$

$$\sigma_{xx}=\frac{2}{3}\lambda(T)\left(2\frac{\partial u}{\partial x}-\frac{\partial v}{\partial y}-\frac{\partial w}{\partial z}\right), \sigma_{yy}=\frac{2}{3}\lambda(T)\left(-\frac{\partial u}{\partial x}+2\frac{\partial v}{\partial y}-\frac{\partial w}{\partial z}\right), \tag{10.4}$$

$$\sigma_{zz}=\frac{2}{3}\lambda(T)\left(-\frac{\partial u}{\partial x}-\frac{\partial v}{\partial y}+2\frac{\partial w}{\partial z}\right) \tag{10.5}$$

$$\sigma_{xy}=\lambda(T)\left(\frac{\partial u}{\partial y}+\frac{\partial v}{\partial x}\right), \sigma_{xz}=\lambda(T)\left(\frac{\partial u}{\partial z}+\frac{\partial w}{\partial x}\right),$$
$$\sigma_{yz}=\lambda(T)\left(\frac{\partial w}{\partial y}+\frac{\partial v}{\partial z}\right) \tag{10.6}$$

Equation (10.1) can be displayed as,

$$\frac{\partial \widehat{Q}}{\partial t}+\frac{\partial \widehat{A}}{\partial \xi}+\frac{\partial \widehat{B}}{\partial \eta}+\frac{\partial \widehat{C}}{\partial \xi}=\frac{1}{Re}\left(\frac{\partial \widehat{A}_v}{\partial \xi}+\frac{\partial \widehat{B}_v}{\partial \eta}+\frac{\partial \widehat{C}_v}{\partial \zeta}\right) \tag{10.7}$$

$$q_x=\frac{\mu}{(\gamma-1)M_\infty^2 Pr}\frac{\partial T}{\partial x}, q_y=\frac{\mu}{(\gamma-1)M_\infty^2 Pr}\frac{\partial T}{\partial y},$$
$$q_z=\frac{\mu}{(\gamma-1)M_\infty^2 Pr}\frac{\partial T}{\partial z}, p=\frac{1}{\gamma M_\infty^2}\rho T,\ Pr=0.72 \tag{10.8}$$

where

$$\widehat{Q} = J^{-1}Q, \tag{10.9}$$

$$\widehat{A} = J^{-1}(\xi_x A + \xi_y B + \xi_z C), \tag{10.10}$$

$$\widehat{B} = J^{-1}(\eta_x A + \eta_y B + \eta_z C), \tag{10.11}$$

$$\widehat{C} = J^{-1}(\zeta_x A + \zeta_y B + \zeta_z C), \tag{10.12}$$

$$\widehat{A}_v = J^{-1}(\xi_x A_v + \xi_y B_v + \xi_z C_v), \tag{10.13}$$

$$\widehat{B}_v = J^{-1}(\eta_x A_v + \eta_y B_v + \eta_z C_v), \tag{10.14}$$

$$\widehat{C}_v = J^{-1}(\zeta_x A_v + \zeta_y B_v + \zeta_z C_v) \text{ and} \tag{10.15}$$

$$J^{-1} = \begin{vmatrix} 1 & 0 & 0 & 0 \\ 0 & x_\xi & x_\eta & x_\zeta \\ 0 & y_\xi & y_\eta & y_\zeta \\ 0 & z_\xi & z_2 & z_\zeta \end{vmatrix} \tag{10.16}$$

$$\begin{pmatrix} \xi_x & \xi_y & \xi_z \\ \eta_x & \eta_y & \eta_z \\ \zeta_x & \zeta_y & \zeta_z \end{pmatrix} = J \begin{pmatrix} y_\eta z_\zeta - y_\zeta z_\eta & z_\eta x_\zeta - z_\zeta x_\eta & x_\eta y_\zeta - x_\zeta y_\eta \\ y_\zeta z_\xi - y_\xi z_\zeta & z_\zeta x_\xi - z_\xi x_\zeta & x_\zeta y_\xi - x_\xi y_\zeta \\ y_\xi z_\eta - y_\eta z_\xi & z_\xi x_\eta - z_\eta x_\xi & x_\xi y_\eta - x_\eta y_\xi \end{pmatrix} \tag{10.17}$$

In the dimensionless form, the reference values for length, density, velocity, and temperature are $\delta_{in}, p_\infty, U_\infty,$ *and* T_∞, respectively. δ_{in} is the displacement thickness of inflow and the Mach number M_∞ and Reynolds number Re are expressed as follows:

$$M_\infty = \frac{U_\infty}{\sqrt{\gamma R T_\infty}}, \text{Re} = \frac{\rho_\infty U_\infty \delta_{in}}{\mu_\infty}, \tag{10.18}$$

where R is the ideal gas constant, γ is the ratio of specific heats, and μ_∞ *is the viscosity.*

Sutherland's equation gives the dynamic viscosity

$$\mu = \frac{T^{\frac{3}{2}}(1+s)}{T+S}, S = \frac{110.3K}{T} \tag{10.19}$$

10.2.2 Numerical Methods

The sixth-order compact scheme is used for spatial derivatives in the streamwise, spanwise, and wall-normal directions to eliminate numerical oscillations caused by the central difference scheme; spatial filtering is used instead of artificial dissipation. Adding to the methods, a third-order TVD Runge–Kutta method is used for time generation.

$$Q^{(0)} = Q^{(n)}, \tag{10.20}$$

$$Q^{(1)} = Q^{(0)} + \Delta t R^{(1)}, \tag{10.21}$$

$$Q^{(2)} = \frac{3}{4}Q^{(0)} + \frac{1}{4}Q^{(1)} + \frac{1}{4}\Delta t R^{(1)}, \tag{10.22}$$

$$Q^{(n+1)} = \frac{1}{3}Q^{(0)} + \frac{2}{3}Q^{(2)} + \frac{2}{3}\Delta t R^{(2)}, \tag{10.23}$$

10.2.3 Case Setup

The axes x, y, and z in the streamwise, spanwise, and wall-normal has the grid number 1920 $\times$128 $\times$ 241. These grids are stretched in the normal direction, while it is uniform in the streamwise and spanwise directions. x_{in} is the distance between the inlet and the leading edge of the flat plate. δ_{in} is the inflow displacement thickness. The physical domain of DNS simulation is shown in Fig. 10.1, and geometry parameters are listed in Table 10.1. In Table 10.1, M_∞ is Mach number, Re is Reynolds number and, T_∞ and T_w are free and wall stream temperature, respectively. L_x, L_y, *and* L_z are the lengths of computational domain in x, y, and z directions. Figure 10.2 shows the domain composition along the streamwise direction.

In the study, a subzone in the middle of the whole domain of DNS simulation is selected, which is halved in the spanwise direction. The main reason for choosing this subzone is that the flow is well developed of complex structures and relatively much data for different time steps. The power spectra of pressure and Liutex are obtained at 22 $\times$22 $\times$ 68 = 32,912 points evenly distributed in the selected subzone. The x-coordinate varies from 624 to 944, y-coordinate varies from 0.52 to 10.83, and z-coordinate varies from 0.6 to 13.8. Figure 10.3 gives the location of the subzone where the power spectra of pressure and Liutex were analyzed.

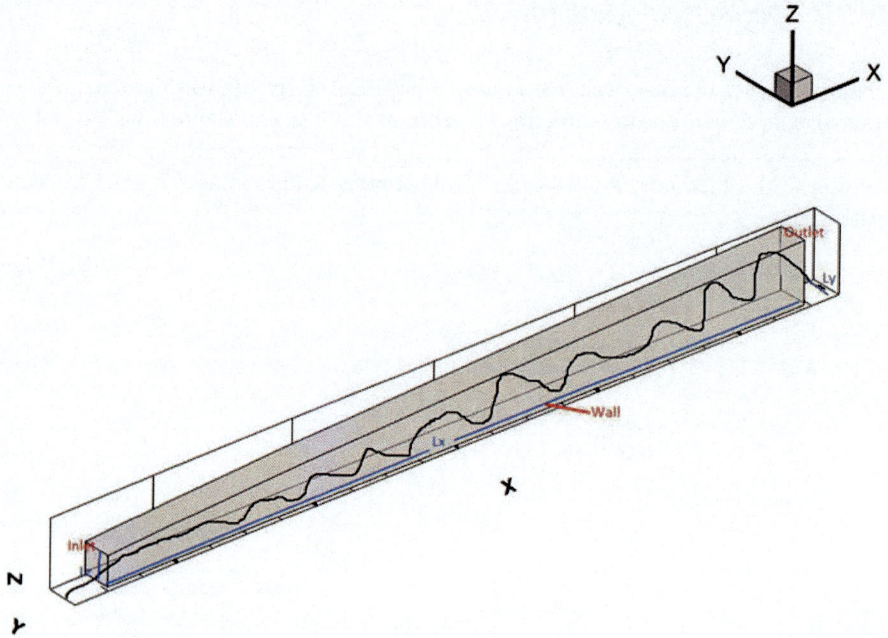

Fig. 10.1 Computation domain

Table 10.1 DNS parameters

M_∞	Re	x_{in}	Lx	Ly	Lz_{in}	T_w	T_∞
0.5	1000	300.79 δ_{in}	798.03 δ_{in}	22 δ_{in}	40 δ_{in}	273.15 K	273.15 K

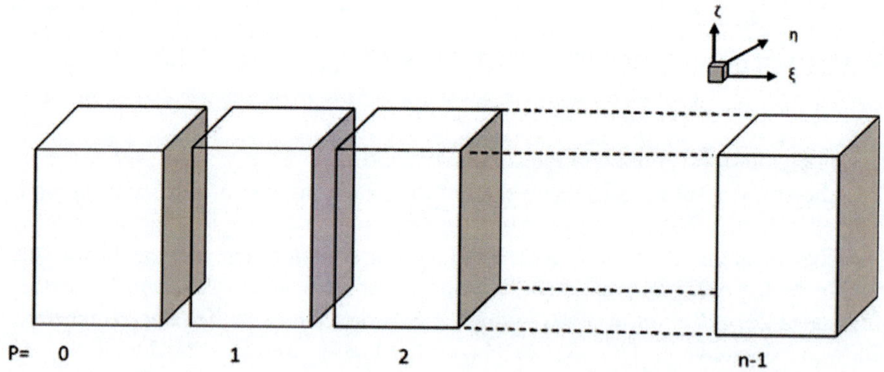

Fig. 10.2 Domain decomposition along the streamwise direction in the computational space

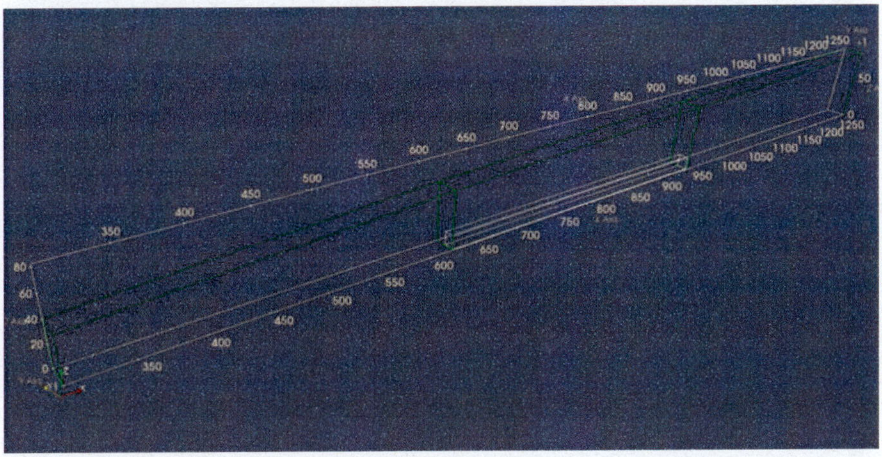

Fig. 10.3 Domain of analysis

10.2.4 Code Validation

The computational domain is displayed in Fig. 10.1. The flat plate flow with Re = 1000 and Ma = 0.5 is selected as the testing case. In the study, the DNSUTA code proposed by the University of Texas at Arlington is adopted [8, 28]. It has been cross-verified by NASA Langley [29] with Parabolized Stability Equation results for wave propagation incompressible boundary layers past curvilinear surfaces to make sure that the DNS results are correct. Readers are encouraged to refer to our papers [11, 30] for further details.

10.3 Liutex Definition

Liutex is a third-generation vortex visualization method discovered by Liu et al. [13, 25] at the University of Texas at Arlington in 2018. The ultimate purpose of Liutex is to extract the grid rotation part from fluid motion to represent the vortex. Mathematically, it is defined as a vector that uses the real eigenvector of velocity gradient tensor as its direction and twice the local fluid angular speed as its magnitude. Liutex is a vector method defined as [15]

$$\boldsymbol{R} = R\boldsymbol{r} \tag{10.24}$$

where R is the magnitude of Liutex and $\boldsymbol{r}$ is the local rotation axis. An explicit expression of R [16] given as

$$R = \vec{\omega} \cdot \vec{r} - \sqrt{(\vec{\omega} \cdot \vec{r})^2 - 4\lambda_{Ci}^2} \tag{10.25}$$

where $\vec{\omega}$ represents the vorticity vector, $\vec{r}$ is the direction of the Liutex vector, λ_{Ci} is the imaginary part of the complex eigenvalue of the velocity gradient tensor.

Thus, the Liutex vector is obtained by $\vec{R} = R\vec{r}$. The Liutex gradient vector is defined by the gradient of the Liutex magnitude

$$\nabla R = \begin{bmatrix} \frac{\partial R}{\partial x} \\ \frac{\partial R}{\partial y} \\ \frac{\partial R}{\partial z} \end{bmatrix} \tag{10.26}$$

10.4 Results and Discussion

Complex vortex structures, such as lambda-vortices, hairpin vortices, and ring-like vortices, are a prominent characteristic of the transitional flow. In the previous work on the transition flow, the large-scale vortices are generated in the upper boundary layer by the multiscale shear layer in the flow [17, 18]. The structures are quite robust and will gradually travel toward the downstream and interact with the strong ramp shock. As shown in Fig. 10.4, the ring-like vortices are visible from x = 420 δ_{d0} and the series of vortices are observed from x = 500 δ_{d0}. These structures are staggered in the streamwise direction. From x > 600 δ_{d0}, the structures of the ring-like vortices are broken, which leads to the generation of small-scale structures.

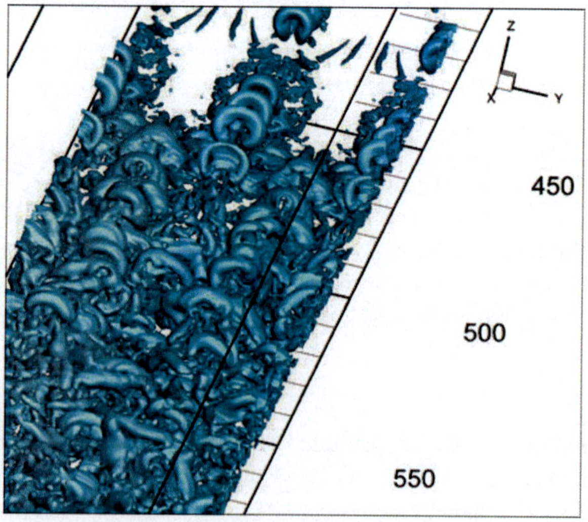

Fig. 10.4 Ring-like vortex structures

Figure 10.5a–d shows the correlation coefficients of pressure and Liutex spectra on YZ cross-sections (spanwise and normal directions). The white curve in each figure shows the locations of boundary layer thickness ($u/U_\infty = 99\%$), and the blue curves correspond to the location with $y^+(z^+) = 30$ and 100. It can be clearly observed that there are variously scaled vortex structures below the boundary layer thickness and extremely high correlation (colored in red).

The correlation coefficient of pressure and Liutex spectra on XZ cross-sections (streamwise and normal directions) is shown in Fig. 10.6. The white curve in each figure shows the locations of boundary layer thickness ($u/U_\infty = 99\%$), the blue curves correspond to the locations with $y^+(z^+) = 30$ and 100. The extremely high correlation is observed below the boundary layer thickness, where variously scaled vortex structures exist.

The distribution of correlation coefficients of pressure and Liutex spectra on XY cross-sections (streamwise and spanwise directions) is given in Fig. 10.7. As we got closer to the bottom surface (above the viscous sublayer), the more significant the correlation is. Also, as the flow develops, the correlation gradually increases.

Definition 10.1 The correlation coefficient is determined by dividing the covariance by the product of the two variables' standard deviations. Let x and y be the random variables with the sample size n, r(x, y) is the correlation coefficient of x and y if:

$$r(x, y) = \frac{cov(x, y)}{\sigma(x)\sigma(y)} \tag{10.27}$$

where,

$$cov(x, y) = E[(x - E(x))(y - E(y))], \tag{10.28}$$

$$E(x) = \frac{1}{n}\sum_{i=1}^{n} x_i; \quad x_i \text{ is the } i \text{ th entry of } x, \tag{10.29}$$

$$Var(x) = E\left[(x - E(x))^2\right], \tag{10.30}$$

$$\sigma(x) = \sqrt{Var(x)} \tag{10.31}$$

The Correlation Coefficient is a statistical concept revealing the extent to which two data groups are related. If these two data groups are entirely correlated, their correlation is 1, and if they are irrelevant, their correlation coefficient is near 0.

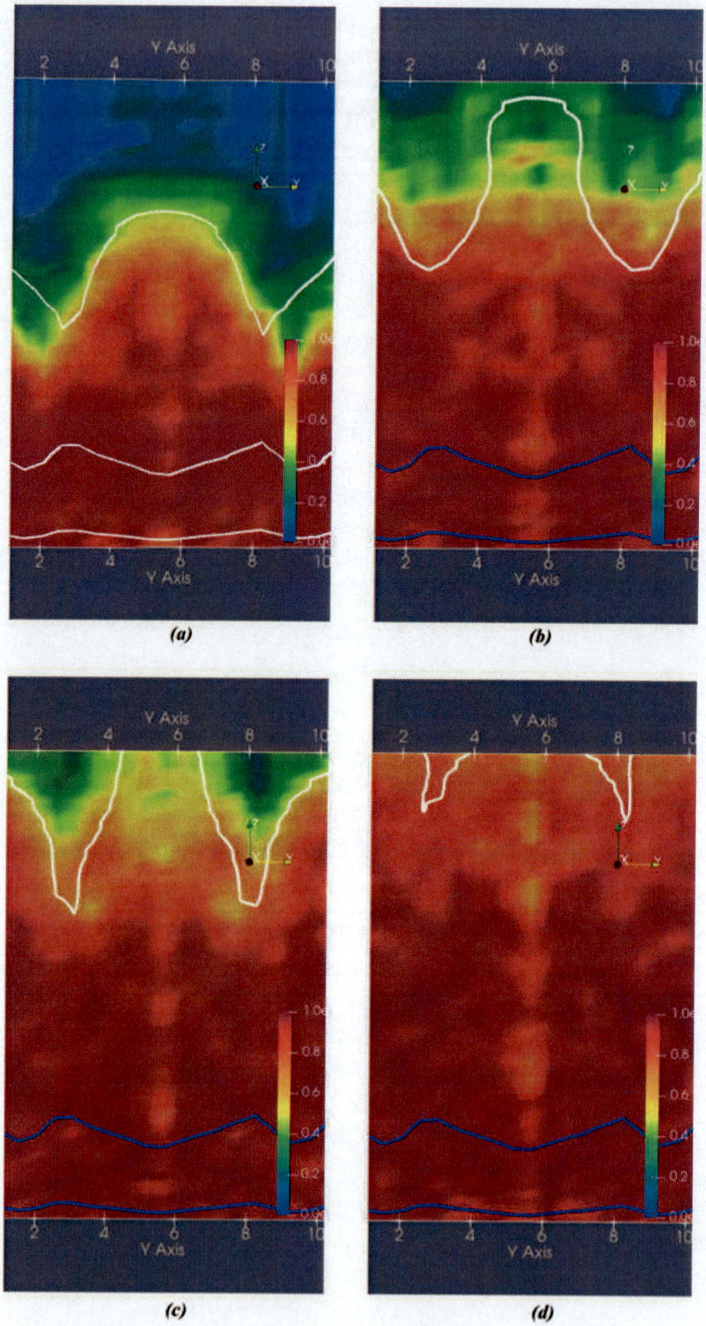

Fig. 10.5 Distribution of correlation coefficients of pressure and Liutex spectra (0.0–1.0) on YZ cross sections **(a)**x = 624δin **(b)** x = 704δin **(c)** x = 784δin **(d)** x = 864δin

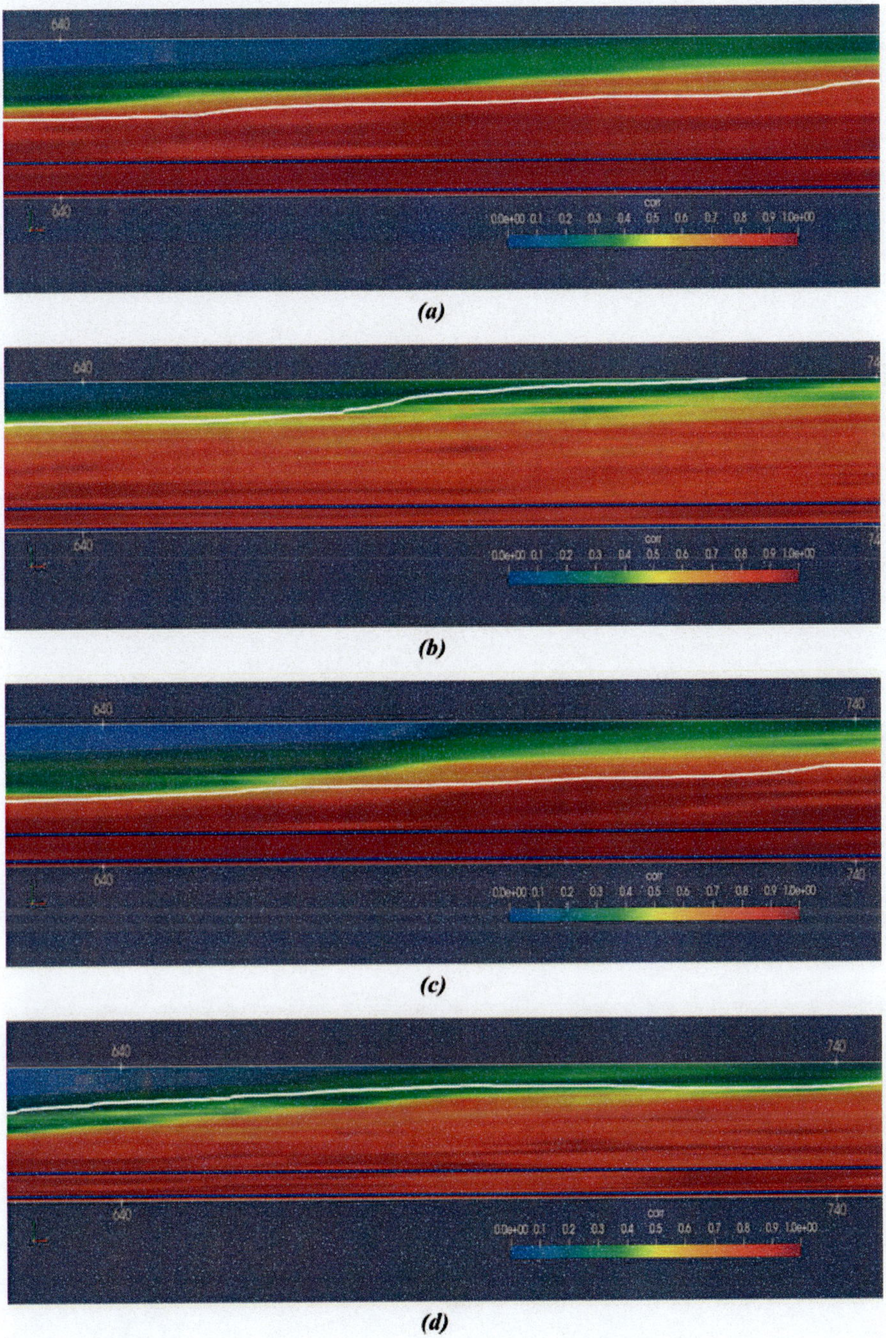

Fig. 10.6 Distribution of correlation coefficients of pressure and Liutex spectra (0.0–1.0) on XZ cross sections **(a)**y = 3.09 **(b)** y = 5.67 **(c)** y = 8.25 **(d)** x = 10.83 (y varies from 0.52 to 10.83 from the subzone selected)

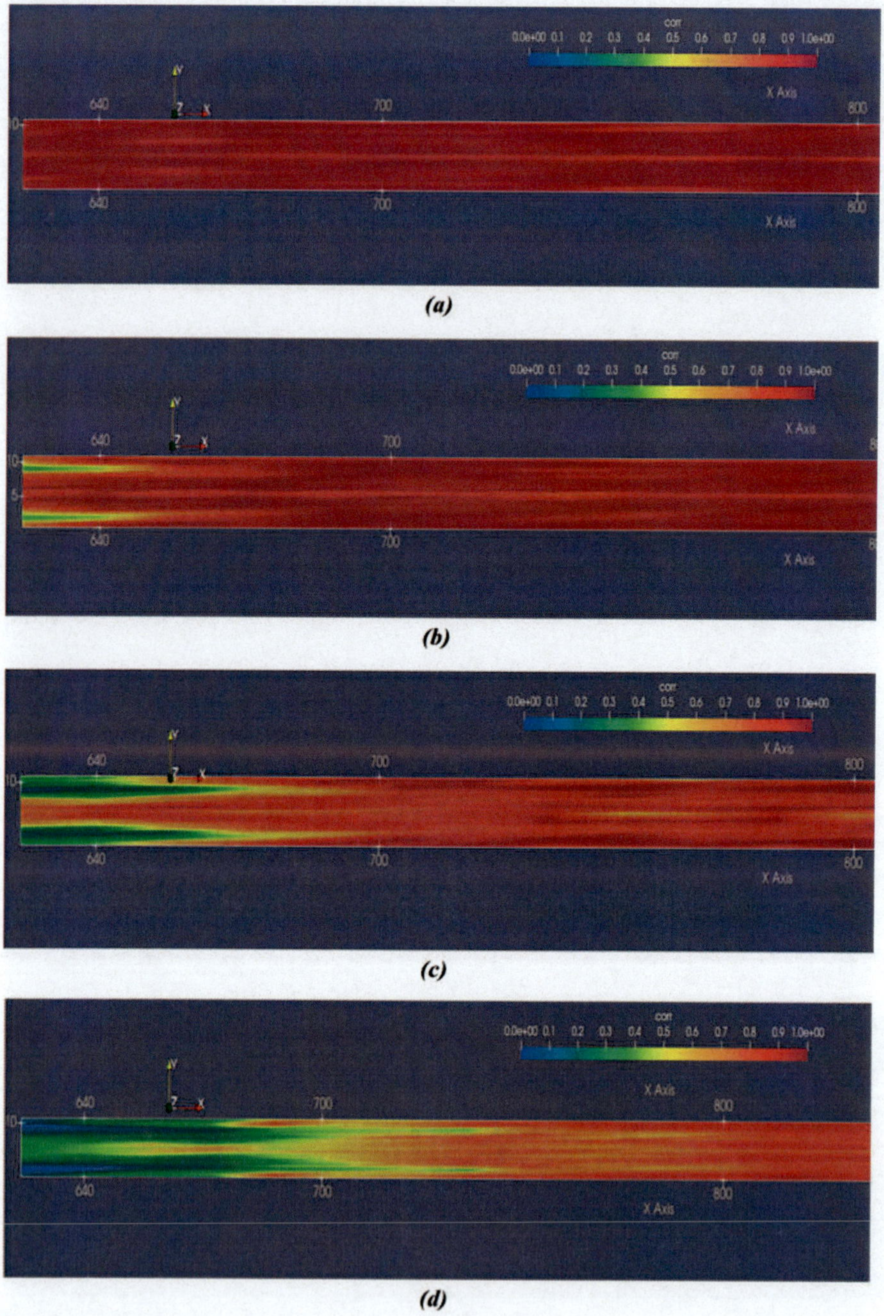

Fig. 10.7 Distribution of correlation coefficients of pressure and Liutex spectra (0.0–1.0) on XY cross sections **(a)**z = 4.6 **(b)** z = 6.6 **(c)** y = z = 8.6 **(d)** z = 10.6 (z varies from 0.6 to 10.8 from the subzone selected)

Table 10.2 Correlations Coefficients of power spectra

Points	R
1	0.88
2	0.86
3	0.96
4	0.95
5	0.94
6	0.92
7	0.89
8	0.90
9	0.94
10	0.90

Table 10.2 gives the corresponding correlations. By the qualitative analysis, the correlation coefficient is above 85%. Overall, a strong correlation between pressure and Liutex is observed at every point. At the first 2 points, the correlation is relatively low because the points are in the upper boundary layer, where the pressure fluctuation is less. On the other side, points 3–6 and 8–10 have a correlation of 90% due to the fact that they are located in the lower part of the boundary layer. Thus, by the power spectrum analysis of the vortex ring motion and Liutex, the results clearly show that the pressure fluctuation is closely related to the Liutex. The power spectra at 10 points of Liutex and pressure fluctuation is shown in Fig. 10.8.

10.5 Conclusions

1. A significantly higher correlation was observed in the middle and the lower boundary layer positions.
2. The low-frequency noises caused by pressure fluctuation are observed as mainly dominated by the vortex structure or the spectrum of Liutex.
3. The vortex rings are weakened because of the decrease in the pressure gradient between upstream and downstream of the shock.
4. The correlation analysis shows that the pressure fluctuation is closely related to the Liutex. The frequency of pressure fluctuation is the same as the Liutex frequencies.

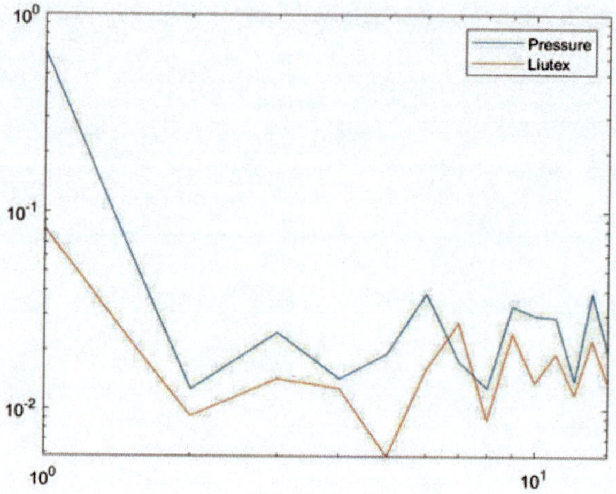

Point 1

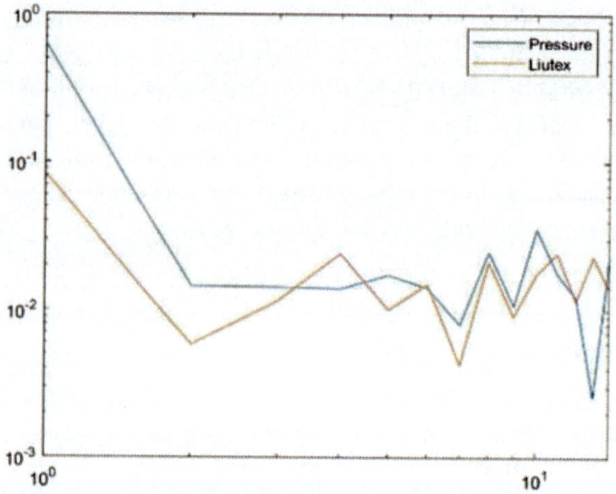

Point 2

Fig. 10.8 Power spectra of 10 points

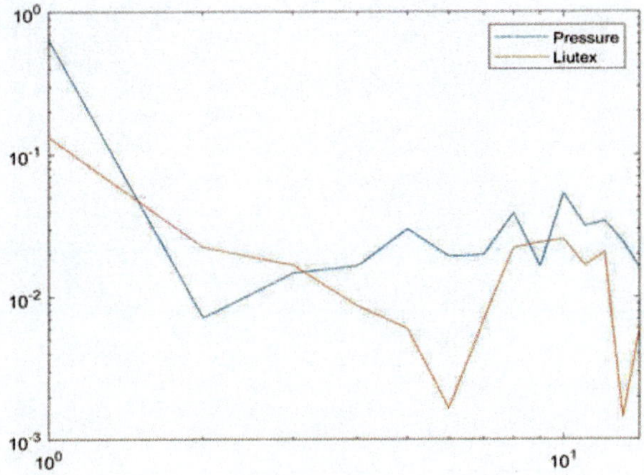

Point 3

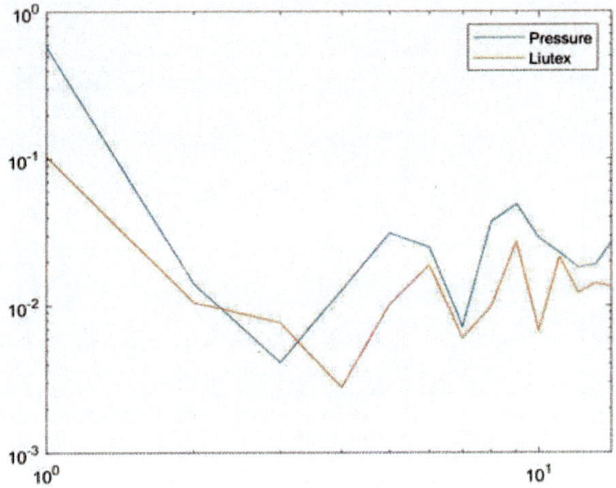

Point 4

Fig. 10.8 (continued)

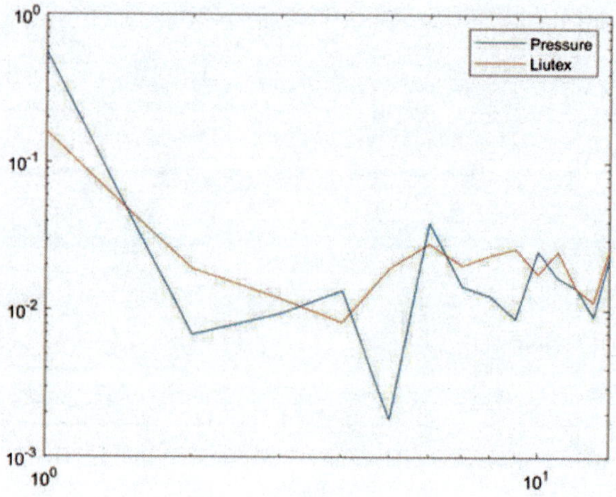

Point 5

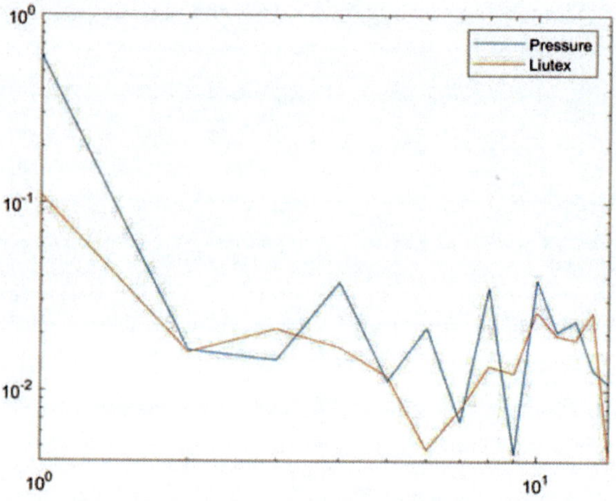

Point 6

Fig. 10.8 (continued)

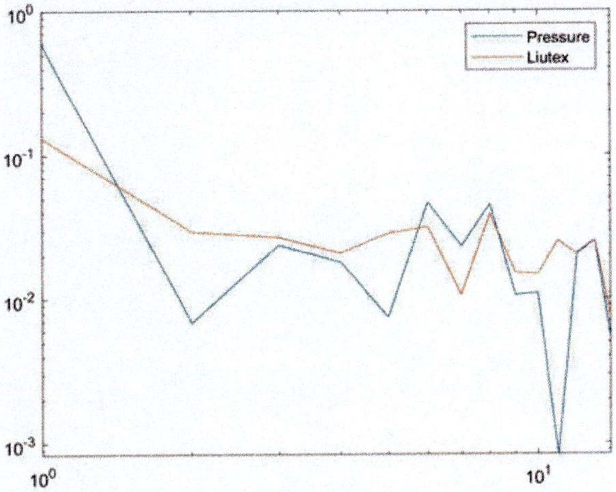

Point 7

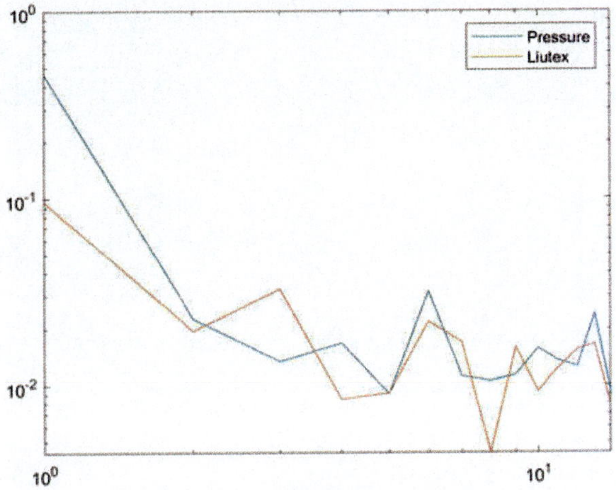

Point 8

Fig. 10.8 (continued)

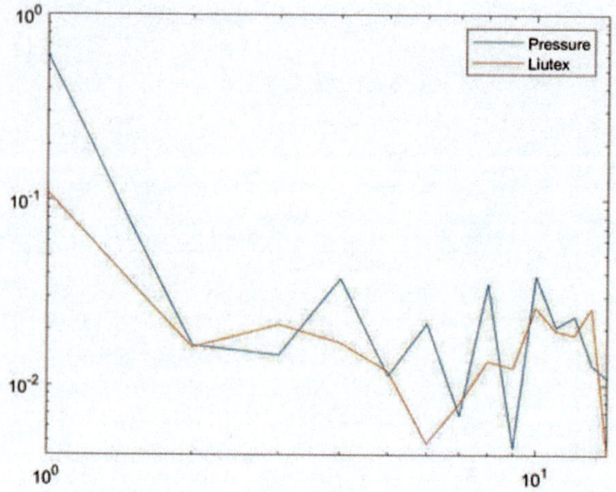

Point 9

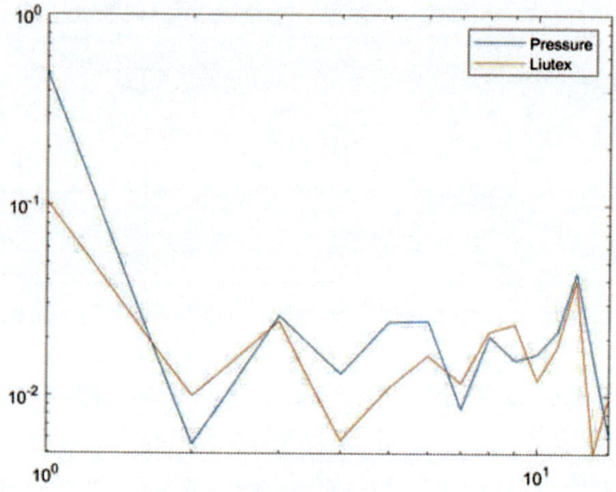

Point 10

Fig. 10.8 (continued)

References

1. D.S. Dolling, Fifty years of shock-wave/boundary-layer interaction research: what next? AIAA J. **39**, 1517–1531 (2001)
2. J. Freund, Noise sources in a low-Reynolds-number turbulent jet at Mach 0.9. J. Fluid Mech. **438**, 277–305 (2001)
3. Z. Hu, C. Morfey, N. Sandham, Sound radiation in turbulent channel flows. J. Fluid Mech. **475**, 269–302 (2003)
4. O. Marsden, C. Bogey, C. Bailly, Direct noise computation of the turbulent flow around a zero incidence airfoil. AIAA J. **46**(4), 874–883 (2008)
5. W.R. Wolf, J.L.F. Azevedo, S.K. Lele, Convective effects and the role of quadrupole sources for aerofoil aeroacoustics. J. Fluid Mech. **708**, 502–538 (2012)
6. R.D. Sandberg, N.D. Sandham, Direct numerical simulation of turbulent flow past a trailing edge and the associated noise generation. J. Fluid Mech. **596**, 353–385 (2008)
7. D.G. Crighton, Goals for computational aeroacoustics, in *Computational Aeroacoustics: Algorithms and Applications, Proceedings of the 1st IMACS Symposium on Computational Aeroacoustics* (Elsevier Science Publishers, Amsterdam, 1986)
8. J. Li, C. Liu, M. Choudhari, and C. L. Chang, Cross-validation of DNS and PSE results for instability-wave propagation in compressible boundary layers past curvilinear surfaces, in the 16th AIAA Computational Fluid Dynamics Conference, Orlando, 2003
9. C. Liu and L. Chen, Parallel DNS for vortex structure of late stages of flow transition. Comput. Fluids **45**(1), 129–137 (2011)
10. L. Jiang, C. L. Chang, M. Choudhari, and C. Liu, Cross-validation of DNS and PSE results for instability—Wave propagation in compressible boundary layers past curvilinear surfaces. AIAA Paper 2003–3555 (2003)
11. C. Liu, Y. Yan, and P. Lu, Physics of turbulence generation and sustenance in a boundary layer. Comput. Fluids **102**, 353–384 (2014)
12. Y. Yan, C. Chen, H. Fu, and C. Liu, DNS study on -vortex and vortex ring formation in flow transition at Mach number 0.5. J. Turbul. **15**(1), 1–21 (2014)
13. C. Liu, Y, Gao, X. Dong, Y. Wang, J. Liu, Y. Zhang, X. Cai, N. Gui, Third generation of vortex identification methods: Omega and Liutex/Rortex based systems. J. Hydrodyn. **31**(2), 774–781 (2019)
14. J. Liu, C. Liu, Modified normalized Rortex/vortex identification method. Phys. Fluids **31**, 061704 (2019)
15. X. Dong, S. Tian, C. Liu, Correlation analysis on volume vorticity and vortex in late boundary layer transition Xiangrui Dong, Shuling Tian, and Chaoqun Liu Citation. Phys. Fluids **30**, 014105 (2018)
16. T. Wang, Y, Gao C, Liu, Physical meaning of vorticity based on the Liutexshear decomposition and explicit formula for the Liutex vector [J]. arXiv:1812.10672 (also submitted to Physics of Fluids)
17. Y. Yan, L. Chen, Q. Li, C. Liu, Numerical study on microramp vortex generator for supersonic ramp flow control at Mach 2.5", shock waves. Int. J. Shock Waves Detonations Explos. (2016). https://doi.org/10.1007/s00193-016-0633-4
18. Y. Gao, C. Liu, Rortex and comparison with eigenvalue-based vortex identification criteria. Phys. Fluids **30**(8), 85107 (2018)

Chapter 11
Statistical Analysis for Liutex Growth in Flow Transition

Check for updates

Charles Nottage, Yifei Yu, and Chaoqun Liu

Abstract In computational fluid dynamics, many researchers and textbooks accepted that vorticity is vortex. However, this is a misunderstanding of the tensors derived from the Cauchy-Stokes decomposition of the velocity gradient tensor. It was believed that the symmetric tensor $\boldsymbol{A}$ and antisymmetric tensor $\boldsymbol{B}$ (vorticity tensor) represented stretching/compression and rotation, respectively. Decomposing the vorticity tensor yields $\boldsymbol{R}$ (rotation part) and $\boldsymbol{S}$ (antisymmetric shear deformation part). Liutex, on the other hand, represents rigid rotation and the Liutex magnitude represents twice the angular speed. We analyze three flow areas in boundary-layer transition: laminar, transitional, and turbulent. In laminar flow, there is no vortex structure. In transitional flow, the formation of hairpin vortex rings will begin. Finally, in turbulent flow, many vortex rings have formed. In this paper, a DNS simulation of boundary transition is conducted, then statistical analysis is performed on the recorded results for Liutex, shear, and vorticity. The resulting values for Liutex followed the proper growth trend, starting at zero in laminar flow and steadily increasing through the transitional and turbulent flows. On the other hand, the vorticity values were much greater and remained consistent with little change throughout the flow transition periods. The analysis also revealed that the shear component negatively relates with Liutex, i.e., as Liutex increases, shear decreases. Since shear substantially impacts the vorticity value where it can be misrepresented as rotation in laminar flow, vorticity, in general, should not be considered vortex.

11.1 Introduction

A vortex is recognized as the rotational motion of fluids. Many vortex identification methods have been developed within the last several decades to track the vortical structure in a fluid flow; however, we still lacked unambiguous and universally accepted vortex identification criteria. This obstacle caused a lot of confusion and misunderstandings in turbulence research [1]. These methods are characterized

C. Nottage (✉) · Y. Yu · C. Liu
Department of Mathematics, The University of Texas at Arlington, Arlington, TX 76019, USA
e-mail: charles.nottage@mavs.uta.edu

Y. Wang et al. (eds.), *Liutex and Third Generation of Vortex Identification*,
Springer Proceedings in Physics 288, https://doi.org/10.1007/978-981-19-8955-1_11

into three generations starting with vorticity-based methods as the first generation, eigenvalue-based such as Δ [2, 3], Q [4], λ_2 [5], λ_{ci} [6], and Ω [7, 8], as the second generation, and the recently developed Liutex methods as the third generation of vortex identification methods [9]. The Liutex method is a novel eigenvector-based method that is local, accurate, unique, and systematic [10].

In Computational fluid dynamics, many researchers and textbooks accept that vorticity is vortex. This is due to a misunderstanding from the Cauchy-Stokes decomposition of the velocity gradient tensor. It was understood that the symmetric tensor $\boldsymbol{A}$ represented stretching/compression, and the antisymmetric tensor $\boldsymbol{B}$ (vorticity tensor) represented rotation [11]. In this paper, we will show that antisymmetric tensor $\boldsymbol{B}$ (vorticity tensor) does not represent only rotation and investigate the behavior of shear, Liutex, and vorticity in the boundary layer from laminar flow to turbulent flow.

This manuscript is split into sections, where Sect. 11.2 is a review of vorticity and Liutex, Sect. 11.3 shows that vorticity is not strictly rotation, Sect. 11.4 shows the data structure for the DNS study, and Sect. 11.5 is the results of the investigation of the behavior of shear Liutex, and vorticity.

11.2 Review of Related Vortex Identification Methods

The vortex identification methods that we will analyze in this paper are vorticity and Liutex. We will review these methods in this section.

11.2.1 Vorticity

In 1858, Helmholtz introduced the concept of the vorticity tube/filament [12]. Since then, many researchers have believed that vortices consist of small vorticity tubes called vortex filaments, and the magnitude of vorticity gives the vortex strength.

11.2.1.1 Vorticity Vector

The vorticity vector is mathematically defined as the curl of the velocity. i.e.,

$$\text{vortivity} = \nabla \times \vec{\mathbf{v}} = \begin{vmatrix} \mathbf{i} & \mathbf{j} & \mathbf{k} \\ \frac{\partial}{\partial x} & \frac{\partial}{\partial y} & \frac{\partial}{\partial z} \\ u & v & w \end{vmatrix}$$

$$= \mathbf{i}\left(\frac{\partial w}{\partial y} - \frac{\partial v}{\partial z}\right) - \mathbf{j}\left(\frac{\partial w}{\partial x} - \frac{\partial u}{\partial z}\right) + \mathbf{k}\left(\frac{\partial v}{\partial x} - \frac{\partial u}{\partial y}\right) \tag{11.1}$$

The vorticity vector is also derived as follows [5]:

$$\vec{\omega} = \nabla \times \vec{\mathbf{v}} = \left(\frac{\partial}{\partial x}, \frac{\partial}{\partial y}, \frac{\partial}{\partial z}\right)^{T} \times (\mathrm{u}, \mathrm{v}, \mathrm{w})^{T}$$

$$= \left(\frac{\partial w}{\partial y} - \frac{\partial v}{\partial z}, \frac{\partial u}{\partial z} - \frac{\partial w}{\partial x}, \frac{\partial v}{\partial x} - \frac{\partial u}{\partial y}\right)^{T}. \tag{11.2}$$

11.2.1.2 Vorticity Magnitude

The vorticity magnitude is defined as

$$\| \vec{\omega} \| = \sqrt{\left(\frac{\partial w}{\partial y} - \frac{\partial v}{\partial z}\right)^2 + \left(\frac{\partial u}{\partial z} - \frac{\partial w}{\partial x}\right)^2 + \left(\frac{\partial v}{\partial x} - \frac{\partial u}{\partial y}\right)^2}. \tag{11.3}$$

11.2.1.3 Vorticity Tensor

The vorticity tensor is the antisymmetric tensor $\boldsymbol{B}$ from the traditional Cauchy-Stokes decomposition of the velocity gradient tensor $\nabla \vec{\mathbf{v}}$ [13]:

$$\boldsymbol{B} = \frac{1}{2}\left(\nabla\vec{\mathbf{v}} - \nabla\vec{\mathbf{v}}^{\mathbf{T}}\right) = \begin{bmatrix} 0 & \frac{1}{2}\left(\frac{\partial u}{\partial y} - \frac{\partial v}{\partial x}\right) & \frac{1}{2}\left(\frac{\partial u}{\partial z} - \frac{\partial w}{\partial x}\right) \\ \frac{1}{2}\left(\frac{\partial v}{\partial x} - \frac{\partial u}{\partial y}\right) & 0 & \frac{1}{2}\left(\frac{\partial v}{\partial z} - \frac{\partial w}{\partial y}\right) \\ \frac{1}{2}\left(\frac{\partial w}{\partial x} - \frac{\partial u}{\partial z}\right) & \frac{1}{2}\left(\frac{\partial w}{\partial y} - \frac{\partial v}{\partial z}\right) & 0 \end{bmatrix}. \tag{11.4}$$

11.2.2 Liutex

Liutex [9, 10] is a vector defined as $\vec{R} = R\vec{r}$. R represents the Liutex magnitude defined as twice the angular velocity, and $\vec{r}$ represents the directional unit vector of Liutex. According to Wang [14], $\vec{r}$ is the real eigenvector of the velocity gradient tensor, and the explicit formula of R is

$$R = \vec{\omega} \cdot \vec{r} - \sqrt{(\vec{\omega} \cdot \vec{r})^2 - 4\lambda_{ci}^2} \tag{11.5}$$

Liutex, as a vector, overcomes the drawbacks of the scalar methods, e.g., the threshold requirement when creating and analyzing graphics.

11.3 Vorticity Versus Rotation

In this section, we will decompose vorticity in the principal coordinate to show that vorticity is not strictly rotational.

11.3.1 Vorticity Tensor in the Principal Coordinate

The Vorticity tensor in the Principal Coordinate has the following form:

$$\boldsymbol{B}_{\boldsymbol{PC}} = \begin{bmatrix} 0 & \frac{1}{2}\left(\frac{\partial U}{\partial Y} - \frac{\partial V}{\partial X}\right) & \frac{1}{2}\left(\frac{\partial U}{\partial Z} - \frac{\partial W}{\partial X}\right) \\ \frac{1}{2}\left(\frac{\partial V}{\partial X} - \frac{\partial U}{\partial Y}\right) & 0 & \frac{1}{2}\left(\frac{\partial V}{\partial Z} - \frac{\partial W}{\partial Y}\right) \\ \frac{1}{2}\left(\frac{\partial W}{\partial X} - \frac{\partial U}{\partial Z}\right) & \frac{1}{2}\left(\frac{\partial W}{\partial Y} - \frac{\partial V}{\partial Z}\right) & 0 \end{bmatrix}$$

$$\boldsymbol{B}_{\boldsymbol{PC}} = \begin{bmatrix} 0 & -\frac{R+\varepsilon}{2} & -\frac{\xi}{2} \\ \frac{R+\varepsilon}{2} & 0 & -\frac{\eta}{2} \\ \frac{\xi}{2} & \frac{\eta}{2} & 0 \end{bmatrix} \tag{11.6}$$

and can be decomposed further into this form:

$$\boldsymbol{B}_{\boldsymbol{PC}} = \begin{bmatrix} 0 & -\frac{R}{2} & 0 \\ \frac{R}{2} & 0 & 0 \\ 0 & 0 & 0 \end{bmatrix} + \begin{bmatrix} 0 & -\frac{\varepsilon}{2} & -\frac{\xi}{2} \\ \frac{\varepsilon}{2} & 0 & -\frac{\eta}{2} \\ \frac{\xi}{2} & \frac{\eta}{2} & 0 \end{bmatrix} = \boldsymbol{R} + \boldsymbol{AS}. \tag{11.7}$$

where $\boldsymbol{R}$ &$\boldsymbol{AS}$ represent the rotation and antisymmetric shear deformation part, respectively. This implies that the vorticity tensor is not strictly rotation [15].

11.3.2 Vorticity Vector in the Principal Coordinate

Using Eq. 11.2 from Sect. 11.2.1.1 and applying the Principal Coordinate yields:

$$\vec{\omega} = \nabla \times \vec{\mathbf{v}_\theta} = \left(\frac{\partial}{\partial X}, \frac{\partial}{\partial Y}, \frac{\partial}{\partial Z}\right)^{\boldsymbol{T}} \times (U, V, W)^{\boldsymbol{T}}$$

$$= \left(\frac{\partial W}{\partial Y} - \frac{\partial V}{\partial Z}, \frac{\partial U}{\partial Z} - \frac{\partial W}{\partial X}, \frac{\partial V}{\partial X} - \frac{\partial U}{\partial Y}\right)^{\boldsymbol{T}}$$

$$= \left(\eta, -\xi, \frac{R}{2} + \varepsilon - \left(-\frac{R}{2}\right)\right)^T = (\eta, -\xi, R + \varepsilon)^T$$

$$= (\eta, -\xi, \varepsilon)^T + (0, 0, R)^T = \vec{S} + \vec{R} \quad (11.8)$$

where $\vec{R}$ is a rotational vector and $\vec{S}$ is a non-rotational shear vector. This implies that $\vec{\omega}$ contains shearing and rotation components. Therefore, the vorticity vector is not strictly rotational [5, 11, 16].

11.3.3 Vorticity Magnitude in the Principal Coordinate

The vorticity magnitude in the Principal Coordinate is

$$\| \vec{\omega} \| = \sqrt{(\eta)^2 + (\xi)^2 + (R + \varepsilon)^2}. \quad (11.9)$$

The vorticity magnitude contains rotation R and shearing components η, ξ, ε; therefore, the vorticity magnitude does not only represent rotational strength [11].

11.4 Data Structure

The computational domain has the grid number 1920 ×128 × 241, representing the number of grids in streamwise (x), spanwise (y), and wall-normal (z) directions. In normal direction, these grids are stretched, while in streamwise and spanwise directions, they are uniform. The length of the first grid interval in the normal direction at the entrance is 0.43 in wall units (Z+ = 0.43) (see Fig. 11.1).

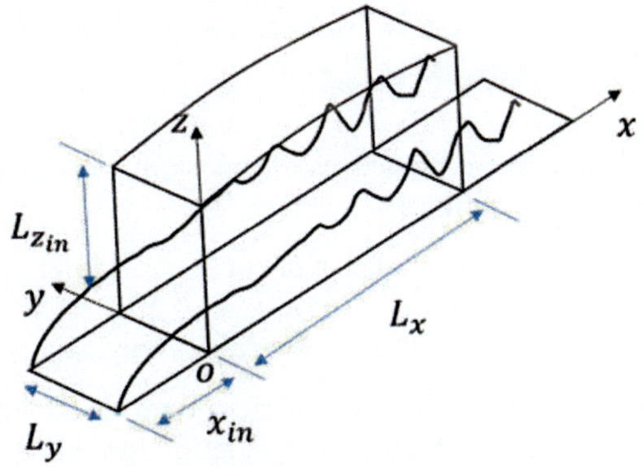

Fig. 11.1 Computation domain

Table 11.1 DNS parameters

M_∞	Re	x_{in}	Lx	Ly	Lz_{in}	T_w	T_∞
0.5	1000	300.79 δ_{in}	798.03 δ_{in}	22 δ_{in}	40 δ_{in}	273.15 K	273.15 K

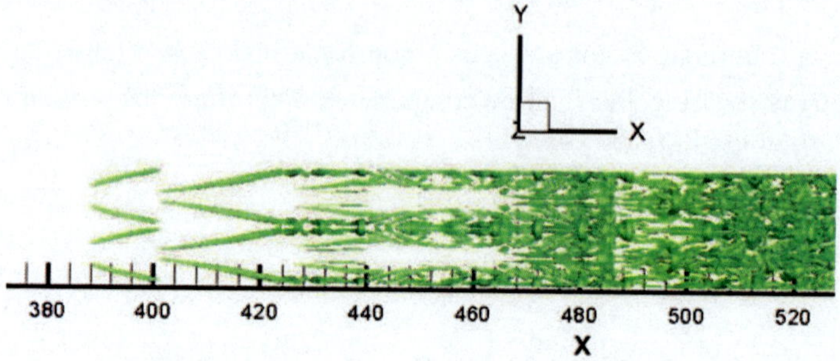

Fig. 11.2 Vortex structures by modified Liutex-Omega with $\tilde{\Omega}_L = 0.52$ at t = 13.00 T

The flow parameters are listed in Table 11.1. Here, M_∞ is Mach number, Re is Reynolds number and, T_w *and* T_∞ are wall and free stream temperature, respectively. Likewise, x_{in} represents the distance between the leading edge and the inlet of the flat plate. Lx, Ly and Lz_{in} are the lengths of computational domain in x, y, and z directions. δ_{in} is the inflow displacement thickness [1, 17].

Figure 11.2 shows the formation of the vortex structures from the Y direction view in laminar, transitional, and turbulent flow. In laminar flow, little to no vortex activity is detected. In transitional flow, the formation of hairpin vortex rings begins, and this is where weak to strong vortex activity starts to be detected. Then in turbulent flow, many vortex rings have formed. This area is more chaotic and complicated. This means the strength of the vortex should increase from laminar flow to turbulent flow. According to Dong et al., the spanwise Y direction is the most prominent since it contributes the most to the value of the magnitudes [8].

11.5 Results

There are two objectives of this study; by using the DNS simulation of boundary layer transition we:

1. Investigate the behavior of shear, Liutex, and Vorticity from laminar flow to turbulent flow.
2. Analyze the effect of shear on vorticity.

Statistical analysis is performed over the whole grid domain and across time domain T. The Statistical integration formula is

$$\sum_i \sum_j \sum_k \tau_{ijkt} vol_{ijk} = \iota_t, \tag{11.10}$$

where τ_{ijkt} is a point in the grid, $i = 1$ to 1920, $j = 1$ to 128, $k = 1$ to 241 and t represents the step-in time. vol_{ijk} is the volume of the space around the point τ_{ijkt}. ι_t is the integration output value at time step t.

The results of the Direct Numerical Simulation are recorded, Statistical Analysis is performed over the whole grid domain, and the data is plotted across time domain **T** for:

- $\boldsymbol{l_{mag}}$ = Liutex magnitude component
- $\boldsymbol{\omega_{mag}}$ = Vorticity magnitude component
- $\boldsymbol{s_{mag}}$ = Shear magnitude component
- $\boldsymbol{l_y}$ = Liutex component in the y direction
- $\boldsymbol{\omega_y}$ = Vorticity component in the y direction
- $\boldsymbol{s_y}$ = Shear component in the y direction

The difference in the values of vorticity and Liutex is significantly high, so the relative values are used to compare the change in the values over period **T**.

Figures 11.3 and 11.4 show the behavior of vorticity, shear, and Liutex from the y direction. It is observed that the relative $\boldsymbol{l_y}$ values increased significantly over period **T**, which is the T-S wave period. While $\boldsymbol{\omega_y}$ showed no significant change throughout period **T**. The $\boldsymbol{s_y}$ values were observed to be decreasing. This change is negative in nature and coincides with the increase in $\boldsymbol{l_y}$ and the behavior of $\boldsymbol{\omega_y}$. The period **T** travels from laminar flow to turbulent flow. There should be minuscule rotation or vortex activity in laminar flow and an increasing trend of vortex activity as we move into transitional flow, where hairpin vortex rings are formed, and on to turbulent flow. This behavior only coincides with the behavior of the $\boldsymbol{l_y}$ relative values.

It can be observed from Figs. 11.5 and 11.6 that the relative change in $\boldsymbol{l_{mag}}$ across time is much greater than $\boldsymbol{\omega_{mag}}$. As we moved from laminar flow to turbulent flow in time, the values of $\boldsymbol{l_{mag}}$ continually increased, showing that $\boldsymbol{l_{mag}}$ picked up the formation of vortex rings, whereas $\boldsymbol{\omega_{mag}}$ barely changed.

11.6 Conclusion

The effect of shear on vorticity can be substantial, leading to a misrepresentation of vortex indication in the laminar flow where there is basically no rotation. The Y direction graphs show that vorticity had little to no change in value over time, while Liutex increased as time progressed. The shear was observed to be decreasing as Liutex increased. This shows that Liutex has a negative relation with Shear deformation. The magnitude graphs show that the increase in Liutex as time progressed is more significant than the increase in vorticity and shear. Therefore, vorticity should not be considered as Vortex.

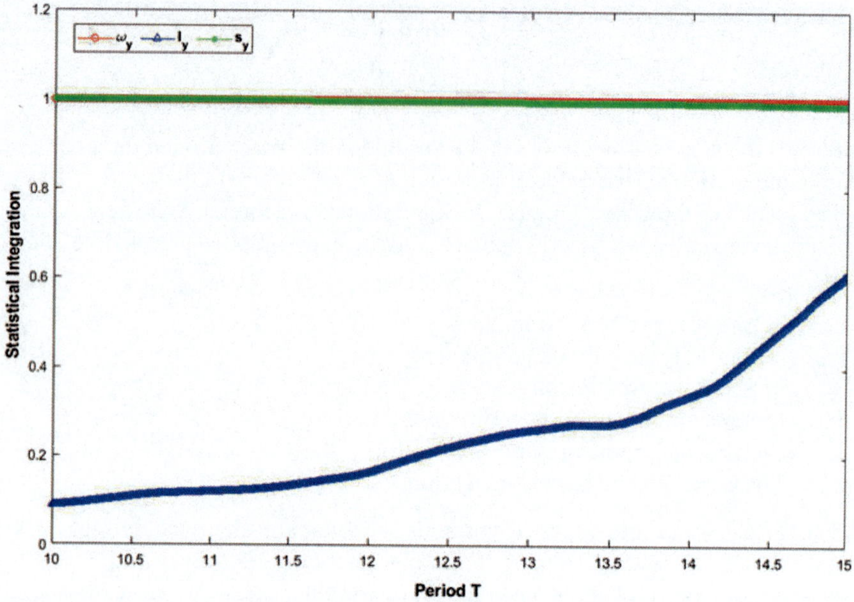

Fig. 11.3 Relative integration values for ω_y, l_y & s_y plotted over time domain **T**

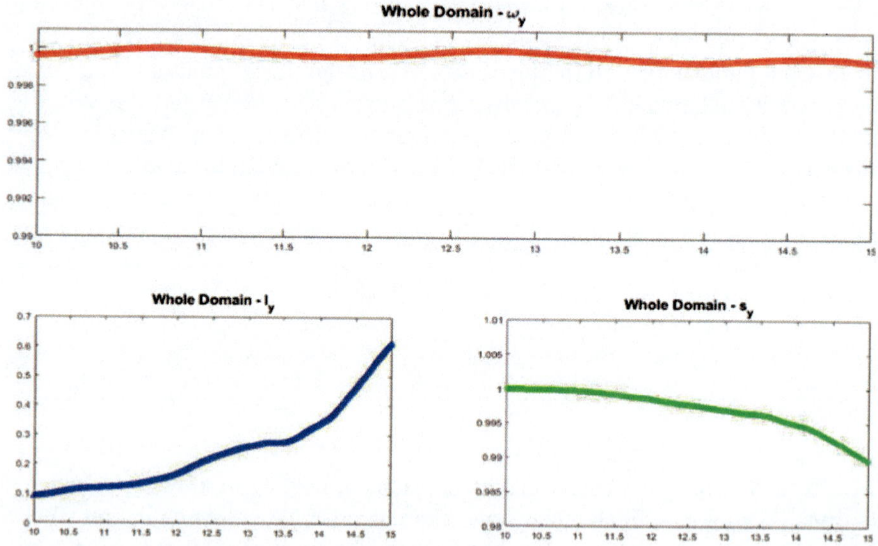

Fig. 11.4 Relative integration values for ω_y, l_y & s_y plotted over time domain **T - individual graphs**

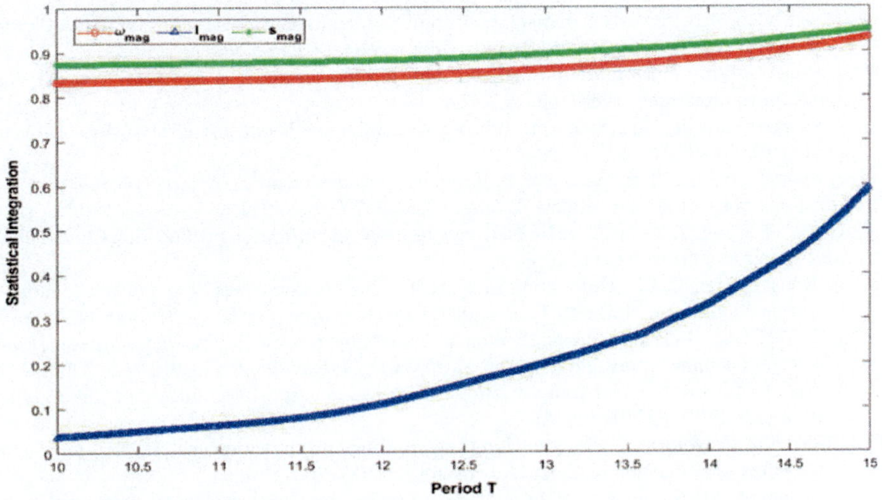

Fig. 11.5 Relative integration values for ω_{mag}, l_{mag} & s_{mag} plotted over time domain **T**

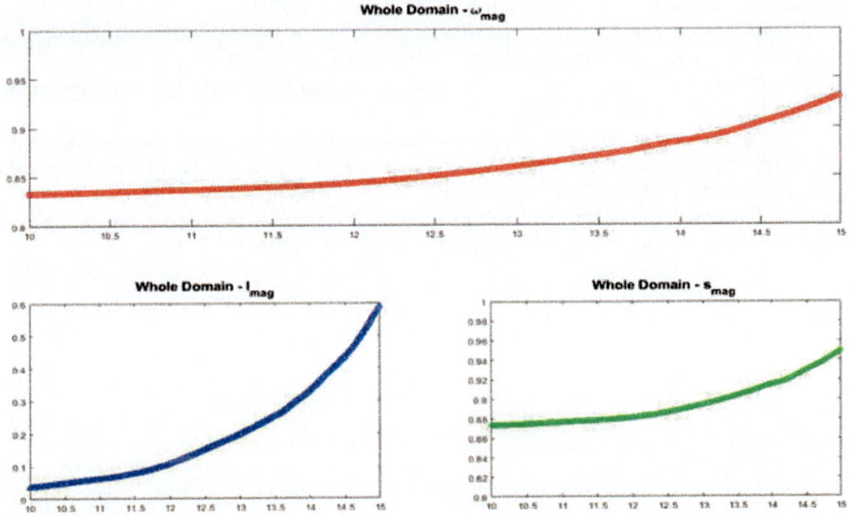

Fig. 11.6 Relative integration values for ω_{mag}, l_{mag} & s_{mag} plotted over time domain **T** - **individual graphs**

References

1. C. Liu, Y. Yun, P. Lu, Physics of turbulence generation and sustenance in a boundary layer. Comput. Fluids **102**, 353–384 (2014)
2. A. Perry, M. Chong, A Description of eddying motions and flow patterns using critical-point concepts. Annu. Rev. Fluid Mech. **19**(1), 125–155 (1987)

3. M.S. Chong, A.E. Perry, A general classification of three-dimensional flow fields. Phys. Fluids A: Fluid Dyn. **2**(5), 765–777 (1990). https://doi.org/10.1063/1.857730
4. J. Hunt, A. Wray, P. Moin, Eddies, stream, and convergence zones in turbulent flows (Center for Turbulent Research, 1988), pp. 193–208
5. J. Jeong, F. Hussain, On the identification of a vortex. J. Fluid Mech. **285**, 69–94 (1995). https://doi.org/10.1017/s0022112095000462
6. J. Zhou, R. Adrian, S. Balachandar, T. Kendall, Mechanisms for generating coherent packets of hairpin vortices in channel flow. J. Fluid Mech. **387**, 353–396 (1999)
7. C. Liu, Y. Wang, Y. Yang, Z. Duan, New omega vortex identification method. Sci. China Phys. Mech. Astron. **59**(8), 684711 (2016)
8. X. Dong, S. Tian, C. Liu, Correlation analysis on volume vorticity and vortex in late boundary layer transition. Phys. Fluids **30**(1), 014105 (2018). https://doi.org/10.1063/1.5009115
9. C. Liu, Y. Gao, S. Tian, X. Dong, Rortex-a new vortex vector definition and vorticity tensor and vector decompositions. Phys. Fluids **30**(3) (2018). https://doi.org/10.1063/1.5023001
10. Y. Gao, C. Liu, Rortex and comparison with eigenvalue-based vortex identification criteria. Phys. Fluids **30**(8), 085107 (2018)
11. P. Shrestha, C. Nottage, Y. Yu, Stretching and shearing contamination analysis for Liutex and other vortex identification methods. Adv. Aerodyn. **3**(8) (2021)
12. H. Helmholtz, On the integrals of the hydrodynamic equations corresponding to vortex motions. Journal für die reine und angewandte Mathematik (Crelles Journal) **55**, 22–25 (1858)
13. C. Liu, Y.-S. Gao, X.-R. Dong, Y.-Q. Wang, J.-M. Liu, Y.-N. Zhang, N. Gui, et al. Third generation of vortex identification methods: omega and Liutex/Rortex based systems. J. Hydrodyn. **31**, 205–223 (2019). https://doi.org/10.1007/s42241-019-0022-4
14. J. Wang, Y. Gao, C. Liu, Explicit formula for the Liutex vector and physical meaning of Vorticity based on the Liutex-Shear decomposition. J. Hydrodyn. **31**(3), 464–474 (2019)
15. Y. Yu, P. Shrestha, C. Nottage, C. Liu, Principal coordinates and principal velocity gradient tensor decomposition. J. Hydrodyn. **32**, 441–453 (2020)
16. C. Nottage, Y. Yu, P. Shrestha, C. Liu, Dimensional and theoretical analysis of second-generation vortex identification methods, in *Liutex and Third Generation of Vortex Definition and Identification.* ed. by C. Liu, Y. Wang (Springer, Cham, 2021), pp. 57–70
17. Y. Yan, C. Chen, F. Huankun, C. Liu, DNS study on Λ-vortex and vortex ring formation in the flow transition at Mach number 0.5. J. Turbul. **15**(1), 1–21 (2014)

Part II
Liutex and Third Generation of Vortex Identification Methods in Engineering Applications

Chapter 12
Three-Dimensional Vortex Structure Identification of Fluid Coupling and Analysis of Spatial-Temporal Evolution Mechanism

Bosen Chai, Dong Yan, Jin Zhang, Wenjie Zuo, and Guangyi Wang

Abstract The interior of fluid coupling is full of turbulent and multi-scale vortex flow. The generation, development and interaction of the multi-scale vortex structure dominate the overall internal flow. Accurate identification of unsteady multi-scale vortex structure characteristics is extremely important for revealing the law of flow and the mechanism of energy conversion and loss mechanism. The flow field of fluid coupling under braking condition is numerically simulated based on the advanced turbulence model. The temporal and spatial evolution characteristics of the three-dimensional vortex structure inside the pump wheel and turbine are extracted based on different vortex recognition methods, and the results are compared with the flow field visualization test. The accuracy and applicability of the extraction results for different vortex recognition methods are analyzed. The results show that: the threshold selection of Q method extraction vortex structure is blind. It is difficult to capture both strong and weak vortices at the same time. Ω method can accurately capture the weak vortex structure within a large threshold range. Ω-Liutex method is not sensitive to threshold selection, and it is the most effective in extracting vortex structure. From the perspective of the temporal and spatial evolution of the three-dimensional vortex structure, Ω-Liutex method can reveal the mechanism of energy conversion and loss best.

B. Chai (✉) · D. Yan · J. Zhang · W. Zuo · G. Wang
School of Mechanical and Aerospace Engineering, Jilin University, Changchun, Jilin 130025, China
e-mail: chaibs2012@jlu.edu.cn

B. Chai · W. Zuo
State Key Laboratory of Automotive Simulation and Control, Jilin University, Changchun, Jilin 130025, China

B. Chai
Sinotest Equipment Co., Ltd., Changchun, Jilin 130103, China

Y. Wang et al. (eds.), *Liutex and Third Generation of Vortex Identification*, Springer Proceedings in Physics 288, https://doi.org/10.1007/978-981-19-8955-1_12

12.1 Introduction

With the increasing demand for energy in society, energy conservation and energy efficiency improvement have become key areas of national concern driven by the national development strategy of "carbon peaking and carbon neutrality". As a hydraulic transmission element with excellent performance of speed regulation, energy saving and emission reduction, the fluid coupling is widely used in power plants, petroleum, chemical industry, metallurgy, mining and other high energy-consuming industries. Especially in the speed regulation drive of heavy-duty and high-power equipment, the fluid coupling not only has excellent transmission performance, but also has significant energy-saving benefits [1]. The structure of the fluid coupling includes: pump wheel, turbine, and pump wheel housing, as shown in Fig. 12.1a. The pump impeller and turbine form a closed working chamber. The pump wheel rotates under the drive of the prime mover. Under the dual action of centrifugal force and blade push, The flowing medium from the pump impinges the turbine blades. The flow medium from the pump impinges the turbine blades to drive the turbine to rotate, and the energy transfer and conversion is realized in this process, as shown in Fig. 12.1b.

The working performance of the fluid coupling is determined by its internal flow characteristics. When it works, it is full of turbulent vortex flow at different scales. The generation, development and interaction of unsteady multiscale vortices govern the overall flow inside the fluid coupling, and affect its external performance. At present, there are still many key scientific problems in the accurate prediction method of the interaction relationship between the vortex flow in the fluid coupling and the flow performance. Accurately identifying the multi-scale vortex structure features inside the fluid coupling has important scientific significance for analyzing the evolution law of the internal flow field.

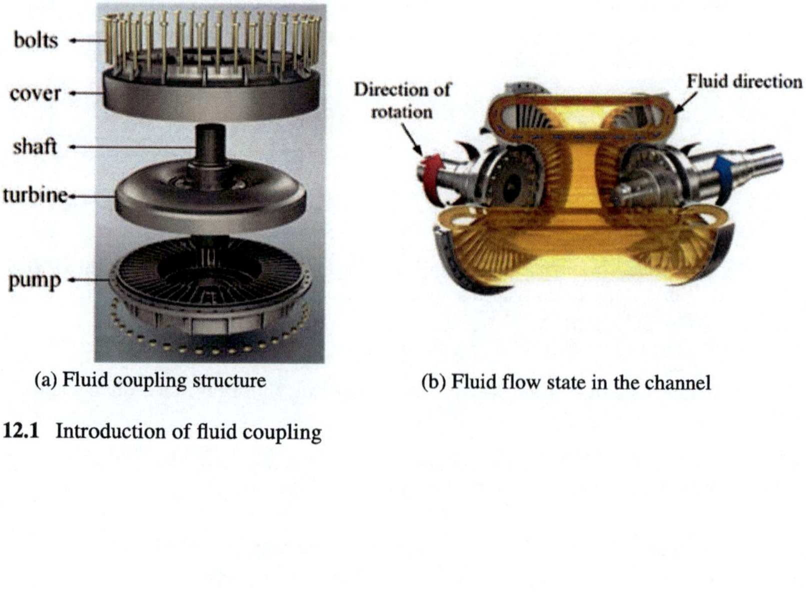

(a) Fluid coupling structure (b) Fluid flow state in the channel

Fig. 12.1 Introduction of fluid coupling

In recent years, scholars at home and abroad have carried out a lot of work in the field of vortex identification methods. They proposed a series of vortex structure identification methods and judgment criteria [2, 3]. Among them, the Q criterion is widely used [4]. However, the accuracy of the vortex structure extracted by the Q criterion depends heavily on the threshold. The physical expression of the vortex structure of this method has ambiguous meaning. In order to solve this problem, since 2014, the team of Professor Chaoqun Liu from the University of Texas at Arlington has carried out a series of research. Subsequently, a dimensionless and regularized Ω criterion and Liutex vortex vector method are proposed to accurately identify the vortex structure in the boundary layer transition flow [5, 6]. Vishwa [7] successfully applied the Liutex vortex vector method to the identification of complex vortex structures in supersonic boundary layer separation flows, which promoted the study of vortex motion and low-frequency shock wave oscillations. Charkrit [8] used Liutex method to accurately identify the hairpin vortex structure in natural flow, and the effect is more obvious. Wang [9] comprehensively applied vortex identification methods such as the Ω criterion and the Q criterion. It was found that the Burgers vortex and Sullivan vortex Ω criterion had better identification results. Guinan [10] analyzed the results of vortex motion in a visualized rotating jet based on direct numerical simulation and different vortex identification criteria. It was found that the Ω criterion and the Liutex method can identify additional small-scale vortices, while other criteria are difficult to achieve. Sun [11] used Liutex method to explore the flow characteristics in the draft tube of the turbine.

Although the third-generation vortex identification method with the Ω criterion and Liutex vortex vector method as the core has been applied in boundary layer flow and some fluid machinery fields. This is the first scientific application of the identification and extraction of unsteady multi-scale vortex structure features in the fluid coupling. In this paper, different vortex identification methods are used to extract the multi-scale vortex structure inside the hydraulic coupling based on the interactive research method of high-precision numerical simulation and flow visualization test measurement. From the perspective of the analysis of the spatiotemporal evolution mechanism of the flow field and the accurate identification of the vortex structure characteristics, the accuracy and reliability of different vortex identification aspects are evaluated. This paper highlights the practical effect of the newly developed Ω method and Ω-liutex method on the accurate identification of the internal vortex structure characteristics of the hydraulic coupling.

12.2 Simulation Model

12.2.1 Calculation Model

The effective circulation circle diameter of the square cavity fluid coupling model is 230 mm, and the blade thickness is 4 mm. Among them, the number of impeller

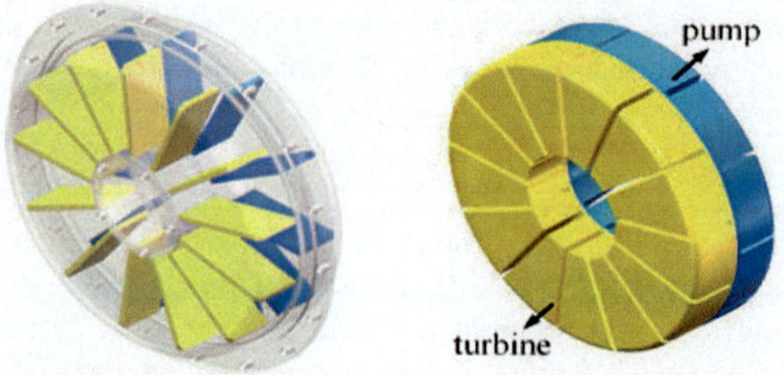

(a) 3D model (b) flow channel model

Fig. 12.2 Simulation model

(a) Overall grid (b) Turbine grid (c) Pump grid

Fig. 12.3 Grid model

blades is 13, and the number of turbine blades is 12, as shown in Fig. 12.2a. The runner model is extracted as the computational domain, as shown in Fig. 12.2b. Based on ICEM software, the hexahedral structured mesh is used to mesh the flow channel model. The number of meshes in the pump wheel flow channel is 579956, and the number of turbine flow channel meshes is 545688, with a total of 1,125,644 meshes, as shown in Fig. 12.3.

12.2.2 Turbulence Model

Based on the Large Eddy Simulation (LES) turbulence model, the improved algebraic wall model (WMLES S-Ω) is used for numerical simulation. Filtering the N-S equation, the governing equation is obtained:

$$\frac{\partial}{\partial t}(\rho\overline{u_i}) + \frac{\partial}{\partial x_j}(\rho\overline{u_i u_j}) = -\frac{\partial\overline{p_t}}{\partial x_i} + \frac{\partial}{\partial x_j}\left(\mu\frac{\partial\overline{u_i}}{\partial x_j}\right) + \frac{\partial\tau_{ij}}{\partial x_j} \tag{12.1}$$

$$\frac{\partial \overline{u_i}}{\partial x_i} = 0 \tag{12.2}$$

$$\tau_{ij} = \rho\overline{u_i u_j} - \rho\overline{u_i}\,\overline{u_j} \tag{12.3}$$

where the sub-grid stress term τ_{ij} represents the momentum conversion between the solvable large-scale vortices and the unsolvable small-scale vortices. The WMLES S-Ω uses the large eddy simulation method outside the near-wall region, and the turbulent eddy viscosity is:

$$\mu_t = \min\left[(\kappa d_\omega)^2, \left(C_{\text{Smag}}\Delta\right)^2\right] \cdot |\text{S} - \Omega| \cdot \left\{1 - \text{e}^{-(y^+/25)^3}\right\} \tag{12.4}$$

$$\Delta = \min(\max(C_\omega \cdot d_\omega, C_\omega \cdot h_{\max}, h_{\omega n}), h_{\max}) \tag{12.5}$$

where d_ω represents the distance between the calculated point and the wall; S represents the strain rate, the Ω represents the vortex intensity, constant $\kappa = 0.41$, $C_{\text{Smag}} = 0.2$, $C_\omega = 0.15$. Filter size Δ is selected according to flow field conditions, $h_{\max}$ represents the maximum side length of the hexahedral element, $h_{\omega n}$ represents the grid spacing in the normal direction of the wall.

12.2.3 Simulation Parameters

Table 12.1 shows the basic parameters of fluid coupling under braking condition. Pure water is used as the flow medium of this model. Due to the slow speed, the influence of temperature rise and leakage is ignored. The flow field structure distribution in the mainstream area of the flow channel is complex. The large eddy simulation turbulence model is selected for simulation analysis [12].

Table 12.1 Parameter setting

Parameter type	Parameters
Time step	0.01 s
Iteration steps	300
Convergent residual	10^{-4}
Medium viscosity	0.001003 Pa·s
Algorithm selection	SIMPLE algorithm
Boundary conditions	Smooth wall without slippage

12.3 Eddy Identification Criterion

12.3.1 *Q Criterion*

Hunt [13] proposed to use the second Galilean invariant of the velocity gradient tensor to represent the vortex structure. The velocity gradient tensor is decomposed into a symmetric part and an anti-symmetric part. Through the Cauchy-Stokes decomposition, the Q expression can be written as:

$$Q = \frac{1}{2}\left(\|B\|_{\mathrm{F}}^{2} - \|A\|_{\mathrm{F}}^{2}\right) \tag{12.6}$$

where $\| \|_{\mathrm{F}}^{2}$ is the Frobenius norm of the matrix. The physical significance of Q lies in the fact that not only the vorticity is required in the vortex structure, but also the anti-symmetric tensor B needs to be able to overcome the deformation effect represented by the symmetric tensor A. In practical applications, the threshold $Q_{threshold}$ must be used to determine the vortex area, and $Q > Q_{threshold}$ is defined as the vortex.

12.3.2 *Omega Vortex Identification Method*

Since the vortex is an area where the vorticity exceeds the deformation, and the vorticity cannot directly represent the fluid rotation. Therefore, the ratio of vorticity to deformation is considered to facilitate the identification of vortices.Vorticity $\boldsymbol{\omega}$ cannot represent the rotational movement of the fluid, The $\boldsymbol{\omega}$ is decomposed into a rotating part and a non-rotating part:

$$\boldsymbol{\omega} = \boldsymbol{R} + \boldsymbol{S} \tag{12.7}$$

where $\boldsymbol{R}$ is the vorticity of the rotating part, $\boldsymbol{S}$ is the non-rotating part vorticity. Generally speaking, the directions of $\boldsymbol{R}$ and $\boldsymbol{\omega}$ are different. The parameter Ω is quoted to represent the ratio of the vorticity of the rotating part to the total vorticity. The Ω can be presented as:

$$\Omega = \frac{\|B\|_F^2}{\|A\|_F^2 + \|B\|_F^2} \tag{12.8}$$

In order to prevent the problem of division by zero, a small positive number is added to the denominator term of formula (12.8). The Ω expression becomes

$$\Omega = \frac{\|B\|_F^2}{\|A\|_F^2 + \|B\|_F^2 + \varepsilon} \tag{12.9}$$

where the value range of Ω is reduced to $0 \leq \Omega \leq 1$, which represents the concentration of vorticity. When the $\Omega = 1$, it means that the fluid does a rigid body rotation.

12.3.3 Omega-Liutex Vortex Identification Method

Proposal based on Ω method and Liutex vector, scholars [14] combined the two to propose the Ω-Liutex vortex recognition method. Denoted as Ω_{L} method. The Ω_{L} is defined as:

$$\boldsymbol{\Omega}_{\mathrm{L}} = \frac{\beta^2}{\alpha^2 + \beta^2 + \varepsilon} \tag{12.10}$$

Introduce small positive number ε into the denominator of Ω_{L} to eliminate non-physical noise, ε is defined by experience as

$$\varepsilon = b \times \left(\beta^2 - \alpha^2\right)_{\max} \tag{12.11}$$

where b is a small positive number, e.g., 0.001. For each case, b is a fixed parameter and the term $\left(\beta^2 - \alpha^2\right)_{\max}$ can be easily obtained at each time step and the manual adjustment of ε is avoided. The α and β are the two-dimensional principal strain rate and vorticity on a plane perpendicular to the local axis of rotation, respectively.

12.4 Vortex Recognition Image and Analysis

12.4.1 Analysis of Threshold Selection Range

Different vortex identification criteria are used for the threshold selection range of the three-dimensional vortex structure extraction inside the fluid coupling channel. The threshold ranges of the Ω method and the Ω-Liutex method are normalized. In order to facilitate the comparison of the threshold range, the threshold range extracted by the Q criterion is proportionally reduced to 0~1, which are listed in Table 12.2.

Table 12.2 Threshold selection range under different vortex identification criteria

Vortex identification criterion	Minimum threshold	Maximum threshold	Optimal threshold
Q Criterion	0	1	0.18
Ω	0	0.992187	0.56
Ω-Liutex	0	0.997305	0.52

The Q criterion is selected to extract the three-dimensional vortex structure of the flow channel. When the threshold value is higher than 0.3, the three-dimensional vortex structure inside the flow channel is obviously broken. And there are weak vortex structures in the flow channel. The three-dimensional vortex structure is complete, but the extracted vortex structures are mixed when the threshold value is too small. The $Q = 0.18$ is finally determined as the optimal threshold after repeated extraction.

The threshold range of the Ω method is selected from 0 to 0.992187. Although the three-dimensional vortex extracted by the recommended threshold [15] of 0.52 is complete. The vortex structure near the wall has the phenomenon of mutual adhesion. When the threshold is gradually increased from 0.52 to 0.6, the overall variation of the three-dimensional vortex in the flow channel is small, and the adhesion between vortex structures of different scales is reduced. After many attempts, the optimal threshold of this method is determined to be 0.56. The threshold range of the Ω-Liutex method is selected from 0 to 0.997305, which is similar to the threshold range of the Ω method. And when the threshold is reduced to between 0.52 and 0.6, the extracted vortex structure changes little. At this time, the three-dimensional vortex structure in the flow channel can not only reflect the overall motion state of the fluid, but also the main vortex structure is divided into clearly visible multi-scale vortex structures.

To sum up, the threshold range can be reduced to 0~0.3 when the Q criterion is used, but the change of the threshold seriously affects the extraction effect of the vortex structure. When the Ω method and the Ω-Liutex method are used, the threshold range is reduced to 0.52~0.6. The extracted vortex structure is clear and accurate, and is less affected by the threshold.

12.4.2 Three-Dimensional Vortex Structure Analysis

Under the braking condition, the internal circulation phenomenon of the turbine flow channel is extremely obvious, the turbulent flow structure is complex, and the three-dimensional vortex structure information is rich, which is shown in Fig. 12.4. In the mainstream area of the turbine, a large-scale vortex-like strong vortex structure is formed, and a small-scale vortex-like weak vortex structure is formed near the blade wall and the outer ring.

Longitudinally, different vortex identification criteria are used to extract the vortex structure of the turbine runner, as shown in Fig. 12.5. Based on the Q criterion, the threshold value $Q = 0.18$ is selected to extract the vortex structure inside the turbine flow channel. As shown in Fig. 12.5a, the large-scale vortex are distributed in the mainstream area, and close to the blades and near the wall surface. The extraction effect of small-scale vortex is obvious.

Based on the Ω method, the recommended threshold $\Omega = 0.52$ is used to extract the vortex structure [15]. As shown in Fig. 12.5a and b, while the Ω method captures the strong vortex structure, the recognition effect of the weak vortex structure near

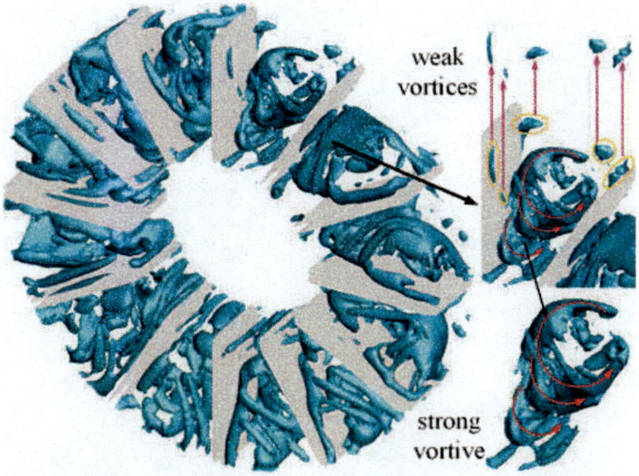

Fig. 12.4 Distribution of multi-scale vortex structure in fluid coupling channel

the wall area is clear, and it can reflect the overall large vortex flow trend in the flow channel. However, most of the strong vortex structures extracted from the mainstream area are strong vortex structures, and there are few weak vortex structures.

On this basis, combined with the advantages of the Liutex method, $\Omega_L = 0.52$ is selected to extract the three-dimensional vortex structure, as shown in Fig. 12.5c. This method has the best effect on feature extraction of 3D vortex structures. And it can capture both strong and weak vortex structures in the flow channel. In particular, the vortex structure extraction in the near-wall area and the runner corner area is more abundant. It can highlight the multi-layered sense of the multi-scale vortex structure inside the flow channel, and present the dynamic evolution process of the stretching, twisting and tearing of the vortex structure at different scales.

Horizontally, the thresholds of the Q criterion are selected as 0.18, 0.22 and 0.26 in Fig. 12.5a. The small-scale adjustment of the threshold has obvious differences in the effect of vortex structure extraction. Although the vortex structure characteristics inside the turbine channel can be displayed under the appropriate threshold selection, the vortex identification method based on the Q criterion is extremely sensitive to the threshold selection. The application of this method is limited by the precise selection of the threshold, and the blind selection is often large, and the efficiency of accurately identifying the vortex structure features is low.

In order to facilitate comparative analysis, in the form of arithmetic progression, the threshold size in the Ω and Ω_L methods is increased. In the Ω method, as the threshold increases, the elongated vortex structure near the wall is fractured. The block vortex structure gradually decreases, the change of the strong vortex structure inside the flow channel is small, and the sensitivity to the threshold decreases, as shown in Fig. 12.5b. In the Ω_L method, as the threshold increases, the vortex structure extraction effect has little effect. Only a small number of small-scale vortex structures change slightly, and the overall change trend is least affected by the threshold. The Ω-Liutex method measures the relative strength of the fluid. In the complex flow field structure, the distribution of strong and weak vortex is different, and it is difficult

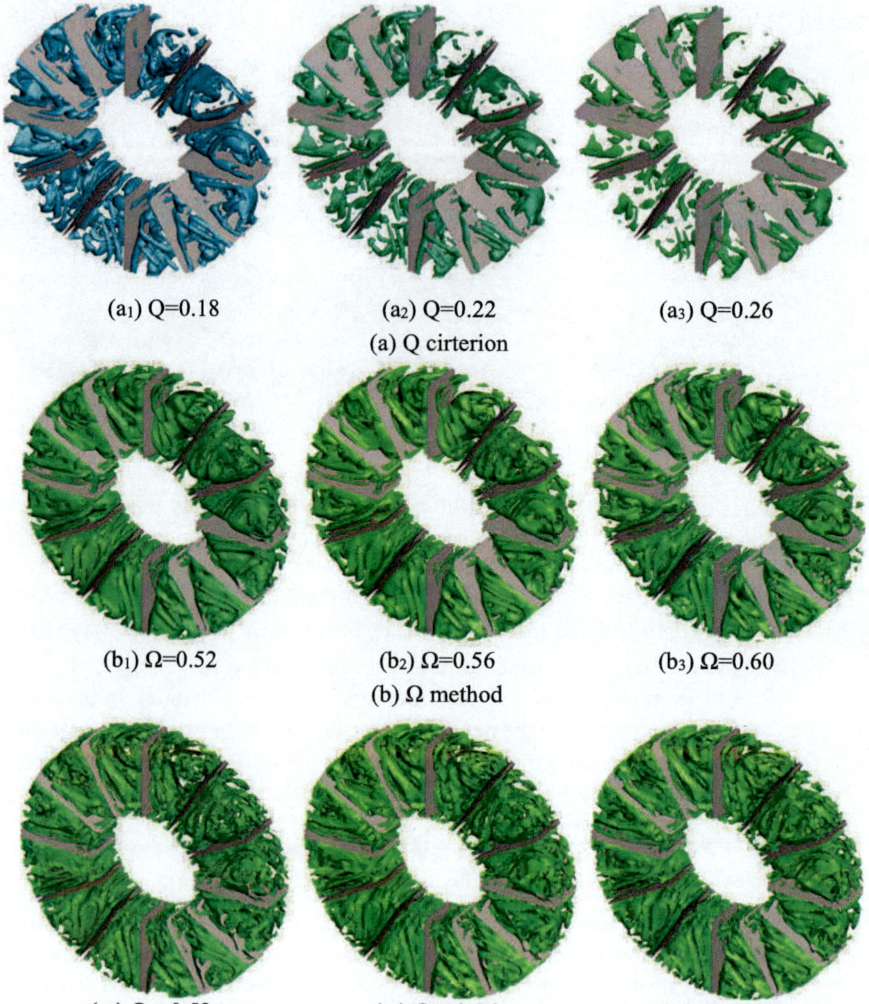

(a1) Q=0.18 (a2) Q=0.22 (a3) Q=0.26

(a) Q cirterion

(b1) Ω=0.52 (b2) Ω=0.56 (b3) Ω=0.60

(b) Ω method

(c1) $Ω_L$=0.52 (c2) $Ω_L$=0.56 (c3) $Ω_L$=0.60

(c) Ω-Liutex method

Fig. 12.5 Multi-scale vortex structure inside the flow channel under different thresholds

to select an appropriate threshold in all fluid structures by absolute strength. The relative intensity dynamically adjusts the threshold range based on the distribution of the vortex structure in the flow field itself, and the effect of the threshold selection on the vortex structure is reduced. Therefore, the Ω-Liutex method is selected to extract the vortex structure inside the fluid coupling channel with the best effect and the highest efficiency.

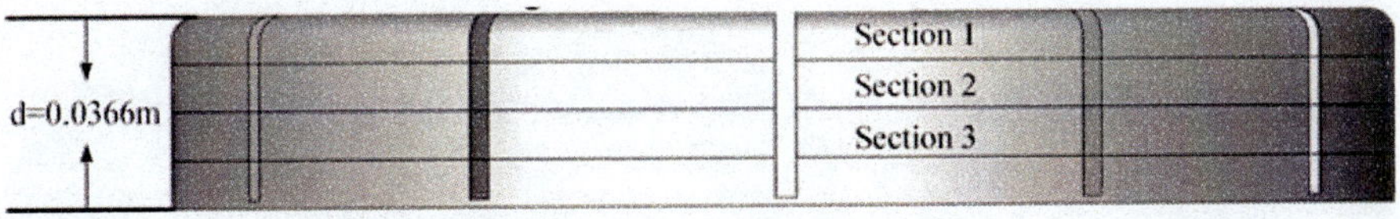

Fig. 12.6 Turbine flow channel isometric diagram

12.4.3 Two-Dimensional Vortex Flow Field Analysis

In order to facilitate the comparison and analysis of the theoretical and two-dimensional flow field visualization test results and compare the ability of different vortex identification methods to accurately capture the vortex structure inside the fluid coupling channel. Three sections of the turbine flow channel are taken along the radial direction, namely the upper section 1, the middle section 2 and the lower section 3, as shown in Fig. 12.6. The distances of each section from the outer wall of the turbine are: section 1 = 9.15 mm, section 2 = 18.3 mm, and section 3 = 27.45 mm. The vorticity information extracted from each section using different vortex identification methods is shown in Fig. 12.7.

12.4.4 Text Verification

Based on the laser section flow field visualization test platform, the flow field image was collected by Particle Image Velocimetry (PIV), as shown in Fig. 12.8. The internal flow medium of the transparent fluid coupling test prototype is pure water. Under the state of full liquid filling, PSP tracer particles with a diameter of 1.5 μm are evenly injected into the test prototype. The continuous laser outputs fan-shaped laser light. The flow field inside the turbine is illuminated [16]. Three groups of flow field visualization experiments were carried out for the turbine flow channel. Adjust the spatial position of the laser sheet light. In the three groups of experiments, the laser sheet light is adjusted to coincide with the upper section 1, the middle section 2 and the lower section 3 respectively. Due to space limitations, this paper only takes the middle section as an example to collect flow field visualization images and post-process the experimental data.

Under braking condition, the pump wheel speed is adjusted to 300r/min, and the turbine speed is 0r/min. Adjust the laser sheet light to coincide with the middle section of the turbine runner. The optical axis of the CCD camera is perpendicular to the outer wall of the turbine of the sample prototype, and two consecutive frames of flow images are collected, as shown in Fig. 12.9a. Based on the PIVlab program [17], the normalized cross-correlation algorithm is used to calculate two consecutive images, and the flow field vorticity information is extracted, as shown in Fig. 12.9b.

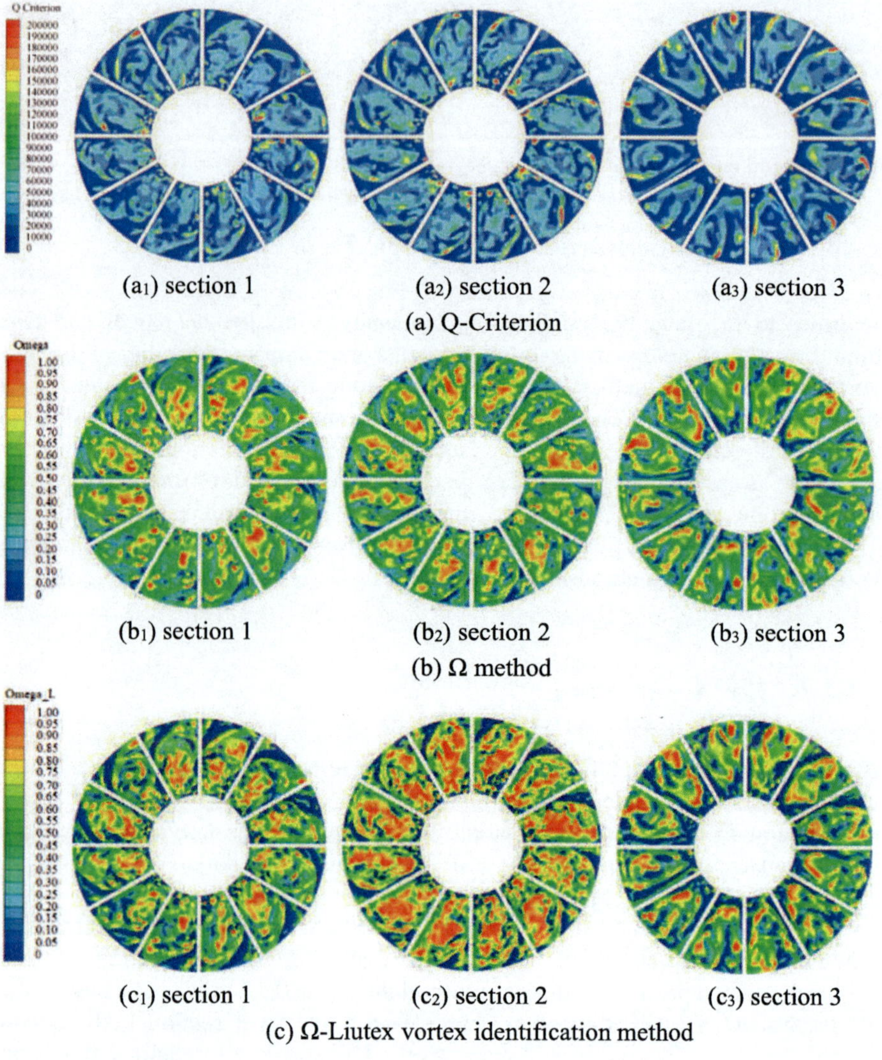

Fig. 12.7 Cross-section view of vorticity in turbine runner

12.5 Analysis and Discussion

12.5.1 Two-Dimensional Cross-Sectional Flow Field

For Sect. 12.2, the small-scale vortex structure cannot be accurately captured by the outer ring region extracted using the Q criterion. The three-dimensional vortex structure of different scales near the inner ring is doped, and the extraction effect

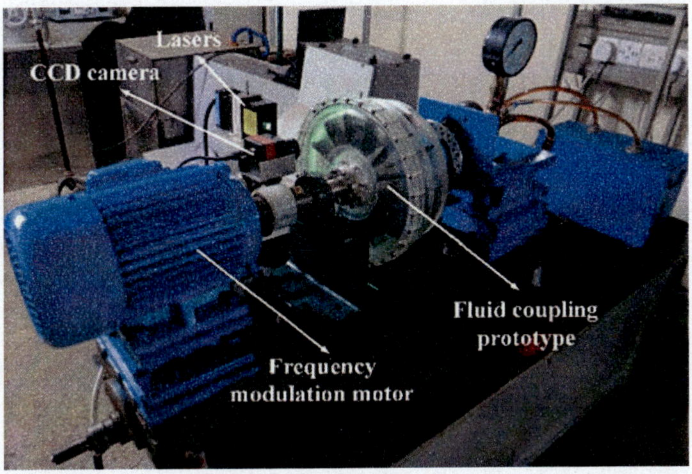

Fig. 12.8 Laser section flow field visualization test platform

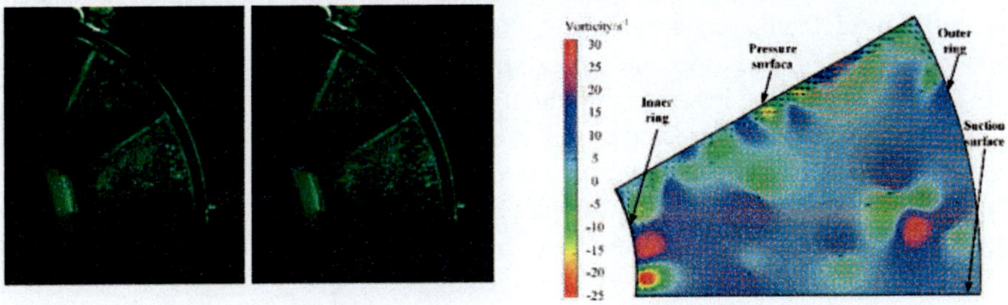

(a) image acquisition (b) image post-processing

Fig. 12.9 Experimental image processing

is blurred, which is difficult to assist the analysis of the test results. When the Ω method is adopted, the high-speed fluid in the pump wheel with the tendency of circular motion flows into the turbine flow channel. At this time, the multi-scale vortex are accelerated and flow into the inside of the flow channel under the action of "lifting and throwing" of the circular arc surface of the outer ring. At the same time, the flow channel changes from wide to narrow. Under the combined action of the two, the fluid accumulates to form a large-scale vortex. The vortex structure near the inner ring is complex, and small-scale vortex can be clearly captured. When the Ω-Liutex method is used, the flow trend of the incident fluid in the flow channel after entering the flow channel can be analyzed through the distribution of the three-dimensional vortex structure in the weak vortex region. The three-dimensional vortex structure in the main flow area completely depicts the diversion of the fluid after impingement through the blades.

For Sect. 12.3, using the Q criterion to extract the three dimensional vortex structure is too dependent on the choice of threshold. While extracting the strong vortex structure in the mainstream area, the capture effect of the weak vortex structure is poor. It is difficult to select an appropriate threshold range to completely capture its overall vortex structure characteristics. When the Ω method is used, the shape of the fluid vortex structure extracted in the mainstream region is large. The fluid velocity near the pressure surface vane is fast, the flow channel begins to fill with liquid, the vortex structure becomes smaller, and the large vortex structure in the flow channel gradually forms. When the Ω-Liutex method is used, large-scale vortex are formed by the accumulation of slow fluid in the middle region of the flow channel. The high-speed fluid at the inlet rotates and rises, and the fluid velocity at the center of the flow channel after the flow channel is filled with liquid drops through the impact of the vane and the wall surface. The high-speed fluid near the upper end of the inner ring merges with the slow flow at the intermediate interface to form a small-scale vortex.

For Sect. 12.4, when the Q criterion is used, the small-scale vortex structure can be accurately captured near the inner ring of the flow channel. However, the overall vortex structure distribution gradient is poor, and the effect is blurred. When the Ω method is used, the fluid near the section flows out from the turbine outlet. The fluid velocity increases, and the scale vortex structure formed after the high-speed flow hits the blades near the inner ring gradually decreases.

The vortex structure extracted by the Ω-Liutex method not only has the advantages of the Ω method, but also the layering effect of the vortex structure is more obvious. It can separate the discontinuous structure in the shear layer from the rotating vortex, and the overall trend of the captured vortex structure is better than the Ω method.

12.5.2 *Visual Analysis of Flow Field in Vortex Structure*

The numerical simulation results of the single flow channel of the fluid coupling extracted by the Ω-Liutex method are compared with the vortex structure extracted from the flow field visualization test, as shown in Fig. 12.10. For the convenience of analysis, the runner area is divided into six main areas, labeled 1~6 respectively. Among them, regions 1 and 2 are close to the outlet of the flow channel, regions 3 and 4 are close to the pressure surface of the flow channel, region 5 is close to the inlet of the flow channel, and region 6 is close to the suction surface of the flow channel.

Regions 1 and 2 near the inner ring of the runner are the outlet of the runner. The high-speed fluid inside the flow channel is returned to the vicinity of the outlet by the impact of the pressure surface. Part of the fluid flows out of the outlet which causes the flow rate to slow down. Close to the inner ring space of the flow channel, the space becomes narrow, and the fluid accumulates into a strong vortex structure here. The high-speed flow from the pump wheel is continuously poured in from the inlet. The fluid inside the turbine rotates around the central area, forming a large-scale

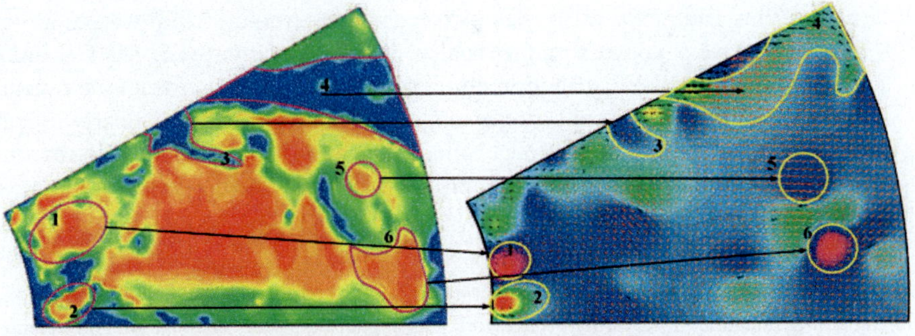

(a) Numerical simulation results (b) PIV processing results

Fig. 12.10 Results comparison

vortex dominated by the center of the flow channel. At this time, the fluid that has not yet impacted the pressure surface has a circular motion tendency, and the flow speed is fast, forming a large-area weak vortex structure in regions 3 and 4. As the fluid flows in the flow channel, the high-speed fluid from areas 3 and 4 pushes the low-speed flow in areas 1 and 2, and pushes the fluid to transfer energy in the flow channel. At this time, part of the slow-moving fluid gradually forms a swirl near the wall of the flow channel, the fluid velocity slows down, and a strong vortex structure is formed in this area.

12.6 Conclusions

The multi-scale vortex structure inside the fluid coupling is extracted by different vortex identification methods. Based on the threshold selection dependence and the vortex structure extraction effect, the applicability of different vortex identification methods is compared in detail.

(1) In terms of threshold selection. The Q criterion takes the absolute strength of the fluid as the threshold. The threshold selection range of vortex structure extraction is large. It is necessary to continuously adjust and determine the appropriate threshold selection range, which leads to the blindness of threshold selection and low vortex extraction efficiency. The Ω method and the Ω-Liutex method use the relative strength of the fluid as the threshold, which effectively reduces the workload of threshold selection and improves the efficiency of vortex extraction. The threshold range is narrowed to 0.52–0.6, which can effectively improve the vortex structure extraction effect, and has little dependence on the threshold selection.

(2) In terms of thc cffcct of vortex structure extraction, The Q criterion can only extract vortex structures with small intensity differences in the flow field. And it is difficult to capture both strong and weak vortex at the same time. The Ω

method can accurately display the strong vortex structure, and the extraction effect of the weak vortex structure is also significantly improved. The Ω-Liutex method improves the distribution effect of the Ω method to extract the overall vortex structure. While capturing strong and weak vortex, it can clearly reflect the trend of fluid flow in the flow channel. The ability of this method to capture the vortex structure of the fluid coupling is obviously better than that of the Q criterion and the Ω method.

Funding The authors disclosed receipt of the following financial support for the research, authorship, and/or publication of this article: This work was supported by the National Natural Science Foundation of China (grant no.52075212, 51405184), Scientific research project of Jilin Provincial Department of Education (grant no. JJKH20220977KJ), Free Exploration Project of the State Key Laboratory of Automotive Simulation and Control, Jilin University (grant no. ascl-zytsxm-202010), Funded by the special fund for the basic scientific research business expenses of the central colleges and universities of Chang'an University(grant no. 300102251511), and General Program of China Postdoctoral Science Foundation (grant no. 2018M641776).

References

1. C. Chu, W. Ma, *Construction Machinery Hydraulic and Hydraulic Transmission System (Hydraulic volume)* (Chemical Industry Press, Beijing, 2015)
2. Y. Zhang, X. Qiu, F. Chen et al., A selected review of vortex identification methods with applications. J. Hydrodyn. **30**(5), 767–779 (2018)
3. B. Epps, *Review of Vortex Identification Methods*. Aiaa Aerospace Sciences Meeting (2017)
4. Y. Wang, N. Gui, A review of the third-generation vortex identification method and its applications. Chin. J. Hydrodyn. **34**(4), 413–429 (2019)
5. C. Liu, Y. Wang, Y. Yang et al., New omega vortex identification method. Mech. Astron. **59**(8), 62–70 (2016)
6. C. Liu, Y. Gao, S. Tian et al., Rortex—a new vortex vector definition and vorticity tensor and vector decompositions. Phys. Fluids **30**(3), 035103 (2018)
7. V. Patel, Y. Yan, Y. Yu, et al., Correlation analysis between low-frequency shock oscillation and Liutex in SBLI, in *Liutex and Third Generation of Vortex Definition and Identification* (2021), pp. 263–277
8. S. Charkri, P. Shrestha, C. Liu, Liutex core line and POD analysis on hairpin vortex formation in natural flow transition. J. Hydrodyn. **32**(6), 1109–1121 (2020)
9. Y. Wang, S. Fu, On the thresholds of vortex identification methods, in *Fluid Structure Sound Interactions and Control* (2019), pp. 45–49
10. N. Gui, L. Ge, P. Cheng et al., Comparative assessment and analysis of rorticity by Rortex in swirling jets. J. Hydrodyn. **31**, 495–503 (2019)
11. L. Sun, P. Guo, X. Luo, Visualization investigation into precessing vortex rope in Francis turbine draft tube based on several vortex identification criterions. Chin. J. Hydrodyn. **34**(06), 779–787 (2019)
12. B. Chai, Y. Xiang, W. Ma et al., Analysis and experimental verification of turbulence models in flow simulation for hydrodynamic coupling under braking condition. Trans. Chin. Soc. Agric. Eng. **32**(03), 34–40 (2016)
13. C. Liu, Y. Gao, X. Dong et al., Third generation of vortex identification methods: omega and Liutex/Rortex based systems. J. Hydrodyn. **31**(2), 205–223 (2019)

14. X. Dong, Y. Gao, C. Liu, New normalized Rortex/vortex identification method. Phys. Fluids **31**(1), 011701 (2019)
15. C. Liu, Liutex-third generation of vortex definition and identification methods. Acta Aerodyn. Sin. **38**(3), 413–431 (2020)
16. B. Chai, Study on Visualization of Internal Flow in Hydrodynamic Coupling and Recognition Method of Flow Velocity (2012)
17. M. Ochowiak, S. Włodarczak, A. Krupińska, et al., Particle image velocimetry based on Matlab and PIVlab for testing flow disturbing elements. Lect. Notes Mech. Eng. **9**(1), 268–276 (2021)

Chapter 13
Numerical Simulation and Analysis of Two-Phase Flow Around Cylinder Using Pseudo-Potential Model and Liutex Method

Zeliang Chen, Pengxin Chen, Xiaoli Huang, Xiaoxi Zhang, Nan Gui, Xingtuan Yang, Jiyuan Tu, and Shengyao Jiang

Abstract Two-phase systems have a wide range of applications in nuclear energy, chemical industry, petroleum, refrigeration and other industrial processes. In this paper, the lattice Boltzmann method that is apt for multiphase flow simulation is adopted, and the third-generation vortex identification method Liutex that is able to distinctly identify rotational vortex is utilized to analyze the vortex field. The two-phase cross flow around columns or tube bundles is widely used in industrial equipment with heat exchange. Based on the practical engineering background, this paper presents a comparative numerical analysis on the two-phase flow around single cylinder under different Reynolds numbers and investigates the evolution of vortex field. The 2D and 3D numerical simulation results has shown the flow field, the vortex shedding pattern, as well as the drag and lift of each column. In particular, the conditions of Re = 30 and Re = 120 in 2D are calculated respectively, corresponding to the conditions of wake vortex stability and laminar vortex street. The drag and lift forces are mainly affected by continuous phase.

Nomenclature

$\mathbf{c_i}$	Discrete particle speed
f	Distribution function
f_i	Particle distribution function
f^{eq}	Equilibrium distribution function
F_i	Discrete external force
F_{SC}	Shan-Chen force

Z. Chen · P. Chen · X. Huang · X. Zhang · N. Gui (✉) · X. Yang · J. Tu · S. Jiang
Institute of Nuclear and New Energy Technology, Collaborative Innovation Center of Advanced Nuclear Energy Technology, Key Laboratory of Advanced Reactor Engineering and Safety, Ministry of Education, Tsinghua University, Beijing 100084, China
e-mail: guinan@mail.tsinghua.edu.cn

J. Tu
School of Engineering, RMIT University, Melbourne, VIC 3083, Australia

Y. Wang et al. (eds.), *Liutex and Third Generation of Vortex Identification*, Springer Proceedings in Physics 288, https://doi.org/10.1007/978-981-19-8955-1_13

lu	Lattice unit
lt	Lattice time
LBM	Lattice Boltzmann method
Q	Q vortex identification
$\boldsymbol{R}$	Liutex
Re	Reynolds number
$\boldsymbol{S}$	Shear
St	Strouhal number
$\mathbf{u}$	Macroscopic velocity

Greek letters

α	Void fraction
ρ	Density
τ	Dimensionless relaxation time
τ_c	Relaxation time
υ	Collision frequency
ψ	Pseudo-potential function
$\boldsymbol{\omega}$	Vorticity
ω_i	Weight coefficient correspond with $\mathbf{c_i}$
$\Omega(f)$	Collision operator
Ω	Omega vortex identification
Ω_R	Omega-Liutex vortex identification

Superscript

c	Continuous phase
d	Dispersed phase
t	Total force
σ	'σ' Component of multiphase system

Subscripts

D	Drag
L	Lift

13.1 Introduction

The gas–liquid two-phase flow around a cylinder or tube bundle is widely used in industrial heat exchange equipment, such as evaporator, steam generator, condenser, reboiler and so on. It is difficult to study the relevant problems, due to the complexity of gas–liquid-solid three-phase interaction in two-phase flow, for example, the bubbles impact on the columns, the bubbles are entrained into the vortices after cylinder, the bubbles influence on vortex shedding and boundary layer, etc. Inoue [1] studied the flow characteristics near and behind a cylinder which was located in a vertical upward air–water bubbly flow. It revealed that the vortex flow and the change of the static pressure and liquid velocity distributions around the cylinder resulted in the large distortion of the void fraction distribution around the cylinder. Joo [2] studied the drag coefficient and the void fraction around a tube subjected to two-phase cross flow for a single tube and for a tube placed in an array. Lu [3, 4] studied the gas–liquid two-phase flow around a cylinder, and analyzed the changes of parameters, such as cylinder surface pressure distribution, fluctuating lift and Strouhal number, with void fraction and Reynolds number. Zhou [5, 6] analyzed changes of fluctuating lift, vortex shedding frequency and vortex street characteristics in the two-phase flow around a single cylinder with different void fraction and Reynolds number by experiments and simulations.

The lattice Boltzmann method (LBM) has outstanding advantages in the numerical simulation of the two-phase flow around single cylinder which has extensive engineering applications. Li [7] used LBM to simulate the 2D flow field around a single cylinder under the Reynolds number equals to 20 and 100. The LBM numerical results showed good precision in space and time, when compared to the experiment results and the finite volume method (FVM) numerical simulation results. Kou [8] computed the velocity and vorticity distribution of flow with three different Reynolds numbers around obstacles in six different shapes by using LBM. Based on the LBM and large eddy simulation (LES), Chen [9] investigated the flow past a circular cylinder under different flow conditions, including steady uniform flow, sinusoidally oscillating flow and linear shear flow. Most of the above LBM simulations focused on single-phase flow, with the traditional vortex identification method determining the vortex field.

In this paper, the two-dimensional and three-dimensional numerical simulation of the flow around single cylinder are analyzed by using the multiphase Shan-Chen pseudo-potential single relaxation time lattice Boltzmann model. Besides, a variety of vortex identification methods are used to identify the vortex field in the calculation results.

13.2 Numerical Simulation Models

13.2.1 Lattice Boltzmann Method

13.2.1.1 Single Relaxation Time Lattice Boltzmann Method

The lattice Boltzmann method originated from the lattice gas cellular automaton models. Ludwig Boltzmann proposed the governing equation describing the temporal and spatial evolution law of the non-equilibrium velocity distribution, which is named the following Boltzmann equation:

$$\frac{\partial f}{\partial t} + \mathbf{v} \cdot \frac{\partial f}{\partial \mathbf{x}} + \mathbf{a} \cdot \frac{\partial f}{\partial \mathbf{v}} = \Omega(f) \tag{13.1}$$

where $f(\mathbf{x}, \mathbf{v}, t)$ is the particles distribution function, representing the particle number density at time t and the spatial position $\mathbf{x} = (x, y, z)$ with a velocity along the direction $\mathbf{v} = (\mathbf{v}_x, \mathbf{v}_y, \mathbf{v}_z)$. $\Omega(f)$ is the collision operator, which describe the interaction between micro fluid particles, plays a key role in whether the model can accurately reappearance the internal physical laws of fluid.

Because of the complex nonlinear integral calculation in the Boltzmann equation, the single relaxation time model proposed by Bhatnagar et al. [10] was widely used to simplify the calculation. They make a linear approximation to the collision operator, consider the f converge to Maxwell equilibrium distribution function f^{eq} after the collision. Assuming the collision operator is direct proportional to $f^{eq} - f$, the scale coefficient is introduced υ, the collision operator is

$$\Omega(f) = \upsilon\left(f^{eq} - f\right) = -\frac{f - f^{eq}}{\tau_c} \tag{13.2}$$

where the υ represents the total number of collisions between particles and other particles in unit time, i.e. collision frequency. $\tau_c = 1/\upsilon$ is the average time interval between two particle collisions, it is also called relaxation time, which is related to the physical properties of fluid. The Boltzmann-BGK equation can be obtained by substituting Eq. (13.2) into Eq. (13.1), whose corresponding model is called single relaxation time lattice Boltzmann model, because only one relaxation parameter in the model. The Boltzmann-BGK equation can be discretized into single relaxation time lattice Boltzmann equation (LBGK equation):

$$f_i(\mathbf{x} + \mathbf{c_i}\delta t, t + \delta t) - f_i(\mathbf{x}, t) = -\frac{f_i(\mathbf{x}, t) - f_i^{eq}(\mathbf{x}, t)}{\tau} + \left(1 - \frac{1}{2\tau}\right) F_i(\mathbf{x}, t)\delta t \tag{13.3}$$

where $\mathbf{c_i}$ is discrete particle speed, $f_i(\mathbf{x}, t)$ is the particle density distribution function along the direction $\mathbf{c_i}$, τ is the dimensionless relaxation time.

This LBM classification scheme was proposed by Qian [11] and is in widespread use. This model is known as DdQm as it is d dimensional and contains m velocities. In this passage, the D2Q9 and D3Q19 model are used to simulate the flow around columns.

13.2.1.2 Multiphase Pseudo-Potential Lattice Boltzmann Model

The Shan-Chen [12] pseudo-potential model is the most widespread used in multi-phase simulation. The Shan-Chen force equation at $\mathbf{x}$ can be transformed into summation form after discretization.

$$\begin{aligned} F_{SC}(\mathbf{x}) &= -\int (\mathbf{x}' - \mathbf{x})G(\mathbf{x}, \mathbf{x}')\psi(\mathbf{x})\psi(\mathbf{x}')\mathrm{d}^3\mathbf{x}' \\ &= -G\psi(\mathbf{x})\sum_i \omega_i \psi(\mathbf{x} + \mathbf{c}_i\delta t)\mathbf{c}_i\delta t \end{aligned} \tag{13.4}$$

where $\psi(\rho)$ is pseudo-potential function,

$$\psi(\rho) = \rho_0\left(1 - e^{-\rho/\rho_0}\right) \tag{13.5}$$

the reference density is generally set to 1. When $\rho \ll \rho_0$, $\psi(\rho)$ close to ρ. Some studies directly take $\psi(\rho) = \rho$. $G(\mathbf{x}, \mathbf{x}')$ is the Green's function to determine the intensity and range of interaction, and its simplified form generally only considers the interaction between adjacent lattice points.

$$G(\mathbf{x}, \mathbf{x}') = \begin{cases} \omega_i G, & \mathbf{x}' = x + \mathbf{c}_i\delta t \\ 0, & \text{others} \end{cases} \tag{13.6}$$

For a multiphase system composed of n components, the single relaxation time lattice Boltzmann equation of each component is,

$$\begin{aligned} f_i^\sigma(\mathbf{x} + \mathbf{c}_i\delta t, t + \delta t) - f_i^\sigma(\mathbf{x}, t) = &-\frac{f_i^\sigma - f_i^{eq(\sigma)}(\rho, \mathbf{u}^{eq})}{\tau^\sigma} \\ &+ \left(1 - \frac{1}{2\tau^\sigma}\right)F_i^\sigma(\mathbf{x}, t) \end{aligned} \tag{13.7}$$

where f_i^σ, $f_i^{eq(\sigma)}$, τ^σ, F_i^σ are 'σ' component distribution function, equilibrium distribution function, relaxation time and external force, respectively.

The inter particle force applied to the 'σ'component can be accumulated by other components forces,

$$F_{SC}^\sigma(\mathbf{x}) = -\psi_\sigma(\mathbf{x})\sum_{\sigma'} G_{\sigma\sigma'}\sum_i \omega_i \psi_{\sigma'}(\mathbf{x} + \mathbf{c}_i\delta t)\mathbf{c}_i\delta t \tag{13.8}$$

The numerical simulation model in the paper is a two-component multiphase system without considering phase transition, which means $n = 2$, $G_{\sigma\sigma} = 0$. $G_{\sigma\sigma'}$ sets to positive, to achieve repulsion and separation between different components. The solid surface is generally set as non-wettability, which means the interaction force between fluid and solid is 0.

Setting a certain density random disturbance during the density field initialization can automatically realize the separation of two phases by using the Shan Chen force. The initialization can obtain the bubble flow field with uniform dispersion distribution, which can simulate the two-phase flow problem simply and efficiently.

13.2.1.3 Bounce-Back Boundary Conditions and Momentum Exchange Method

In the lattice Boltzmann model, the bounce-back boundary condition is often used to realize the non-slip wall. The bounce-back boundary condition means the incident particle distribution function will bounce back at the original value along the opposite direction of incidence, which can ensure the mass and momentum conservation of the system.

In this paper, the drag and lift in the flow around the cylinder can be calculated by the momentum exchange method. The drag force equals to the component force on the solid opposite to the flow direction, while the lift force is the component force on the solid perpendicular to the flow direction.

13.2.1.4 The Unit of LBM

In the LBM, lattice units are used for calculation, such as the lattice length unit *lu* and the lattice time unit *lt*. Based on the similarity principle, the key to convert the lattice units to actual units, is the dimensionless criterion numbers, which play an important role in the control system physical laws, need to be consistent. In multicomponent multiphase systems, the dimensionless parameters, such as Weber number (We), Reynolds number (Re), viscosity ratio, density ratio, and so on, play a great role in the process of droplets fragmentation and polymerization.

$$\mathrm{We} = \frac{\rho_l D U^2}{\sigma}, \mathrm{Re} = \frac{DU}{\upsilon}, \upsilon_r = \frac{\upsilon_l}{\upsilon_g}, \rho_r = \frac{\rho_l}{\rho_g} \tag{13.9}$$

The units in this paper are lattice units, including lattice length *lu*, lattice time *lt*, and lattice speed *lu*/*lt*. The units are omitted below for simplification.

13.2.2 Vortex Identification

The vortex identification criteria selected in this paper are as follows, ω, Q, Ω, Liutex R, Ω-Liutex Ω_R.

Vorticity criterion is the first generation vortex identification method, which originates from the concept proposed by Helmholtz [13]. It is twice the average angular velocity of the fluid micro element rotation. Generally, the vortex is recognized by the concentration area of the vorticity, means the area of $\omega > 0$.

$$\omega = \|\boldsymbol{\omega}\| = \|\nabla \times \mathbf{u}\| \tag{13.10}$$

Q vortex identification [14] is widely used in the second generation vortex identification method. However, there is a few errors in judging rotating vortex because some terms in Q formula represent tensile, compression and shear deformation.

$$Q = \frac{1}{2}\left(\|\boldsymbol{B}\|_F^2 - \|\boldsymbol{A}\|_F^2\right) \tag{13.11}$$

where $\|\boldsymbol{A}\| = \frac{1}{2}\left(\nabla\mathbf{u} + (\nabla\mathbf{u})^T\right)$, $\|\boldsymbol{B}\| = \frac{1}{2}\left(\nabla\mathbf{u} - (\nabla\mathbf{u})^T\right)$ are symmetric and anti-symmetric tensors of velocity gradient tensors, respectively. $\|\cdot\|_F^2$ is Frobenius norm.

Ω vortex identification [15] identify the proportion of rotating vorticity to the total vorticity, which can be understood as the vorticity concentration, and the value range is [0,1]. Ω vortex identification is also the second-generation vortex identification method, with the symmetric and anti-symmetric tensors based on Cauchy-Stokes decomposition. It can better capture strong vortices and weak vortices at the same time, because it represents the relative intensity of vortex and has low sensitivity to threshold.

$$\Omega = \frac{\|\boldsymbol{B}\|_F^2}{\|\boldsymbol{A}\|_F^2 + \|\boldsymbol{B}\|_F^2} \tag{13.12}$$

In practical application of Ω vortex identification, a small positive number ϵ will be added to the denominator, to ensure the stability.

$$\Omega = \frac{\|\boldsymbol{B}\|_F^2}{\|\boldsymbol{A}\|_F^2 + \|\boldsymbol{B}\|_F^2 + \epsilon} \tag{13.13}$$

Liu et al. [16, 17] developed and proposed the third generation vortex identification method. This method decomposes the vorticity into two parts, $\boldsymbol{\omega} = \mathbf{S} + \mathbf{R}$, Liutex **R** representing the rigid rotating part and **S** representing the non-rotating part (antisymmetric shear). **RS** decomposition gives a unique tensor decomposition with Galileo invariance. In most cases, the main component of ω is **S**, and the proportion of **R** is smaller than **S**. Therefore, in traditional methods, using ω to characterize fluid rotation (**R**) will cause large errors.

The direction **r** of Liutex represents the local rotation axis direction, which is determined by the direction of the real eigenvector of the velocity gradient tensor. According to the explicit formula proposed by Wang [18], the Liutex module value *R*, which represents the intensity of local rotation, can be calculated by removing the non-rotating part from the module value of vorticity.

$$\mathbf{R} = R\mathbf{r} = \left(\langle \omega, \mathbf{r} \rangle - \sqrt{\langle \omega, \mathbf{r} \rangle^2 - 4\lambda_{ci}^2}\right)\mathbf{r} \tag{13.14}$$

where λci is the imaginary part of complex eigenvalues in $\nabla \mathbf{u}$, **r** is the real eigenvector of $\nabla \mathbf{u}$.

The Ω-Liutex [19] vortex identification method is developed on the basis of Liutex to determine the relative intensity of vorticity.

$$\Omega_R = \frac{\beta^2}{\alpha^2 + \beta^2 + \epsilon} \tag{13.15}$$

where $\alpha = \frac{1}{2}\sqrt{\left(\frac{\partial V}{\partial Y} - \frac{\partial U}{\partial X}\right)^2 + \left(\frac{\partial V}{\partial X} + \frac{\partial U}{\partial Y}\right)^2}$, $\beta = \frac{1}{2}\left(\frac{\partial V}{\partial X} - \frac{\partial U}{\partial Y}\right)$. Under the XYZ coordinate system, the rotation matrix $\boldsymbol{Q}$ transform $\nabla \mathbf{u}$ into $\nabla \mathbf{V} = \boldsymbol{Q}\nabla \mathbf{u}\boldsymbol{Q}^T$. *U, V, W* are the velocity components meeting the conditions $[\nabla \mathbf{V}]_{1,3} = [\nabla \mathbf{V}]_{2,3} = 0$.

13.3 Method Verification

In this section, the multiphase computational model is validated, which is illustrated in Fig. 13.1a for multi-component Poiseuille flow [20]. The left and right sides of the flow channel are solid wall boundaries, and the center of the flow channel is occupied by liquid *a* with viscosity υ_a and density ρ_a. The sides of flow channel are occupied by liquid *b* with viscosity υ_b and density ρ_b. From Fig. 13.1b, it can be seen that the calculated velocity distribution fits well with the theoretical equation, and there is only a certain velocity error at the gas–liquid interface.

In addition, the correlation coefficients are also verified for the flow around two-dimensional circular cylinder. As shown in Table 13.1, the accuracy of the numerical method is verified by comparing the calculation with previous results [21], which are in good agreement with the reference values.

(a)

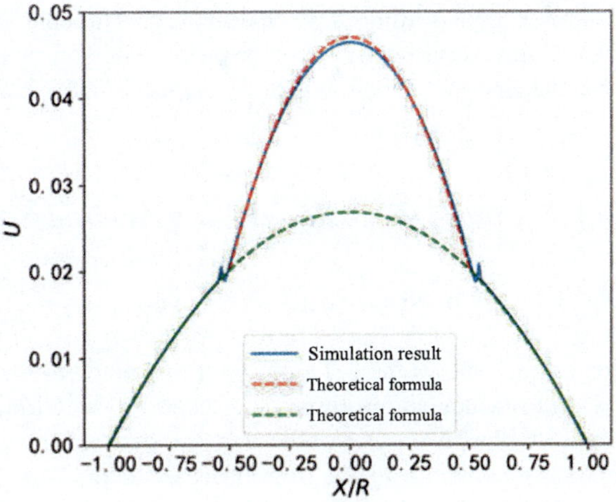

(b)

Fig. 13.1 Validation of multi-component Poiseuille flow

Table 13.1 Verification of two-dimensional flow around circular cylinder

Reynold number	20	20	100	100	100
Target	C_D	C_L	C_{Dmax}	C_{Lmax}	S_t
Numerical result	5.55	0.0118	3.29	0.98	0.306
Lower limit of benchmark value	5.57	0.0104	3.22	0.99	0.295
Upper limit of benchmark value	5.59	0.0110	3.24	1.01	0.305

13.4 Numerical Simulation of Flow Around Single Cylinder

13.4.1 Simulation Initial Settings

In this paper, two-dimensional and three-dimensional flow around single cylinder are simulated respectively. The settings are as follows: the calculation domain of two-dimensional example is $\left(L_x, L_y\right) = (256, 2048)$, and the cylinder diameter $D = 100$. Three-dimensional example calculation domain is $\left(L_x, L_y, L_z\right) = (160, 160, 450)$, and the cylinder diameter $D = 80$. The solid wall surface set as non-wettability, and the surrounding boundaries set as periodic boundaries. By selecting different vertical gravity values to drive the fluid movement, the flow simulation under different Reynolds numbers is realized. In 2D simulations, the conditions of Re = 30 and Re = 120 are calculated respectively, corresponding to the conditions of wake vortex stability and laminar vortex street. As the Ref. [22] shows that the generation of vortex and the stability of vortex street would be disturbed at high void fraction ($\alpha > 0.15$) because the bubbles are entrained into the vortices, the void fraction in the simulation sets to 0.13.

The example settings have passed the grid independence test. The dispersed phases are bubbles, signified by superscript 'd'. The continuous phases are fluids, signified by superscript 'c'. The total forces are signified by superscript 't'.

13.4.2 Two-Dimensional Flow Around Single Cylinder

13.4.2.1 Re = 30

Figure 13.2 shows the 2D simulation results with the Re = 30 when the flow field velocity reaches equilibrium. The dispersed small bubbles converge into medium-sized bubbles in the vortex field at the back of the cylinder, and then the bubbles gradually converge in the flow process, leading to the bubbles decreases to form discrete large bubbles. Most of the bubbles remain spherical and flow downward at a low speed, but the larger bubbles formed by coalescence have certain deformation. Because of the low-speed flow, the bubbles always gather together. There is no bubbles burst during flowing or impacting with cylinder, and some bubbles contacting with the cylinder would slip across the surface.

$\boldsymbol{R}$, $\boldsymbol{S}$ and ω have positive and negative discrete distribution areas on both sides of the cylinder, respectively. And the distribution diffuses to both sides behind the cylinder with the movement of bubbles. A high $\boldsymbol{R}$ value distribution can also be observed (as shown in Fig. 13.2h) near the interface of bubbles coalescence and large bubbles deformation. Besides, there is no vortex separation behind the cylinder at low Reynolds number.

13.4.2.2 Re = 120

Figure 13.3 shows the 2D simulation results with the Re = 120 when the flow field velocity reaches equilibrium. The dispersed small bubbles converge into medium-sized bubbles in the vortex field at the back of the cylinder, and then bubbles move downward rapidly, forming a region with less local void fraction behind the cylinder. Because of high-speed flow, some bubbles are stretched into ellipsoid, strip and other irregular shapes, and burst to small satellite bubbles under the shear. In addition, after impacting with cylinder at high speed, the bubbles also burst to large bubbles and some small satellite bubbles. The quantities and shapes of bubbles are more than the low-speed flow, because the more bubbles coalescence and bursting in the high-speed flow field.

In the distribution of $\boldsymbol{R}$, $\boldsymbol{S}$ and ω, there are positive and negative discrete distribution areas on both sides of the cylinder at the initial time (as shown in Fig. 13.3e). Due to the larger gravity, the bubble moves downward faster, and the areas of $\boldsymbol{R}$, $\boldsymbol{S}$ and ω diffusion are more extensive than the low-speed flow results in the previous section. Behind the cylinder, there are periodic vortices falling off with clockwise rotation on the left and counterclockwise rotation on the right, forming two rows of

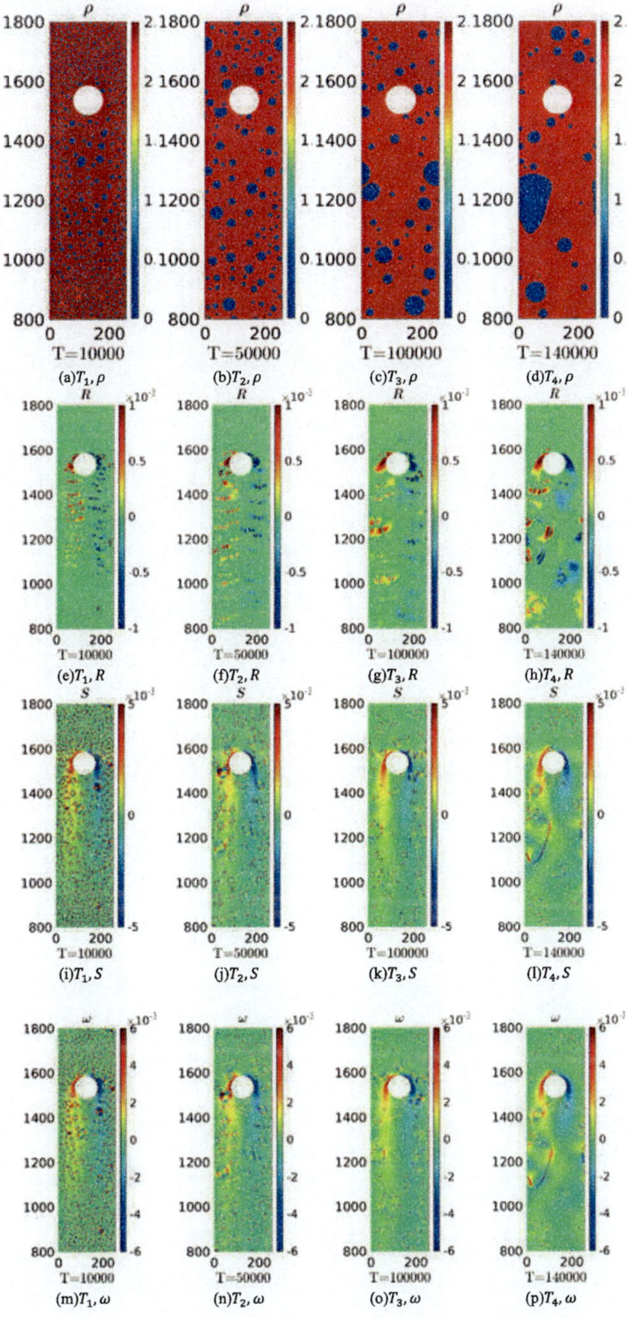

Fig. 13.2 The evolution of two-phase flow around a single cylinder (Re = 30)

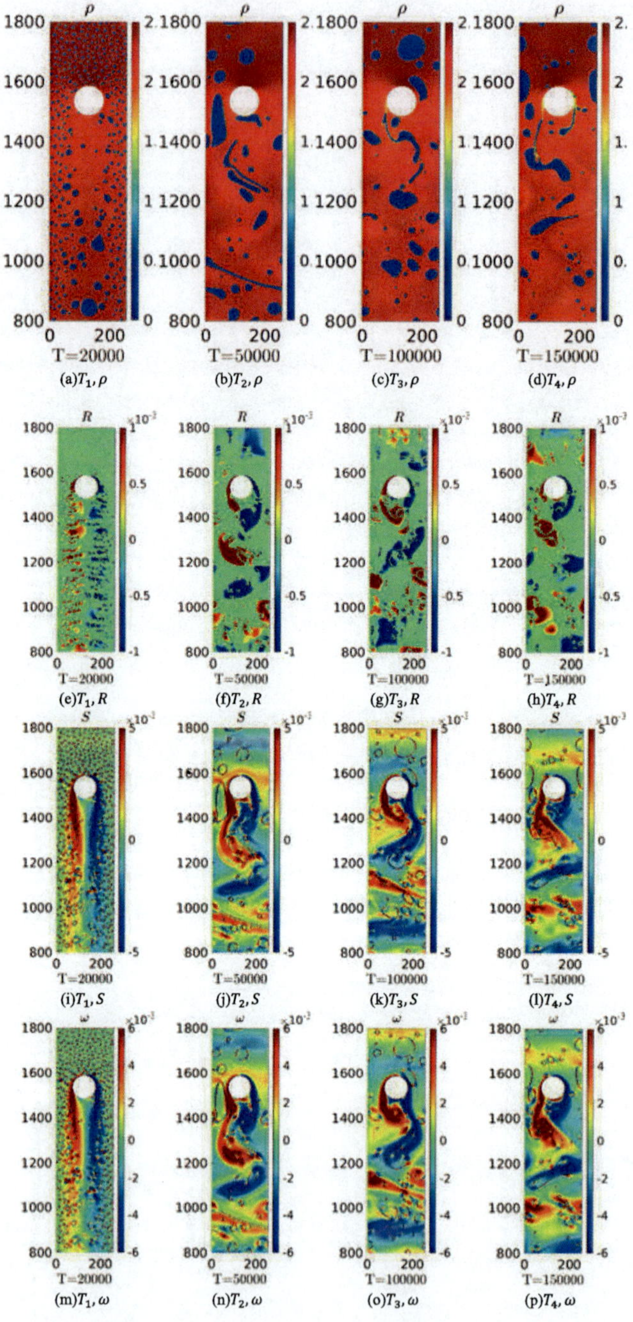

Fig. 13.3 The evolution of two-phase flow around a single cylinder (Re = 120)

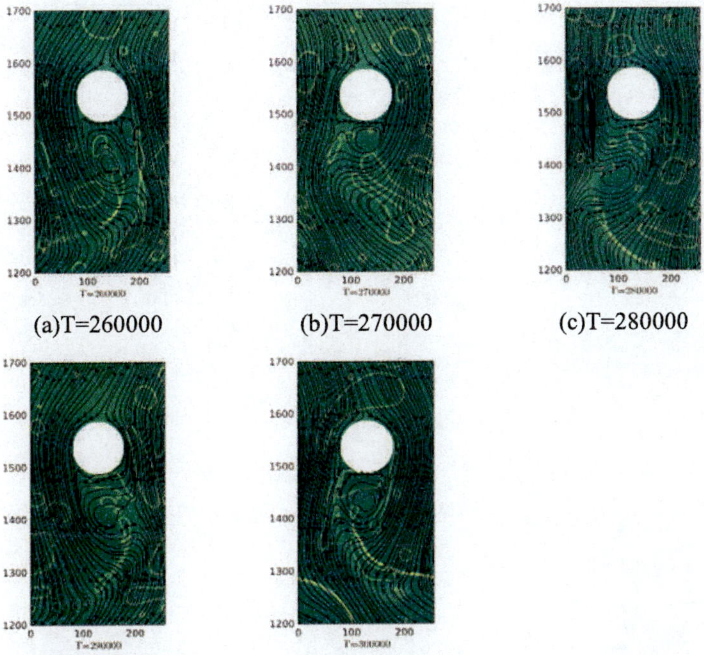

(a)T=260000 (b)T=270000 (c)T=280000

(d)T=290000 (e)T=300000

Fig. 13.4 The streamline evolution of two-phase flow around a single cylinder (Re = 120)

opposite-rotation staggered vortices. The vortices have the characteristics of Carmen vortex street, with the obvious unsteady characteristics of flow field and vortex field. Figure 13.3f–h have shown the positive and negative distribution areas of rotation, which show periodic changes of stretching, deformation, detachment and diffusion.

Figure 13.4 shows the evolution of the streamline, in which the black arrow curves represent the streamline and the yellow curves represent the bubble boundary in the two-phase flow.

Figure 13.5 shows the comparison of different vortex identification criteria. It can be observed that each vortex identification criterion can capture the rotating vortices falling off periodically behind the cylinder. The vorticity is mainly disturbed by shear. Besides, Q, Ω, Ω_R could not identify the positive and negative rotating vortices. Meanwhile, the rotating vortices distribution identified by Liutex $\boldsymbol{R}$ is the most clear and intuitive.

13.4.2.3 The Drag and Lift Forces Variation

Figures 13.6 and 13.7 show the changes of drag and lift forces in the two-phase flow with Re = 30 and Re = 120. F_D^c, F_D^d, F_D^t are drag forces of continuous phase, discrete phase and total, respectively. The drag and lift force are mainly influenced by the

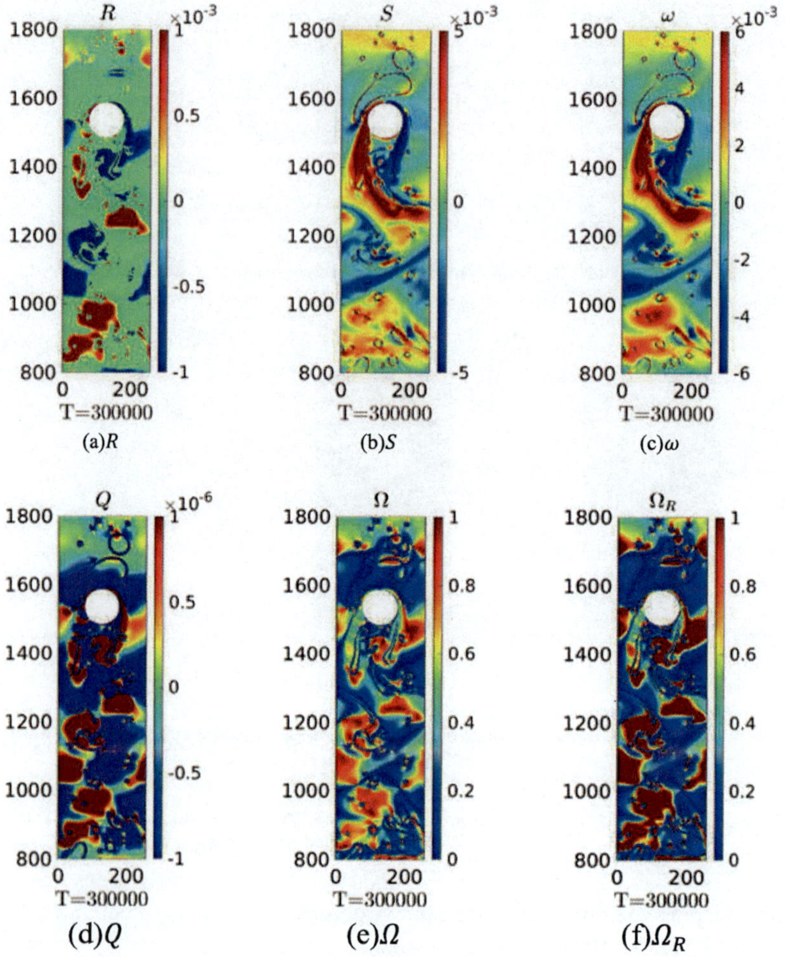

Fig. 13.5 Vortex identification of two-phase flow around a single cylinder (Re = 120)

continuous phase. In the low-speed flow (Re = 30), the drag and lift forces variation are relatively stable, except that a pulsation occurs with the bubbles impacting or contacting with the cylinder. In the high-speed flow (Re = 120), the drag and lift force variation are significantly higher than those in low-speed flow, and its pulsation is also significantly enhanced.

There are three parts in the fluctuating lift force of the two-phase flow around single cylinder. First, the fluctuating lift force caused by the periodic staggered shedding of vortices on both sides behind the cylinder, which has large amplitudes and the main energy of vibration. Second is the fluctuating lift force caused by the turbulence intensity random fluctuation. The two-phase flow has larger turbulence intensity than the single-phase flow, but the turbulence intensity has little effects under the low

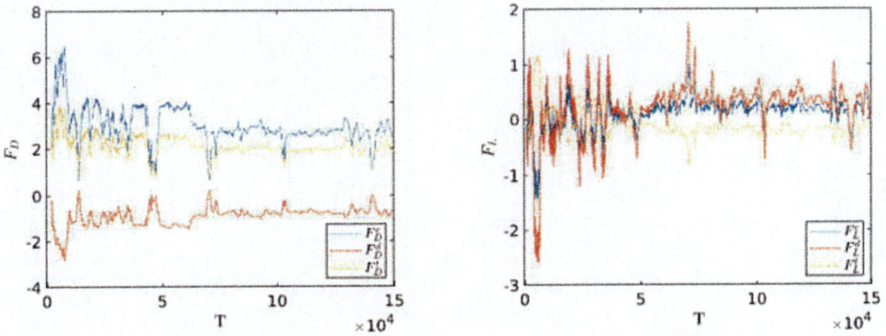

(a)F_D (b)F_L

Fig. 13.6 The drag and lift forces variation of two-phase flow around a single cylinder (Re = 30)

Reynolds number in this paper. The third is the fluctuating lift force caused by the continuous bubbles impacting on the cylinder. The fluctuating lift caused by vortex shedding in the first part has periodicity, and the other two parts have randomness. As shown in Fig. 13.7e and f, the fast Fourier transform (FFT) is carried out on the lift force. It can be found that the pulsation frequency of discrete phase and continuous phase are coincide, which is $4.3869 \times 10^{-4} lt^{-1}$, i.e. corresponding to vortex frequency. The peak amplitude of discrete phase is lower than continuous phase. The equilibrium velocity of the flow field is $0.2lu/lt$, and the corresponding Strouhal number is 0.219.

13.4.3 Three-Dimensional Flow Around Single Cylinder

Figure 13.8 shows the three-dimensional simulation results of bubble flow around a single cylinder. $\|R\|$, $\|S\|$ have high amplitude distribution area on the side of cylinder. Figure 13.9 shows the drag and lift forces variation. As shown in Fig. 13.9a, the discrete phase drag force F_D^d is represented by coordinate scale on the right. In the beginning, the discrete phase drag force is high when the bubbles distribution is dense, then it decreases at first and gradually increases with the flow velocity increasement. The total drag force F_D^t and continuous phase drag force F_D^c are almost the same. With the acceleration of the fluid under the gravity, F_D^t and F_D^c continuous increases and gradually reaches equilibrium. The lift forces are influenced by the random bubbles impacting on the wall, resulting in pulsating fluctuations.

In this work, the multiphase lattice Boltzmann method is used to simulate the two-phase flow around a single cylinder under different Reynolds numbers. In the laminar vortex street, the effects of periodic shedding vortices on the flow field, the drag forces and the lift forces are analyzed, respectively. Besides, different vortex identification methods are used to identify the shedding vortices in the flow field.

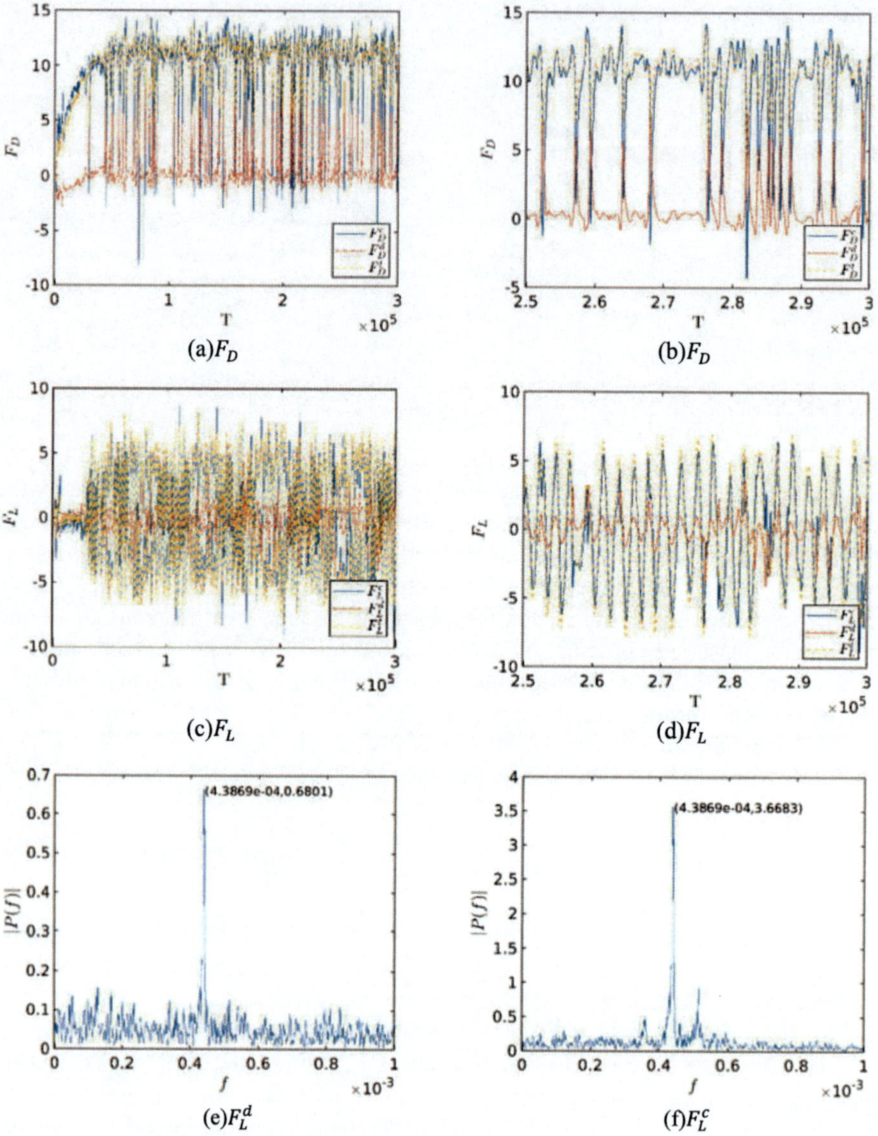

Fig. 13.7 The drag and lift forces variation of two-phase flow around a single cylinder and the lift force frequency (Re = 120)

13.5 Conclusions

Multiphase Shan-Chen pseudo-potential single relaxation time lattice Boltzmann model can simulate the evolution process of two-dimensional and three-dimensional two-phase flow around cylinders under different Reynolds numbers. The simulation

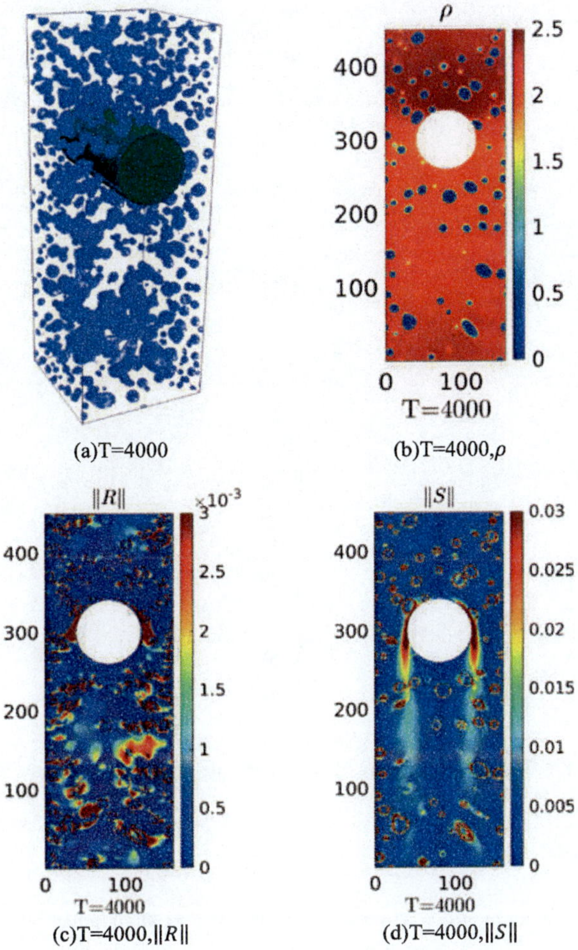

Fig. 13.8 3D two-phase flow around a single cylinder

results can show the development and changes of flow field in two-phase flow, and the vortex field can be identified and analyzed by vortex identification criteria. In different vortex identification criteria, each vortex identification method can capture the periodical rotating vortices falling off behind the cylinder, and the distribution of rotating vortices identified by Liutex R is the most clear and intuitive.

In the low-speed flow (Re = 30), the drag and lift forces variation are relatively stable, except that a pulsation occurs with the bubbles impacting or contacting with the cylinder. In the high-speed flow (Re = 120), the drag and lift force variation are significantly higher than those in low-speed flow, and its pulsation is also significantly enhanced. In 3D simulations, the variation of drag and lift forces also show similar phenomena.

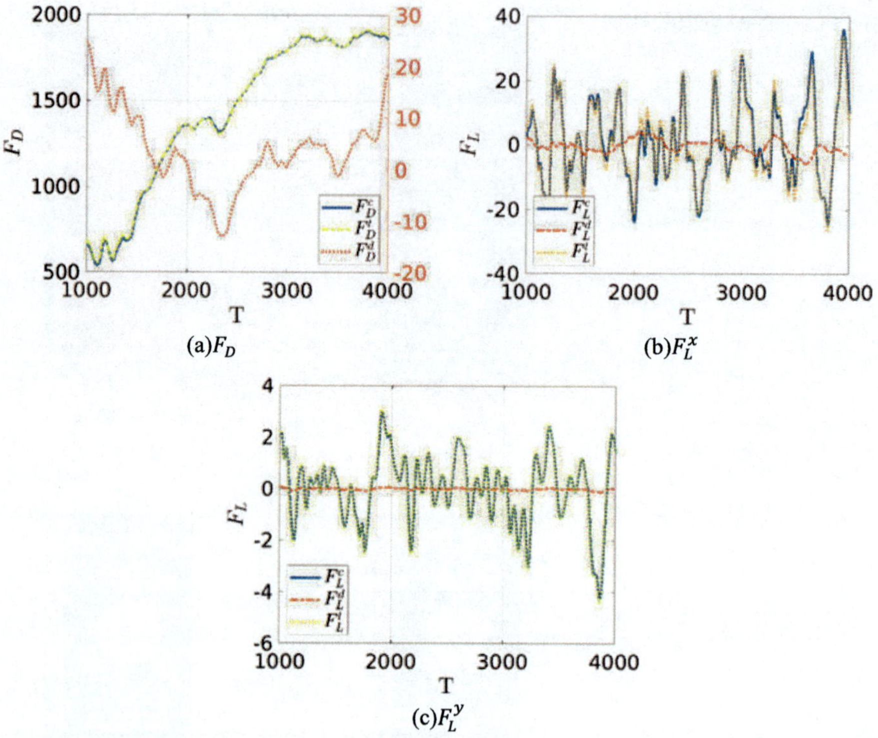

(a)F_D (b)F_L^x

(c)F_L^y

Fig. 13.9 The drag and lift forces variation of 3D two-phase flow around a single cylinder

References

1. A. Inoue, Y. Kozawa, M. Yokosawa et al., Studies on two-phase cross flow. Part I: flow characteristics around a cylinder. Int. J. Multiph. Flow **12**(2):149–167 (1986)
2. Y. Joo, V.K. Dhir, An experimental study of drag on a single tube and on a tube in an array under two-phase cross flow. Int. J. Multiph. Flow **20**(6), 1009–1019 (1994)
3. J.C. Lu, Z.W. Xie, Y.P. Wang, Z.H. Lin, M.K. Wang, Pressure distributions around circular cylinder for gas-liquid two-phase cross flow. J. Xi'an Jiaotong Univ. **01**, 58–61 (1999)
4. J.C. Lu, Z.W. Xie, Y.P. Wang, Z.H. Lin, M.K. Wang, Study on the fluctuating lift acting on a circular cylinder under two-phase cross-flow. Chin. J. Appl. Mech. **1**, 76–80 (2000)
5. Y.L. Zhou, W.P. Hong, G.P. Wang, Numerical simulation of vortex shedding frequency and the oscillation lift based on gas-liquid two-phase flow around circular cylinders in the vertical pipeline. J. Hydrodyn. **02**, 194–201 (2007)
6. Y.L. Zhou, C.D. Diao, R. Cao, Study of gas-liquid two-phase vortex street characteristics of a square cylinder and a circular one. J. Eng. Thermal Energy Power **24**(06), 746–749+817–818 (2009)
7. X.B. Li, Lattice Boltzmann method and its application on the numerical simulation of flows past cylinders. (Tianjin University, 2006)
8. W.Y. Kou, Z.G. Bai, Numerical simulation of flow around multi-type obstacles by the lattice Boltzmann method. J. Tianjin Univ. Technol. **31**(02), 1–5 (2015)
9. T.Q. Chen, Lattice Boltzmann simulation of the flow around a circular cylinder. (Tianjin University, 2007)

10. P.L. Bhatnagar, E.P. Gross, M. Krook, A model for collision processes in gases. I. Small amplitude processes in charged and neutral one-component systems. Phys. Rev. **94**(3), 511–525 (1954)
11. Y.H. Qian, D. d'Humières, P. Lallemand, Lattice BGK models for Navier-Stokes equation. EPL (Europhys. Lett.) **17**(6), 479 (1992)
12. X. Shan, H. Chen, Lattice Boltzmann model for simulating flows with multiple phases and components. Phys. Rev. E **47**(3), 1815–1819 (1993)
13. H. Helmholtz, Über integrale der hydrodynamischen gleichungen, welche den wirbelbewegungen entsprechen. Journal für die reine und angewandte Mathematik **55**, 25–55 (1858)
14. J.C.R. Hunt, A. Wray, P. Moin, Eddies, stream, and convergence zones in turbulent flows. Center for turbulence research report CTR-S88, pp. 193–208 (1988)
15. C.Q. Liu, Y.Q. Wang, Y. Yang et al., New omega vortex identification method. Sci. China **59**(8), 684711 (2016)
16. C. Liu, Y. Gao, S. Tian, et al., Rortex a new vortex vector definition and vorticity tensor and vector decompositions. Phys. Fluids **30**(3) (2018)
17. Y. Gao, C. Liu, Rortex and comparison with eigenvalue-based vortex identification criteria. Phys. Fluids **30**(8), 085107 (2018)
18. Y.Q. Wang, Y.S. Gao, J.M. Liu, et al., Explicit formula for the liutex vector and physical meaning of vorticity based on the liutex-shear decomposition. J. Hydrodyn. **31**(3), 464–474 (2019)
19. X. Dong, Y. Gao, C. Liu, New normalized Rortex/vortex identification method. Phys. Fluids **31**(1) (2019)
20. A. Banari, C. Janßen, S.T. Grilli, M. Krafczyk, Efficient GPGPU implementation of a lattice Boltzmann model for multiphase flows with high density ratios. Comput. Fluids **93**, 1–17. (2014). https://doi.org/10.1016/j.compfluid.2014.01.004.
21. P. Cheng, N. Gui, X. Yang, J. Tu, S. Jiang, H. Jia, Liutex-based analysis of drag force and vortex in two-phase flow past 2-D square obstacle using LBM on GPU. J. Hydrodyn. **32**(5), 820–833. (2020) https://doi.org/10.1007/s42241-020-0058-5
22. Z.H. Lin, Y.G. Li, J.C. Lu, et al., Vortex shedding characteristics of gas-liquid two-phase flow and its engineering application (in Chinese). (Chemical Industry Press, 2001)

Chapter 14
Analysis of Vortices Shed by Generic Submarines Based on Liutex

Fenglai Huang, Liushuai Cao, and Decheng Wan

Abstract Submarines in towed conditions shed a wealth of vortical structures from the main body and appendages, such as casing, fins, rudders, and diving planes. The interaction of these coherent structures is detrimental to its hydrodynamic performance and is a source of vibrations and emission of noise, having a negative impact on its stealth capabilities. Studies on flow control and vortex manipulation are of great importance for submarines. However, most of the current vortex identifications based on Eulerian criteria, including the Q criterion and λ_2 criterion are fail to identify the swirl axis or orientation. To address the problem, we analyzed vortices shed by generic submarines by using an alternative eigrnvector-based definition of Liutex. The platform under study is the DARPA suboff submarine model and the Joubert BB2 submarine, which is now as international benchmarks for underwater vehicles. This paper is aimed at analyzing vortices shed by a submarine by using the Liutex and the traditional Q criterion.

14.1 Introduction

Submarines in towed conditions shed a wealth of vortical structures from the main body and appendages, such as casing, fins, rudders, and diving planes. There are various vortical structures from the main body and appendages, such as casing, fins, rudders, and diving planes [1]. The interaction of these coherent structures is detrimental to its hydrodynamic performance and it a source of vibrations and emission of noise, which will have a negative impact on its stealth capabilities. Thus, it is important to study the flow control and vortex manipulation of submarines.

As stated before, the key to studying the flow control and vortex manipulation of submarines is to identify vortex structure correctly. The history of vortex

F. Huang · L. Cao · D. Wan (✉)
Computational Marine Hydrodynamics Lab (CMHL), School of Naval Architecture, Ocean and Civil Engineering, Shanghai Jiao Tong University, Shanghai 200240, China
e-mail: dcwan@sjtu.edu.cn

D. Wan
Ocean College, Zhejiang University, Zhoushan 316021, China

Y. Wang et al. (eds.), *Liutex and Third Generation of Vortex Identification*, Springer Proceedings in Physics 288, https://doi.org/10.1007/978-981-19-8955-1_14

identification methods can be classified into three generations so far. The first-generation methods are based on vorticity, with the implication that the antisymmetric part of the velocity gradient tensor obtained by Cauchy-Stokes decomposition represents the rigid-body rotation [2]. However, it has been found that this method cannot distinguish between swirling motions and shearing motions. The second generation of vortex identification methods, such as Q [3] and lamada2 [4] criteria, has been proposed to overcome the problems associated with vorticity-based methods. However, the second generation methods are often plagued with case-related threshold problems in actual use and are prone to contamination by shearing. Another thing that confuses many engineers is that with the different values of Q or lamada2, the vortex images are often different. The "correct value of Q" seems to be an empirical value for every engineer.

Until Liu [5] and Gao [6, 7] proposed the third generation of vortex identification method called Liutex which can provide a mathematical and systematical definition of the local rigid rotation part of the fluid motion, including both the local rotational axis and the rotational strength. Compared to the first and the second generation methods, Liutex has three advantages when referring to engineering use. First, engineers do not need to hesitate about the value of Q, the threshold for the iso-surface is always 0.52. Second, Liutex can identify vortex excludes shearing motions and shear boundary from vorticity which will help engineers capture true vortex structure. Last but not least, Liutex can figure out the strong and weak vorticities simultaneously. Therefore, Liutex can be a strong tool for researchers to analyze engineering problems.

In this paper, the aim of the present study is to analyze vortices shed by a generic submarine, and compare the results between the Liutex and the traditional Q criterion. We chose two different types of generic submarines and two different environmental conditions are selected. The first case is the DARPA Suboff submarine moving in the constant density environment in Sect. 14.2, and the second case is the Joubert BB2 submarine moving in the linearly stratified environment in Sect. 14.3. Since there are various vortical structures from the main body and appendages, during the simulation, we will concentrate on the positions where the strong vortices are generated, such as the root of the sail (the well-known horse-shoe vortices), the tip of the sail, the forward hydroplanes, and from the aft rudders.

14.2 The DARPA Suboff Submarine

The first benchmark is the DARPA suboff submarine model. In this part, 6 different velocity conditions are computed with RANS methods. Before we checked the vortex structures with liutex of the suboff, we must ensure we get a reasonable hydrodynamic result. Thus, we introduce the geometry of the DARPA suboff submarine model in Sect. 14.2.1. Then the computational setup procedures are presented in Sect. 14.2.2. Finally, we discussed the resistance and coherent vortical structures of the suboff in Sect. 14.2.3.

Table 14.1 Main particulars of the DARPA Suboff model

Item	Value
Length of overall, L_{oa}	4.356 m
Length between perpendiculars, L_{pp}	4.261 m
Maximum hull diameter, $D_{\max}$	0.508 m
Center of buoyancy, LCB	$0.4621L_{oa}$
Volume of displacement, ∇	0.718m^3
Wetted surface area, S	6.338m^2

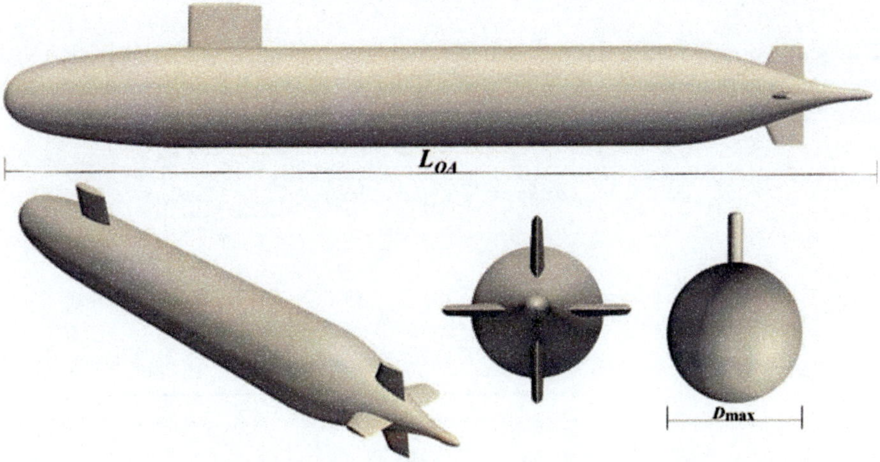

Fig. 14.1 The fully appendage DARPA Suboff model

14.2.1 *The DARPA Suboff Submarine Model*

In this part, a generic fully appended submarine model, DARPA Suboff AFF8 is chosen as the test case [8]. Table 14.1 presents the main particulars of the DARPA suboff submarine model, and Fig. 14.1 shows the side and front view of the DARPA Suboff submarine.

14.2.2 *Computational Setup*

14.2.2.1 Governing Equations

The incompressible Navier–Stokes equations are used and the continuity and momentum equations are as follows:

$$\nabla \cdot \mathrm{u} = 0 \tag{14.1}$$

$$\frac{\partial \boldsymbol{u}}{\partial t} + \nabla \cdot (\boldsymbol{u}\boldsymbol{u}) = -\nabla \frac{p}{\rho} + \nabla \cdot (\nabla \boldsymbol{u}) \tag{14.2}$$

where $\boldsymbol{u}$ is the velocity, ρ is the weighted average density, p is the dynamic pressure, and υ is the effective dynamic viscosity.

14.2.2.2 Turbulence Modeling

The whole problem is simulated with Reynolds averaged Navier–Stokes (RANS) equation, and the Shear-Stress Transport (SST) k-ω turbulence model is adopted to close the equation. The governing equations of k and ω can be written as follows.

$$\frac{\partial}{\partial t}(\rho k) + \nabla \cdot (\rho \boldsymbol{v} k) = \nabla \cdot \left(\mu_{eff,k} \nabla k\right) + \underbrace{P_k - \beta^* \rho k \omega}_{S^k} \tag{14.3}$$

$$\frac{\partial}{\partial t}(\rho \omega) + \nabla \cdot (\rho \boldsymbol{v} \omega) = \nabla \cdot \left(\mu_{eff,\omega} \nabla \omega\right) + \underbrace{\widetilde{C}_\alpha \frac{\omega}{k} P_k - \widetilde{C}_\beta \rho \omega^2 + 2(1 - F_1)\sigma_{\omega 2} \frac{\rho}{\omega} \nabla k \cdot \nabla \omega}_{S^\omega} \tag{14.4}$$

where all their coefficients depend on the Menter and Kuntz [9].

14.2.2.3 Computational Domain and Mesh

The entire domain, which includes $1L$ in front of the submarine, $3L$ behind the submarine, and $1L$ on each broadside, is used to calculate the hydrodynamic performance of the submarine. To prevent wall effects, the simulation's vertical dimension is set to $2L$. In addition, four different boundary condition types are used in this paper. The DARPA suboff submarine has a no-slip wall boundary condition established; velocity inflow and outflow boundary conditions are used for the inlet and outlet, respectively; and symmetry boundary conditions are used on the remaining portion of the domain. The dimensions stated above are better visualized in Fig. 14.2.

The whole calculation grid for the entire domain is displayed in Fig. 14.3. The ICEM CFD is used to adapt the structured mesh. A topology of O-grids is used to surround the submarine. We established an inner zone that is close to the suboff and an outer zone for resolving the submarine's flow field in order to obtain a reasonable hydrodynamic result. To ensure the proper resolution of the shear layer and lower force and moment errors, more cells are added to the inner zone in the port, starboard,

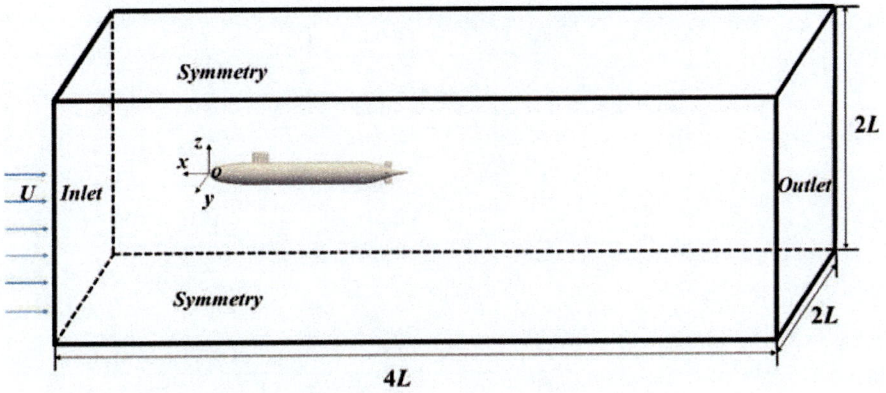

Fig. 14.2 Domain features and dimensions

rudder, and stern sections. A suitable mesh density is offered in the outer zone to provide a balance between resolution and computational expense. There are roughly 6 million cells.

In this part, we have computed 6 different velocity conditions with the Reynolds number is up to 10,000,000. The computational matrix is shown as Table 14.2.

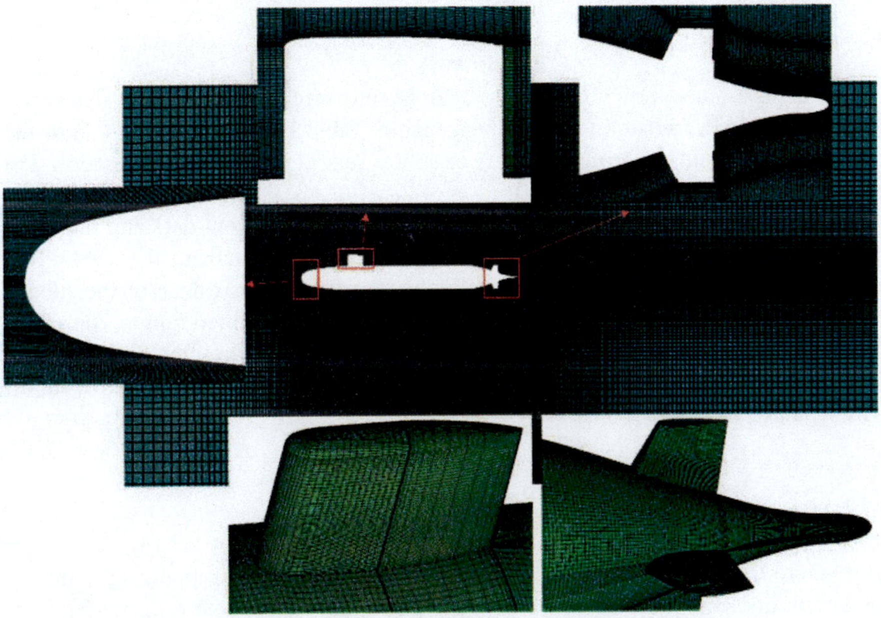

Fig. 14.3 Computational grids for the DARPA Suboff submarine

Table 14.2 Computational Matrix

Cases	v (m/s)	Re
1	3.051	1.17×10^7
2	5.144	1.96×10^7
3	6.096	2.33×10^7
4	7.161	2.74×10^7
5	8.231	3.15×10^7
6	9.152	3.50×10^7

Table 14.3 The resistance of suboff at different speeds

v (m/s)	R_T (N) (CFD)	R_T (N) (EFD)	Relative errors (%)
3.051	102.6	102.3	0.29
5.144	288.9	283.8	1.80
6.096	398.4	389.2	2.36
7.161	530.3	526.6	0.70
8.231	678.6	675.6	0.44
9.152	829.5	821.1	1.02

14.2.3 Results and Discussions

14.2.3.1 Resistance

Before using Liutex and Q methods to analyze the vortex structures, the first step is to ensure the hydrodynamic result is reasonable. Table 14.3 and Fig. 14.4 show the compared CFD and experiment results of suboff resistance at different speeds. The experiment data is from Huang [10].

In Fig. 14.4, the white square points stand for the experiment data and the black circle point represent the present CFD data. It can be seen from the Table 14.3 and Fig. 14.4, the CFD method always higher predict the resistance of the suboff, however, the relative errors are within 3% which is in an accepted range. Therefore, it can be concluded that the CFD method can accurately predict the hydrodynamic performance of the suboff. In Sect. 14.2.3.2, the coherent vortical structures of suboff will be analyzed.

14.2.3.2 Coherent Vortical Structures

Since there are various vortical structures from the main body and appendages, during the simulation, we will concentrate on the positions where the strong vortices are generated, such as the root of the sail (the well-known horse-shoe vortices), and the aft rudders. We compare the Q criteria result and the Liutex result in different locations of the suboff. The first location is the head of the suboff, and the second is the sail of the suboff and the last location is the aft rudders.

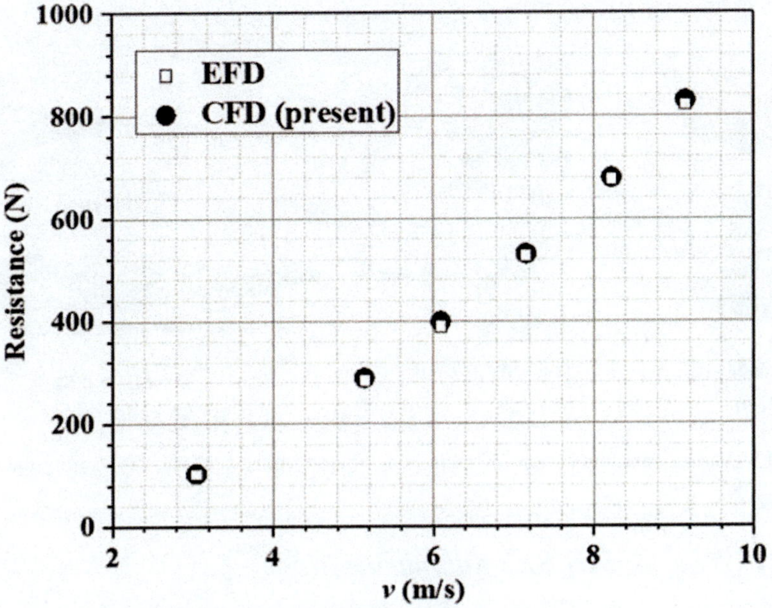

Fig. 14.4 The comparison of the resistance of the DARPA Suboff submarine at different speeds between experiments [10] and the present CFD results

In Fig. 14.5a, when using the *Q* criteria to identity the vortex structure of the head of the submarine, it would include shearing motions and shear boundary from vorticity, which means there are fake vortex structure in the head of the submarine. However, when using the Liutex method, it filters out false vortex structures in the head, and there are only several true vortex structure appear in the curvature change of the head. This also happened in the sail of the suboff and the rudders. In Fig. 14.5b and c, Liutex method can filter out false vortex structures on the sail and rudder, so that the true sail horseshoe vortex and the rudder vortex are presented.

14.3 The Joubert BB2 Submarine

The second benchmark is the Joubert BB2 submarine case. In this part, we compute a submarine case in a real ocean environment which completely different from the suboff case. The test geometry is provided in Sect. 14.3.1. Then we set a linearly stratified environment based on the thermocline model in Sect. 14.3.2. Finally, we discussed the hydrodynamic performance and coherent vortical structures of the Joubert BB2 submarine in the linearly stratified environment.

Fig. 14.5 Vortical structures from the main body and appendages of suboff by using the Q and Liutex criteria

14.3.1 The Joubert BB2 Submarine Model

The Joubert BB2 submarine model is the common submarine that is being studied in this section. It is a modified version of the underwater vehicle standard known as the Joubert BB1, which was created by Joubert [11]. Several research teams have tested the hydrodynamic studies on BB2, including experiments using free sailing maneuvering models at the Maritime Research Institute Netherlands (MARIN) and various CFD solvers (Torunski [12]; Donadei, [13] and Bettle [14]).

The Joubert BB2 submarine and the coordinate systems used in this section are seen in a side and front perspective in Fig. 14.6. The Joubert BB2's complete and model scale geometric major characteristics are shown in Table 14.4, and the scaling ratio of 18.348 which is in line with international conventions and makes it easier to compare following resistance results.

14.3.2 Computational Setup

14.3.2.1 The Thermocline Model

Based on Cao [15] and Huang [16], we solved the Navier–Stokes equations under Boussinesq approximation and proposed a thermocline model to achieve the linearly stratified environment. The governing equations are shown as follows:

$$\nabla \cdot \boldsymbol{u} = 0 \tag{14.5}$$

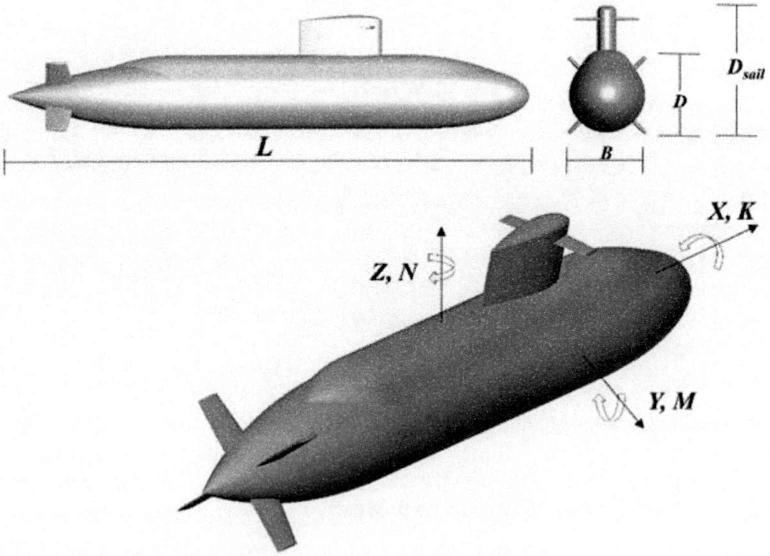

Fig. 14.6 The Joubert BB2 geometry

Table 14.4 Main parameters of BB2 (model scale 1:18.348)

Description	Symbol	Scale	
		Full	Model
Length	L (m)	70.2	3.8260
Beam	B (m)	9.6	0.5232
Draft to Duck	D (m)	10.6	0.5777
Draft to Sail top	D_{sail} (m)	16.2	0.8829
Wetted surface area	S_{wa} (m^2)	2142.3	6.3635

$$\frac{\partial \boldsymbol{u}}{\partial t} + (\boldsymbol{u} \cdot \nabla)\boldsymbol{u} = -\frac{1}{\rho}\nabla P + \upsilon\nabla^2\boldsymbol{u} + \beta \boldsymbol{g}T' + \boldsymbol{f} \tag{14.6}$$

$$\frac{\partial T'}{\partial t} + (\boldsymbol{u} \cdot \nabla)T' + \gamma u_z = \alpha\nabla^2 T' \tag{14.7}$$

where $\boldsymbol{u}$, ρ, P, υ, α, β, $\boldsymbol{g}$, T', $\boldsymbol{f}$ denote the velocity, density, modified pressure, kinematic viscosity, thermal diffusivity, volumetric expansion coefficient, gravitational acceleration, temperature fluctuation, and large-scale force required to maintain the turbulence, respectively.

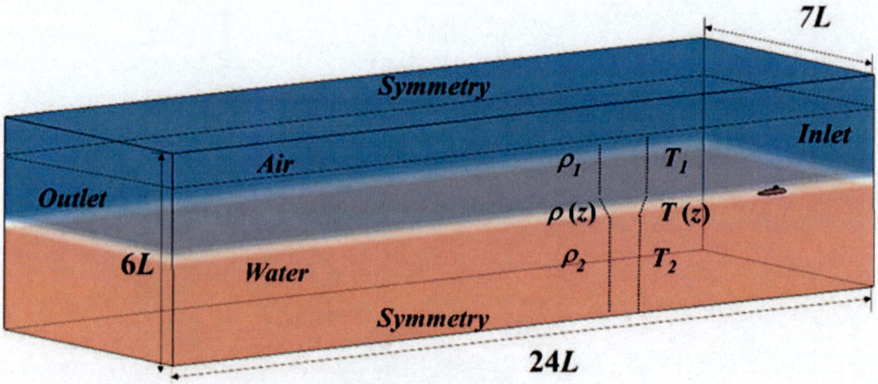

Fig. 14.7 Domain features and dimensions

14.3.2.2 Computational Domain and Mesh

With the submarine located at the front of the block, the computational domain is a cuboid. The entire domain is $4L$ in front, $20L$ behind, and $7L$ on either broadside of the submarine. To prevent wall effects, the simulation's vertical dimension is set to $3L$. A sponge layer with a length of $1L$ is inserted in the inlet and broadside to prevent wave reflection. In addition, four different boundary condition types are used in this paper. The Joubert BB2 submarine has a no-slip wall boundary condition established; velocity inflow and outflow boundary conditions are used for the inlet and outlet, respectively; and symmetry boundary conditions are used on the remaining portion of the domain. Figure 14.7 helps to visualize the dimensions listed above.

The whole computing grid for the entire domain is displayed in Fig. 14.8. The original mesh is a half mesh with symmetry plane $y = 0$. Then it is reflected. To ensure the proper resolution of the shear layer and lessen force and moment mistakes, more cells are added to the port, starboard, rudder, and stern regions. For the stratified case, there are roughly 16 million total cells.

In this part, we calculated a model scale submarine model with all appendages near the free surface ($z = 0.25L$) with an onset velocity of $U = 1.201$ m/s in the specific stratification.

14.3.2.3 Initial Conditions for Stratification

As to the stratified case, we simulated a weakly stratified condition with $Fr \approx 3$. The linear density stratification visualized in Fig. 14.9. While the temperature is a function of depth z, which is measured from the submarine to the calm water surface. As depicted in Fig. 14.9, $T_0 = 287.15$ K and varies from 286.15 to 288.15 K at the top and bottom, and $\rho_0 = 1011$ kg/m^3 and varies from 998 to 1024 kg/m^3.

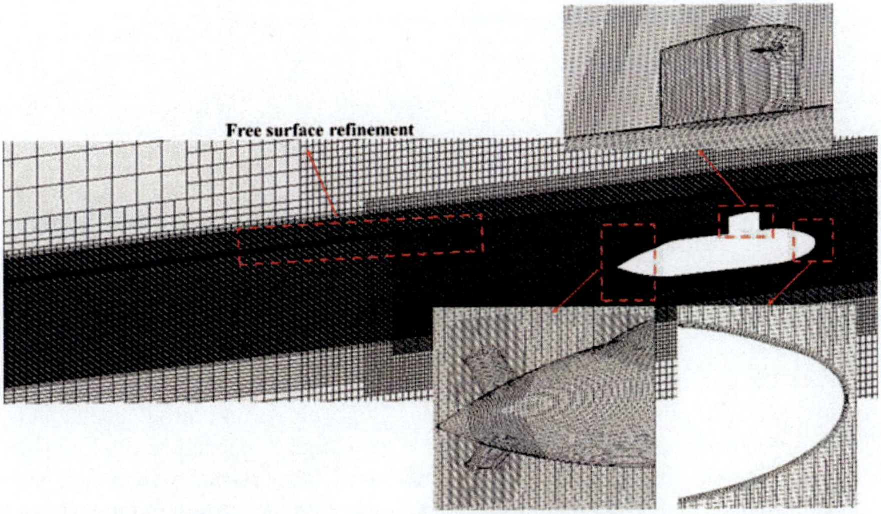

Fig. 14.8 Computational grids for the Joubert BB2 submarine with grids details on the main BB2 submarine control surfaces

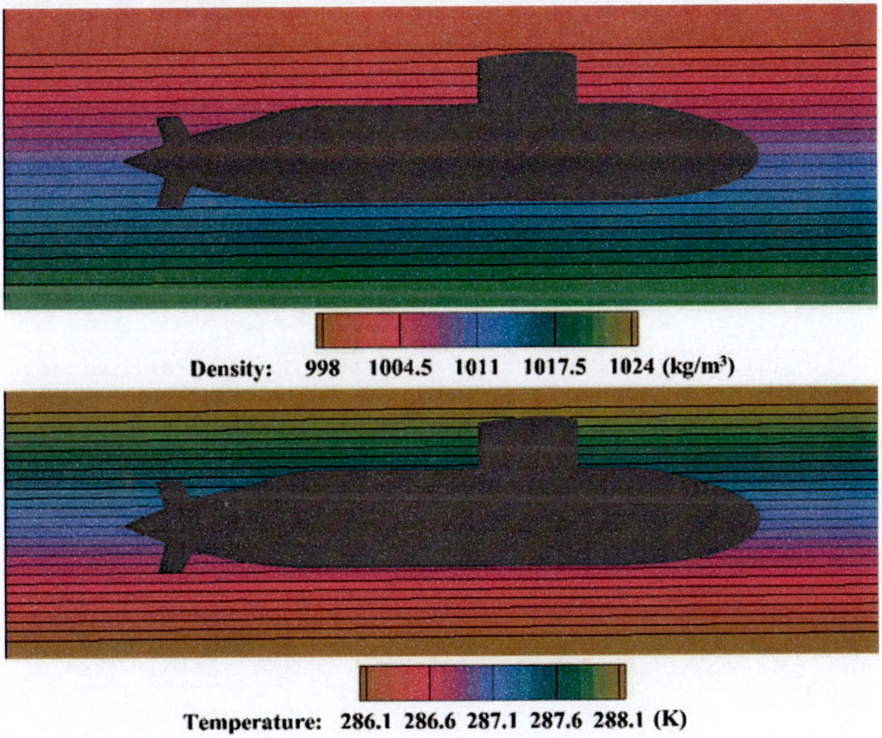

Fig. 14.9 Initial density and temperature profile in the median vertical plane ($y = 0$)

14.3.3 Results and Discussions

14.3.3.1 Forces and Moments

The validation study is carried out for the deeply homogeneous condition because there are not enough stratified data for submarines. Forces (X, Y, Z) and moments (K, M, N) are converted non-dimensional for comparison using Eqs. (14.8) and (14.9). The normalized force X' represents the resistance coefficient, positive Z' here shows an attempt to push the submarine to the surface, and M' is the non-dimensional pitch moment which equals positive indicating a bow-up moment. The current analysis of the hydrodynamic data is displayed in Table 14.5. The stratified example is the main case, and the current homogeneous case is utilized to compare the hydrodynamic performance with other findings from various CFD solvers and turbulence models. The non-dimensional force and moment value very slightly depend on the Reynolds number, as seen in Table 14.5, across all simulations. The resistance force and pitch moment are marginally less than in the prior simulation, according to the results of this study. The current findings and those of Torunski [12] differ by 0.55% in the case of X', while a 6.45 and 20.2% difference was found for Z' and M', respectively. At a Reynolds number of 3.9×10^6, the resistance and pitch moment likewise exhibit strong agreement with the supplementary data reported by Bettle [14] and Donadei [13]. Additionally, the resistance increases between the stratified and homogeneous cases by approximately 5.5%, which is a common occurrence in the stratified situation [16].

$$X', Y', Z' = \frac{X, Y, Z}{\frac{1}{2}\rho U^2 L^2} \tag{14.8}$$

$$K', M', N' = \frac{K, M, N}{\frac{1}{2}\rho U^2 L^3} \tag{14.9}$$

where ρ is fluid density, U is the submarine speed, and L is the length of the submarine.

Table 14.5 Comparison of the BB2 forces and moments in deeply submerged conditions

Simulation Type	$Re \cdot 10^{-6}$	$X' \cdot 10^3$	$Z' \cdot 10^4$	$M' \cdot 10^4$
STAR-CCM+ —SST homogeneous (Present)	4.6	1.82	1.16	0.79
STAR-CCM+ —SST stratified (Present)	4.6	1.92	–	–
ReFRESCO KSKL—(Torunski, [12])	4.6	1.83	1.24	0.99
ANSYS CFX—SST (Bettle, [13])	3.9	1.87	1.06	1.03
STAR-CCM+ —SST (Donadei, [14])	3.9	1.85	1.13	1.04

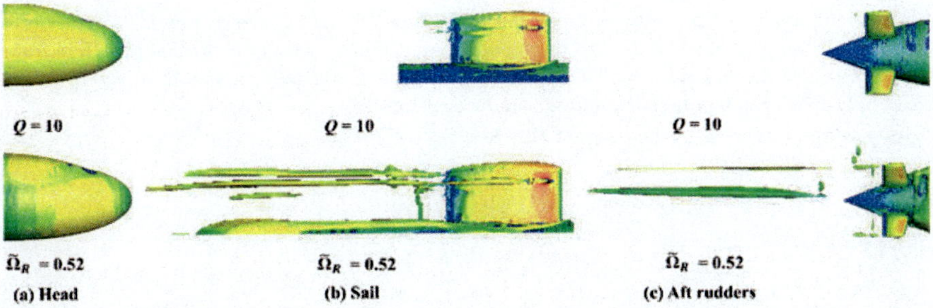

Fig. 14.10 Vortical structures from the main body and appendages of the Joubert BB2 by using the Q and Liutex criteria

14.3.3.2 Coherent Vortical Structures

As stated before, we will concentrate on the positions where the strong vortices are generated, such as the root of the sail (the well-known horse-shoe vortices), and the aft rudders. We compare the Q criteria result and the Liutex result in different locations of the Joubert BB2 submarine. The first location is the head of the submarine, and the second is the sail of the submarine and the last location is the aft rudders.

In Fig. 14.10a, when using the Q criteria to identity the vortex structure of the head of the submarine, it would include shearing motions and shear boundary from vorticity, which means there are fake vortex structure in the head of the submarine. However, when using the Liutex method, it filters out false vortex structures in the head, and there are only several true vortex structure appear in the curvature change of the head. This also happened in the sail of the submarine and the rudders. In Fig. 14.10b and c, Liutex method can filter out false vortex structures on the sail and rudder, so that the true sail horseshoe vortex and the rudder vortex are presented.

14.4 Conclusions

The present work is devoted to a preliminary research on identifying vortex structure of a three-dimensional fully appended submarine at high Reynolds number. The main conclusions are as follows:

1. Liutex can avoid case-related thresholds used by the second generation methods and capture both strong and weak vortices simultaneously.
2. Liutex can offer reasonable answers for vortex definition and identification even in the simulation of complex submarines at high Re numbers.
3. Last but not least, Liutex can accurately represent vortices which will help understand the submarine's hydrodynamic performance.

Further efforts can be devoted to applying to submarine maneuvering conditions and try to determine the vortex core center lines by using Liutex.

Acknowledgements This work is supported by National Natural Science Foundation of China (Nos. 52001210 and 52131102), the Oceanic Interdisciplinary Program of Shanghai Jiao Tong University (SL2020PT104), the National Key Research and Development Program of China (2019YFB1704200), to which the authors are most grateful.

References

1. C. Fureby, B. Anderson, D. Clarke, Experimental and numerical study of a generic conventional submarine at 10° yaw. Ocean Eng. **116**, 1–20 (2016)
2. H. Helmholtz, Uber Integrale der hydrodynamischen Gleichungen, welche den Wirbelbewegungen entsprechen. J. Reine Angew. Math. **55**, 25–55 (1858)
3. J. Hunt, A. Wray, and P. Moin, Eddies, streams, and convergence zones in turbulent flows, Report CTR-S88, Center For Turbulence Research, (1988)
4. J. Jeong, F. Hussain, On the identification of a vortices. J. Fluid Mech. **285**, 69–94 (1995)
5. C. Liu, Y. Gao, S. Tian, X. Dong, Rortex—a new vortex vector definition and vorticity tensor and vector decompositions. Phys. Fluids **30**, 035103 (2018)
6. Y. Gao, C. Liu, Rortex based velocity gradient tensor decomposition. Phys. Fluids **31**, 011704 (2019)
7. Y. Gao, C. Liu, Rortex and comparison with eigenvalue-based vortex identification criteria. Phys. Fluids **30**, 085107 (2018)
8. N.C. Groves, T.T. Huang, M.S. Chang, Geometric characteristics of DARPA Suboff Models (DTRC Model Nos. 5470 and 5471) (No. DTRC/SHD-1298–01) (1989)
9. F.R. Menter, M. Kuntz, Adaptation of eddy-viscosity turbulence models to unsteady separated flow behind vehicles, in *The Aerodynamics of Heavy Vehicles: Trucks, Buses, and Trains.* ed. by R. McCallen, F. Browand, J. Ross. Lecture Notes in Applied and Computational Mechanics. (Springer, Heidelberg, 2004), pp.339–352
10. H.-L. Liu, T.T. Huang, Summary of DARPA Suboff experimental program data (No. CRDKNSWC/HD-1298–11). Naval Surface Warfare Center Carderock Division (NSWCCD), West Bethesda, MD, USA (1998)
11. P.P.N. Joubert, Some aspects of submarine design-part 2. Shape of a submarine 2026 (2006)
12. B. Torunski, Computational analysis of the free surface effects on a BB2 submarine undergoing horizontal maneuvers. M.Sc. thesis, The University of New Brunswick (2016)
13. C. Donadei, Computational fluid dynamics predictions of submarine hydrodynamics. M.Sc. thesis, ENSTA Bretagne (2019)
14. M.C. Bettle, Development of computational fluid dynamics (CFD) meshes for the BB2 generic submarine. Scientific Report, Defence Research and Development Canada (2020)
15. L. Cao, F. Huang, C. Liu, D. Wan, Vortical structures and wakes of a sphere in homogeneous and density stratified fluid. J. Hydrodyn. **33**, 207–215 (2021)
16. F. Huang, Q. Meng, L. Cao, D. Wan, Wakes and free surface signatures of a generic submarine in the homogeneous and linearly stratified fluid. Ocean Eng. **250**, 111062 (2022)

Chapter 15
Vortex Identification Methods Applied to Complex Viscous Flow Field of Ship in Restricted Waters

Hongjie Cao, Chao Guo, and Decheng Wan

Abstract Vortex plays a vital role in the generation and maintenance of turbulence. The flow mechanism around the ship and marine structures can be deeply analyzed through the vortex structure, which is of great significance to the study of turbulence problems. The performance of ships in restricted waters is much different from ships in open waters. The issue of the ship entering a lock is one of the most typical situations of the restricted waters. In this paper, different generations of vortex identification methods are applied to complex viscous flow field of an 8,000t bulk carrier entering a lock. The turbulent flows are modeled by Reynolds averaged Navier–Stokes (RANS) simulations based on the two-equation k-ω shear stress transport (SST) model. With the help of the overset grid method, the surface pressure of the ship, wave height of the free surface and the wave field are analyzed numerically to explain the ship hydrodynamic performance. Then, based on Liutex vortex identification, the vortex structures in the viscous flow field are captured. Compared with the traditional vortex identification methods, the third-generation methods show a better ability of capturing the vortex structures and gives a more accurate definition of vortex, which can be used in the study focused on the flow mechanism, especially on the problem of the ship-lock interaction. All the analysis aspects of the flow field in this work above provide a reference for the post-processing method of the complex viscous flow field of ship in restricted waters.

H. Cao · D. Wan (✉)
Computational Marine Hydrodynamic Lab (CMHL), School of Naval Architecture, Ocean and Civil Engineering, Shanghai Jiao Tong University, Shanghai, China
e-mail: dcwan@sjtu.edu.cn

C. Guo
Key Laboratory of Navigation Structure Construction Technology, Ministry of Transport, Nanjing Hydraulic Research Institute, Nanjing, China

Y. Wang et al. (eds.), *Liutex and Third Generation of Vortex Identification*, Springer Proceedings in Physics 288, https://doi.org/10.1007/978-981-19-8955-1_15

15.1 Introduction

With the development of modern shipping industry, infrastructures such as ports, docks, rivers, and lock chambers are increasingly difficult to meet the navigation needs of large ships. The originally open channels such as coastal, inland rivers, ports and locks have become restricted waters. In 2021, the 400-m vessel Ever Given was stuck in the Suez Canal. More than one week blockage resulted in 422 vessels waiting to pass through the canal and caused huge economic losses.

As a result, the issue of restricted waters has been given more attention. In the restricted waters, the ship is in a state of low-speed navigation and is vulnerable to the influence of bottom and bank of the channel, which seriously threatens the navigation safety of the ship. Generally, the differences between the restricted waters and open waters can be concluded into three main aspects: the pressure distribution around the hull, the components of resistance and the recirculation speed of water. These differences may influence the maneuvering performance of the ship, which may lead to the ship collision or grounding in the restricted waters.

Currently, there are many researchers have work on the issue of restricted waters. Toxopeus, S. L investigated the forces and moments acting on a ship as a function of different drift angle, yaw rate and water depth [1]. Liu and Wan analyzed the pure yaw motion of a ship and investigated its forces, moments, and local flow physics in shallow water [2]. Zou et al. focused on the bank effects on a tanker moving straight ahead in a canal and further investigated the "squat" phenomenon [3].

The issue of a ship sailing in a lock is a typical case of restricted waters. Generally, the space of the lock is shallow and narrow, which means the ship in a lock may face both shallow water effect and bank effect. Consequently, the ship-lock interaction is complicated. Most of researchers take the Panama lock as the research target. Based on the third group of locks in Panama, Vantorre et al. conducted a series of experimental studies to analyze the effects of speed, water depth, eccentricity and drift angle on the hydrodynamic characteristics of ships passing through the lock [4, 5]. Meng and Wan used RANS equations in combination with the *k-w* SST turbulence model to simulate a 12 thousand TEU ship model entering the Third set of Panama locks. His work introduced three important instants to analyze the work, focusing on wave pattern, dynamic pressure [6, 7].

Vortex plays a vital role in the generation and maintenance of turbulence. In marine hydrodynamics, the vortex structure can be generated by large drift angle maneuvering, deep-draft column stabilized floaters and appendages of a fully appended ship [8]. The mechanism around the ship and marine structures can be deeply analyzed through the vortex structure. For a better study on the vortex structures, currently, there are three generations of vortex identification methods. The first-generation method is simply based on the definition of vorticity proposed by Helmholtz in 1858. However, many researchers found that this method cannot accurately reflect the vortex structure of the flow field. Then, by modifying the first generation vortex identification method, the scholars proposed the second generation represented by Q-criterion, $\lambda 2$, etc. Unfortunately, the second generation vortex identification method is plagued by unclear physical meaning and artificial threshold selection, and cannot accurately reflect the vortex structure in the flow field. Therefore, the latest third generation vortex recognition method represented by Ω_R and Liutex raised by Liu's team [9, 10] initially has been gradually favored by people. Then, Ren, Wang and Wan use four different vortex identification methods to capture vortex structures to analyze the flow mechanism in the viscous large separated flow field, and find the third generation vortex identification methods are more suitable for displaying the vortex structures in a complex viscous flow field [11, 12].

In the present work, a new lock chamber is studied. The complex viscous flow field and the hydrodynamics of the hull are investigated to explore the ship-lock interaction. The RANS equations in combination with the *k-w* SST turbulence model is applied to simulate the complex viscous flow field on the self-developed CFD solver, naoe-FOAM-SJTU. And the third generation of vortex identification methods are applied to analyze the complex viscous flow field of the restricted waters. The paper is organized as follows: first the governing equations and main numerical methods are introduced, then the settings of numerical domain are established, and the hydrodynamics performance, free surface, dynamic pressure, vortex field are analyzed. Finally, the conclusion of this paper is drawn.

15.2 Numerical Methods

15.2.1 Governing Equations

In this paper, the whole work is based on the CFD solver, naoe-FOAM-SJTU, which is developed on the open-source code platform OpenFOAM, mainly composed of a dynamic overset grid module, a 6 degrees of freedom module and a numerical wave tank module. Then, in the numerical computations, the governing equations are the unsteady Reynolds averaged Navier–Stokes equations (URANS). Which are shown as follows:

$$\nabla \cdot \boldsymbol{U} = 0 \tag{15.1}$$

$$\frac{\partial \rho \boldsymbol{U}}{\partial t} + \nabla \cdot \left(\rho\left(\boldsymbol{U} - \boldsymbol{U}_g\right)\boldsymbol{U}\right) = -\nabla p_d - \boldsymbol{g} \cdot x \nabla \rho + \nabla \cdot \left(\mu_{eff} \nabla \boldsymbol{U}\right) + (\nabla \boldsymbol{U}) \cdot \nabla \mu_{eff} + \boldsymbol{f}_\sigma + \boldsymbol{f}_s \tag{15.2}$$

where **U** and $\mathbf{U}_g$ represent the velocity and the velocity of grid respectively; $p_d = p - \rho g x$ represents the dynamic pressure, ρ represents the density of the fluid, g is the gravity acceleration, μ_{eff} is the effective dynamic viscosity; f_σ and f_s are the surface tension term and source term for the wave elimination region.

15.2.2 Turbulence Model

In order to enclosure the URANS equations, the shear stress transport turbulence model, SST *k-w* model is selected. The SST *k-w* model is one of the widely used turbulence models, which can not only deal with the near-wall region efficiently, but also solve the free surface well. The equations of SST k-w model in OpenFOAM are shown as follows:

$$\frac{\partial k}{\partial t} + \nabla \cdot (Uk) = \tilde{G} - \beta^* k\omega + \nabla \cdot [(v + a_k v_t)\nabla k] \tag{15.3}$$

$$\frac{\partial \omega}{\partial t} + \nabla \cdot (U\omega) = \gamma S^2 - \beta\omega^2 + \nabla \cdot [(v + a_\omega v_t)\nabla \omega] + (1 - F_1)CD_{k\omega} \tag{15.4}$$

where, the k represents the turbulent kinetic energy; ω represents the turbulent dissipation rate; F_1 is a mixed function, which can be applied to switch between SST *k-w* model (near-wall region) and SST *k-ε* model (far-filed region).

15.2.3 Free Surface

For the purpose of capturing the change of free surface, the high precision volume of fluid (VOF). The core concept of this method is to calculate the volume fraction of different fluids in the grid element α to determine the interface. The relative proportion of different fluids in the grid cell is shown as follows:

$$\begin{cases} \alpha = 0 & air \\ \alpha = 1 & water \\ 0 < \alpha < 1 & interface \end{cases} \tag{15.5}$$

where, the α represents the relative proportion. When it equals to 0, it means the grid cell is completely filled with air while when it equals to 1, it means the grid cell is full of water.

15.2.4 Overset Grid Method

The dynamic overset grid technology is the key point for direct simulating the complex ship motions. Generally, it includes two or more blocks of overlapping structured or unstructured grids. Then, the decomposed parts will be nested into a uniform and orthogonal background grid. Besides, there will be some overlap between the grids of each subpart, so that the information of the flow field can be coupled and matched by interpolation, which is favor of realizing the calculation of the whole filed. Therefore, the overset grid method can not only retain the advantages of structured grids, but also make up for its deficiency on dealing with complex objects.

15.2.5 Vortex Identification Method

The third generation of vortex identification method generally has two kinds: $\boldsymbol{\Omega}_{\boldsymbol{R}}$ and Liutex/Rortex. Firstly, the vorticity ω is obtained by Cauchy Stokes decomposition. And it can be further decomposed into rotating part R and non-rotating pure shear S, which is show as:

$$\boldsymbol{\omega} = \boldsymbol{R} + \boldsymbol{S} \tag{15.6}$$

The $\boldsymbol{\Omega}$ is defined as a ratio of the vorticity tensor norm squared over the sum of the vorticity tensor norm squared and deformation tensor norm squared.

$$\boldsymbol{\Omega} = \frac{\|\boldsymbol{B}\|_{\mathrm{F}}^{2}}{\|\boldsymbol{A}\|_{\mathrm{F}}^{2} + \|\boldsymbol{B}\|_{\mathrm{F}}^{2} + \varepsilon} \tag{15.7}$$

where, ε is a small positive number to prevent division by zero. The value of ε can be determined by the Eq. (15.8), which avoids the influence of manual selection on vortex structure.

$$\varepsilon = 0.001 \times \left(\|\boldsymbol{B}\|_{F}^{2} - \|\boldsymbol{A}\|_{F}^{2}\right) \tag{15.8}$$

As a result, the value range of $\boldsymbol{\Omega}$ is between 0 and 1. When $\boldsymbol{\Omega} = 1$, it indicates that the fluid is rotating in a rigid body; When $\boldsymbol{\Omega} > 0.5$, the antisymmetric tensor B is dominant. In practical application, generally, $\boldsymbol{\Omega} = 0.51$ or 0.52 can be used as a fixed threshold to identify the vortex structure in the flow field.

For Liutex method, the rotational part of the vorticity is defined as Liutex/Rortex vector. Consequently, the physical quantity Liutex, which is a vector, has its direction and magnitude. It contains both the local rotation axis and the rigid-body angular speed. The Liutex can be defined as:

$$\boldsymbol{R} = \left(\langle \omega, r\rangle - \sqrt{\langle \omega, r\rangle^{2} - 4\lambda_{ci}^{2}}\right) r \tag{15.9}$$

where ω is the vorticity vector; r is the real eigenvector of ∇V; λ_{ci} is a Galilean invariant; ∇V is the velocity gradient tensor, which can be expressed as:

$$\nabla V = Q \nabla v Q^T = \begin{bmatrix} \frac{\partial U}{\partial X} & \frac{\partial U}{\partial Y} & 0 \\ \frac{\partial V}{\partial X} & \frac{\partial V}{\partial Y} & 0 \\ \frac{\partial W}{\partial X} & \frac{\partial W}{\partial Y} & \frac{\partial W}{\partial Z} \end{bmatrix} \tag{15.10}$$

where Q is a rotation matrix while U, V, W represent the velocity components in the XYZ coordinate respectively.

15.3 Numerical Setup

15.3.1 Geometry Model

In the present work, an 8000t ship is adopted as the research object. The geometry model of the ship is shown in Fig. 15.1 and the principal particulars are presented in Table 15.1. In the numerical simulations, the reduced model scale is adopted, which is 1:28.97, so that the length between the perpendiculars of the hull is regarded as 4.35 m.

The geometry model of the lock is symmetrical and at the entrance of the lock, there is an obvious rise in the bottom. In the top view, the channel changes from wide to narrow. And the water depth in the entrance of the lock has a sharp drop. The effective size of the full lock is 280.0 m × 40.0 m × 8.0 m (length by width by the height of the water in the lock). The geometry model of the ship is shown in Fig. 15.1. The draft of the ship is 5.5 m while the depth of the water in the lock is only 8 m. The ratio of water depth to draft is less than 1.5. According to the 23rd International Towing Tank Conference (ITTC), a ship's behavior and maneuverability depends on the depth h of the navigation area. When the ratio of water depth to draft is less than 1.5, it means the ship sailing in the shallow water and the effect of depth restrictions can be significant [13] (Fig. 15.2).

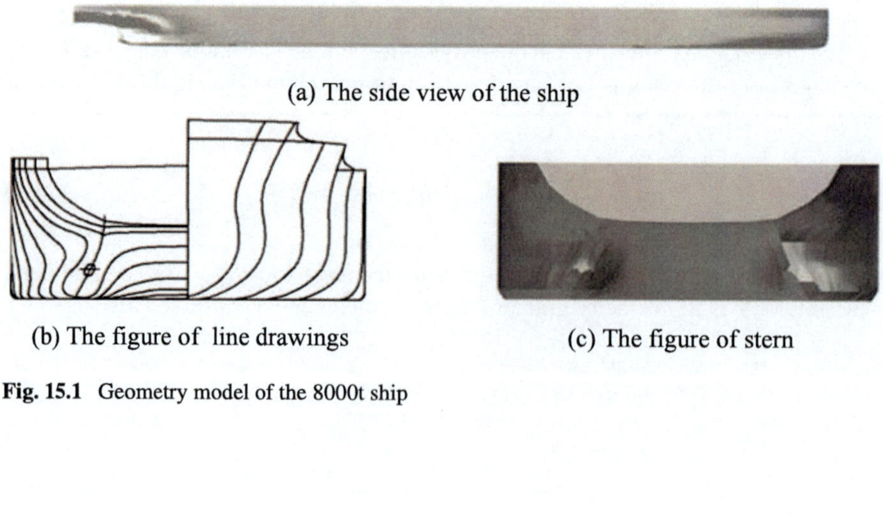

(a) The side view of the ship

(b) The figure of line drawings

(c) The figure of stern

Fig. 15.1 Geometry model of the 8000t ship

Table 15.1 Principal particulars of the 8000t ship

Main particulars	Symbols (Unit)	Model scale	Full scale
Length between perpendiculars	L_{pp}(m)	4.35	126
Beam	B_{WL} (m)	0.663	19.2
Draft	T(m)	0.190	5.5
Depth	D(m)	0.242	7
Displacement volume	∇(m^3)	0.495	12,045.124
Wetted surface area	S(m^2)	4.373	3670.293
Block coefficient	C_B	0.872	0.877

(a) The top view of the lock

(b) The side view of the lock

Fig. 15.2 Geometry model of the lock

15.3.2 Computational Domain

In the numerical simulation, the ship is towed into the lock from open water. The length between the perpendiculars of the ship model is 4.35 m and the speed of ship model is 0.162 m/s, corresponding to Fr = 0.0248. The computational domain and boundary are shown in Fig. 15.3. In the computational domain, the x-axis points stern, the y-axis points starboard, and the z-axis points upward. The dimension of the calculation domain is (taking the length between LPP vertical lines as the reference): $-2.46\ L_{pp} \leqslant x \leqslant\ 3.47\ L_{pp}$ in length; the open water: $-0.23\ L_{pp} \leqslant z \leqslant 0.23\ L_{pp}$ in height; the lock: $-0.063\ L_{pp} \leqslant z \leqslant 0.23\ L_{pp}$ in height; the open water: $-0.556\ L_{pp} \leqslant y \leqslant 0.556\ L_{pp}$ in width; the lock: $-0.159\ L_{pp} \leqslant y \leqslant 0.159\ L_{pp}$ in width. Plane $z = 0$ is the free surface, with air above and water below. The origin of the coordinates is located at the intersection of the free surface, the mid-ship section, and the lock entrance section.

In the case, the dynamic overset mesh is selected to support the simulation. The grids of lock domain and the hull domain are generated respectively in order to fit

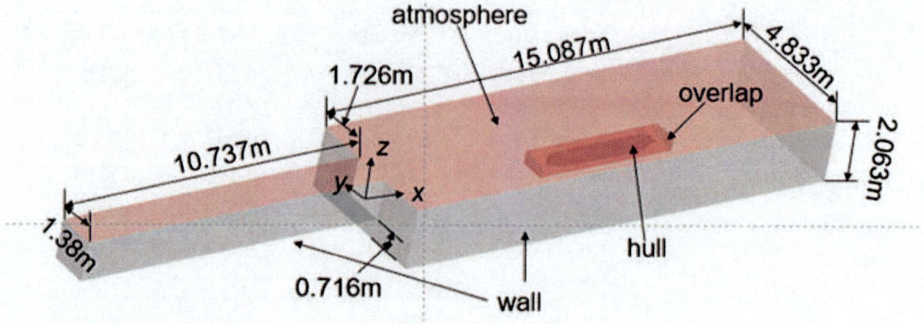

Fig. 15.3 Computational domin

the geometry well. The lock grids are made by the commercial software, Pointwise. And the grids of the hull are made by SnappyHexMesh module in OpenFOAM. Due to the special structure at the entrance of the lock, which is an obvious water level rise, the area of the free surface and the shallow water region are refined. The total number of grids is 6.41 M (M means million.), including 5.26 M background grids and 1.15 M hull grids. Figure 15.4. presents the grid distribution and Table 15.2. Shows the details of the mesh generation.

15.3.3 Typical Instant

In this work, 3 instants are considered as the analysis targets for a better analysis of the flow field and hydrodynamic characteristics of the ship, which represent three conditions of the ship in the lock. As shown in Fig. 15.5. Time 0 is the ship still in the open water; Time 1 represents the condition that the bow of the ship just reaches the entrance of the lock; Time 2 represents the condition that half of the ship is in the lock while Time 3 represents the condition the ship is fully in the lock.

15.4 Hydrodynamic Results

15.4.1 Hydrodynamic Characteristics

Figure 15.6 shows the time history curve of the hydrodynamic characteristics of the ship. First, it can be seen from the curve of drag force that, when the ship enters the lock from the far-field, the magnitude of the drag force shows a rising trend. Before the ship gets close to the lock, the resistance presents a regularity change. When the ship reaches the gate of the lock at Time 1, the value of the drag force decreases. This is because the backwater phenomenon inside the lock is unobvious in the beginning.

(a) The global grid

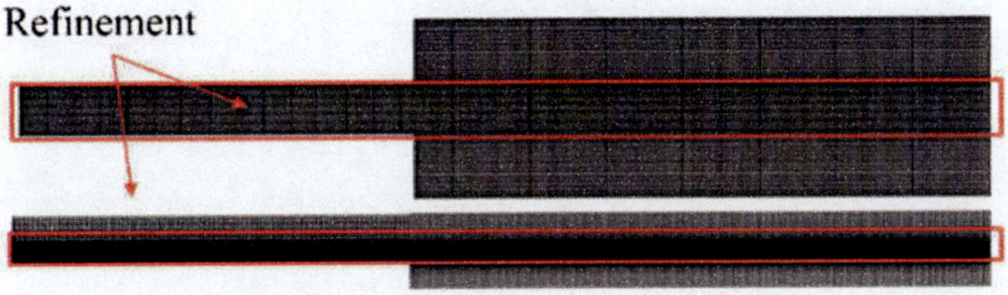

(b) The grid distribution and refinement

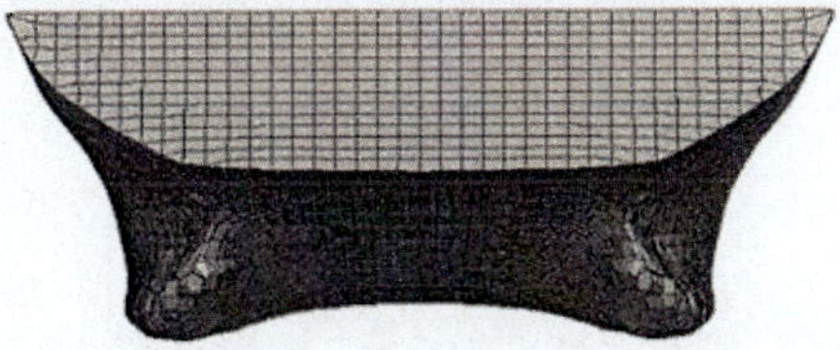

(c) The grid distribution on hull

Fig. 15.4 Grid distribution

Table 15.2 Details of the grid distribution

Background mesh (Million)	Hull mesh (Million)	Total (Million)
5.26	1.15	6.41

After Time 1, with the ship gradually enters in the lock, the water in the lock is accumulated in front of the ship, and due to the narrowing of the channel and the shallowing of the navigation water area, the blockage coefficient increases and the backwater phenomenon gets worse, result in the increase of the pressure resistance and friction resistance. On the contrary, since the ship enters the lock from the middle of the channel, the flow filed is symmetrical. So, the lateral force and yaw moment of the ship is also symmetrical, reflecting on the curve is their values are almost equal to zero. These results are in line with the facts.

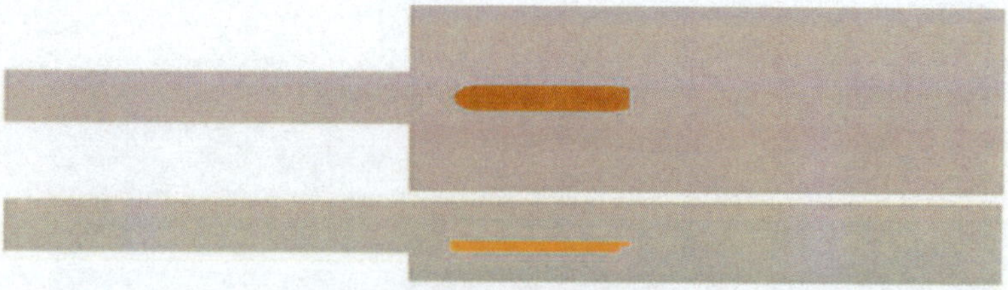

(a) T0: Open water

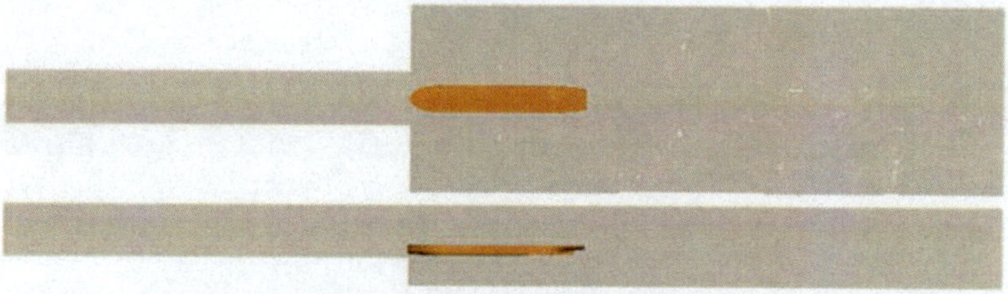

(b) T1: Reaching the entrance of the lock

(c) T2: Half in the lock

(d) T3: All in the lock

Fig. 15.5 Typical instants

15.4.2 Free Surface and Dynamic Pressure

The free surface distributions of the typical instants are shown in Fig. 15.7. When the ship gradually enters in the lock, there is an increase of the free surface in front of the ship. The flow section decreases rapidly from a wide-open channel to the

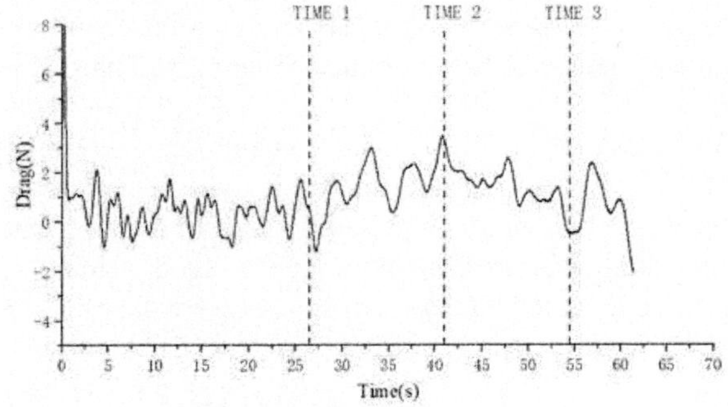

(a) Drag force

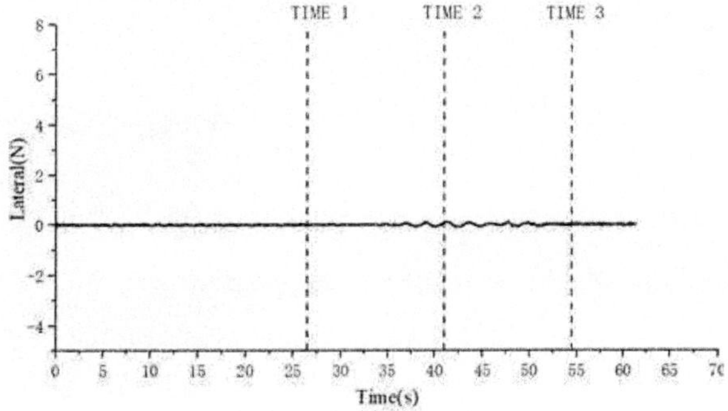

(b) Lateral force

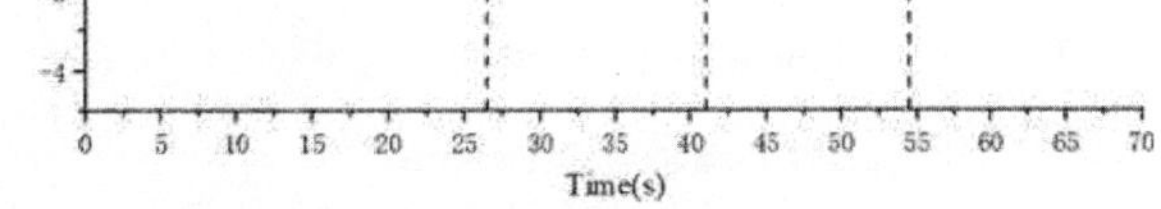

(c) Yaw moment

Fig. 15.6 Time history of the hydrodynamic forces and moment

narrow space between the ship and the lock, especially the space between the ship's bottom and the lock's bottom. At Time 2, when the half of the ship is in the lock, the backwater is most severe. In the picture, the lock is almost occupied by the red, which means the high level of the free surface while outside the entrance, the level is lower, colored by blue. At the last station, thc ship has entered the lock completely. The area of the high-level free surface reduces due to the accumulated water before evacuated out of the lock. The level of the free surface around the ship is lower than the other area. Besides, the blackwater effect causes a faster velocity of the return flow and a sinkage of the water level around the ship, leading to a general sinkage of the ship.

The same results can be obtained in the dynamic pressure distribution on the bottom of the hull. Overall, the distribution of the dynamic pressure is symmetrical because the flow field is symmetrical. When the half of the ship enters in the lock, the part in the lock has a lower pressure owing to the higher velocity according to the Bernoulli principle. At that time, the lower pressure pare will be pulled downward. As a result, the ship moves vertically downward, which is called the 'squat' phenomenon. And, with the ship entering the lock little by little, the area of negative pressure zone becomes larger. At Time 3, due to the return water evacuated out of the lock, the pressure on the stern is higher than that on the bow, causing a trim and threating the safety (Fig. 15.8).

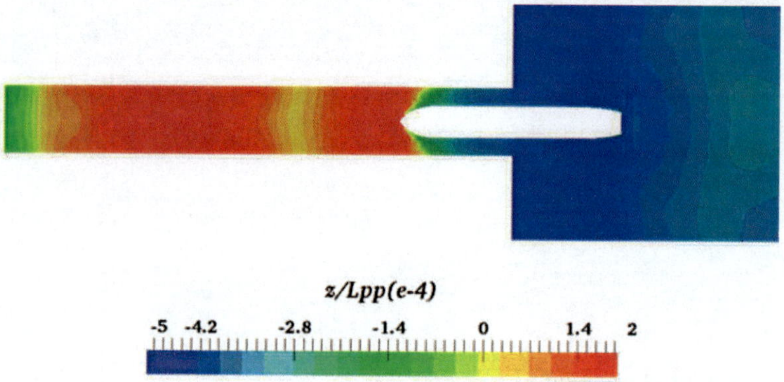

(a) T2: Half in the lock

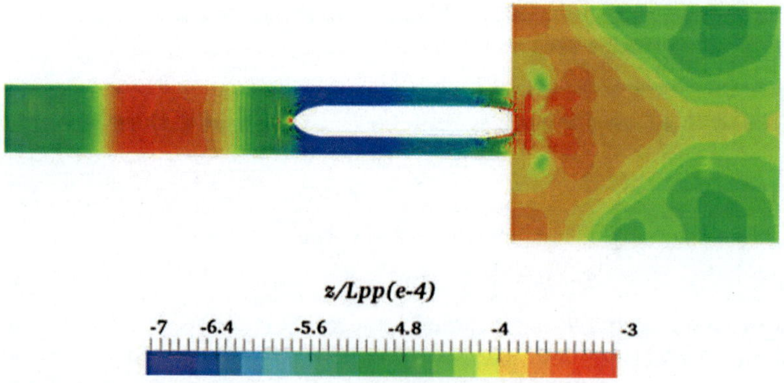

(b) T3: All in the lock

Fig. 15.7 The free surface distributions of the typical instants

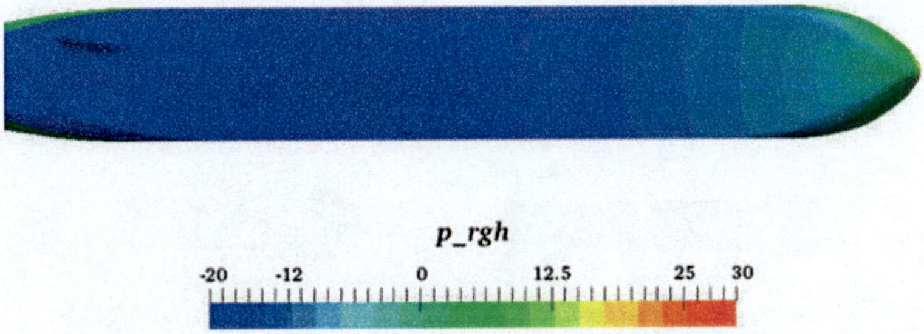

(a) T2: Half in the lock

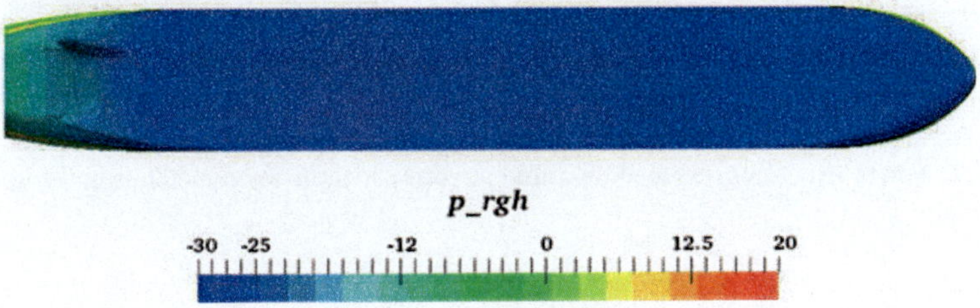

(b) T3: All in the lock

Fig. 15.8 The dynamic pressure distributions of the typical instants

15.4.3 Vortex Structures

Vortex structures around the hull are in different shapes and intensities. Four kinds of vortex identification methods from three generation are applied to capture the vortex structures, as is shown in Fig. 15.9. The Fig. 15.9a shows the different instants' vortex structures captured by the first-generation method: vorticity. From the figure, the hull is covered fully by the vortex structures in all instants. Obviously, these results are incredibility. The results obtained by Q criterion seem to be more convinced, which is shown in the Fig. 15.9b. Then, as a third-generation vortex identification method, the $\mathbf{\Omega}_{\mathrm{R}}$ method captures larger and more vortex structures in the flow field. However, the results of Q criterion might be easily affected by artificial threshold. Figure 15.10. shows the vortex structures captured by Q criterion with different artificial thresholds. In Time 3, when the ship has entered the lock, the vortex structures in the aft of the ship shows a winding shape and diffuses outwardly form the entrance of the lock, due to the sharp narrowing at the entrance. If the Q is larger than 0.2, the winding vortex cannot be capture. On the contrary, if the Q equals to 0.1 or much smaller, there will be more surplus disturbing the correct identification of the vortex.

On the whole, the second generation method Q criterion and the third generation methods $\boldsymbol{\Omega}_R$ and Liutex can capture the stream-wise vortex structures caused by the

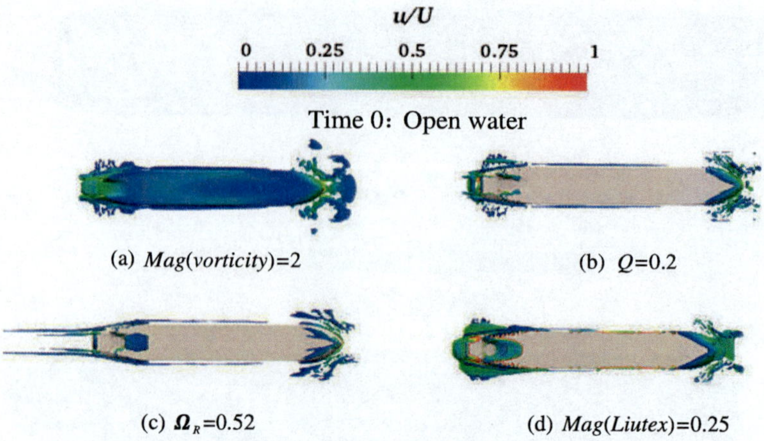

Fig. 15.9 Vortex structures obtained by different vortex identification methods in the open water

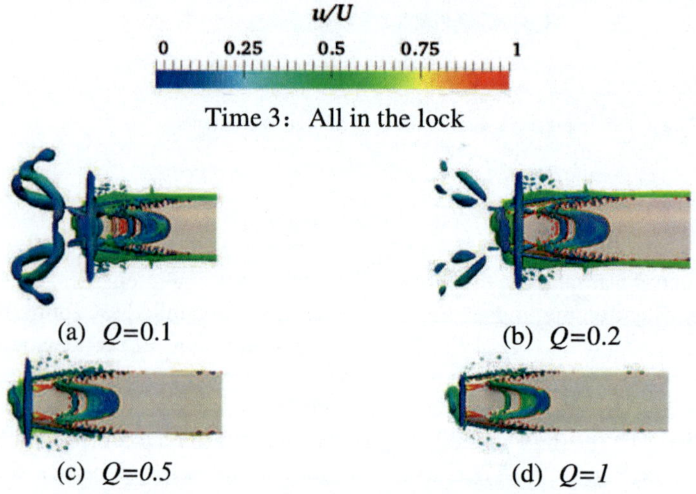

Fig. 15.10 Vortex structures obtained by Q criterion with different threshold in Time 3

upward of the lock bottom at the entrance and the span-wise vortex structures caused by the narrowing channel successfully. Between the two third generations of vortex identification methods, the wake vortex structures are obvious and the Liutex method shows more broken vortex structures (Fig. 15.11).

In addition, as the third generation vortex identification method, Liutex vector represents the rotational angular velocity of the rotating part of the local fluid rigid body, which eliminates the shear effect in various previous vortex identification methods. Consequently, the complex viscous flow field of low-speed ships will be further analyzed by using the Liutex vector method. Figure 15.12 shows the vertical

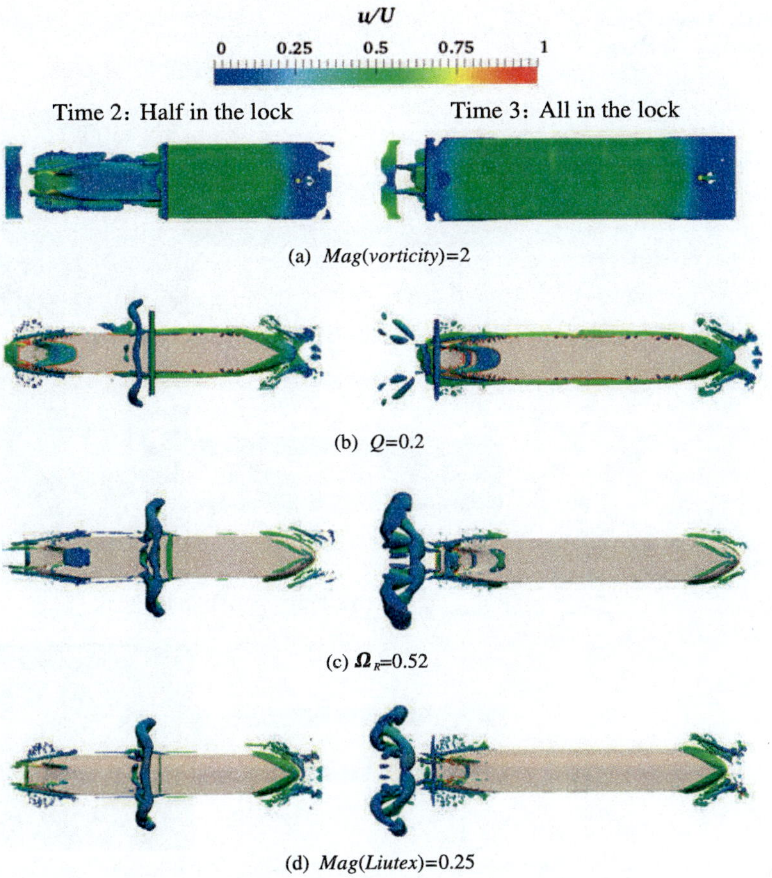

Fig. 15.11 Vortex structures obtained by different vortex identification methods

Liutex vorticity distribution at different times. When the ship enters the gate, there is a region of sharp change in vortex structures at the entrance. This is because the ship wake is blocked by the narrowing lock wall, and the wake velocity changes dramatically at the entrance of the lock. At the same time, after the ship completely enters the lock, a positive and negative vortex pair like Karmen vortex appears in the wake flow field of the ship.

15.5 Conclusion and Prospect

In this paper, the numerical simulations of a towing naked hull in restricted waters are carried out by coupling with overset grid technology. For a better analysis of the ship-lock interaction when the ship enters the lock, three typical instants are selected. The

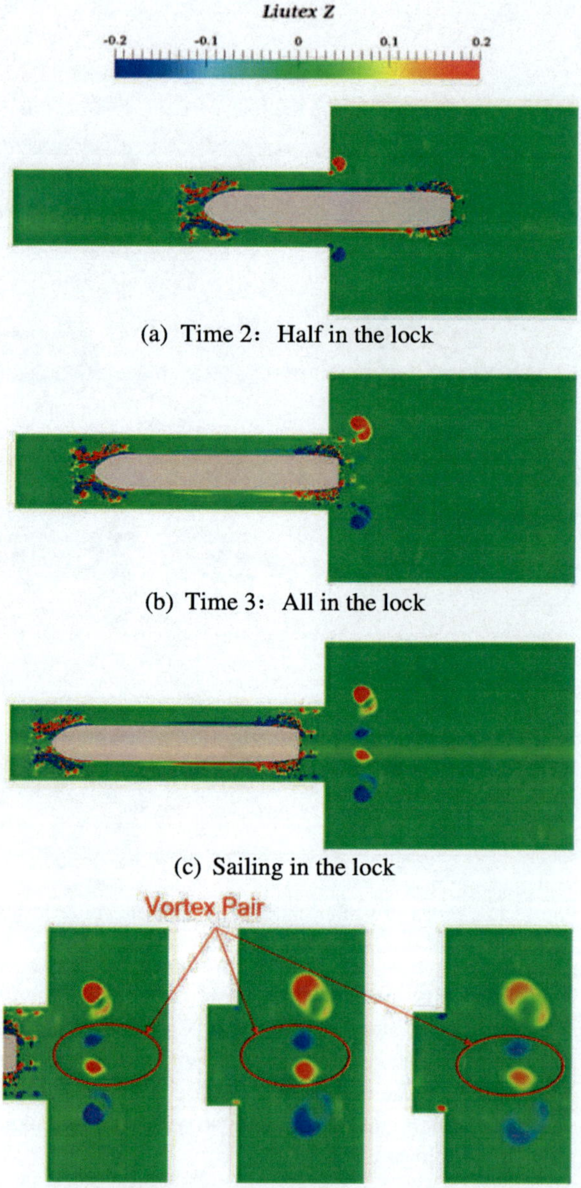

(a) Time 2: Half in the lock

(b) Time 3: All in the lock

(c) Sailing in the lock

(d) Vortex pair in the wake flow

Fig. 15.12 Vertical Liutex distribution on the plane

free surface and dynamic pressure of these instants are obtained. Besides, combined with the four different vortex identification methods, the vortex structures of the hull are gotten. Overall, three conclusions are summarized:

- As the dynamic pressure of the hull and the free surface shown, the results of the CFD simulations show the most dangerous time is half of the ship entering the lock. At that time, the magnitude of the darg force is the largest and the backwater phenomenon in the lock is the most severe.
- Through analysis, it is found that the third-generation vortex identification method can more accurately give the vortex structure around the hull. Due to the narrowing of the channel and the shallowing of the water depth, there will be a span-wise vortex structure and a stream-wise vortex structure at the entrance of lock, where the changes is the most significantly.
- Compared with the other vortex identification methods, Liutex vector taking advantage of its directionality, can be used to analyze the vorticity evolution process.

The future work will focus on the heavy and pitch motion of the hull while in the present work, the hull is fixed expect moving forward. And more towing velocity will be tested to study the impacts of different velocity on the case.

Acknowledgements This work is supported by National Natural Science Foundation of China (52131102, 51909160) and Foundation of Key Laboratory of Navigation Structure Construction Technology, to which the authors are most grateful.

References

1. S.L. Toxopeus, *Viscous-Flow Calculations for KVLCC2 in Deep and Shallow Water* (Springer, Netherlands, 2013)
2. X. Liu, D. Wan, Numerical simulation of ship yaw maneuvering in deep and shallow water (2015)
3. L. Zou, CFD prediction and validation of ship-bank interaction in a canal. 2nd International conference on ship manoeuvring in shallow and confined water, Trondheim, Norway (2011)
4. M. Vantorre, G. Delefortrie, Behaviour of ships approaching and leaving locks: open model test data for validation purposes. Flanders Hydraulic Research (2013), pp. 337–352
5. T. Vergote, K. Eloot, M. Vantorre, et al., Hydrodynamics of a ship while entering a lock, in *3rd International Conference on Ship Manoeuvring in Shallow and Confined Water: with non-exclusive focus on Ship Behaviour in Locks, R.I.N.A* (2012), pp. 281–289
6. Q. Meng, D. Wan, Numerical simulations of viscous flows around a ship while entering a lock with overset grid technique, in *Proceedings of the 25th International Ocean and Polar Engineering Conference* (2015), pp. 989–996.
7. Q. Meng, D. Wan, W. Huang, Numerical investigation of influence of eccentricity on the hydrodynamics of a ship maneuvering into a lock, in *Proceedings of the 6th International Conference on Computational Methods*, 962-3341-1-PB (2015)
8. W.W. Zhao, J.H. Wang, D.C. Wan, Vortex identification methods in marine hydrodynamics. J. Hydrodyn. **32**(2) (2020)

9. C. Liu, Y.S. Gao, X.R. Dong, Y.Q. Wang, J.M. Liu, Y.N. Zhang, X.S. Cai, N. Gui, Third generation of vortex identification methods: omega and Liutex/Rortex based systems. J. Hydrodyn. **31**(2), 205–223 (2019)
10. C. Liu, Y. Gao, S. Tian, X. Dong, Rortex—a new vortex vector definition and vorticity tensor and vector decompositions. Phys. Fluids **30**(3), 035103 (2018)
11. Z. Ren, J. Wang, D. Wan, Investigation of the flow field of a ship in planar motion mechanism tests by the vortex identification method. J. Marine Sci. Eng. **8**(9), 649 (2020)
12. Z. Ren, J. Wang, D. Wan, Investigation of fine viscous flow fields in ship planar motion mechanism tests by DDES and RANS methods. Ocean Eng. **2022**, 243 (2022)
13. ITTC, The Manoeuvring Committee, Final Report and Recommendations to the 23rd ITTC, in *Proceeding of the 23rd ITTC* (2002)

Chapter 16
The Applicability of Third Vortex Identification Methods on Atmospheric Boundary Layer and Wind Turbine Wakes

Shun Xu, Weiwen Zhao, and Decheng Wan

Abstract To better understand the interaction between complex atmospheric boundary layer (ABL) inflow and wind turbine, the third vortex identification methods including the Liutex and OmegaR methods are implemented to identity and visualize the vortex structures, and the results are compared with those of Q criterion method. Two stages of numerical simulation are employed, including precursor stage for generation of ABL inflow and successor stage for wind turbine subjected to it. The wind turbine blades are modeled by actuator line model, and large eddy simulations are performed for the two stages. Compared with Q criterion method, the Liutex method can better identify vortices in ABL inflow and wind turbine wakes. However, the vortices in wind turbine wakes are not clearly visualized by OmegaR method due to its ability of capturing weak vortices. But the small-scale vortices in wind turbine wakes and large-scale vortices in ambient atmosphere are evident based on OmegaR method. Generation and fragmentation of blade tip vortices, interaction between vortices induced by wind turbine and ambient atmospheric vortices and wake expansion effect at middle wake region are clearly visualized by Q criterion and Liutex methods. The unphysical effect that sheet-like vortex structures upstream the wind turbine is wrongly captured by Q criterion method, whereas this error is eliminated by Liutex method.

16.1 Introduction

In recent years, the wind energy has received a rising attention due to its properties of no pollution and renewable [1], which makes a significant advance in wind turbine technology. According to the 2021 Global Wind Energy Report, the installed capacity of wind turbine in 2020 reached up to 93GW [2], resulting a growth rate of 53% compared to last year. The wind turbines are gradually developing to large scales with aim of maximizing the captured wind energy. And a lot of researches have

S. Xu · W. Zhao · D. Wan (✉)
Computational Marine Hydrodynamics Lab (CMHL), School of Naval Architecture, Ocean and Civil Engineering, Shanghai Jiao Tong University, Shanghai, China
e-mail: dcwan@sjtu.edu.cn

Y. Wang et al. (eds.), *Liutex and Third Generation of Vortex Identification*, Springer Proceedings in Physics 288, https://doi.org/10.1007/978-981-19-8955-1_16

investigated the operation performance of wind turbine under external environmental conditions.

Cheng et al. [3] conducted a fully coupled aero-hydrodynamic numerical simulation of a semi-submersible floating wind turbine (FOWT) under uniform inflow wind condition. Huang et al. [4, 5] performed numerical simulation for a spar type FOWT, and shear inflow wind was adopted to consider the shear property of atmospheric boundary layer (ABL) inflow caused by friction of the land or sea surface. For a more complex inflow wind condition, Li et al. [6] studied the FOWT's aerodynamic performance under turbulence inflow wind condition, and they pointed out the generated power was becoming unstable because of the distribution of turbulence in wind fields.

With the increase of wind turbine scale, its dynamic responses maybe significantly impacted by the ABL inflow. The inflow wind conditions of above literatures were simplified, because of the presence of two-phase issue. However, for the one-phase issue of onshore wind turbine study, the wind turbine subjected to the ABL inflow was widely studied. Churchfield et al. [7] performed a large eddy simulation (LES) to study the effects of atmosphere and turbine wakes on wind turbine dynamic responses, and the wind turbine blades were modeled by actuator line model (ALM). Lee et al. [8] investigated the effects of atmospheric stability and surface roughness on wind turbine fatigue load. Ning and Wan [9] studied the wake meandering and its effects on wind turbine aerodynamics. The inflow wind conditions of the above references were generated using the LES with sufficient simulation duration to obtain the quasi-equilibrium atmospheric turbulence.

The vortices in ABL inflow and wind turbine wakes need more attention to further investigate the interaction between atmosphere and wind turbine. The third vortex identification methods can reasonably answer the six elements problem of vortex, and quantify the properties of the vortex [10, 11]. Consequently, Huang et al. [12] employed the Liutex method to identify vortices in wind turbine wakes with the purpose of further understanding the complicated wake characteristics of wind turbine. They pointed out that the Liutex vector can better describe the complex wake characteristic of wind turbine in various conditions, compared with the Vorticity method and Q method.

In this study, the third vortex identification methods are implemented to capture the vortices in ABL inflow and turbine wakes. The precursor stage strategy with sufficient simulation duration is used to generate the quasi-equilibrium ABL inflow. In the successor stage, the wind turbine is subjected to ABL inflow and its blades are modeled by ALM. What's more, the LES is employed for both stages.

16.2 Numerical Method

16.2.1 Actuator Line Model

The wind turbine blades are modeled using the ALM, which avoids suffering the expensive computational cost with the solution of blade surface boundary layer and the satisfied numerical results are also obtained by solving Navier-Stokes equations. The ALM was firstly proposed by Sørensen and Shen [13], the basic theory behind ALM is to employ the actuator points of actuator lines representing the radial discreted wind turbine blades. The body forces of actuator points are projected to the flow fields, which reflecting the influence of wind turbine blades to the flow fields. The body forces of actuator points can be calculated by following expression:

$$f = (\boldsymbol{L}, \boldsymbol{D}) = \frac{1}{2}\rho U_{rel}^2 cdr\left(C_L \vec{e_L} + C_D \vec{e_D}\right) \tag{16.1}$$

where, $\boldsymbol{L}$ and $\boldsymbol{D}$ are the lift and drag forces of actuator points located at the blade radius r, ρ is the air density, $\boldsymbol{U_{rel}}$ denotes the relative velocity of two-dimensional airfoil, c represents the chore length of airfoil, dr is the width of blade element, C_L and C_D are the coefficients of lift and drag forces, respectively, which can be determined by the local attack angle, $\overline{\boldsymbol{e_L}}$ and $\overline{\boldsymbol{e_D}}$ are the unit vectors of lift and drag forces.

Figure 16.1 shows the velocity vectors of two-dimensional airfoil, the relative inflow velocity $\boldsymbol{U_{rel}}$ can be calculated by:

$$\boldsymbol{U_{rel}} = \sqrt{\boldsymbol{U_z^2} + (\boldsymbol{\Omega r} - \boldsymbol{U_\theta})^2} \tag{16.2}$$

where, $\boldsymbol{U_z}$ and $\boldsymbol{U_\theta}$ are the axial and tangential inflow wind velocities, $\boldsymbol{\Omega}$ is the rotor speed. The local attack angle $\alpha = \phi - \beta$, ϕ and β are the local inflow angle and pitch angle, respectively.

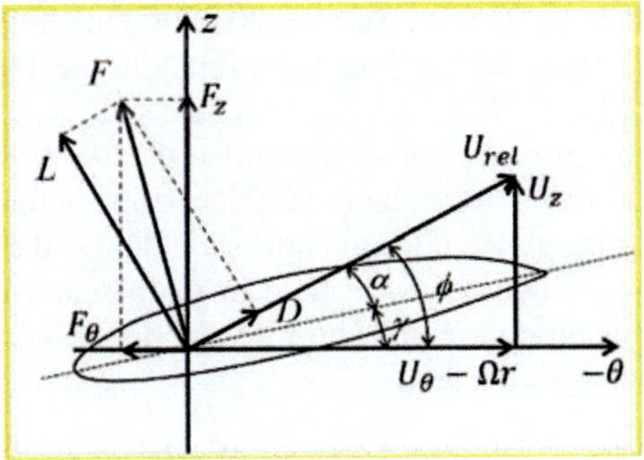

Fig. 16.1 Velocity vectors of two-dimensional airfoil

The body forces of actuator points are smoothly projected to the flow fields to avoid the numerical singularity. Therefore, the Gauss kernel function is adopted for the smooth projection procedure. The body force after smoothing is expressed by:

$$\boldsymbol{f}_\varepsilon = \boldsymbol{f} \otimes \eta_\varepsilon = \sum_{i=1}^{N} \boldsymbol{f}_i(\boldsymbol{x}_i, \boldsymbol{y}_i, \boldsymbol{z}_i, t)\frac{1}{\varepsilon^3 \pi^{\frac{3}{2}}} exp\left[-\left(\frac{\boldsymbol{d}_i}{\varepsilon}\right)^2\right] \tag{16.3}$$

where, N is the amount of actuator points for one wind turbine blade, $(\boldsymbol{x}_i, \boldsymbol{y}_i, \boldsymbol{z}_i)$ is the coordinate of the i-th actuator point, $\boldsymbol{d}_i$ donates the distance between the actuator point and the projection point, ε reflects the projection width, which set to $\varepsilon \approx 2\Delta x$ [14], Δx is the mesh size around the wind turbine blades.

16.2.2 Governing Equations

In this study, the LES method is employed for the numerical simulation of processor and successor stages. The spatially filtered governing equations are shown as follows:

$$\frac{\partial \overline{\boldsymbol{u}}_i}{\partial \boldsymbol{x}_i} = 0 \tag{16.4}$$

$$\begin{aligned}\frac{\partial \overline{\boldsymbol{u}}_i}{\partial t} + \frac{\partial}{\partial \boldsymbol{x}_j}(\overline{\boldsymbol{u}}_j\overline{\boldsymbol{u}}_i) = &-\underbrace{\frac{\partial \hat{p}}{\partial \boldsymbol{x}_i}}_{I} - \underbrace{\frac{1}{\rho_0}\frac{\partial}{\partial \boldsymbol{x}_i}\overline{p}_0(\boldsymbol{x}, \boldsymbol{y})}_{II} - \underbrace{2\boldsymbol{\varepsilon}_{i3k}\boldsymbol{\Omega}_3\overline{\boldsymbol{u}}_k}_{III} \\ &+ \underbrace{\boldsymbol{g}\left(\frac{\overline{\theta} - \theta_0}{\theta_0}\right)\boldsymbol{\delta}_{i3}}_{IV} - \underbrace{\frac{\partial}{\partial \boldsymbol{x}_j}(\boldsymbol{\tau}_{ij}^D)}_{V} + \underbrace{\frac{1}{\rho_0}\boldsymbol{f}_i^T}_{VI}\end{aligned} \tag{16.5}$$

where, as for the momentum equation, the Term I on the right side represents modified pressure gradient, the Term II is background pressure gradient, in order to drive the flow fields from uniform state to complex atmospheric state, the Term III donates Coriolis force due to earth rotation, the Term IV is temperature-induced buoyancy term, the Term V represents fluid stress tensor, which can be calculated based on the Smagorinsky sub-scale model, the Term VI donates the source term of body force of wind turbine blade, which can be obtained by employing the ALM. More details of the governing equations should refer the reference [7].

In addition, as illustrated in temperature-induced buoyancy term of the momentum equation, a temperature transportation equation needs to be solved to obtain the temperature field:

$$\frac{\partial \overline{\theta}}{\partial t} + \frac{\partial}{\partial \boldsymbol{x}_j}(\boldsymbol{u}_j\overline{\theta}) = -\frac{\partial}{\partial \boldsymbol{x}_j}(\boldsymbol{q}_j) \tag{16.6}$$

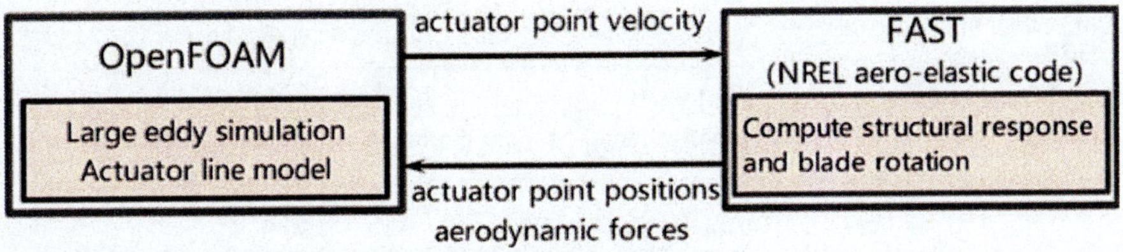

Fig. 16.2 Coupling procedure between OpenFOAM and FAST

where, heat flux $\boldsymbol{q}_j$ can be expressed by:

$$\boldsymbol{q}_j = -\frac{\nu_{SGS}}{Pr_t}\frac{\partial\overline{\theta}}{\partial \boldsymbol{x}_j} \tag{16.7}$$

where, Pr_t is the Prandtl number, set to 1/3 in the neutral and convection heat stability of ABL.

16.2.3 Coupling Procedure

When the wind turbine subjected to atmospheric inflow, the coupling strategy between FAST [15] and OpenFOAM [16] is adopted for the successor stage, which was firstly proposed and implemented by SOWFA [17]. The momentum part of blade element momentum (BEM) in FAST is replaced by the computational fluid dynamics (CFD). The inflow wind velocities at blade elements are sampled in the flow fields based on OpenFOAM, then the inflow information is delivered to FAST for the calculation of aerodynamic forces, as shown in Fig. 16.2. The wind turbine rotor rotation and dynamic responses are also calculated in FAST, and the aerodynamic forces and blade actuator point positions are feedback to OpenFOAM.

16.3 Computational Set up

16.3.1 Model Description

The NREL 5 MW reference wind turbine [18] is employed for the study object, which is a conventional three-bladed upwind turbine with pitch control and torque control systems. As presented in Table 16.1, the rated mechanical power of the wind turbine is 5 MW with rated wind velocity of 11.4 m/s, and rotor diameter, hub diameter and hub height are 126 m, 3 m and 90 m, respectively. The cut-in and cut-out wind velocities are 3 m/s and 25 m/s, and the cut-in and rated rotor speed are 6.9 rpm and 12.1 rpm, respectively.

Table 16.1 Main parameters of NREL 5 MW reference wind turbine

Parameter	Value
Rating	5 MW
Rotor orientation, configuration	Upwind, 3 blades
Rotor, hub diameter	126 m, 3 m
Hub height	90 m
Cut-in, rated, cut-out wind speed	3 m/s, 11.4 m/s, 25 m/s
Cut-in, rated rotor speed	6.9 rpm, 12.1 rpm

16.3.2 Precursor Stage

Muti-scales turbulence are existed in the ABL, the largest turbulent scales are of the order of 1 km, whereas the smallest scales are of the order of 1 mm [19]. Therefore, the computational domain of precursor stage is 3 km × 3 km × 1 km to contain large-scale turbulence as much as possible, as illustrated in Fig. 16.3. For the boundary conditions, the four side boundaries are set to cyclic conditions. The upper surface is slip condition, whereas no-slip wall condition is used for lower surface, and the wall stress is modeled based on the Moeng model [20]. Inflow wind is configured to 240° southwest direction, in order to eliminate the stuck phenomenon that the vortex structures at the same height reach the downstream and enter the upstream at the same time for the cyclic boundary condition.

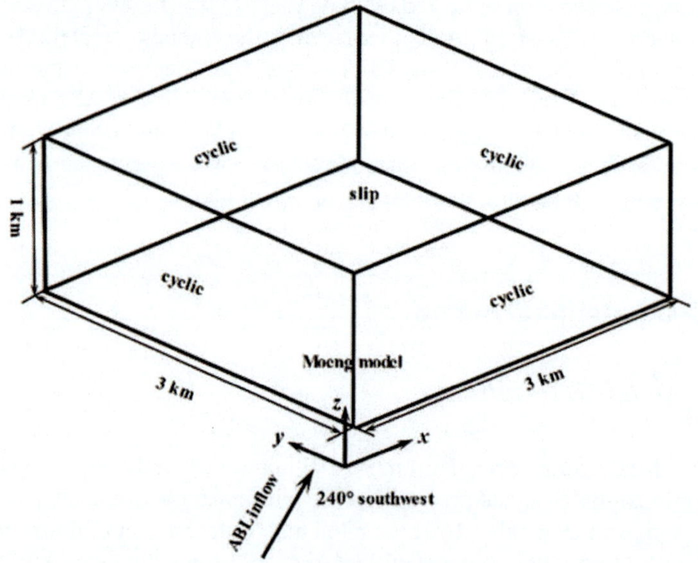

Fig. 16.3 Configuration of computational domain and boundary conditions of precursor stage

For the setting of initial temperature, the 300 K uniform distribution is employed from the lower surface to 700 m height. Temperature inversion layer is adopted from 700 m to 800 m to limit the growth of the ABL, in which the temperature linearly increases to 308 K. For the height from 800 m to upper boundary, the temperature increases linearly with a rate of 0.003 K/m. The time step is 0.2 s, and the simulation time is set to 19000 s, long enough to obtain the quasi-equilibrium atmospheric turbulence. The last 1000 s calculation data of upstream plane are saved for the input of the successor stage. Initial wind velocity in flow field is set to 11.4 m/s, which is consistent with the rated wind speed. The 240° inflow wind direction is adopted, as aforementioned in Fig. 16.3. The neutral atmospheric stability is employed for the precursor stage, and the surface roughness is set to 0.001, corresponding to representative sea surface condition.

16.3.3 Successor Stage

The inflow data saved from the precursor stage are used for the successor stage. Computational domain and background mesh resolution of successor stage is the same as precursor stage, whereas the north and east sidewall boundaries are modified to zero gradient boundary conditions to prevent downstream wind turbine wake circularly entering the upstream boundary. When the wind turbine subjected to the ABL inflow, two level mesh refinement strategy is employed for the accurately capture of wake vortex structures, as shown in Fig. 16.4. The numerical simulation time step of successor stage is 0.02 s, and the simulation time is 1000 s, corresponding to the stored inflow data time of precursor stage.

16.4 Results and Discussions

16.4.1 Precursor Stage

Figure 16.5 shows the vorticity visualization of the complex ABL inflow in hub height level plane (z = 90 m). Because of high turbulence intensity of the complex ABL, the vortices are widely distributed in the flow fields. A few strong vortices are captured based on the Q criterion method, whereas the strong vortices are more evident when the Liutex method is employed. Besides, the muti-scale characteristics of vortices are presented by the Liutex method. Both large- and small-scale vortices are visualized, whereas this phenomenon is not obvious for the Q criterion method. Similar with the Liutex method, the OmegaR method also has the ability to visualize the muti-scale properties of vortices. In addition, more weak vortices are captured and visualized when using the OmegaR method.

Fig. 16.4 Configuration of the calculation region for successor stage

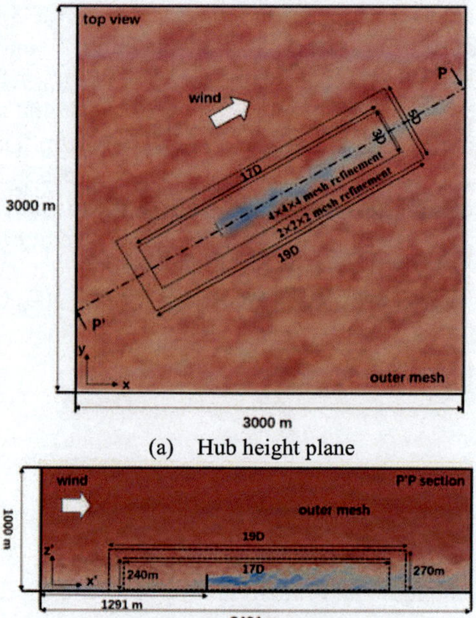

(a) Hub height plane

(b) $P'P$ vertical plane

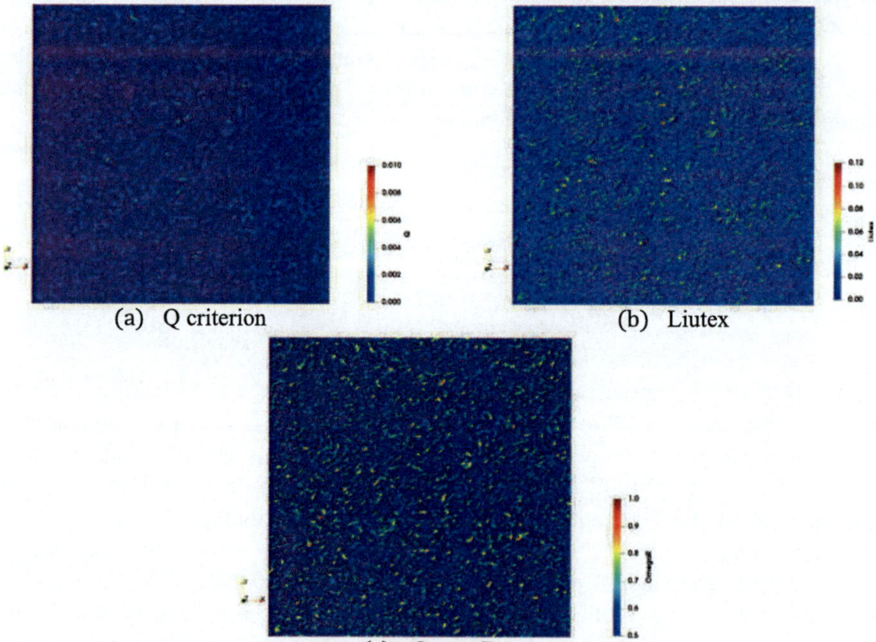

(a) Q criterion

(b) Liutex

(c) OmegaR

Fig. 16.5 Vorticity visualization of ABL inflow in hub height level plane (z = 90 m)

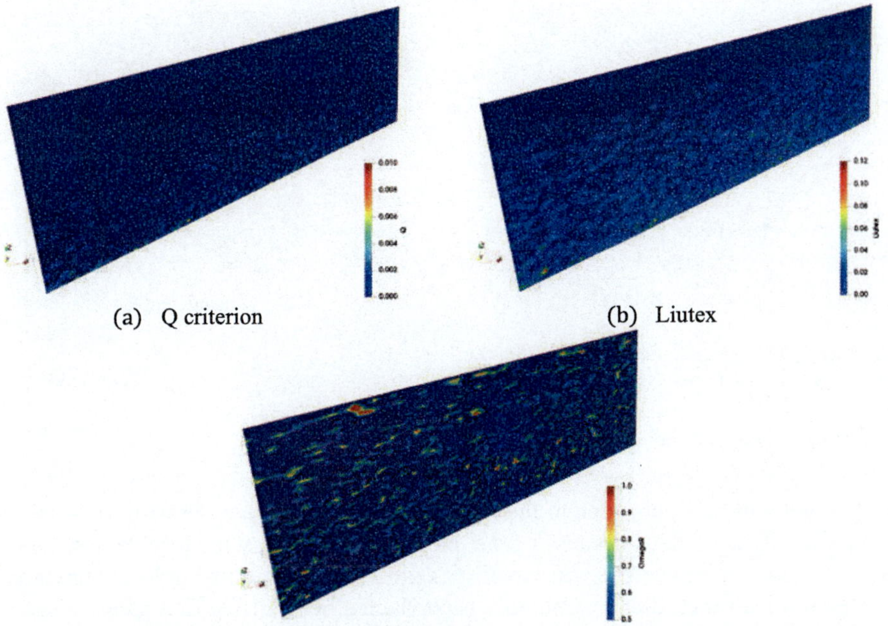

(a) Q criterion (b) Liutex

(c) OmegaR

Fig. 16.6 Vorticity visualization of ABL inflow in 240° inflow wind vertical plane

Figure 16.6 illustrates the vorticity visualization of the 240° inflow wind vertical plane. Because of the lower surface friction, some strong vortices are identified and visualized based on the Q criterion method, whereas the vortices located at middle height of calculation region are clearly shown by the Liutex method, which is consistent with previous analysis. However, the vortices near upper boundary are evident when using the OmegaR method, reflecting the capability of OmegaR method for capturing the weak vortices.

The vortex structures are identified and visualized based on the Q criterion and OmegaR methods, and the vortex structures are colored by wind velocity, as presented in Fig. 16.7. The vortex structures near lower boundary are visualized by the Q criterion method. However, the slender vortex structures near upper boundary surface are identified and presented based on the OmegaR method, which representing the 240° inflow wind direction. Compared with the Q criterion method, the OmegaR method can present the weak vortex structures around the upper boundary surface, which is consistent with the previous analysis.

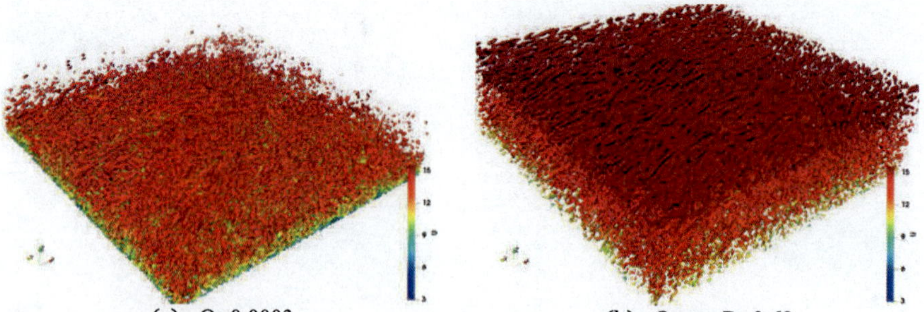

(a) Q=0.0003 (b) OmegaR=0.68

Fig. 16.7 Vortex structures of the ABL inflow

16.4.2 Successor Stage

When wind turbine subjected to the ABL inflow, the turbine wake with high turbulent intensity is obvious because of the interaction between wind turbine and atmospheric ambient inflow. The vorticity visualization in wind turbine wakes of the three different vortex identification methods is conducted in Fig. 16.8. Compared with the ambient turbulence, the vortices in wind turbine wakes are more evident when the Q criterion method is employed, whereas this phenomenon is clearer with the Liutex method. Besides, the blade shaped vortex is induced when the turbine blades passing hub height level plane. As aforementioned, the OmegaR method has the ability of capturing weak vortices, so the vortices in wind turbine wakes is not clearly presented, compared with the other two methods. However, the small-scale vortices behind the wind turbine and large-scale vortices in ambient atmosphere are presented based on the OmegaR method. The interaction between large-scale ambient vortices and wind turbine is responsible for the generation of small-scale vortices in wind turbine wakes. Besides, the mesh refinement strategy also contributes to the capture of small-scale vortices.

According to Fig. 16.9, the vortices induced by wind turbine are visible for a long distance behind the turbine. This phenomenon is more evident based on the Liutex method, compared with the Q criterion. Similar with Fig. 16.8, the turbine induced vortices are not clearly visualized by OmegaR method, because of the capability of OmegaR method of capturing weak vortices. However, due to the combined effects of local mesh refinement and interaction between wind turbine and large-scale atmospheric vortices, the small-scale vortices behind wind turbine are clearly visible, compared with Q criterion and OmegaR methods.

Figure 16.10 illustrates the vortex structures of wind turbine subjected to ABL inflow. The vortex structures are identified and visualized by the Q criterion and Liutex methods, without the OmegaR method, because the induced vortex structures by wind turbine are not easily visible if the weak vortices upper on the wind turbine are captured. A lot of vortex structures upstream the wind turbine are presented, due to the high turbulence intensity of ambient atmospheric inflow. The blade tip vortices

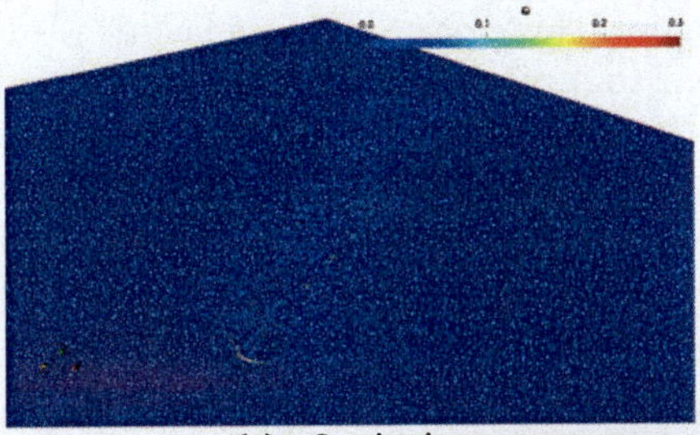

(a) Q criterion

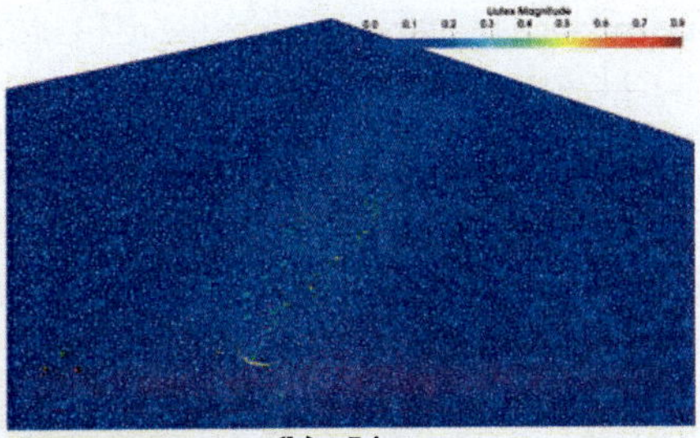

(b) Liutex

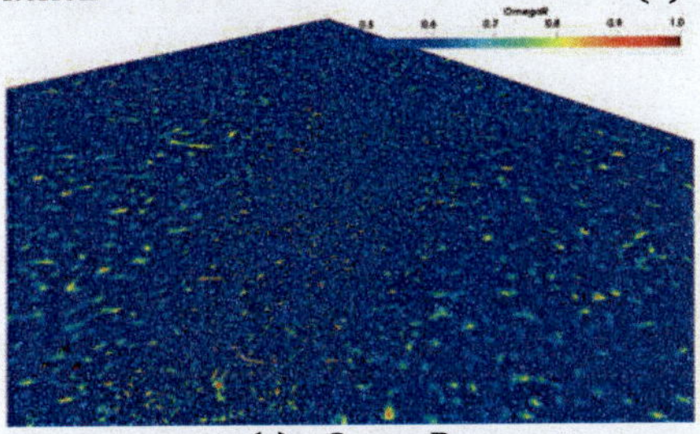

(c) OmegaR

Fig. 16.8 Vorticity visualization of turbine wakes in hub height level plane (z = 90 m)

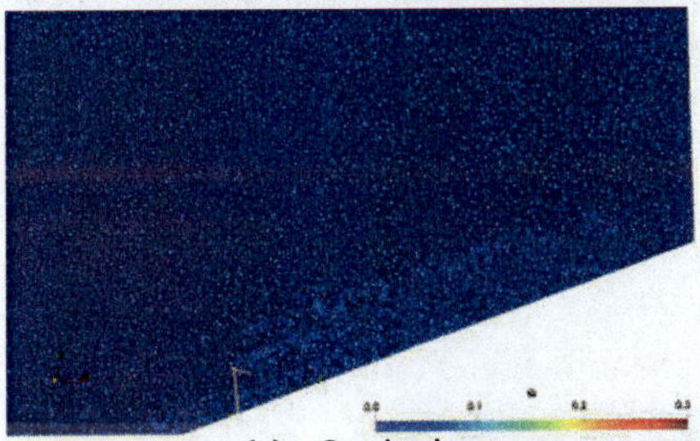

(a) Q criterion

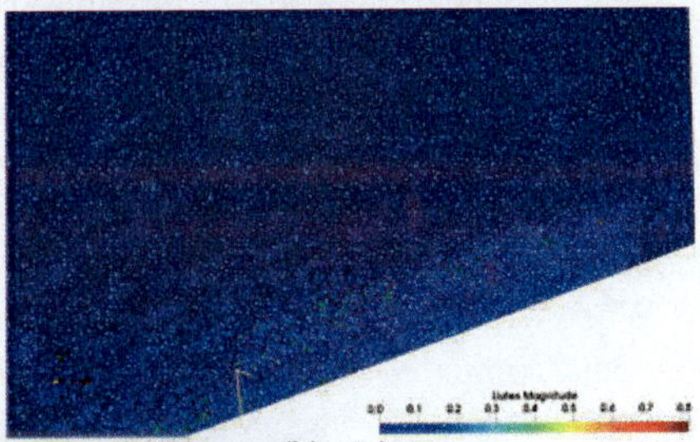

(b) Liutex

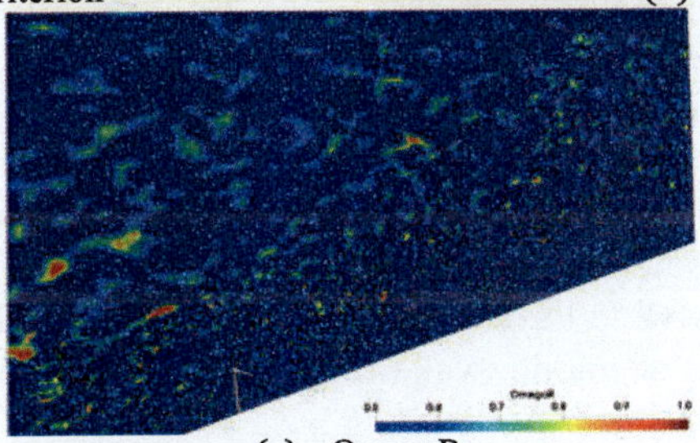

(c) OmegaR

Fig. 16.9 Vorticity visualization of turbine wakes in 240° inflow wind vertical plane are induced by wind turbine due to blade rotation. However, the blade tip vortices are quickly break down when the vortices travel to downstream. Both Q criterion and Liutex methods visualize the break down of blade tip vortices. The broken turbine wake vortices located at middle wake region are tend to expand outward, which called the wake expansion effect. In the far wake region, the wake vortices break into many small-scale vortices. The unphysical effect that the sheet-like vortex structures upstream wind turbine is wrongly presented by the Q criterion method, whereas this error is eliminated when employing the Liutex method.

16.5 Conclusions

In this study, the precursor stage combined with successor stage strategy is adopted for the numerical simulation of wind turbine subjected to ABL inflow. The target of precursor stage is to generate the ABL inflow data for the boundary inflow data of successor stage. Three vortex identification methods are employed for the ABL inflow and wind turbine subjected to it. And some conclusions are obtained.

For the ABL inflow, the vortices are widely distributed in flow fields. And strong vortices are visualized based on the Q criterion method, whereas the strong vortices are clearer when using the Liutex method. The weak vortices near upper boundary surface are presented by OmegaR method, indicating that it has the ability of identifing the weak vortices. For the vortex structures, Q criterion only identify the vortex structures near lower boundary surface, whereas slender vortex structures near upper boundary surface are identified and presented based on the OmegaR method, which representing the 240° inflow wind direction.

When wind turbine subjected to ABL inflow, the vortices in wind turbine wakes are obvious. Compared with the Q criterion, this phenomenon is more evident when using the Liutex method. The vortices in wind turbine wakes are not clearly presented by OmegaR method, because of its ability of capturing weak vortices. However, the small-scale vortices in wind turbine wakes and large-scale vortices in ambient atmosphere are presented based on the OmegaR method.

Because OmegaR method has the ability to identify weak vortices, this method is not employed for the vortex structures visualization when wind turbine subjected to ABL inflow. Generation and fragmentation of blade tip vortices, the interaction of vortices induced by wind turbine with the ambient atmospheric vortices and the wake expansion effect at middle wake region are clearly visualized based on the Q criterion and Liutex methods. The unphysical effect that the sheet-like vortex structures upstream wind turbine is wrongly presented by the Q criterion method, whereas this error is eliminated by Liutex method.

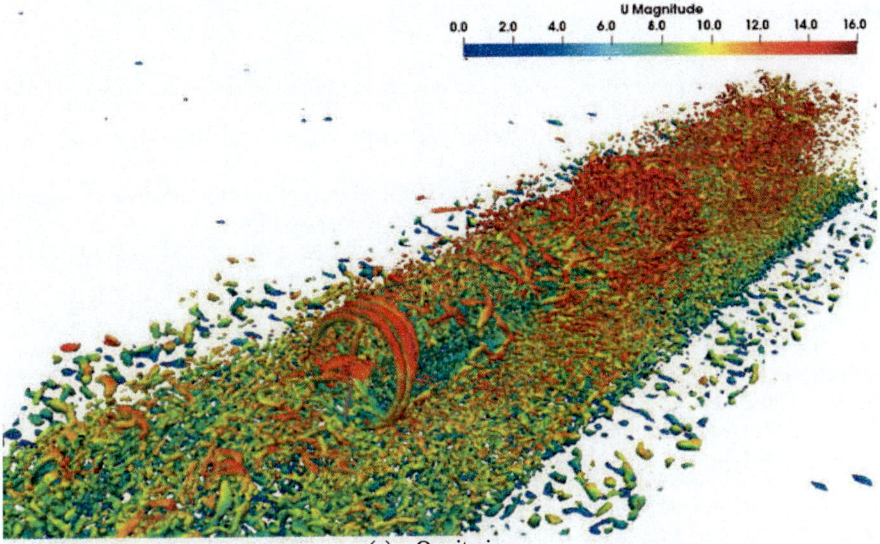

(a) Q criterion

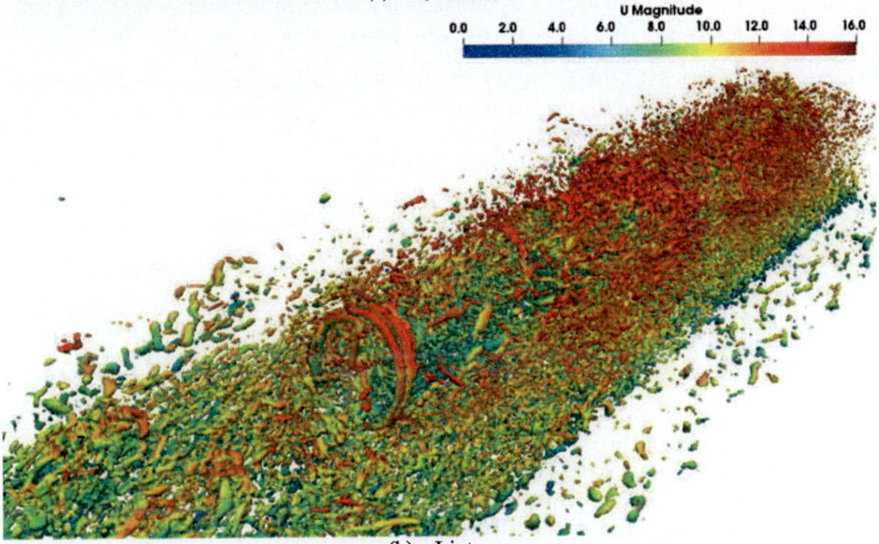

(b) Liutex

Fig. 16.10 Vortex structures of wind turbine subjected to ABL inflow

Acknowledgments This work is supported by National Natural Science Foundation of China (52131102, 51909160), and the National Key Research and Development Program of China (2019YFB1704200), to which the authors are most grateful.

References

1. K.B. Sahu, Wind energy developments and policies in China: a short review. Renew. Sustain. Energy Rev. **81**, 1393–1405 (2018)
2. Council GWE, GWECǀ GLOBAL WIND REPORT 2021. Global Wind Energy Council: Brussels, Belgium (2021)
3. P. Cheng, Y. Huang, D. Wan, A numerical model for fully coupled aero-hydrodynamic analysis of floating offshore wind turbine. Ocean Eng. **173**, 183–196 (2019)
4. Y. Huang, P. Cheng, D. Wan, Numerical analysis of a floating offshore wind turbine by coupled aero-hydrodynamic simulation. J. Mar. Sci. Appl. **18**(1), 82–92 (2019)
5. Y. Huang, D. Wan, Investigation of interference effects between wind turbine and spar-type floating platform under combined wind-wave excitation. Sustainability. **12**(1), 246 (2020)
6. L. Li, Y. Liu, Z. Yuan, Y. Gao, Wind field effect on the power generation and aerodynamic performance of offshore floating wind turbines. Energy **157**(AUG.15), 379–390 (2018)
7. M.J. Churchfield, S. Lee, J. Michalakes, P.J. Moriarty, A numerical study of the effects of atmospheric and wake turbulence on wind turbine dynamics. J. Turbul. **13**, N14 (2012)
8. S. Lee, M. Churchfield, P. Moriarty, J. Jonkman, J. Michalakes, Atmospheric and wake turbulence impacts on wind turbine fatigue loadings, in *50th AIAA Aerospace Sciences Meeting including the New Horizons Forum and Aerospace Exposition* (2012), p. 540
9. X. Ning, D. Wan, LES Study of wake meandering in different atmospheric stabilities and its effects on wind turbine aerodynamics. Sustainability **11**(24), 6939 (2019)
10. C. Liu, H. Xu, X. Cai, Y. Gao, Liutex and third generation of vortex identification methods. Liutex and Its Applications in Turbulence Research (2021)
11. Y.N. Zhang, X.Y. Wang, Y.N. Zhang, C. Liu, Comparisons of and analyses of vortex identification between omega method and Q criterion. J. Hydrodyn. **31**(2), 224–230 (2019)
12. Y. Huang, L. Cao D. Wan, Application of Liutex for analysis of complex wake flows characteristics of the wind turbine, in *Liutex and Third Generation of Vortex Definition and Identification* (Springer, 2021), pp. 353–371
13. J.N. Sorensen, W.Z. Shen, Numerical modeling of wind turbine wakes. J Fluids Eng. **124**(2), 393–399 (2002)
14. N. Troldborg, Actuator line modeling of wind turbine wakes. PhD Thesis, Technical University of Denmark, Lyngby, Denmark (2009)
15. J.M. Jonkman, M.L. Buhl, FAST user's guide. National Renewable Energy Lab. (NREL), Golden, CO (United States) (2005)
16. C.J. Greenshields, OpenFOAM user guide. OpenFOAM Foundation Ltd, version. 3(1), 47 (2015)
17. P. Fleming, P. Gebraad, J.-W. van Wingerden, S. Lee, M. Churchfield, A. Scholbrock, et al., SOWFA super-controller: a high-fidelity tool for evaluating wind plant control approaches. National Renewable Energy Lab.(NREL), Golden, CO (United States) (2013)
18. J. Jonkman, S. Butterfield, W. Musial, G. Scott, Definition of a 5-MW reference wind turbine for offshore system development. National Renewable Energy Lab.(NREL), Golden, CO (United States) (2009)
19. R.B. Stull, An introduction to boundary layer meteorology, vol **13**, Springer Science & Business Media (1988)
20. C.H. Moeng, A large-eddy-simulation model for the study of planetary boundary-layer turbulence. J. Atmos. Sci. **41**(13), 2052–2062 (1984)

Chapter 17
Propeller Wake and Noise Analysis Based on the Third-Generation Vortex

Lianjie Yu, Weiwen Zhao, and Decheng Wan

Abstract Vortex is one of the complex contents of fluid dynamics. The traditional first-generation vortex 'Q-criterion' and second-generation 'vorticity' have various problems in calculation. The third-generation vortex recognition 'Rortex', which is more in line with the actual laws of physics, came into being. Based on the third-generation vortex recognition technology, this paper conducts a detailed analysis of the propeller tail vortex on the OpenFOAM platform. Taking the DTMB4119 propeller as the research object, the wake characteristics of the propeller are studied by comparing the wake vortex considering cavitation and ignoring it. The development process is qualitatively analyzed in axial velocity, turbulent kinetic energy, turbulent viscosity and so on. Besides, since the vortex is an important source of noise, the acoustic characteristics is analyzed by acoustic analogy. Combined with the vortex distribution, this paper is committed to explaining the influence of the vortex on the linear and nonlinear terms of the propeller noise. The results show that the third-generation vortex recognition technology can remove the shear in the flow field and the spurious vorticity near the boundary, so that the continuous vorticity calculated by the Q criterion becomes a separated vorticity, which is more in line with the actual laws of physics. At the same time, the cavitation affects vorticity and nonlinear sound source, which increases the overall sound pressure level.

17.1 Introduction

Vortex is a common phenomenon in turbulent flow, which is closely related to drag, lift, vibration, noise, etc. Accurate capture of vortices is the basis for the simulation of the flow field. In the last 30 years, various fluid dynamics governing equations is expressed in eddies perspective, and they explain many problems in the flow field from another angle [8].

L. Yu · W. Zhao · D. Wan (✉)
Computational Marine Hydrodynamics Lab (CMHL), School of Naval Architecture, Ocean and Civil Engineering, Shanghai Jiao Tong University, Shanghai, China
e-mail: dcwan@sjtu.edu.cn

Y. Wang et al. (eds.), *Liutex and Third Generation of Vortex Identification*, Springer Proceedings in Physics 288, https://doi.org/10.1007/978-981-19-8955-1_17

The earliest definition of vortex was defined by the vorticity tube, which is considered to be equivalent to a vortex. The method is categorized as the first generation of vortex identification. In the past 30 years, scholars proposed the Q-criterion, Δ, λ_2 and so on. These parameters are usually based on the Cauthy-Stokes decomposition of the velocity gradient tensor. They are regarded as the second-generation vortex identifications [8]. However, there are some problems with these methods. In 2014, the Vortex and Turbulence Research Team at the University of Texas at Arlington (UTA Team) focused on developing a new generation of vortex identification method, omega, which defines a vortex as a connected region where the vorticity exceeds the deformation [9]. Meanwhile, Liutex is proposed to represent the local rigid rotation of fluids. Such the third-generation vortex identification technology can describe the absolute strength, the relative strength, the rotational axis, the vortex core position and size, and the vortex boundary. Other vortex identification methods fail to answer these problems, except for the vortex boundary [3, 10].

The simulation of the wake of an open-water propeller is a complex problem in the field of hydrodynamics, not only for the complex geometry of the propeller, but also for the existence of complex vortex systems in the propeller wake, such as hub and tip vortex [6]. For high-speed operating conditions, cavitation may occur. The interaction between cavitation and wake vortices enhances the pulsation of turbulent flow and increases turbulent kinetic energy. The accurate prediction of propeller thrust and torque is dependent on the accuracy of the flow field, and the precise capture of vortex is the basis of all this [14]. The first and second generation vortex identification methods cause shear tensor and false vorticity near the boundary, and the omega identification method helps overcome these problems.

In recent years, the prediction of propeller noise has become a research hotspot. Due to the underwater incompressibility, it is not feasible to use direct numerical simulation (DNS) to predict the noise. Therefore, acoustic analogy becomes the main method [1, 4, 6]. In the acoustic analogy, the area near the propeller is regarded as the near field, and all nonlinear effects are within this range. With the assumption of compact sound sources, the sound pressure can be obtained by radiating the sound source obtained by CFD calculation to the far field. This method was first proposed by Lighthill [7] and developed by Curle [2] and Ffowcs Williams et al. [5] They classified sound sources into three types: monopoles, dipoles, and quadrupoles. Among them, the quadrupole is considered to be related to the vortex, so the accurate capture of the vorticity is the basis of noise prediction.

In this paper, the wake of the DTMB4119 propeller are studied, and the third-generation vortex identification method is used to analyze the noise in cavitation and non-cavitation conditions. The sound pressure distribution law is explained from the perspective of vortex. The structure of this paper is arranged as follows: Chap. 2 is the mathematical foundation, including omega expression, DDES turbulence model, Schnerr-Sauer cavitation model and FW-H acoustic equation. The Chap. 3 introduces the settings of numerical examples. Chapter 4 presents the hydrodynamic results, including the verification of hydrodynamic results and the analysis of vortex structure. Chapter 5 presents the sound field results, which are divided into near-field and far-field results. Finally, Chap. 6 gives the conclusion.

17.2 Mathematical Formulations

17.2.1 Omega Vortex Identification Method

Vorticity does not represent fluid rotation because the effects of deformation should also be considered. The relationship between the two is like a mixture of salt and water. The salinity does not depend on the mass of salt. It depends on the ratio of salt and water. Larger vorticity does not necessarily cause strong rotations, and small vorticity may also cause strong rotations. The Blasius boundary layer is a typical example. The third generation of vortex identification technology uses the ratio of vorticity and deformation to define the vortex boundary. Ω is defined as a ratio of the vorticity tensor norm squared over the sum of the vorticity tensor norm squared and deformation tensor norm squared.

$$\boldsymbol{\Omega}=\frac{\|\boldsymbol{B}\|_F^2}{\|\boldsymbol{A}\|_F^2+\|\boldsymbol{B}\|_F^2}=\frac{b}{a+b}$$

where, $a=\|\boldsymbol{A}\|_F^2$, $b=\|\boldsymbol{B}\|_F^2$. To prevent the appearance of unphysical vortices, a small value ϵ is added to the denominator, as shown below.

$$\boldsymbol{\Omega}=\frac{b}{a+b+\epsilon}$$

Many scholars examined the robustness of $\boldsymbol{\Omega}$ and found that the range of vortices hardly changes between the thresholds of 0.52–0.6. This is one of the advantages of the third-generation vortex identification technology over the previous methods, which are often sensitive to threshold values. For $\boldsymbol{\Omega}$, increasing or decreasing the threshold values only make the vortex structure thinner or fatter. But a higher threshold value is needed if the position of the vortex core is desired, such as 0.8, 0.9, etc.

17.2.2 Turbulence Model

The problem of propeller wake is still an open issue, and the interaction between turbulence and cavitation is particularly complex. Previous studies proved that the RANS model cannot obtain the small-scale vortices for the underwater propeller, which has a great impact on the acoustic prediction [13].

For this reason, this paper uses the improved Spalart–Allmaras DES as the turbulence model, namely SA-DDES [12]. Its governing equation is,

$$\frac{\partial \widetilde{\nu}}{\partial t}+div(\widetilde{\nu}\,u)=\frac{1}{C_\sigma}\{div[(\nu+\widetilde{\nu})grad\,\widetilde{\nu}]+C_{b2}\frac{\partial \widetilde{\nu}}{\partial x_i}\frac{\partial \widetilde{\nu}}{\partial x_j}\}+C_{b1}\tilde{S}\,\widetilde{\nu}-C_{w1}f_w\left(\frac{\widetilde{\nu}}{\tilde{d}}\right)^2$$

The first term on the left side is the transient term, and the second term is the convection term; the first term on the right side is the diffusion term, the second term is the source term, and the third term is the dissipation term. The DDES equation modifies $\tilde{d}$ in the formula to

$$\tilde{d} = d_w - f_d \cdot \max(d_w - C_{DES}\Delta, 0)$$

$$f_d = 1 - \tanh[(8r_d)^3]$$

$$r_d = \frac{\nu_t + \nu}{\sqrt{U_{ij}U_{ij}}(\kappa d_w)^2}$$

17.2.3 Cavitation Model

The cavitation model, or mass transport model, is derived from the commonly-used Rayleigh-Plesset cavity dynamics equation. The process of condensation and evaporation in cavitation describes the changes in each phase by adding source terms. This paper adopts the Schnerr-Sauer cavitation model [11]. The mass conversion equation of the model is as follows:

$$\frac{\partial(\rho_v \alpha_v)}{\partial t} + \frac{\partial(\rho_v \alpha_v u_j)}{\partial x_j} = \dot{m}^+ - \dot{m}^-$$

where α_v is the vapor volume fraction, and the source terms $\dot{m}^+$ and $\dot{m}^-$ represent the evaporation and condensation processes, respectively. When the phase change occurs, there is

$$\dot{m}^+ = \frac{\rho_v \rho_l}{\rho}\alpha_v(1-\alpha_v)\frac{3}{R_b}\sqrt{\frac{2}{3}\frac{\max(p_v - p, 0)}{\rho_l}}$$

$$\dot{m}^- = \frac{\rho_v \rho_l}{\rho}\alpha_v(1-\alpha_v)\frac{3}{R_b}\sqrt{\frac{2}{3}\frac{\max(p - p_v, 0)}{\rho_l}}$$

Among them, $\dot{m}^+$ represents the evaporation process, and $\dot{m}^-$ represents the condensation process. R_b is the radius of the cavity, and p_v is the saturation vapor pressure at the local temperature. The calculation formula of the cavity radius is:

$$\mathrm{R_b} = \left(\frac{\alpha_v}{(1-\alpha_v)}\frac{3}{4\pi}\frac{1}{N_b}\right)^{\frac{1}{3}}$$

Among them, $N_b = 10^{13}m^{-3}$ is the cavity number density.

17.2.4 FW-H Formulation

Here, we use the most widely used and universally significant FW-H equation, as is shown below.

$$\left(\frac{1}{c^2}\frac{\partial^2}{\partial t^2} - \nabla^2\right)p' = \frac{\partial}{\partial t}[\rho_0 v_n \delta(f)] - \frac{\partial}{\partial x_i}[pn_i\delta(f)] + +\frac{\partial^2}{\partial x_i \partial x_j}[H(f)T_{ij}]$$

The surface of an object can be represented by $f(x,t) = 0$, where $\nabla f = \vec{n}$ points to the normal direction outside the object surface. According to the linear assumption, $p' = c^2\rho' = c^2(\rho - \rho_0)$, where c and ρ_0 respectively represent the fluid sound velocity and density under the undisturbed fluid medium. $\delta(f)$ represents the Dirac function. The National Aeronautics and Space Administration (NASA) summarized the different integral solutions of the FW-H equation and systematically named these solutions, such as Farassat Formulations 1 and Farassat Formulations 1A. The derivation process of Farassat Formulation 1A is omitted here, and the integral result expression is directly given as follows:

$$4\pi p'_T(x,t) = \int_{f=0}\left[\frac{\rho_0 \dot{v}_n}{r(1-M_r)^2} + \frac{\rho_0 v_n \hat{r}_i \dot{M}_i}{r(1-M_r)^3}\right]_{ret} dS + \\ + \int_{f=0}\left[\frac{\rho_0 c v_n (M_r - M^2)}{r^2(1-M_r)^3}\right]_{ret} dS$$

$$4\pi p'_L(x,t) = \int_{f=0}\left[\frac{\dot{p}\cos\theta}{cr(1-M_r)^2} + \frac{\hat{r}_i \dot{M}_i p\cos\theta}{cr(1-M_r)^3}\right]_{ret} dS \\ + \int_{f=0}\left[\frac{p(\cos\theta - M_i n_i)}{r^2(1-M_r)^2} + \frac{(M_r - M^2)p\cos\theta}{r^2(1-M_r)^3}\right]_{ret} dS$$

Here, p'_T stands for thickness noise. p'_L stands for load noise, $(x,t)(y,t)$ are the space–time variables of the observation point and the sound source respectively; $r = |x - y|$ is the norm of the vector radius from the observation point to the sound source; $\hat{r_i} = \frac{(x-y)}{r}$ represents the normalization of the vector radius; $\mathrm{M_r} = \frac{\hat{r_i} v_i}{c}$ is the sound source Mach number in the satellite coordinate system; $1 - M_r$ is called the Doppler factor; $\dot{v_n} = \frac{\partial}{\partial \tau}(v \cdot n)$ represents the derivative of the speed to the sound source time; $\cos\theta = n_i\hat{r_i}$ is the vector path from the observation point to the sound source; $[]_{\mathrm{ret}}$ is retarded time, which represents the satellite coordinates after considering the Doppler effect.

17.3 Numerical Setup

The DTMB 4119 propeller (three-blade) is selected as the research object. As a well-known model widely used both in hydrodynamic and hydroacoustic fields, this propeller has a simple geometric structure and relatively detailed hydrodynamics experimental data [4]. The basic parameters of the propeller are shown in Table 17.1.

The numerical domain is cylindrical, as shown in the Fig. 17.1. The propeller is arranged 2D downstream of the inflow patch, and the outflow patch is 4D away from the disk, so that the wake can be fully developed before reaching the outlet. The diameter of the outer cylinder is set as 3D to avoid interference with the flow of the propeller.

The calculation example uses unstructured grids, and the total number of grids is 3.45 million. The y + value at the boundary layer is less than 10, which meets the requirements of the DDES turbulence model. Figure 17.1 shows the mesh distribution on the propeller surface.

Table 17.1 DTMB 4119 propeller geometry model parameters

Diameter (m)	0.1
Blades number	3
Skew (°)	0
Rake (°)	0
Blade section	NACA66, a = 0.8
Rotation direction	Right

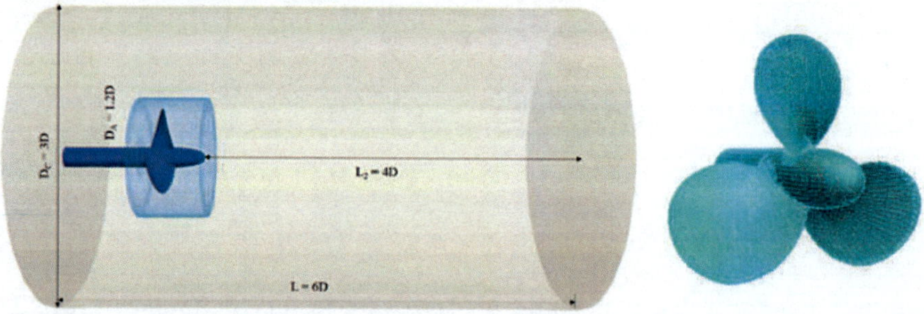

Fig. 17.1 The left panel: Diagram of the calculation domain; The right panel: The propeller mesh

17.4 Hydrodynamic Results

17.4.1 Hydrodynamic Verification

Figure 17.2 shows the comparison between the hydrodynamic coefficient obtained by the computation and the experiment under different advance coefficients. It can be seen that the error is very small. The thrust coefficient, the torque coefficient and the efficiency are defined as below:

$$K_T = \frac{T}{\rho n^2 D^4}, K_Q = \frac{Q}{\rho n^2 D^5}, \eta = \frac{K_T \cdot D}{K_Q}$$

Three sections y/D = -–0.25,0.5,1 are plotted in the time-averaged contours of axial velocity, turbulent kinetic energy (TKE), and Q amplitudes, as Fig. 17.3 shows. The suction force is dominant in the upstream region, which can be seen from the axial velocity and the TKE cloud diagram. Besides, there is almost no vortex upstream of the propeller except near the hub, which is related to the small upstream TKE. The axial velocity downstream is mainly distributed in the range of the disk surface, and this area produces the most of the thrust force. The tip vortex becomes weak rapidly with the increase of the distance, but the hub vortex can continue for a long way. The TKE of the tip part gradually decreases in the downstream region, but the TKE of the hub part is increasing.

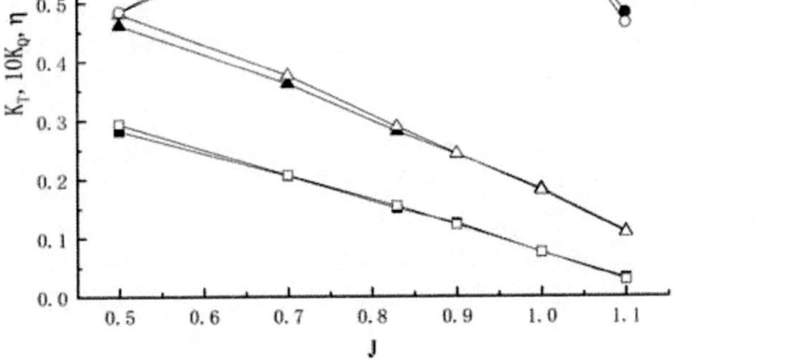

Fig. 17.2 The calculated K_T, K_Q and η under different advance coefficients compared with the experiment results [4]

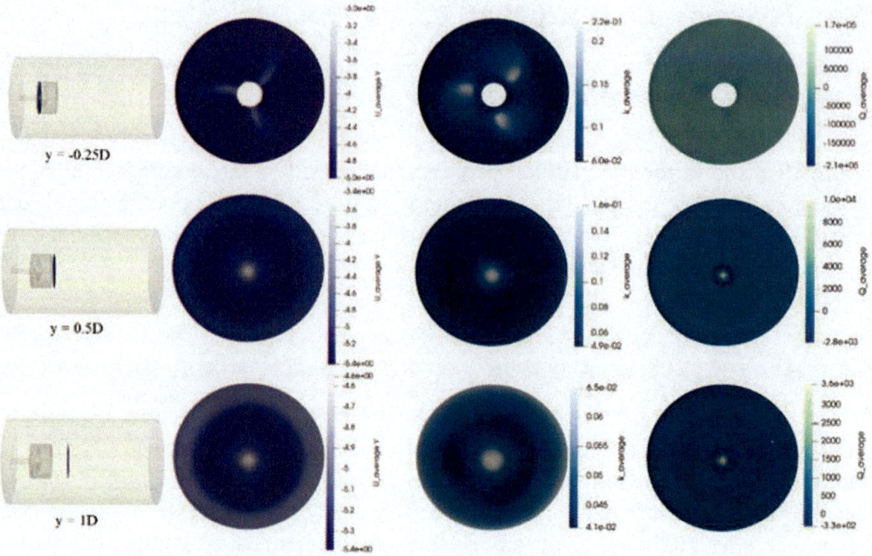

Fig. 17.3 The time-averaged contours of axial velocity, turbulent kinetic energy (TKE), and Q amplitudes plotted in three sections y/D = –0.25,0.5,1

In the Fig. 17.4, the pressure, the epsilon of DDES and the turbulent viscosity are plotted on the three cross-sections. The low-pressure core inside the tip vortex can still be seen until y/D = 1. In the hub vortex, the epsilon is large, which can be considered as the area where the sound source is concentrated. Satisfactory values are also obtained for the viscosity in the tip and hub vortex regions.

17.4.2 Vortex Comparison

$\mathbf{\Omega_R}$ Are compared with vorticity in this subsection. The second-generation vortex 'vorticity' has an obvious drawback, that is, there is false vortex near the AMI surface. The third-generation vortex '$\mathbf{\Omega_R}$' solves the problem of false signal on the AMI surface in the vorticity, but the false signal at the boundary still exists (Fig. 17.5).

For comparison, a cavitation condition is set here, and the relevant parameters of the operating condition are shown in Table 17.2. Here we compare the Q criterion and $\mathbf{\Omega_R}$ in the cavitation state, and find that $\mathbf{\Omega_R}$ can not only capture the vortex structure further downstream, but also eliminate the false vortex on the wall (Fig. 17.6).

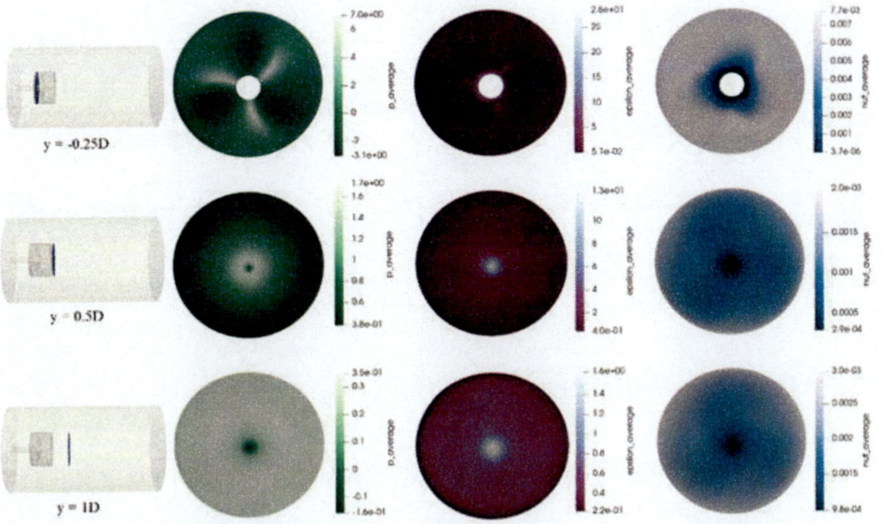

Fig. 17.4 The pressure, the epsilon of DDES and the turbulent viscosity on the three cross-sections y/D = –0.25,0.5,1

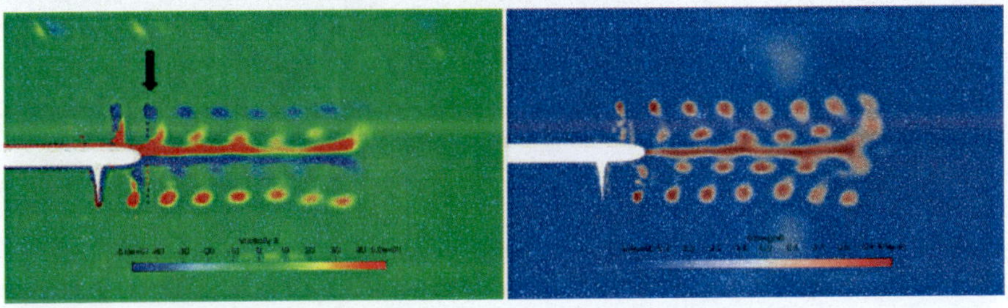

Fig. 17.5 The comparison between vorticity cloud chart (left panel) and omega cloud chart (right panel)

Table 17.2 The cavitation condition parameters

Max inlet velocity	15 m/s
Rotate speed	66.7 r/s
Saturation pressure	2300 Pa
Phase change model	Schnerr Sauer

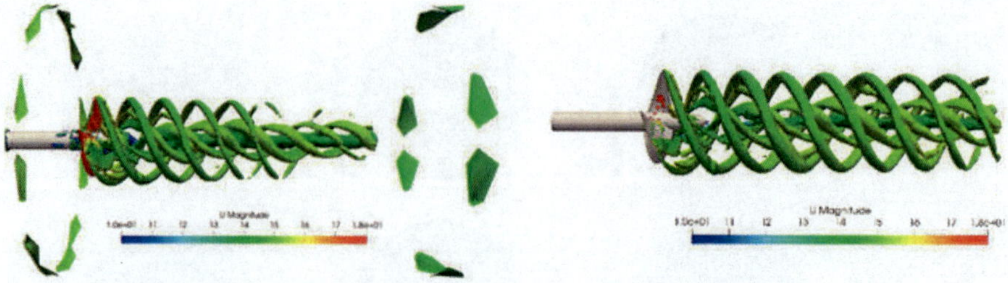

Fig. 17.6 The comparison for cavitation conditions between Q-criterion and Ω_R: The left panel shows the Q = 300 iso surface; The right panel shows $\Omega_R = 0.52$ iso surface

17.5 Hydroacoustic Results

17.5.1 Near-Field Sound Pressure

The near-field sound pressure is analyzed in this section. In the near field of the propeller, six near-field probes are set up in sequence from upstream to downstream, as Fig. 17.7 shows. Taking the object surface as the integral surface, the dipole contribution, monopole contribution and CFD measurement values of 6 test points are plotted respectively, see Fig. 17.8. It is easy to see that the sound pressure at point A2 is the largest, no matter which sound pressure point, the contribution of the dipole is much greater than that of the monopole. And the total sound pressure is relatively consistent with the CFD measurements.

At the same near-field test point, for cavitation, the results of acoustic analogy and CFD measurement are quite different, as Fig. 17.8 shows. This is mainly because the influence of nonlinear sound sources is not considered, and the existence of two-phase flow also has an influence on sound pressure propagation [14].

The difference between the two cases can be explained from the perspective of nonlinear sound sources. The sound source of the cavitation is longer than that of the non-cavitation, and it is more complicated near the object surface. The reason is the vortex distribution.

Fig. 17.7 The six near-field microphones positions: All of these probes are in the x = 0 plane and x = 0.8D plane. A_2 probe is located at y = 0. The spacing of two adjacent microphones is 0.2D

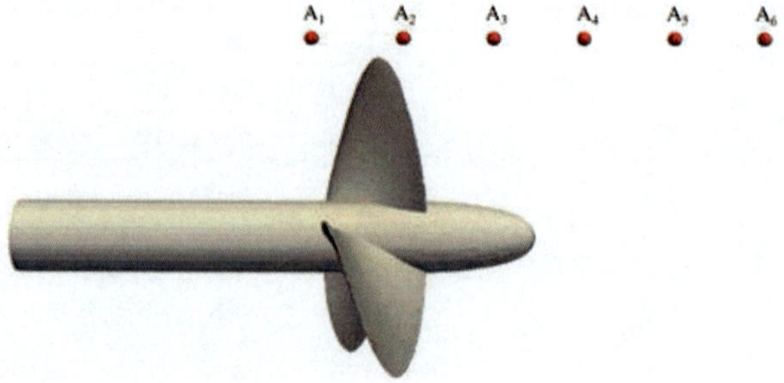

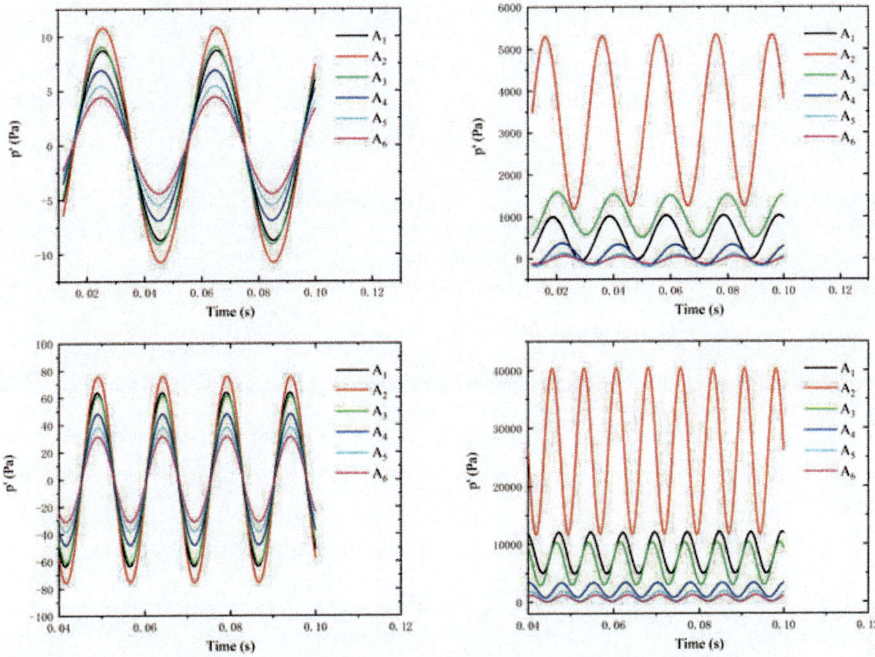

Fig. 17.8 The near-field sound pressure predicted by FW-H equation: The upper row is non-cavitation condition. The lower row is cavitation condition. The left column shows the monopole results. The right column shows the dipole results (Fig. 17.9).

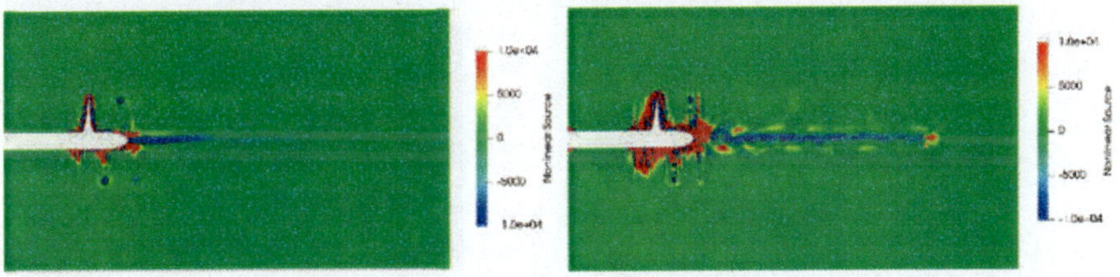

Fig. 17.9 The nonlinear sound source distribution: The left is the non-cavitation condition; The right is the cavitation condition

17.5.2 Far-Field Directivity

As for the far field, 4 planes are selected, and 36 test points are set for each plane, as Fig. 17.10 shows. The penetrable integral surface is used to calculate the acoustic directivity. In this way, nonlinear terms can be considered. Therefore, it can be seen that there are quadrupole characteristics on the two X planes, see Fig. 17.11. This is because the quadrupole is dominant in the far field. On the Y plane, it shows a uniform distribution.

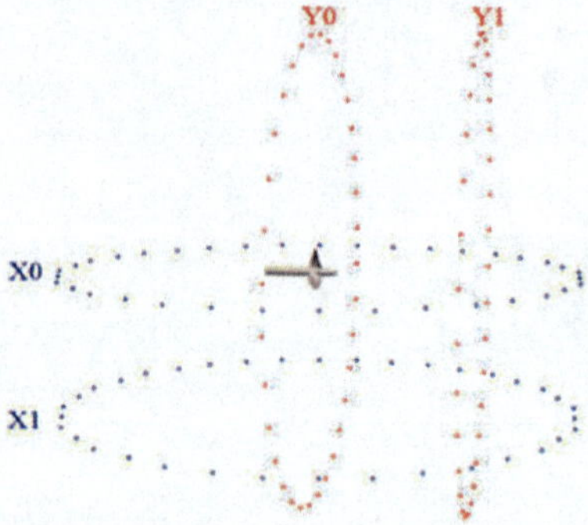

Fig. 17.10 The sketch of far-field microphones distributed on 4 planes: X0 = 0, X1 = –3D, Y0 = 0, Y1 = 3D. The radius of the probes is 10D

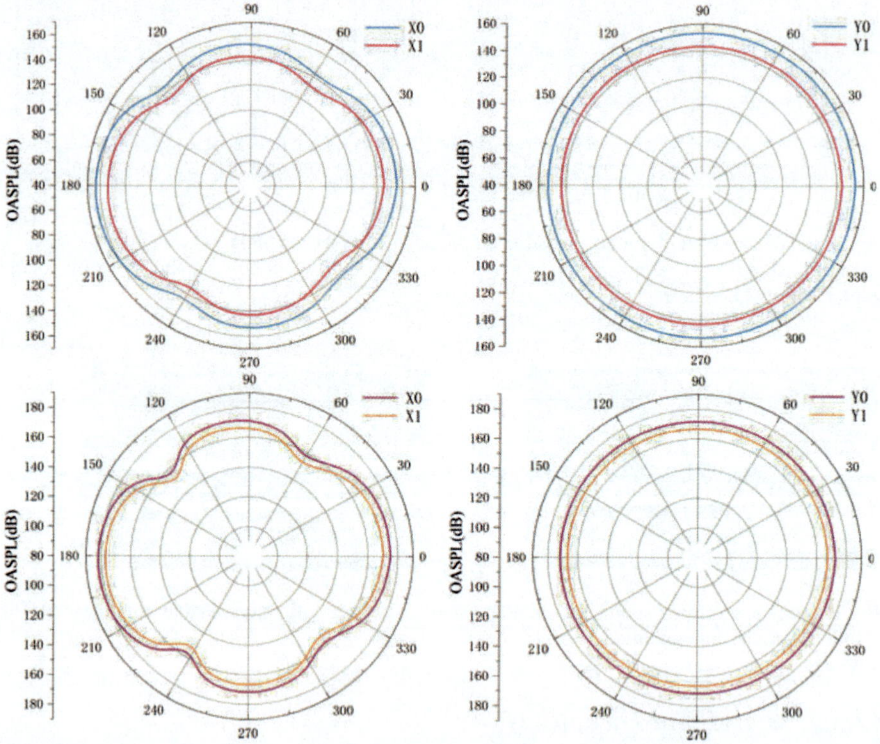

Fig. 17.11 The OASPL distribution on 4 planes for far field: The upper row shows the non-cavitation condition; The lower row shows the cavitation condition

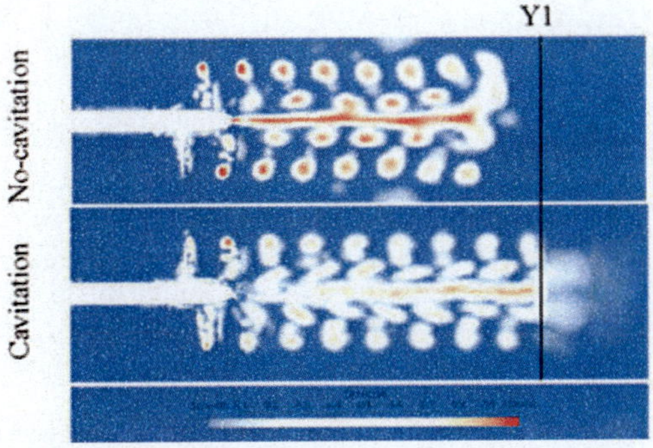

Fig. 17.12 The cloud map for Ω_R: The top panel is for non-cavitation condition; The bottom panel is for cavitation condition. The plane Y1 is marked by the black line

As for the cavitation, the decay speed of cavitation is less than that of non-cavitation, which is caused by different vorticity distribution laws. This can also be explained by the distribution of $\mathbf{\Omega_R}$. As shown in the Fig. 17.12, the vorticity of Y1 is already small for non-cavitation conditions, while for cavitation, the vortex is still considerable. Therefore, the decay of the cavitation state is slower.

17.6 Conclusion

The wake characteristics of the DTMB4119 propeller is analysed using the third-generation vortex identification technology. With the DDES turbulence model and the Schnerr Sauer cavitation model, both non-cavitating and cavitating conditions are simulated. The force coefficients are calculated and compared with the experimental results to verify the accuracy of the simulation. Changes in turbulent kinetic energy and turbulent viscosity et al. are analysed at different sections. The distribution of sound pressure is analyzed from the near-field and far-field perspectives, and an explanation is attempted from the distribution of $\mathbf{\Omega_R}$. Three conclusions are gotten:

- The third-generation vortex recognition technology can capture more detailed vortex structures and eliminate false vortex signals on the AMI surface.
- For the distribution of Lighthill sound sources, the vortex distribution given by Liutex is more consistent with the sound source distribution.
- Nonlinear terms dominate far-field noise. $\mathbf{\Omega_R}$, which predicts the vortex structure more accurately, can explain the far-field decay law better.

Acknowledgements This work is supported by National Natural Science Foundation of China (52131102, 51909160), and the National Key Research and Development Program of China (2019YFB1704200), to which the authors are most grateful.

References

1. M. Cianferra, A. Petronio, V. Armenio, Non-linear noise from a ship propeller in open sea condition. Ocean Eng. **191**, 106474 (2019)
2. N. Curle, The influence of solid boundaries upon aerodynamic sound. Proc. R. Soc. Lond. Ser. A Math. Phys. Sci. **231**(1187), 505–514 (1955)
3. X. Dong, Y. Gao, C. Liu, New normalized Rortex/vortex identification method. Phys. Fluids **31**(1), 011701 (2019)
4. A. Ebrahimi, M.S. Seif, A. Nouri-Borujerdi, Hydrodynamic and acoustic performance analysis of marine propellers by combination of panel method and FW-H equations. Math. Comput. Appl. **24**(3), 81 (2019)
5. J.E. Ffowcs Williams, D.L. Hawkings, Sound generation by turbulence and surfaces in arbitrary motion. Philos. Trans. R. Soc. Lond. Ser. A Math. Phys. Sci. **264**(1151), 321–342 (1969)
6. S. Ianniello, R. Muscari, A. Di Mascio, Ship underwater noise assessment by the acoustic analogy. Part I: nonlinear analysis of a marine propeller in a uniform flow. J. Mar. Sci. Technol. **18**(4), 547–570 (2013)
7. M.J. Lighthill, On sound generated aerodynamically I. General theory. Proc. R. Soc. Lond. Ser. A Math. Phys. Sci. **211**(1107), 564–587 (1952)
8. C. Liu, Y.S. Gao, X.R. Dong, Y.Q. Wang, J.M. Liu, Y.N. Zhang, X.S. Cai, N. Gui, Third generation of vortex identification methods: omega and Liutex/Rortex based systems. J. Hydrodyn. **31**(2), 205–223 (2019)
9. C. Liu, Y. Gao, S. Tian, X. Dong, Rortex—a new vortex vector definition and vorticity tensor and vector decompositions. Phys. Fluids **30**(3), 035103 (2018)
10. C. Liu, Y. Yan, P. Lu, Physics of turbulence generation and sustenance in a boundary layer. Comput. Fluids **102**, 353–384 (2014)
11. G.H. Schnerr, J. Sauer, Physical and numerical modeling of unsteady cavitation dynamics, in *Fourth International Conference on Multiphase Flow* (vol. 1). ICMF New Orleans (2001, May)
12. P. Spalart, S. Allmaras, A one-equation turbulence model for aerodynamic flows, in *30th Aerospace Sciences meeting and Exhibit*. (1992, January), p. 439
13. V.M. Viitanen, A. Hynninen, T. Sipilä, T. Siikonen, DDES of wetted and cavitating marine propeller for CHA underwater noise assessment. J. Mar. Sci. Eng. **6**(2), 56 (2018)
14. Q. Wu, B. Huang, G. Wang, S. Cao, M. Zhu, Numerical modelling of unsteady cavitation and induced noise around a marine propeller. Ocean Eng. **160**, 143–155 (2018)

Chapter 18
Identification of Vortex Around High-Speed Ship Based on Liutex Method

Yunpeng Ma, Weiwen Zhao, and Decheng Wan

Abstract Vortex is ubiquitous in ship navigation. The first-generation vortex identification method is based on vorticity, which is understood as twice the rotational angular velocity around the center. The isosurface is used to explore the vortex structure of the flow field when it comes to second-generation $\boldsymbol{Q}$, Δ and other methods. The third-generation Liutex vector defines the local rigid body rotation of the fluid, including the local rotation axis and the rotation strength. The vorticity can be decomposed into the rigid body rotation and the pure shear, and the rigid rotation part can be accurately extracted from the fluid motion. The vortex identification method overcomes the need to adjust the threshold for specific flows and can capture both strong and weak vortices at the same time; Based on the idea of Liutex, the various characteristics of vortices can be quantitatively described, such as the absolute strength of vortices, relative strength, local axis of rotation, vortex core center, vortex core size, and vortex boundary. As a prototype of a destroyer, many engineers use the DTMB model to do a lot of numerical calculations. In this paper, the third-generation Liutex method is used to identify the vortex around the DTMB, and the vortex structure is analyzed in detail and compared with the Q criterion and so on. The calculation results show that the Liutex method has better accuracy.

18.1 Introduction

Vortexes can be seen everywhere in nature, ranging from tornadoes to turbulence. Turbulence can be defined as the accumulation of eddies. The related research on turbulence is called the problem of the century. Researchers have realized that turbulence is not completely random, but has a pseudo-ordered vortex structure that plays a key role in fluid dynamics and energy transport. How to extract vortex structures accurately from massive data of 3D flow field is increasingly becoming the key to solving turbulent flow problems. The development of vortex identification methods

Y. Ma · W. Zhao · D. Wan (✉)
Computational Marine Hydrodynamic Lab (CMHL), School of Naval Architecture, Ocean and Civil Engineering, Shanghai Jiao Tong University, Shanghai 200240, China
e-mail: dcwan@sjtu.edu.cn

Y. Wang et al. (eds.), *Liutex and Third Generation of Vortex Identification*,
Springer Proceedings in Physics 288, https://doi.org/10.1007/978-981-19-8955-1_18

can be divided into three generations [1]. The first generation method was based on vorticity and the Cauchy-Stokes is decomposed. The antisymmetric part of the velocity gradient tensor (the vorticity tensor) is regarded as a rotation. However, many researchers have found that the correlation between vorticity and vortex is not satisfying. To overcome the defects of the first generation method, the second generation vortex identification method represented by methods, such as Q, emerges. However, the second-generation method depends on the actual threshold value, is also associated with the specific case, and can be affected by shearing. To solve the threshold problem, by the Ω method developed in 2016, the vortex is defined as the vorticity greater than the deformation of the connected area and expressed as a relative value. The Liutex vector, developed in 2017 and 2018, provides a systematic mathematical representation of the local rigid body transformation of fluids including local rotation axis and rotation strength. Ω method, Liutex vector, and a series of related methods are defined as the third generation vortex identification method.

The flow field around the ship is especially important for the performance of the ship, and more and more studies have been done on the vortex structure around the ship. Longo and Stren [2] studies the changes in the flow field around the ship under different drift angles of the Series 60 ship and defines the vortex structures at different positions. In the article, he also compares velocity pressure, etc. Visonneau et al. [3] adopts the ISIS-CFD solver to study the vortex structure near the DTMB sonar dome and at the bottom of the ship. The identification method used is Q. In the article, a structured mesh is used to refine the sonar dome and the bottom of the ship, and the turbulence model used is k-ω models, the anisotropic two-equation Explicit Algebraic Reynolds Stress Model (EARSM), Hybrid LES turbulence models based on Detached Eddy Simulation (DES). Finally, Michel compares calculated vortex structure, turbulent kinetic energy and velocity, according to the comparison, all turbulence models cannot completely reflect the physical reality.

18.2 Numerical Method

18.2.1 Turbulence Model

The most common are the classical two-equation k-ε and k-ω models, where the turbulent kinetic energy k and the turbulent kinetic energy dissipation rate ε and the specific rate of kinetic energy dissipation ω are solved. For the Direct Numerical Simulation, the Navier–Stokes equations are solved numerically directly without any turbulence model. The temporal and spatial scales of turbulence are solved in grids, which require quantitative computing resources. It is only used for low Reynolds number flows currently.

The SST [4] model combines the k-ε and k-ω models. It is kind of a low Reynolds number model. Some of the weaknesses exhibited by the k-ω model and the k-ε model can be eliminated by using this model and it is widely used in engineering.

For large eddy simulation, only large-scale turbulence is calculated and turbulence smaller than the filtering scale, which appears as an additional stress term called the subgrid stress. There is similarity on sufficiently small scales for almost all turbulent flows. It is possible to simulate subgrid stresses with a universal model theoretically, although no such model has yet emerged.

When it comes to the Reynolds-averaged Navier–Stokes equation, which averages the N-S equation in time, the unsteady turbulence problem is transformed into a steady problem study, at the cost of an additional unknown factor, which is also in the same form as the stress, called Reynolds stress. The Reynolds stress also needs to be modeled, which is the so-called turbulent model or turbulence model. However, due to the time averaging, the information contained in the equations themselves has been lost partially. What is more, giving a model for Reynolds stress is very difficult.

As regards Detached Eddy Simulation (DES), RANS is used in the near-wall and LES in the turbulent core region. Delayed DES (DDES) simulation is used to prevent model stress from losing and separation induced by mesh.

DDES is also the model used in this paper. In addition, Improve DDES and DDES models are adopted to solve the problem of mismatching of the logarithmic region.

18.2.2 Naoe-Foam Solver

18.2.2.1 Governing Equation

In this paper, the naoe-Foam solver [5] developed by Computational Marine Hydrodynamics Lab (CMHL) based on OpenFOAM is used. The governing equation used is the two-phase flow incompressible RANS (Reynolds-Averaged Navier–Stokes) equation:

$$\nabla \cdot \boldsymbol{U} = 0 \tag{18.1}$$

$$\begin{aligned} \frac{\partial \rho \mathbf{U}}{\partial t} + \nabla \cdot \left(\rho(\mathbf{U} - \mathbf{U}_g)\mathbf{U}\right) &= -\nabla p_d - g \cdot x \nabla \rho + \nabla \cdot \left(\mu_{eff} \nabla \mathbf{U}\right) \\ &\quad + (\nabla U) \cdot \nabla \mu_{eff} + f_\sigma \end{aligned} \tag{18.2}$$

U is the velocity of the flow field; U_g is the grid velocity, and P_d is the fluid dynamic pressure; $\mu_{eff} = \rho(v + v_t)$ is the effective dynamic viscosity, ν and ν_t are the dynamic viscosity and turbulent eddy viscosity, and ν_t is obtained from the k-ω turbulence model; the Two-phase flow surface tension term of the model is f_σ.

18.2.2.2 DDES Model

As DDES is a combination of RANS and LES, the RANS is used inside the boundary layer and LES is used outside the boundary layer. The problem would be the same as that of LES, the DDES method is particular about the quality of the grid. In addition, since the RANS equation is used inside the boundary layer, the calculation results inside are influenced on the results of the DDES calculation.

As the property of DES relies on the similarity of the turbulent dissipative terms of RANS and LES, the rate of change inside the transition zones of RANS and LES cannot be huge. However, in this region, turbulence viscosity may not be correct due to the switches from RANS to LES. To solve this problem, for the DDES method, introducing a special delay function is brought to modify the turbulent feature length [6]. What is more, DDES is came up with to solve the MSD and GIS in the prior DES [7]. For the SST DDES equation [8], the turbulent length is defined as follows, f_d is a kind of empirical blending function:

$$l_{\mathrm{DDES}} = l_{\mathrm{RANS}} - f_d max(0, l_{\mathrm{BANS}} - C_{\mathrm{DES}}\Delta) \tag{18.3}$$

$$f_d = 1 - \tanh\left((8r_d)^3\right) \tag{18.4}$$

the r_d would be the delay factor, and the r_d is defined as:

$$r_d = \frac{v + v_t}{\sqrt{0.5\left(S^2 + \Omega^2\right)\kappa^2 d^2}} \tag{18.5}$$

In detail, Ω is vorticity, S is strain rate in the definition of turbulence eddy viscosity, ν_t is the eddy viscosity and ν is molecular viscosity; The κ, equals 0.41, is the von Karman factor and d is the distance to the wall. In the near-wall boundary layer, f_d is made to be 0 to deactivate the DES limiter and to make sure that DES runs in RANS way. So the RANS region is limited, it would not translate to LES too early.

18.2.2.3 VOF Boundary Compressible Method

The VOF method (Volume of Fluid) with artificial compression technology is used to capture the change of free surface. It can control the numerical dissipation better, at the same time, has a better surface capture accuracy and its transport equation is defined as:

$$\frac{\partial \alpha}{\partial t} + \nabla \cdot \left[\left(\boldsymbol{U} - \boldsymbol{U}_g\right)\alpha\right] + \nabla \cdot \left[\boldsymbol{U}_r(1-\alpha)\alpha\right] = 0 \tag{18.6}$$

where α is the volume fraction, defined as the fraction of volume occupied by the fluid in a grid cell, ranges from 0 to 1.

18.2.2.4 Discrete Method

In the process of calculation, the RANS Eqs. (18.1), (18.2) and the VOF transport Eq. (18.3) are discretized through the finite volume method. The computational domain is discretized into a series of small cells and the computational flow field information is stored in the center of the grid cell; The value of the cell surface can be obtained by interpolating the value of the cell center. Finally, according to the Gauss theory, the value of the element surface can be accumulated to obtain the volume integral of the element body. In addition, different interpolation methods need to be used for different items.

18.2.3 The Third Generation Vortex Identification Method

Vortex is an unavoidable problem in fluid research. The identification method of vortex was first proposed in the 1980s. After continuous improvement, it has now developed into the third generation.

18.2.3.1 Vorticity and Q

The vorticity method is the easiest way to describe vortices, but it is difficult to distinguish between vortices caused by rotation and vortices caused by shear. Moreover, it is hard to identify tiny and weak vortices.

The Q method, using a small amount of calculation, is the most classic. Based on the definition of the second invariant of velocity gradient tensor, For incompressible flow, the trace of ∇U is 0. Thus, Q can be simplified as

$$Q = \frac{1}{2}\left[\|\boldsymbol{\Omega}\|^2 - \|\boldsymbol{S}\|^2\right] \tag{18.7}$$

The $\| \, \|$ notation represents the Frobenius norm of a tensor. The velocity gradient tensor can be decomposed into two parts, the symmetric part, denoted by S, is usually called the strain rate tensor; the antisymmetric part, denoted by Ω, is usually called the rotation rate or vorticity tensor.

Where

$$\|\boldsymbol{\Omega}\| = \sqrt{\mathrm{tr}\left(\boldsymbol{\Omega}^T \boldsymbol{\Omega}\right)} \tag{18.8}$$

$$\|\boldsymbol{S}\| = \sqrt{\mathrm{tr}\left(\boldsymbol{S}^T \boldsymbol{S}\right)} \tag{18.9}$$

For the Q method, a vortex is defined as a "connected fluid region with a positive second invariant of ∇U", Its physical meaning is that Q represents the balance between the magnitude of the vorticity and the strain rate tensor; It can represent the local fluid rotation when the antisymmetric tensor overcomes the cancellation effect of the symmetric tensor; Finally, the Q method can be understood as a modification of the Cauchy-Stokes decomposition.

18.2.3.2 Liutex Method

To overcome the problems of second-generation identification methods, Professor Chaoqun Liu and his team, at the University of Texas at Arlington, have carried out a series of works. In 2016, Nian Liu et al. [9] proposed the concept of decomposing the vorticity into rotating parts and non-rotating parts. It overcomes the problem that the second-generation vortex identification method needs to adjust the threshold manually. In 2017, Professor Chaoqun Liu proposed how to extract the rotating motion part of the rigid body in the motion systematically, and the rotating part of the vorticity is named Rortex vectors whose direction represents the local axis of rotation, magnitude represents the rotational angular velocity [10]. In 2018, the Rortex vector is renamed the Liutex vector.

The direction of Liutex vector $\boldsymbol{r}$

If there are three real eigenvalues for the velocity gradient tensor ∇U, it means that the fluid motion contains just stretching and deformation. In this case, the magnitude of the Liutex vector is 0. On the contrary, if there are one real and two conjugate complex eigenvalues for ∇U, there exists a vortex in the local region. The direction of the Liutex vector is the same as one of the eigenvector correspondings to the only real eigenvalue. The normalized eigenvector is defined as $\boldsymbol{r}$.

The magnitude of Liutex vector R

By transforming the original coordinate system xyz to a new coordinate one XYZ with a matrix Q, the original Z-axis is rotated to the direction of r. In the new coordinate system, fluid would rotate only in the XY plane and the derivatives of U_x and U_y in Z-direction are zero. Then the velocity gradient tensor is as follows:

$$\nabla \boldsymbol{U} = \boldsymbol{Q}\nabla \boldsymbol{u}\,\boldsymbol{Q}^T = \begin{bmatrix} \frac{\partial U_X}{\partial X} & \frac{\partial U_X}{\partial Y} & 0 \\ \frac{\partial U_Y}{\partial X} & \frac{\partial U_Y}{\partial Y} & 0 \\ \frac{\partial U_Z}{\partial X} & \frac{\partial U_Z}{\partial Y} & \frac{\partial U_Z}{\partial Z} \end{bmatrix} \tag{18.10}$$

Q is the transform matrix. A second transformation through matrix P will be carried out, by which Z-axis is rotated by θ, and the velocity gradient tensor becomes

$$\nabla \boldsymbol{U}_P = \boldsymbol{P}\nabla \boldsymbol{U}\,\boldsymbol{P}^T \tag{18.11}$$

where P is defined as

$$\boldsymbol{P} = \begin{bmatrix} \cos\theta & \sin\theta & 0 \\ -\sin\theta & \cos\theta & 0 \\ 0 & 0 & 1 \end{bmatrix} \tag{18.12}$$

The rotating angle θ is calculated to ensure that the absolute value of $\partial U_X / \partial Y|_\theta$ is going to be the minimum. The magnitude of the Liutex vector is given by

$$R = \begin{cases} 2(\beta - \alpha), \beta^2 > \alpha^2 \\ 0, \alpha^2 \geq \beta^2 \end{cases} \tag{18.13}$$

where

$$\alpha = \frac{1}{2}\sqrt{\left(\frac{\partial U_Y}{\partial Y} - \frac{\partial U_X}{\partial X}\right)^2 + \left(\frac{\partial U_Y}{\partial X} + \frac{\partial U_X}{\partial Y}\right)^2} \tag{18.14}$$

$$\beta = \frac{1}{2}\left(\frac{\partial U_Y}{\partial X} - \frac{\partial U_X}{\partial Y}\right) \tag{18.15}$$

Liutex vector is gained as:

$$\boldsymbol{R} = R\boldsymbol{r} \tag{18.16}$$

Wang et al. [11] made an explicit expression for the Liutex magnitude in the initial coordinate system:

$$\boldsymbol{R} = \left(\langle \boldsymbol{\omega}, \boldsymbol{r} \rangle - \sqrt{\langle \boldsymbol{\omega}, \boldsymbol{r} \rangle^2 - 4\lambda_{ci}^2}\right)\boldsymbol{r} \tag{18.17}$$

where λ_{ci} is the imaginary part of eigenvectors corresponding to the conjugate complex eigenvalues of ∇U.

18.3 Computational Overviews

18.3.1 Geometry and Test Condition

The geometry of DTMB5415 is as follows (Fig. 18.1 and Table 18.1):

The speed is set up to 2.590 m/s, it will be 30Kn when it is full scale. DTMB model is the prototype of a destroyer with high speed. To reduce wave resistance at high speed, a bulbous bow is set at the bow.

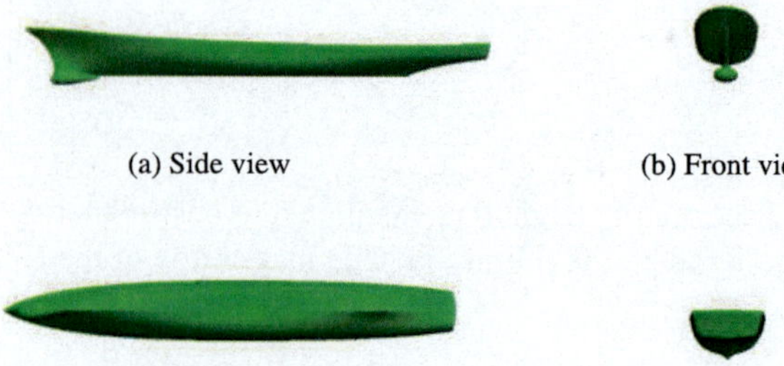

(a) Side view (b) Front view

(c) Upward view (d) Rear view

Fig. 18.1 DTMB5415 geometry

Table 18.1 Main parameters of DTMB5415

α	35.480
Lwl (m)	4.007
Lpp (m)	4.002
B (m)	0.537
T (m)	0.173
∇ (m^3)	0.189
CoG (m)	(1.9218,0,005)
iyy (kg·m^2)	378.6
V (m/s)	2.590

18.3.2 Computational Grids

The compute domain is shown in Fig. 18.2. The origin of coordinates is set at the intersection of the waterline and longitudinal section. Along the mid-longitudinal section, it is positive when pointing to stern, starboard, and upward. Due to the symmetry of shipshape and wave action, a half-domain is used to simulate the slamming of ships against waves. The inlet is arranged at $X = -Lpp$ (Lpp is the distance between perpendiculars), the outlet is at $X = 4Lpp$, the right boundary is at $Y = Lpp$, the bottom boundary is at $Z = -Lpp$, and the upper boundary is at $Z = 0.5Lpp$. The free liquid surface is at $Z = 0$, with air above and water below (Fig. 18.3).

In this paper, snappyHexMesh, a mesh generation tool provided by OpenFOAM, is used to generate mesh, and the number of background mesh is $83 \times 26 \times 26$. Mesh refinement derives from splitting cells, the refinement level (x, y, z) means that one cell is divided into (2^x, 2^y, 2^z) cells. Level (0,0,0) means no splitting in this region. In this paper, four refinements are used, around the ship being the innermost, the refinement levels is (4*4*4), and the remaining levels are 3, 2, and 1 from the inside to the outside. while the total number of cells after refinement is 7 million. Since the refinement scope is large, it covers the entire hull and most of the free surface, so compared with the wave breaking, even compared with the calculation example where only part of the ship is refined, the mesh used in this paper is not very dense.

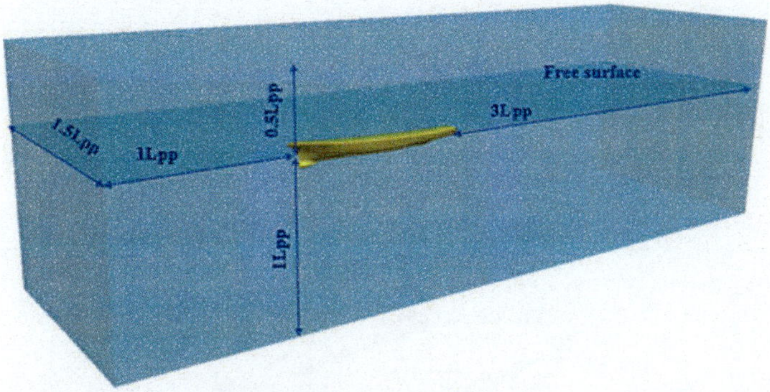

Fig. 18.2 Computational domain for DTMB

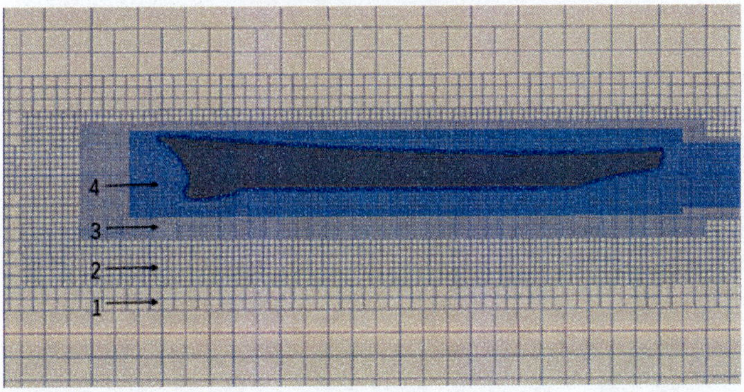

Fig. 18.3 Grid distribution in the computational domain

18.4 Results and Discussions

Isosurfaces identified by different vortex identification methods are shown in Fig. 18.4. For the Vorticity method, it is severely affected by shear pollution, it cannot accurately identify the vortex, For the second-generation Q method, is also affected by shear pollution, but not as much as the vorticity. For the Liutex, it cannot be affected by shear pollution, it can also clearly identify the vortex structure around the ship. Therefore, Liutex is adopted for the following analysis.

The section of the magnitude of Liutex is shown in Fig. 18.5. For the Vorticity method, Vortices near the bilge and bottom of the ship can be identified. In addition, after the stern, many of the identified vortex structures are distributed near the free surface, where the violent motion of the waves would interfere with the identification of the vortex structure.

Isosurface and the pressure near the bow are shown in Fig. 18.6. There are longitudinal vortices at the bottom of the ship, one of which is close to the bottom of the ship, and the other extends from the rear of the sonar dome where the diameter increases

(a) Vorticity = 20

(b) Q = 15

(c) |Liutex| = 0.85

Fig. 18.4 Vortex magnitude upward view

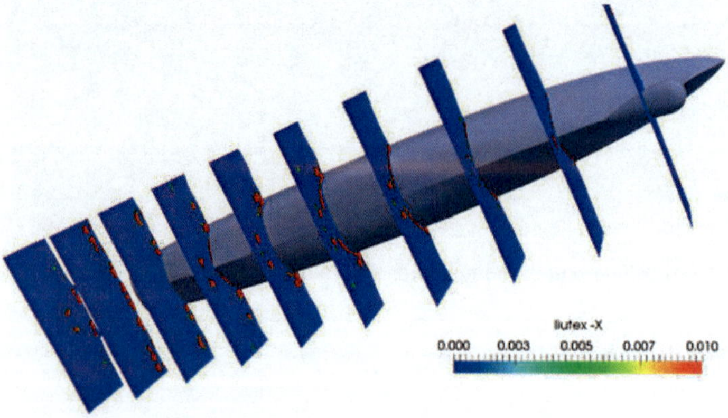

Fig. 18.5 Vortex magnitude in x–z section with |Liutex|=0.85

gradually, which may be related to the high pressure behind the sonar dome. The vortex structure close to the surface can be seen on the bow side.

Isosurface of the whole ship is shown in Fig. 18.7. Two longitudinal vortices at the bottom of the ship extend from the position of the sonar dome to the position before the stern. There are also a large number of vortex structures near the broadside.

(a) Isosurface (b) P of the ship

Fig. 18.6 Vortex structure of the bow

Fig. 18.7 Vortex structure viewed from stern

18.5 Conclusions

In this paper, the self-developed naoe-Foam solver and the DDES turbulence model are used. Different vortex identification methods are adopted to identify the vortex structure in the flow field around the DTMB ship model at high speed. Compared with the first and second-generation vortex identification methods, the third-generation Liutex method is free from the interference of shear contamination, which is an important improvement for vortex identification. The Liutex method can identify the vortex structures near the bow sonar dome and the bottom of the ship clearly. In addition, in this paper, the mesh refinement scope is large. Many vortex structures identified by Liutex are located near the free surface, at this position, the motion amplitude of the wave surface is huge, thus there are still problems with the reliability of the identified vortex structures. In the next, the sonar dome and the bottom of the ship are going to be refined in specialty to study the vortices at that location.

Acknowledgements This work is supported by the National Natural Science Foundation of China (52131102, 51909160) to which the authors are most grateful.

References

1. Y.Q. Wang, N. Gui, A review of the third-generation vortex identification method and its applications[J]. Chinese J. Hydrodyn. **34**(4), 413–429 (2019)
2. J. Longo, F. Stern, Effects of drift angle on model ship flow[J]. Exp. Fluids **32**(5), 558–569 (2002)
3. M. Visonneau, E. Guilmineau, P. Queutey, et al., Assessment of statistic and hybrid LES turbulence closures for complex free-surface flow simulation with combined grid refinement criteria[C], in *30th Symposium on Naval Hydrodynamics*, (2014), pp. 2–7
4. F.R. Menter, M. Kuntz, R. Langtry, Ten years of industrial experience with the SST turbulence model[J]. Turbul. Heat Mass Trans. **4**(1), 625–632 (2003)
5. J. Wang, W. Zhao, D. Wan, Development of naoe-FOAM-SJTU solver based on OpenFOAM for marine hydrodynamics[J]. J. Hydrodyn. **31**(1), 1–20 (2019)
6. Z. Wei-wen, W.A.N. De-cheng, Detached-Eddy simulation of flow past tandem cylinders[J]. Appl. Math. Mech. **37**(12), 1000–0887 (2016)
7. P.R. Spalart, S. Deck, M.L. Shur et al., A new version of detached-eddy simulation, resistant to ambiguous grid densities[J]. Theoret. Comput. Fluid Dyn. **20**(3), 181–195 (2006)
8. M.S. Gritskevich, A.V. Garbaruk, J. Schütze et al., Development of DDES and IDDES formulations for the k-ω shear stress transport model[J]. Flow Turbul. Combust. **88**(3), 431–449 (2012)
9. C. Liu, Y. Wang, Y. Yang, et al., New omega vortex identification method[J]. Sci. China Phys. Mech. Astron. **59**(8), 684711 (2016)
10. C. Liu, Y. Gao, S. Tian et al., Rortex-a new vortex vector definition and vorticity tensor and vector decompositions[J]. Phys. Fluids **30**, 035103 (2018)
11. Y. Wang, Y. Gao, J. Liu et al., Explicit formula for the Liutex vector and physical meaning of vorticity based on the Liutex-Shear decomposition[J]. J. Hydrodyn. **31**(3), 464–474 (2019)

Chapter 19
Identification of Vortical Structures of Flows Past a Surface-piercing Finite Square Cylinder with Rounded Corners

Songtao Chen, Weiwen Zhao, and Decheng Wan

Abstract As one of the typical bluff bodies, square cylinders are widely applied in offshore platforms, such as semi-submersibles and tension leg platforms. In order to reduce the resistance, rounded corners are often adopted instead of the original sharp corners to passively control the flow separation. To further explore its flow mechanism, this paper focuses on the vortical structures of flows past a vertical surface-piercing finite square cylinder with rounded corners at $Re = 2.7 \times 10^5$ and $Fr = 1.1$. The delayed detached-eddy simulation (DDES) is used to resolve the large-scale separation flow, and a piecewise-linear interface calculation (PLIC) based geometric volume of fluid (VOF) method is employed to accurately capture the interface. Due to the co-existence of the free surface and the free end, the wake flow exhibits strong three-dimensional characteristics. To clearly visualize the vortical structures, the third generation of vortex identification method—Liutex—is adopted throughout the present study. The iso-surfaces of the normalized Omega-Liutex $\tilde{\Omega}_R$ represent three-dimensional structures, while the Liutex vector field and integral lines show the direction of the local rotational axis. The results show that the Liutex method and its variant can well characterize the typical vortical structures and reveal more flow details.

19.1 Introduction

As one of the typical bluff bodies, the square cylinder has been widely used in many offshore platforms as a column. For many purposes, the square cylinder with rounded corners instead has become more common in practice. The modification from sharp corners to rounded ones with different radii would greatly change flow characteristics nearby. However, compared with the most common circular cylinders, the flow mechanisms of the square cylinder with rounded corners have not received much attention, especially in marine hydrodynamics. The practical condition where

S. Chen · W. Zhao · D. Wan (✉)
Computational Marine Hydrodynamics Lab (CMHL), School of Naval Architecture, Ocean and Civil Engineering, Shanghai Jiao Tong University, Shanghai, China
e-mail: dcwan@sjtu.edu.cn

Y. Wang et al. (eds.), *Liutex and Third Generation of Vortex Identification*, Springer Proceedings in Physics 288, https://doi.org/10.1007/978-981-19-8955-1_19

the free end and the free surface coexist in engineering will inevitably make the flow characteristics more complicated. Therefore, it is of great importance to explore the internal mechanism of flows past a finite surface-piercing square cylinder with rounded corners, with emphasis on the correlated vortical structures.

In previous studies, many researchers have carried out wind-tunnel experiments to study the aerodynamic behaviors of square cylinders with rounded corners. Tamura and Miyagi [1] studied the effect of inflow conditions on the drag and lift forces of square cylinders with different corner shapes. Their results suggest that the reduction in wake width is responsible for the reduced drag force. Carassale et al. [2] considered two rounded-corner radii (1/15 and 2/15), and studied their performances of global forces and surface pressure in the range of $1.7 \times 10^4 \leq \text{Re} \leq 2.3 \times 10^5$. In addition, their flow behaviors under two inflow conditions (smooth and turbulent flows) and two angles of incidence (0° and 45°) were also studied. They found that these differences could be attributed to the flow reattachment on the lateral surfaces induced by rounded corners. Hinsberg et al. [3] systematically performed a series of wind-tunnel experiments to investigate the variation of two different rounded-corner radii (0.16 and 0.29) under three angles of incidence (0°, −22.5° and −45°). They focused on spanning all flow regimes of interest, in which the studied Reynolds number was as high as $Re = 12 \times 10^6$ for the first time. By measuring global forces and surface pressure, they summarized the variation of aerodynamic coefficients with corner roundness and angle of incidence.

On the other hand, numerical simulations have become another essential research method, with the unique advantage of revealing more flow details. Zhang and Samtaney [4] carried out three-dimensional direct numerical simulations to study five rounded-corner radii (0, 0.125, 0.25, 0.375, and 0.5) at $Re = 1000$ using their in-house finite difference code. They found that the rounded corners could significantly reduce the time-averaged drag force as well as the fluctuations. Cao and Tamura [5] used the dynamic mixed model (DMM) to perform large-eddy simulations on the flows ($Re = 2.2 \times 10^4$ and 1.0×10^6) around a square cylinder with rounded corners of $r/D = 0.167$. In the supercritical regime, the delayed flow separation at the leeward corners resulted in a much smaller recirculation region, which was attributed to the flow reattachment near the frontal corners. On this basis, Cao and Tamura [6, 7] further numerically investigated the effects of shear inflow on the flow characteristics in these two regimes. In the subcritical regime, the asymmetry of the flow pattern due to the difference in flow separation between high-velocity and low-velocity lateral sides was evident, especially under high shear inflows. However, this is not the case in the supercritical regime, where the asymmetry was much lower in the near-wake. Dai et al. [8] used a modified $\text{k} - \varepsilon$ turbulence model to predict the flows past a square cylinder with rounded corners in OpenFOAM and succeeded in capturing the delayed flow separation on lateral sides.

In this paper, as an extension of [9], the flows past a finite surface-piercing square cylinder with rounded corners at $Re = 2.7 \times 10^5$ and $Fr = 1.1$ are numerically investigated. To better understand such bluff body flows, the primary objective of our study is to analyze the involved vortical structures using the third generation of vortex identification method—Liutex. The remainder of this paper is organized as follows.

Numerical methods are first introduced, especially the Liutex vortex identification method and its normalized variant—Omega-Liutex $\tilde{\Omega}_R$. Then, the numerical setup is presented, including the computational geometry, mesh, and boundary conditions. In the next section, the Liutex method is used to extract instantaneous and time-averaged vortical structures from the complex flow field. Finally, the main conclusions are drawn.

19.2 Numerical Methods

19.2.1 Governing Equations

In the present study, our simulation is performed by the open source CFD toolbox OpenFOAM v8. The two-phase incompressible Navier–Stokes equations are used to describe the flow in the present study. The continuity and momentum equations are as follows:

$$\nabla \cdot \mathbf{U} = \mathbf{0} \tag{19.1}$$

$$\frac{\partial \rho \mathrm{U}}{\partial t} + \nabla \cdot (\rho \mathrm{UU}) = -\nabla \mathrm{p_d} - \mathbf{g} \cdot \mathrm{x} \nabla \rho + \nabla \cdot (\mu_{\mathrm{eff}} \nabla \mathrm{U}) + \mathrm{f}_\sigma \tag{19.2}$$

where $\mathbf{U}$ is the velocity, ρ is the weighted average density, $\mathrm{p_d}$ is the dynamic pressure, g is the acceleration of gravity, x is a Cartesian coordinate vector, μ_{eff} is the effective dynamic viscosity, and f_σ is the surface tension term.

19.2.2 Turbulence Model

In the present study, a hybrid RANS/LES approach SST-DDES model is employed to resolve this large-scale separation flow. Compared with the pure RANS method, this scale-resolving strategy could enhance the solution accuracy of the detached region. At the same time, it could avoid the shortcomings that may be caused by the insufficient near-wall mesh layers when the pure LES method is applied. In general, the DDES technique achieves a good balance between accuracy and computational efficiency. In the current version of the SST-DDES model, the transport equations for the turbulence kinetic energy k and the turbulence specific dissipation ω are as follows:

$$\frac{\partial \rho k}{\partial t} + \nabla \cdot (\rho \mathrm{U} \mathrm{k}) = \nabla \cdot [(\mu + \sigma_k \mu_t) \nabla \mathrm{k}] + \mathrm{P_k} - \frac{\rho \mathrm{k}^{3/2}}{l_{\mathrm{DDES}}} \tag{19.3}$$

$$\frac{\partial \rho\omega}{\partial t} + \nabla \cdot (\rho \mathbf{U}\omega) = \nabla \cdot [(\mu + \sigma_\omega \mu_t)\nabla\omega] + (1 - \mathrm{F}_1)\mathrm{CD}_{\mathrm{k}\omega} + \alpha \frac{\rho}{\mu_t}\mathrm{P}_\mathrm{k} - \beta\rho\omega^2 \tag{19.4}$$

$$\mu_\mathrm{t} = \min\left(\frac{\rho k}{\omega}, \frac{\rho a_1 \mathrm{k}}{\mathrm{SF}_2}\right) \tag{19.5}$$

where P_k is the turbulence kinetic energy production term, l_{DDES} is the turbulence length scale, F_1 and F_2 are blending functions, $\mathrm{CD}_{k\omega}$ is a limit function, μ_t is the eddy viscosity, S is the magnitude of the strain rate tensor, σ_k, σ_ω, α, and β are constants. The specific expression of each function and the value of each constant can be found in [10, 11].

19.2.3 Interface-Capturing Method

To accurately capture the violent free surface at a high Froude (*Fr*) number, a geometric volume of fluid (VOF) method is used in our simulation. The transport equation for the phase fraction α_v is as follows:

$$\frac{\partial \alpha_\mathrm{v}}{\partial \mathrm{t}} + \nabla \cdot (\alpha_\mathrm{v}\mathrm{U}) = 0 \tag{19.6}$$

Compared with the original interface-compression algebraic method proposed by Weller [12], (19.6) does not include the additional compression term based on the "counter-gradient" transport. Instead, it improves the accuracy of the surface interpolated phase fraction α_{vf} in advection based on the piecewise-linear interface calculation (PLIC). A detailed description of the whole procedure was presented in [9].

19.2.4 Liutex Vortex Identification Method

For a good representation of the vortical structures involved, the third generation of vortex identification method—Liutex [13–15] proposed by Prof. Liu is used throughout our study. Through coordinate transformation, the pure rigid-body rotation can be extracted from the fluid motion, thus giving the local rotational axis and strength of the vortical structures. Therefore, compared with other existing methods such as vorticity, *Q*, and λ_2 criterions, it can well exclude the possible shearing contamination. Furthermore, since Liutex is in vector form, it can not only characterize three-dimensional structures with iso-surfaces of proper thresholds but also depict the development of rotational axes. The definition of the Liutex vector **R** is given as below:

$$\mathrm{R} = R\mathrm{r} \tag{19.7}$$

$$R = \begin{cases} 2(\beta - \alpha), & \alpha^2 - \beta^2 < 0 \text{ and } \beta > 0 \\ 2(\beta + \alpha), & \alpha^2 - \beta^2 < 0 \text{ and } \beta < 0 \\ 0, & \alpha^2 - \beta^2 \geq 0 \end{cases} \tag{19.8}$$

$$\nabla \mathrm{u} = \mathrm{Q} \nabla \mathrm{U} \mathrm{Q}^{\mathrm{T}} = \begin{bmatrix} \frac{\partial u_x}{\partial x} & \frac{\partial u_x}{\partial y} & 0 \\ \frac{\partial u_y}{\partial x} & \frac{\partial u_y}{\partial y} & 0 \\ \frac{\partial u_z}{\partial x} & \frac{\partial u_z}{\partial y} & \frac{\partial u_z}{\partial z} \end{bmatrix} \tag{19.9}$$

$$\begin{cases} \alpha = \frac{1}{2}\sqrt{\left(\frac{\partial u_y}{\partial y} - \frac{\partial u_x}{\partial x}\right)^2 + \left(\frac{\partial u_y}{\partial x} + \frac{\partial u_x}{\partial y}\right)^2} \\ \beta = \frac{1}{2}\left(\frac{\partial u_y}{\partial x} - \frac{\partial u_x}{\partial y}\right) \end{cases} \tag{19.10}$$

where R is the magnitude of the Liutex vector, **r** is the local rotational axis, **u** is the velocity in the transformed coordinate system, and **Q** is a rotational matrix.

On this basis, Prof. Liu further proposed a normalized Omega-Liutex $\tilde{\Omega}_R$ method [16, 17] to overcome the problem of threshold selection. Therefore, the Omega-Liutex $\tilde{\Omega}_R$ method can capture both strong and weak vortices in the entire flow field simultaneously, and a threshold of 0.52 is recommended to have an overall view. The expression of the normalized Omega-Liutex $\tilde{\Omega}_R$ is as follows:

$$\tilde{\Omega}_{\mathrm{R}} = \frac{\beta^2}{\beta^2 + \alpha^2 + \lambda_{cr}^2 + \frac{1}{2}\lambda_r^2 + \varepsilon} \tag{19.11}$$

Where λ_{cr} is the real part of the complex-conjugate eigenvalues of ∇u, λ_r is the real eigenvalue, and $\varepsilon = b_0(\beta^2 - \alpha^2)$ is a small parameter. Here, according to a sensitivity study by Zhao et al. [18], b_0 is recommended to be 10^{-6} in marine hydrodynamics.

In the following analysis, the iso-surfaces of the normalized Omega-Liutex $\tilde{\Omega}_R$ with different thresholds are used to characterize three-dimensional vortical structures, and the Liutex vector field and integral lines are used to reveal their detailed behaviors inside.

19.3 Numerical Setup

As an extension of [9], a surface-piercing finite square cylinder with rounded corners is adopted in our study, as shown in Fig. 19.1. It can be seen from its cross-section that the diameter D is still 0.2 m while the radius ratio r/D of corners is set to 0.15. Its aspect ratio L/D is 6, and the immersed length ratio h/D is 4. To explore the effect of cylinder shape on flow characteristics, the flow conditions remain the same, i.e., $Re = 2.7 \times 10^5$ and $Fr = 1.1$.

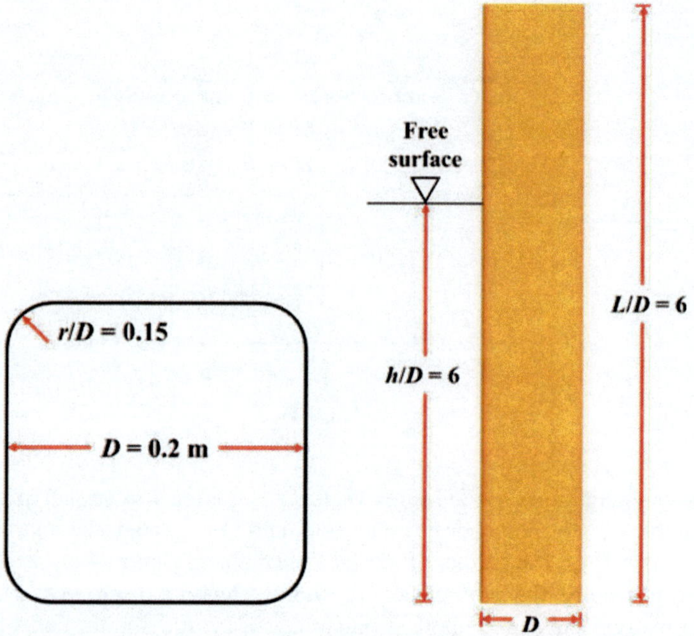

Fig. 19.1 Computational geometry

Figure 19.2 shows the computation domain, which is set to $-20 \leq x/D \leq 20$, $-15 \leq y/D \leq 15$, and $-8 \leq z/D \leq 2$ (length × width × depth). The cylinder is located at the origin of the coordinate system, i.e., the intersection of mid-longitudinal and mid-transverse planes at the free surface. Figure 19.3 shows the body-fitted structured computation mesh from different views, consisting of about 14.8×10^6 cells. The mesh configuration has been verified with quantitative convergence studies and validated with the experimental data, whose specific parameters can be found in [9].

At the inlet, a uniform inflow (U_∞, 0, 0) is applied, which is also set as the initial condition for the entire domain. The no-slip boundary condition is imposed on the

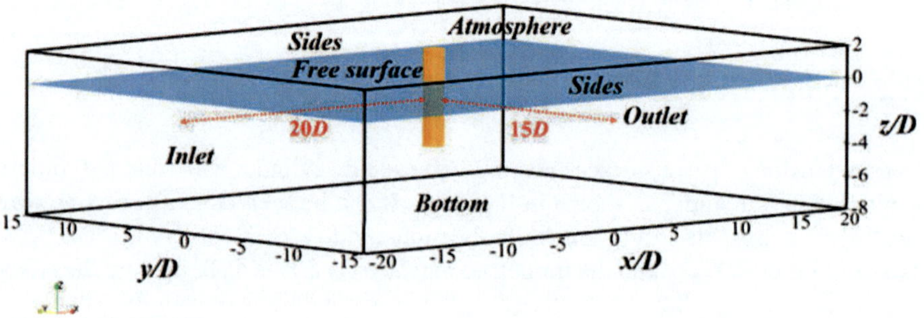

Fig. 19.2 Computational domain

Fig. 19.3 Computational mesh: **a** near free surface and **b** near free end

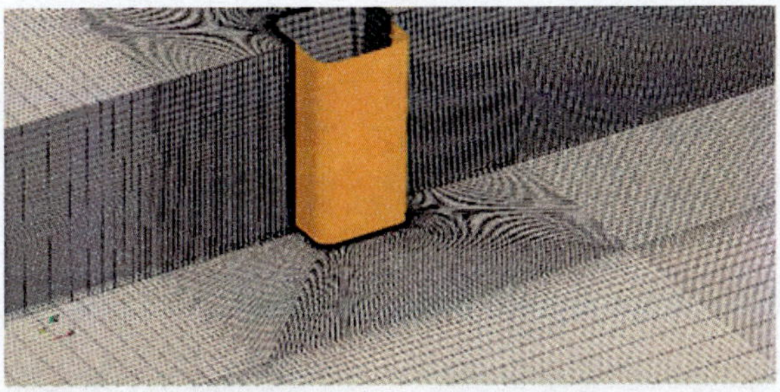

(a)

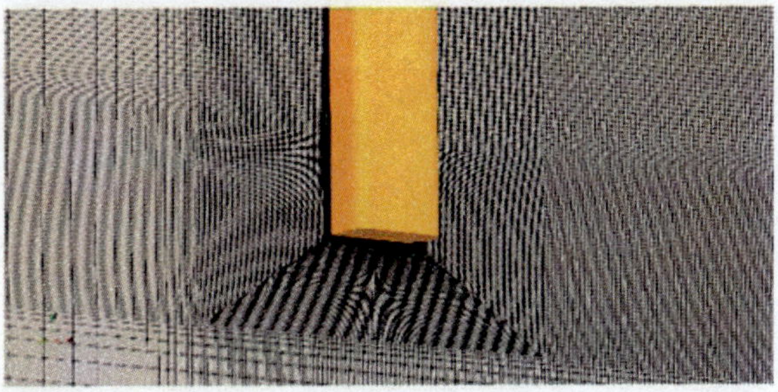

(b)

cylinder surface, and the Neumann boundary condition is used for the outlet and atmosphere. Apart from these, the symmetry boundary condition is set on the sides and bottom. The time step $\Delta tU_{\infty}/D$ is set to 1.5×10^{-3} to ensure the maximum Courant number is below 1, and the total simulation time TU_{∞}/D is set to 100 to obtain enough periodic stable results for statistical analysis.

19.4 Results and Discussion

19.4.1 Instantaneous Flow Field

To first have a straightforward understanding of the instantaneous flow field, Fig. 19.4 shows the three-dimensional vortical structures using the iso-surfaces of $\tilde{\Omega}_R = 0.52$ at $tU_{\infty}/D = 100$. From the top view, Fig. 19.4a reveals the existence of necklace vortices consisting of several slender structures in front of the square cylinder, which can be attributed to the sustained run-up (incoming flow) and wash-down (reversed flow) near the frontal surface. On the other hand, a large number of small-scale structures near the free surface termed wave-induced vortices can be observed in the wake. The violent free surface deformation and distortion at a high *Fr* lead to their wide-spreading in the near and even far wake. From the side view, several large-scale structures known as Karman vortex shedding can

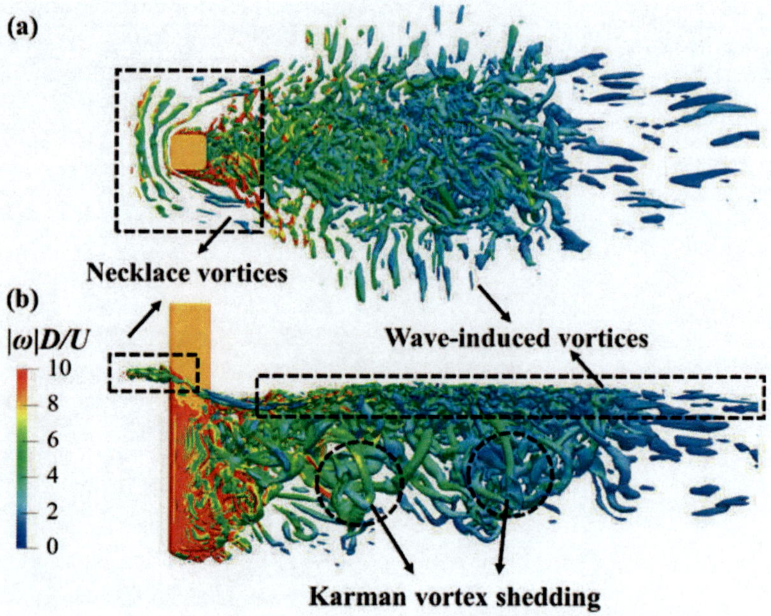

Fig. 19.4 Instantaneous iso-surfaces of $\tilde{\Omega}_R = 0.52$, colored by normalized magnitude of vorticity: **a** top view and **b** side view

be captured at medium depth in Fig. 19.4b, which is the most prominent feature of flows around bluff bodies. When approaching the free end, the Karman vortex shedding appears to be inhibited. Instead, the inclined upward structures generated by the free end dominate the near wake, which will be identified later. In conclusion, the pattern of the instantaneous vortical structures in this study is very similar to those of flows past a circular cylinder [9].

Figure 19.5 further gives the distribution of instantaneous spanwise vorticity at six representative depths, which are interface, $z/D = -0.75$, -1, -2, -3, and -3.5, respectively. Using the vorticity method, the behaviors of free shear layers can be observed. At the interface, the free shear layers bend outward with a large angle rather than being parallel or towards the centerline, accompanied by the dispersed distribution of small vortices between them. This phenomenon is correlated with the pattern of the depression region behind and the above-mentioned violent free surface. Below the interface, as an example of $z/D = -1$, the free shear layers are basically parallel with the lateral surfaces, and Karman vortex shedding are shed from their ends alternatively. However, when approaching the free end, the free shear layers become short and concentrated, and only a few dispersed vortices can be roughly observed in the far wake.

For comparison, Fig. 19.6 gives the instantaneous spanwise Liutex at the same depths. Since the Liutex method can exclude the possible shearing contamination, the free shear layers almost disappear in these contours. As a result, compared with the

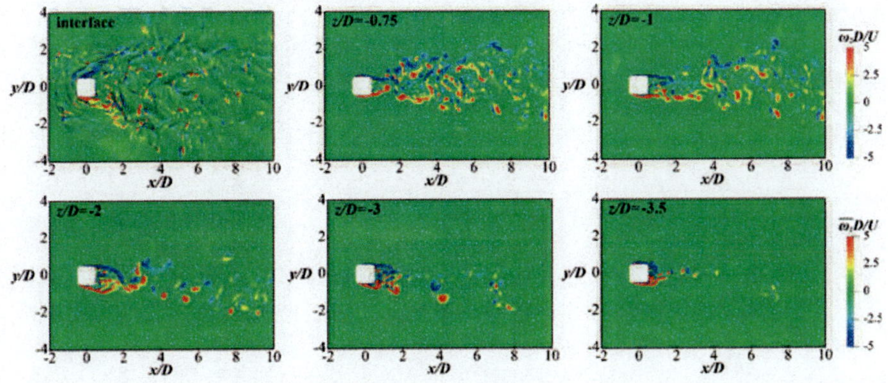

Fig. 19.5 Contours of instantaneous normalized spanwise vorticity

"diffusive" ones using the vorticity method, the vortices even in the far wake become more visible now. With this advantage, it can be clearly observed that the width of the vortex field varies significantly with the depth. When near the free surface, the vortices occupy a large area in the wake, especially in the cross-stream direction. However, as the depth gradually increases, the vortices shrink first in the cross-stream direction and subsequently in the streamwise direction. Therefore, when finally approaching the free end, most of the vortices are concentrated near the leeward surface of the square cylinder.

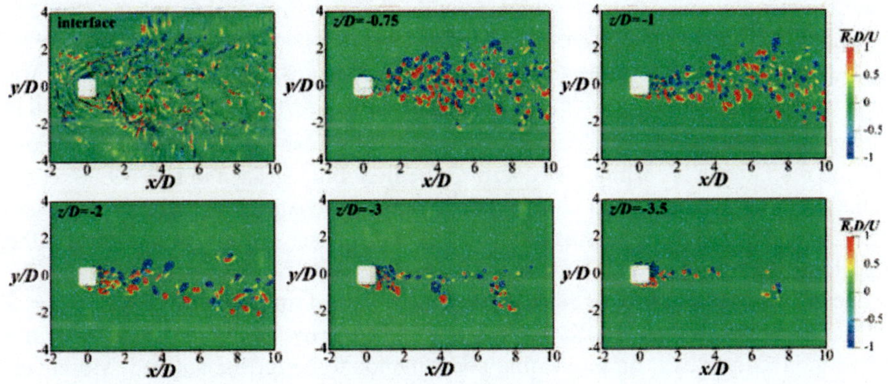

Fig. 19.6 Contours of instantaneous normalized spanwise Liutex

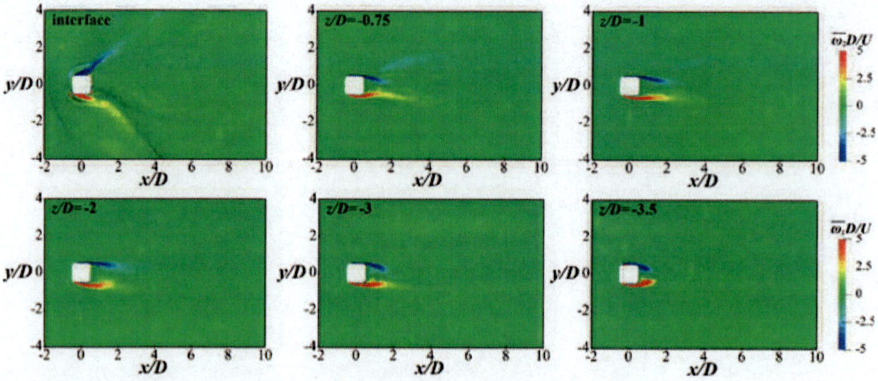

Fig. 19.7 Contours of time-averaged normalized spanwise vorticity

19.4.2 Time-Averaged Flow Field

In this section, the time-averaged flow field from $50 < tU_\infty/D < 100$ is further analyzed to characterize the behaviors of primary flow structures. On the basis of Figs. 19.5, 19.7 further shows the time-averaged spanwise vorticity to study the variation of free shear layers with depth. By excluding the vortices in the wake, the shape of free shear layers and the angle between them can be easily observed. The free shear layers start from a large angle between their ends at the interface, then become parallel to each other, and finally concentrate towards the centerline. By comparing with the parallel condition of medium depth, this variation fully demonstrates the effects of the free surface and the free end.

Figure 19.8 gives the three-dimensional time-averaged vortical structures for the upper part of the square cylinder. Here, the normalized Omega-Liutex $\tilde{\Omega}_R$ method is applied to the time-averaged velocity field, and the threshold $\overline{\tilde{\Omega}_R}$ is increased from 0.52 to 0.65 to extract primary structures. The light blue surface in Fig. 19.8b represents the time-averaged free surface. Upstream of the square cylinder, a complete primary necklace vortex N_1 surrounds the frontal surface and is symmetry about the mid-longitudinal plane $y/D = 0$. At the same time, a pair of counter-rotating wave-induced vortices W_l and W_r originate from the leeward rounded corners. Compared with the ones from the leeward surface in the circular cylinder [9], the origin of the wave-induced vortex is an interesting phenomenon to be further studied. Moreover, as indicated by the streamwise vorticity, the rotations of W_l and W_r deviate from the mid-longitudinal plane $y/D = 0$.

For the lower part of the cylinder, the three-dimensional time-averaged vortical structures are presented in Fig. 19.9. Given that the flows near the free end are more complicated, the threshold $\overline{\tilde{\Omega}_R}$ is increased from 0.65 to 0.8. On the leeward surface, a pair of large-scale counter-rotating vortices A_{vl} and A_{vr}, identified as arch vortices, incline upward. Their behaviors are consistent with the observation of instantaneous vortical structures in Fig. 19.4b, which can be regarded as the dominant factor of

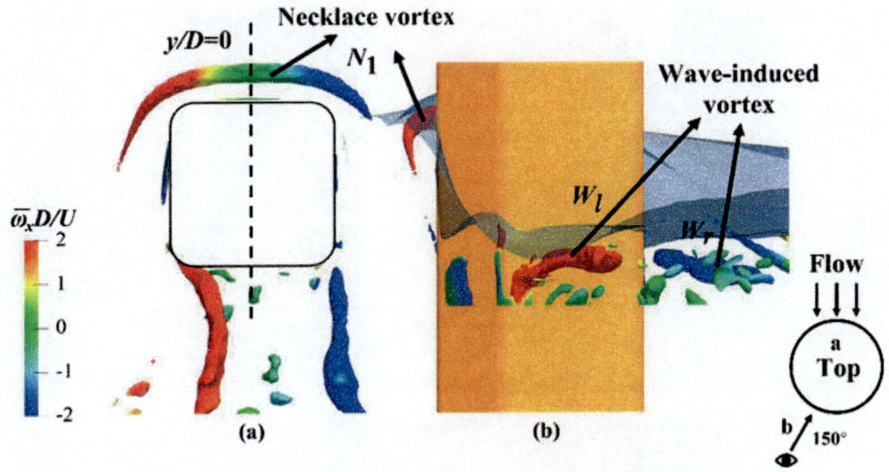

Fig. 19.8 Time-averaged iso-surfaces of $\overline{\tilde{\Omega}}_R = 0.65$ for upper part of cylinder, colored by normalized time-averaged streamwise vorticity: **a** top view and **b** side view. (The views are marked by the sketch in the lower right corner)

the flows near the free end. Besides, similar to the wave-induced vortices described above, their rotations are still away from the mid-longitudinal plane $y/D = 0$. Between A_{vl} and A_{vr}, another λ-type arch vortex A_b can be captured near the bottom of the leeward surface, possibly due to the flow separation caused by the trailing edge of the free end. At the bottom of the free end, a pair of counter-rotating tip vortices T_{vl} and T_{vr} are identified, and their rotations are opposite to that of the arch vortices A_{vl} and A_{vr}. Unlike the origin of tip vortices in circular cylinders [9], T_{vl} and T_{vr} are always attached to the bottom surface rather than originating from the lateral edges.

Apart from the iso-surfaces of normalized Omega-Liutex $\tilde{\Omega}_R$, the Liutex method itself can also provide many information using the vector field and the integral lines. In the present study, the necklace vortices upstream are taken as an example for a detailed analysis. Figure 19.10 shows the time-averaged streamlines near the free surface at $y/D = 0$ and the iso-surfaces of $\overline{\tilde{\Omega}}_R = 0.52$. The reason for choosing $\overline{\tilde{\Omega}}_R = 0.52$ again is that it gives the completest necklace vortices. As shown in Fig. 19.10, three necklace vortices N_1, N_2, and N_3 can be well identified with both streamlines and iso-surfaces. Among them, N_1 corresponds to the primary necklace vortex in Fig. 19.8, which proves that the strength of N_2 and N_3 is slightly weaker. As indicated by streamlines, it is clear that the interaction of the incoming flow and the reverse flow, i.e., run-up and wash-down, results in their generation. Moreover, the diameter of these structures can also reveal their relative strength to a certain extent.

To explore their internal behaviors, the Liutex vector field is given first in Fig. 19.11. As shown in the enlargements, the Liutex vectors are almost tangent to the iso-surfaces, which is consistent with the conclusions of [15]. By integrating these vectors, the Liutex lines can be obtained accordingly, as shown in Fig. 19.12. It can be seen from the directions of Liutex lines that the rotational axes of N_1

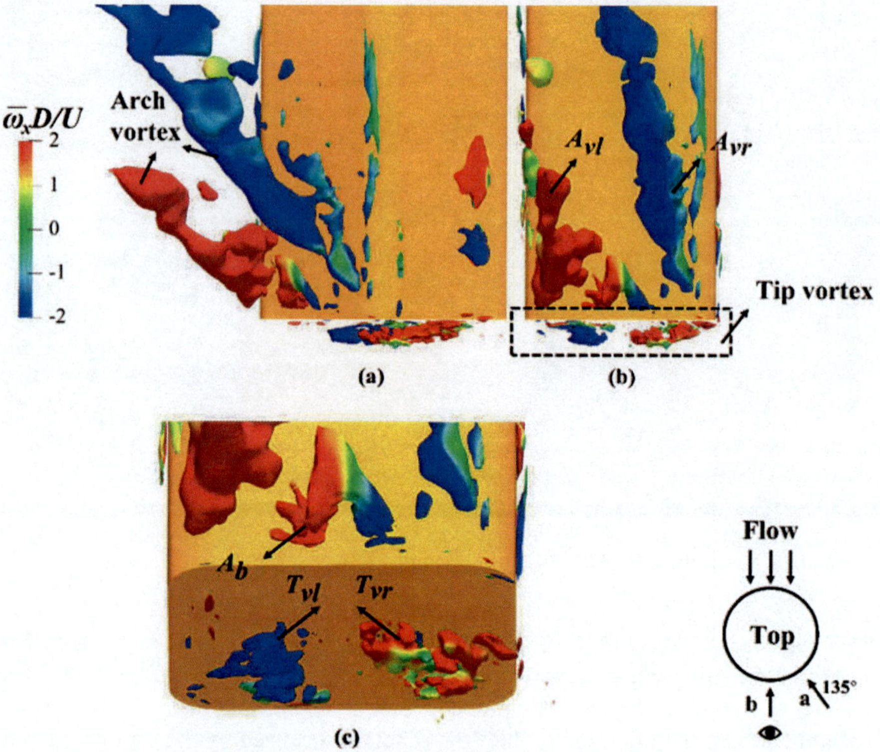

Fig. 19.9 Time-averaged iso-surfaces of $\overline{\tilde{\Omega}}_R = 0.8$ for lower part of cylinder, colored by normalized time-averaged streamwise vorticity: **a** side view, **b** back view, and **c** magnified view near free end. (The views are marked by the sketch in the lower right corner)

and N_3 are opposite to that of N_2, which can also be confirmed by the streamlines in Fig. 19.10. Moreover, when near the frontal surface, the Liutex lines basically remain straight and parallel to each other. However, as they develop laterally downstream, they quickly twist into a spiral, indicating the increasing instability of local rotational axes. The enlargement shows the Liutex lines of N_2, where the above phenomenon is particularly evident. Even without the indication of Liutex lines, the shape of N_2 can clearly reveal the behavior of twisting at both ends.

19.5 Conclusions

In this paper, the flows past a vertical surface-piercing finite square cylinder with rounded corners (r/D = 0.15) at Re = 2.7 × 10^5 and Fr = 1.1 is numerically investigated using an SST-DDES turbulence model and a PLIC-based geometric VOF method. On this basis, the instantaneous and time-averaged vortical structures

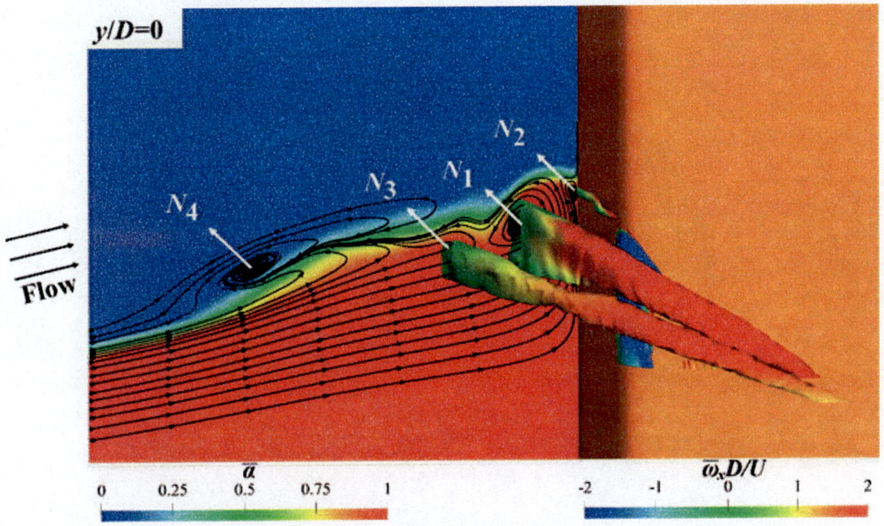

Fig. 19.10 Time-averaged streamlines near free surface at $y/D = 0$

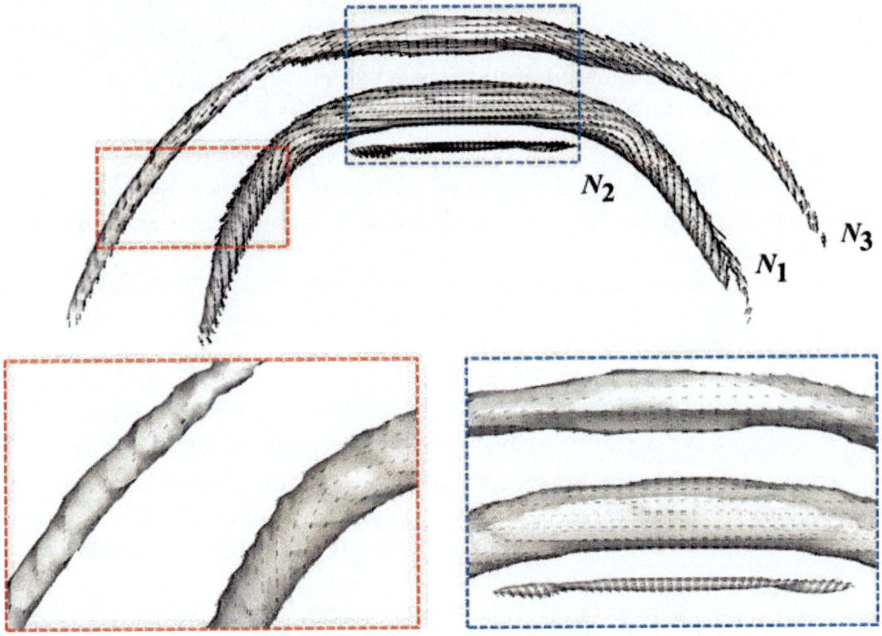

Fig. 19.11 Liutex vector field for necklace vortices. (The dashed rectangles show local enlargements)

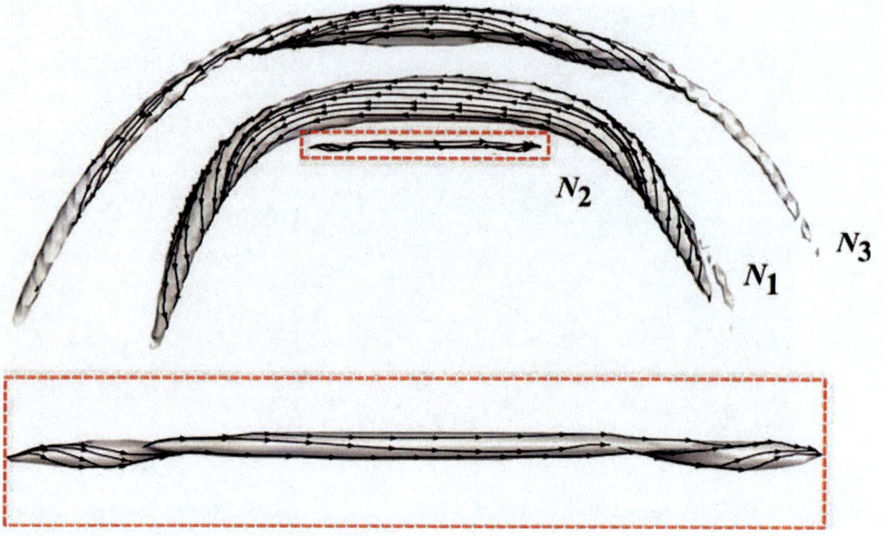

Fig. 19.12 Liutex lines for necklace vortices. (The red dashed rectangle shows the local enlargement of N_2)

involved in this complex flow are analyzed with the vorticity, Liutex, and normalized Omega-Liutex $\tilde{\Omega}_R$ methods. The main conclusions are as follows:

1. Using the iso-surfaces of $\tilde{\Omega}_R = 0.52$, the instantaneous vortical structures can be well identified, including the necklace vortices upstream near the free surface, wave-induced vortices even in the far wake, and Karman vortex shedding at medium depth.
2. For two-dimensional contours, the vorticity method can illustrate the behaviors of free shear layers, while the Liutex method can better characterize the alternative vortices in the wake.
3. Using the iso-surfaces of $\overline{\tilde{\Omega}_R} = 0.65$, the time-averaged vortical structures for the upper part of the square cylinder consist of a primary necklace vortex N_1 upstream and a pair of counter-rotating wave-induced vortices W_l and W_r downstream. Using the iso-surfaces of $\overline{\tilde{\Omega}_R} = 0.8$, the time-averaged vortical structures for the lower part consist of a pair of counter-rotating arch vortices A_{vl} and A_{vr}, a λ-type arch vortex A_b, and another pair of counter-rotating tip vortices T_{vl} and T_{vr}.
4. Using the Liutex vector field and integral lines, the twisting behaviors of necklace vortices at ends can be further revealed, especially for N_1 and N_2.

Acknowledgements This work is supported by National Natural Science Foundation of China (52131102, 51909160), and the National Key Research and Development Program of China (2019YFB1704200), to which the authors are most grate-ful.

References

1. T. Tamura, T. Miyagi, The effect of turbulence on aerodynamic forces on a square cylinder with various corner shapes. J. Wind Eng. Ind. Aerodyn. **83**, 135–145 (1999)
2. L. Carassale, A. Freda, M. Marre-Brunenghi, Experimental investigation on the aerodynamic behavior of square cylinders with rounded corners. J. Fluids Struct. **44**, 195–204 (2014)
3. N.P. van Hinsberg, G. Schewe, M. Jacobs, Experiments on the aerodynamic behaviour of square cylinders with rounded corners at Reynolds numbers up to 12 million. J. Fluids Struct. **74**, 214–233 (2017)
4. W. Zhang, R. Samtaney, Low-Re flow past an isolated cylinder with rounded corners. Comput. Fluids **136**, 384–401 (2016)
5. Y. Cao, T. Tamura, Supercritical flows past a square cylinder with rounded corners. Phys. Fluids **29**, 085110 (2017)
6. Y. Cao, T. Tamura, Shear effects on flows past a square cylinder with rounded corners at Re = 2.2×10^4. J. Wind Eng. Ind. Aerodyn. **174**, 119–132 (2018)
7. Y. Cao, T. Tamura, Aerodynamic characteristics of a rounded-corner square cylinder in shear flow at subcritical and supercritical Reynolds numbers. J. Fluids Struct. **82**, 473–491 (2018)
8. S.S. Dai, B.A. Younis, H.Y. Zhang, Prediction of turbulent flow around a square cylinder with rounded corners. J. Offshore Mech. Arct. Eng. **139**
9. S. Chen, W. Zhao, D. Wan, Turbulent structures and characteristics of flows past a vertical surface-piercing finite circular cylinder. Phys. Fluids **34**, 015115 (2022)
10. M.S. Gritskevich, A.V. Garbaruk, J. Schütze, F.R. Menter, Development of DDES and IDDES formulations for the k-ω shear stress transport model. Flow Turbul. Combust. **88**, 431–449 (2012)
11. W. Zhao, D. Wan, Detached-Eddy simulation of flow past tandem cylinders. Appl. Math. Mech. **37**, 1272–1281 (2016)
12. H.G. Weller, A new approach to VOF-based interface capturing methods for incompressible and compressible flow. (2008)
13. C. Liu, Y. Gao, S. Tian, X. Dong, Rortex—a new vortex vector definition and vorticity tensor and vector decompositions. Phys. Fluids **30**, 035103 (2018)
14. Y. Gao, C. Liu, Rortex based velocity gradient tensor decomposition. Phys. Fluids **31**, 011704 (2019)
15. Y. Gao, C. Liu, Rortex and comparison with eigenvalue-based vortex identification criteria. Phys. Fluids **30**, 085107 (2018)
16. X. Dong, Y. Gao, C. Liu, New normalized Rortex/vortex identification method. Phys. Fluids **31**, 011701 (2019)
17. J. Liu, C. Liu, Modified normalized Rortex/vortex identification method. Phys. Fluids **31**, 061704 (2019)
18. W. Zhao, J. Wang, D. Wan, Vortex identification methods in marine hydrodynamics. J. Hydrodyn. **32**, 286–295 (2020)

Chapter 20
Comparison of Vortex Identification Methods Based on the Liutex Decomposition and Application in a Compressor Cascade

Weibo Zhong, Yumeng Tang, and Yangwei Liu

Abstract Gaining a deeper insight into the complex vortex-dominated flow in aero-engine compressors is of great meaning for improving the aerodynamic performance of compressors. However, vortex regions vary depending on the vortex identification methods by which they are extracted. The Liutex based velocity gradient decomposition provides an effective approach to quantify the rigid-body rotation, pure shearing, and compression-stretching effects in fluid motion. The LT-plane constructed based on the local trace criterion provides a tool to geometrically represent the relationship of different vortex identifications. In this study, the commonly used Q series vortex criteria and the normalized Liutex method $\tilde{\Omega}_L$ are analytically compared under zero threshold and non-zero threshold conditions. A typical corner separation flow in a highly loaded compressor cascade is analyzed. Vortex regions extracted by different vortex criteria with equivalent threshold values are compared. Results show that proper exclusion of pure shearing and compressing-stretching contamination helps to capture entangled vortex structures and indicate the pressure loss regions.

20.1 Introduction

Vortices are ubiquitous in nature and in engineering flow, they are major components of coherent turbulent structures [1–3]. A deeper insight into the nature of vortex is critical when studying turbulence flow mechanisms. Although a universally-accepted definition of vortex has not yet been determined [4–6], advanced requirements for accurate vortex identification are raised due to the development of researches on

W. Zhong · Y. Tang · Y. Liu (✉)
National Key Laboratory of Science and Technology On Aero-Engine Aero-Thermodynamics, Beihang University, Beijing 100191, China
e-mail: liuyangwei@126.com

W. Zhong
Research Institute of Aero-Engine, Beihang University, Beijing 100191, China

Y. Tang · Y. Liu
School of Energy and Power Engineering, Beihang University, Beijing 100191, China

Y. Wang et al. (eds.), *Liutex and Third Generation of Vortex Identification*, Springer Proceedings in Physics 288, https://doi.org/10.1007/978-981-19-8955-1_20

natural and engineering flow. A remarkable example is the corner separation flow in the compressor of modern aeroengine. The corner separation flow is characterized by complex vortical structures in the blade endwall regions, and it significantly affects the performance of compressor. To further analyze these complex vortical flow features in compressor, and thus improving the accuracy of predictions [7, 8] and effectiveness of flow control approaches [9], it is important to distinguish vortices in tangled complexity using proper vortex identification methods.

In the past decades, various vortex identification methods have been proposed and applied in different flow conditions. For example, the Q criterion [10] is one of the most popular methods, its extensions such as the Q_D criterion [11] and the Q_M criterion [12] are developed. As a foundation of eigenvalue-based vortex criteria, the Δ criterion [5], identifies vortical flow in swirling flow zone according to the eigenvalue condition of the velocity gradient tensor. The λ_{ci} criterion [13] and $\lambda_{cr}/\lambda_{ci}$ criterion [6] enhance the Δ criterion from perspective of swirling strength and compactness. These vortex criteria have been shown to be effective in capturing certain signatures of vortices. However, limitations still exist because vortices are quite different in various flow situations, and confusions occur when different vortex criteria reveal unique structures in the same flow of complexity. To avoid confusions and to use vortex criteria properly, the relationships between them need further explanation. The advanced Liutex method [14–18] decomposes the velocity gradient tensor into rigid-body rotation, pure shearing, and compressing-stretching components [19, 20]. It provides an effective approach to better classify the vortical flow motion. Additionally, the recently proposed local trace criterion (LT_{cri}) and LT_{cri}-based elliptical region (LTER) indication method [21, 22] contribute to analyzing the evolutionary process of complex vortices, and provide a geometric view to examine the relationships between the vortical regions confined by different vortex criteria.

In this paper, the relationships between different vortex criteria are developed based on the Liutex decomposition and compared in the projection plane constructed by LT_{cri} (LT-plane) under different threshold conditions. Applications of these vortex criteria are conducted on the delayed detached eddy simulation (DDES) flow field of the corner separation flow in a highly loaded compressor cascade. The analysis aims at obtaining a better understanding and proper use of different vortex criteria, thus providing insights into the flow mechanism, control schemes, and turbulence modeling in engineering.

20.2 Review of Liutex Based Velocity Gradient Decomposition

According to the theory of Liutex [14–18], vortex exists in the swirling flow region defined by the well-known $\Delta > 0$ criterion [5]. Δ acts as the discriminant of the velocity gradient tensor $\nabla \boldsymbol{u}$. When $\Delta > 0$, $\nabla \boldsymbol{u}$ has one real eigenvalue λ_r, and a pair of conjugate complex eigenvalues $\lambda_{cr} \pm i\lambda_{ci}$, which implies the existence of local swirling flow. While, $\nabla \boldsymbol{u}$ have three real eigenvalues when $\Delta \leq 0$. Under positive-Δ condition, the velocity gradient tensor in the transformed principal coordinate system is obtained as [19, 20]

$$\nabla \boldsymbol{u}_P = \begin{pmatrix} \lambda_{cr} & -\phi & 0 \\ \phi + s_1 & \lambda_{cr} & 0 \\ s_2 & s_3 & \lambda_r \end{pmatrix}. \tag{20.1}$$

where ϕ represents the rigid-body rotation strength, the magnitude of Liutex is expressed as $R_L = 2\phi$. The Liutex vector is defined as $\boldsymbol{R}_L = R_L \boldsymbol{r}_L$, where $\boldsymbol{r}_L$ is the local rotation axis, which is aligned to the real eigenvector $\boldsymbol{v}_r$ under the original frame. s_1 is the pure shearing component along $\boldsymbol{r}_L$. s_2 and s_3 represent the pure shearing components in the rotation plane in the principal coordinate system. λ_r and λ_{cr} are the compressing-stretching factors. Thus, the swirling flow motion can be divided into three parts by decomposing $\nabla \boldsymbol{u}_P$ into rigid-body rotation part ($\boldsymbol{R}$), pure shearing part ($\boldsymbol{SS}$), and compressing-stretching part ($\boldsymbol{CS}$), which can be expressed as

$$\nabla \boldsymbol{u}_P = \boldsymbol{R} + \boldsymbol{SS} + \boldsymbol{CS} = \begin{pmatrix} 0 & -\phi & 0 \\ \phi & 0 & 0 \\ 0 & 0 & 0 \end{pmatrix} + \begin{pmatrix} 0 & 0 & 0 \\ s_1 & 0 & 0 \\ s_2 & s_3 & 0 \end{pmatrix} + \begin{pmatrix} \lambda_{cr} & 0 & 0 \\ 0 & \lambda_{cr} & 0 \\ 0 & 0 & \lambda_r \end{pmatrix} \tag{20.2}$$

20.3 Review of Local Trace Criterion Based Elliptical Region

The LT_{cri} and the corresponding LTER indication proposed by Liu and Tang (2019) [21] provide an approach to extract vortical flow region and to distinguish different spiral patterns of vortices in tangled complexity. The magnitude of LT_{cri} is defined as the first invariant of $\boldsymbol{\Psi}$, which is defined as half of the quadratic velocity gradient tensor, $\boldsymbol{\Psi} = (\nabla \boldsymbol{u})^2/2$. I_Ψ reflects the gathering and spreading trait of $\boldsymbol{\Psi}$ in an infinitesimal volume around the local flow point, and thereby to estimate its behavior. When I_Ψ obtains positive values, the local flow point tends to appear as a spin sink, which conforms to the notion of a vortex core. It should be stricter to employ the positive I_Ψ to express the exceedance of local vorticity (rotation rate) to the deformation. Thus, the local method which captures vortical region with $I_\Psi > 0$ is termed as LT_{cri}.

$$LT_{cri} = I_\Psi = -tr(\boldsymbol{\Psi}) = \begin{cases} -\frac{1}{2}\left(\lambda_1^2 + \lambda_2^2 + \lambda_3^2\right) \ (when\ \Delta \leq 0) \\ -\frac{1}{2}\lambda_r^2 - \lambda_{cr}^2 + \lambda_{ci}^2 \ (when\ \Delta > 0) \end{cases} \tag{20.3}$$

From Eq. (20.3), the vortical region with $LT_{cri} > 0$ only exists when $\Delta > 0$. Thus, to avoid superfluous calculation, the condition when $\Delta \leq 0$ is simplified as $LT_{cri} = 0$.

In a space defined by $\Delta > 0$, a plane with its abscissa axis defined as $\xi = \lambda_r/|\lambda_{ci}|$ and ordinate axis defined as $\eta = \lambda_{cr}/|\lambda_{ci}|$ is constructed based on LT_{cri}, it is termed as LT-plane. As shown in Fig. 20.1, any flow point satisfying $\Delta > 0$ can find unique position on the LT-plane. The region encompassed by $LT_{cri} > 0$ is equivalently projected as an ellipse at the center of the LT-plane, which is expressed as

$$\xi^2/2 + \eta^2 < 1. \tag{20.4}$$

The expansion or compression state of fluid motion is informed by the sign of the first invariant of $\nabla \boldsymbol{u}$, $P = -tr(\nabla \boldsymbol{u})$. In the LT-plane, $P = 0$ can be expressed as

$$-\xi - 2\eta = 0. \tag{20.5}$$

The spiraling patterns are classified according to the schematics of the spiraling trajectory of local flow point, which can be expressed in the locally curvilinear coordinate system (c_1, c_2, c_3) based on the unit eigenvectors $(\boldsymbol{v}_r, \boldsymbol{v}_{cr}, \boldsymbol{v}_{ci})$ as

$$\begin{bmatrix} c_1(t) \\ c_2(t) \\ c_3(t) \end{bmatrix} = \begin{bmatrix} c_1(0)e^{\lambda_r t} \\ [c_2(0)cos(\lambda_{ci}t) + c_3(0)sin(\lambda_{ci}t)]e^{\lambda_{cr}t} \\ [c_3(0)cos(\lambda_{ci}t) - c_2(0)sin(\lambda_{ci}t)]e^{\lambda_{cr}t} \end{bmatrix}. \tag{20.6}$$

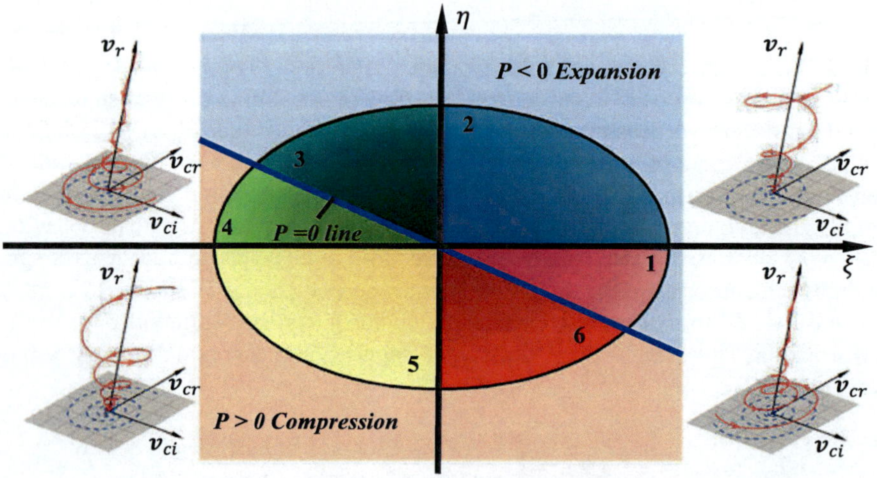

Fig. 20.1 The elliptical region with six divisions in the LT-plane. The $P = 0$ iso-line (blue) divides the LT-plane into a compression state region and an expansion state region. The schematics of trajectories under different spiralling patterns are shown in the corresponding quadrant

Table 20.1 Compressibility information and spiraling patterns in different divisions of the elliptical region in the LT-plane

No	Compressibility	Orbital development	
		Along $\boldsymbol{v}_r$	In $\boldsymbol{v}_{ci}$-$\boldsymbol{v}_{cr}$ plane
1	Expansion	Stretching	Damping
2	Expansion	Stretching	Amplification
3	Expansion	Attenuation	Amplification
4	Compression	Attenuation	Amplification
5	Compression	Attenuation	Damping
6	Compression	Stretching	Damping

From Eq. (20.6), it is clear that the trajectory of the local fluid point spirals with an orbit time period of $T = 2\pi/\lambda_{ci}$. Higher value of λ_{ci} infers a shorter swirling period; thus, λ_{ci} is taken as a measurement of the swirling strength [13]. The positive and negative value of λ_r reflects the stretching and compressing rate of the spiral trajectory along the direction of $\boldsymbol{v}_r$, respectively. In the swirling plane constructed by $\boldsymbol{v}_{cr}$ and $\boldsymbol{v}_{ci}$, the orbital radius is amplified by a positive value of λ_{cr}, or damped by a negative value of λ_{cr}. Thus, the sign of ξ reflects stretching or attenuation along $\boldsymbol{v}_r$, and the sign of η reflects the amplification or declination of the orbital radius in the swirling plane. In Fig. 20.1, schematics of four different spiral patterns are present in the corresponding quadrants of the LT-plane. The $P = 0$ iso-line together with the coordinate axes divide the elliptical region into six parts to indicate compressibility information and spiraling patterns. These six divisions on the LT-plane are numbered, their spiral patterns and compressibility are listed in Table 20.1.

20.4 Relationships between Different Vortex Criteria

In this section, some commonly used vortex criteria are briefly reviewed and are rewritten with decomposed components of swirling flow motion. By comparing the vortical regions confined by these criteria under different threshold conditions in the LT-plane [22], the influences of different swirling flow components on the vortex identification are revealed.

20.4.1 Q Series Criteria

The well-known Q criterion defines vortices as flow regions with positive Q, which acts as the second invariant of $\nabla \boldsymbol{u}$,

$$Q = \frac{1}{2}[tr(\nabla \boldsymbol{u})]^2 - \frac{1}{2}tr\left[(\nabla \boldsymbol{u})^2\right]. \tag{20.7}$$

The analytical expression of Q in the positive-Δ space is obtained as [14]

$$Q = \phi(\phi + s_1) + \lambda_{cr}^2 + 2\lambda_r\lambda_{cr}, \tag{20.8}$$

where the swirling strength λ_{ci} can be defined analytically by the ϕ and s_1 as $\lambda_{ci} = \sqrt{\phi(\phi + s_1)}$ [14]. The region satisfies $Q > 0$ in the LT-plane is expressed as [21]

$$\eta^2 + 2\eta\xi + 1 > 0. \tag{20.9}$$

As shown in Fig. 20.2, the gray region represents where $Q < 0$ in the LT-plane. The $Q = 0$ iso-line and the boundary of ellipse defined by $LT_{cri} = 0$ are tangential at points $a(-2/\sqrt{3}, 1/\sqrt{3})$ and $b(2/\sqrt{3}, -1/\sqrt{3})$.

However, the Q criterion was formulated assuming incompressible flow, in which case the first invariant is $P = 0$. Thus the Q criterion is notate as Q_{cri} here to distinguish it from the second invariant Q. The Q_{cri} can be expressed as

$$Q_{cri} = \frac{1}{2}\left(\|\boldsymbol{\Omega}\|^2 - \|\boldsymbol{S}\|^2\right). \tag{20.10}$$

Where $\boldsymbol{S} = (\nabla \boldsymbol{u} + \nabla \boldsymbol{u}^T)/2$ is the strain rate tensor and $\boldsymbol{\Omega} = (\nabla \boldsymbol{u} - \nabla \boldsymbol{u}^T)/2$ is the rotation tensor. Such that $Q_{cri} > 0$ provides a measure of the excess local rotation rate relative to the local strain rate. As $Q_{cri} > 0$ has been proved to be a subset of $\Delta > 0$ [22], the formula of Q_{cri} can be rewritten as Eq. (20.11) and its projection onto the LT-plane is equivalent to Eq. (20.5). Thus, $Q_{cri} > 0$ is a special case of $LT_{cri} > 0$ under incompressible assumption, vortices defined by positive Q_{cri} value are projected along the line segment ab in Fig. 20.2.

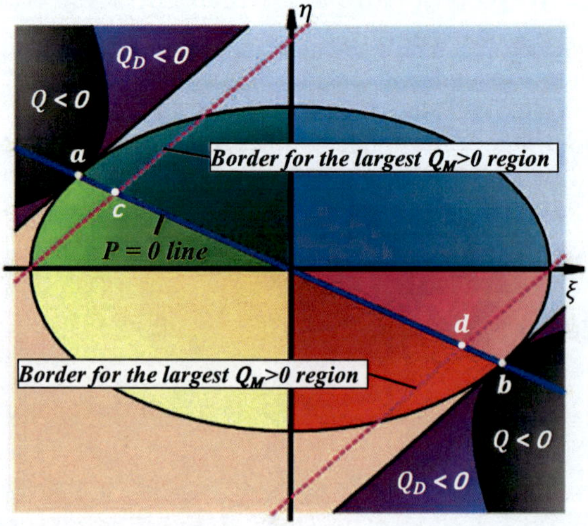

Fig. 20.2 Projections of Q series criteria in the LT-plane under zero threshold value condition

$$Q_{cri} = \phi(\phi + s_1) - \frac{1}{2}\lambda_r^2 - \lambda_{cr}^2. \quad (20.11)$$

As an extension of Q_{cri} to compressible condition, the Q_D criterion [11] excludes the effect of isentropic expansion and compression of fluid, which is performed in the strain rate tensor $\boldsymbol{S}$. The deviatoric part of $\boldsymbol{S}$ is $\boldsymbol{S}_D = \boldsymbol{S} - k\boldsymbol{I}$, where $k = \mathrm{tr}(\nabla \boldsymbol{u})/3$ denotes uniform dilatation. Hence, Q_D is defined as

$$Q_D = \frac{1}{2}\left(\|\boldsymbol{\Omega}\|^2 - \|\boldsymbol{S}_D\|^2\right). \quad (20.12)$$

The formula of Q_D can be rewritten as Eq. (20.13), and the projection of $Q_D > 0$ in the LT-plane is expressed as Eq. (20.14) [22]. As shown in Fig. 20.2, the $Q_D = 0$ iso-lines in the LT-plane are a pair of parallel lines that are tangent to the $LT_{cri} = 0$ and the $Q = 0$ iso-lines at points a and b.

$$Q_D = \phi(\phi + s_1) - \frac{1}{3}(\lambda_{cr} - \lambda_r)^2, \quad (20.13)$$

$$-\sqrt{3} < \eta - \xi < \sqrt{3}. \quad (20.14)$$

The Q_{cri} is further modified as the Q_M criterion which is derived by comparing the magnitudes of the vorticity vector and the principal strain-rate difference vector [12, 23]. Q_M is expressed as

$$Q_M = \frac{1}{2}\left(\|\boldsymbol{\Omega}\|^2 - \frac{3}{2}\|\boldsymbol{S}_D\|^2\right). \quad (20.15)$$

The condition $Q_M > 0$ is stricter than $Q_D > 0$, thus $Q_M > 0$ is a subset of $\Delta > 0$. Q_M in positive-Δ space can be written as Eq. (20.16), and $Q_M > 0$ in the LT-plane is equivalent to Eq. (20.17), where m is a positive number as expressed in Eq. (20.18) [22].

$$Q_M = \phi(\phi + s_1) - \frac{1}{2}(\lambda_{cr} - \lambda_r)^2 - \frac{1}{8}(s_1^2 + s_2^2 + s_3^2), \quad (20.16)$$

$$-m < \eta - \xi < m, \quad (20.17)$$

$$m = \sqrt{2 - \frac{s_1^2 + s_2^2 + s_3^2}{4\lambda_{ci}^2}}. \quad (20.18)$$

Therefore, the stripped region confined by $Q_M > 0$ reaches its maximum width when there is no pure shearing and $m = \sqrt{2}$, the borders for the largest $Q_M > 0$ region is marked by the magenta dashed lines in Fig. 20.2. For compressible flow, swirling flow satisfying $Q_M > 0$ is not guaranteed to be regarded as vortex from

$LT_{cri} > 0$ perspective, when it is projected out of the ellipse. While under incompressible condition, $Q_M > 0$ is becomes a line segment cd in the LT-plane which is stricter than $LT_{cri} > 0$ (and $Q_{cri} > 0$). Compared with other Q series criteria, only Q_M accounts for the negative influence of the local pure shearing contamination on vortex identification.

20.4.2 Normalized Liutex Method

The normalized Liutex method $\tilde{\Omega}_L$ [24] is a nondimensionalized criterion that can extract both strong and weak vortices through the same threshold. By comparing the norms of symmetric and antisymmetric portions of $\nabla \boldsymbol{u}_P$, and $\tilde{\Omega}_L$ is formulated as

$$\tilde{\Omega}_L = \frac{\beta^2}{\beta^2 + \alpha^2 + \lambda_{cr}^2 + \frac{1}{2}\lambda_r^2 + \varepsilon}, \tag{20.19}$$

where $\beta = \phi + s_1/2$ is the vorticity component aligned to $\boldsymbol{r}_L$, and $\alpha = s_1/2$ is the corresponding strain rate. The small positive parameter ε is employed to prohibit computational noise and to avoid division by zero, which is proposed as $\varepsilon = b_0\left(\beta^2 - \alpha^2\right)_{max}$, where b_0 is a small coefficient. To extract the vortex, the threshold value n_L for $\tilde{\Omega}_L$ ranges from 0.5 to 1. The projection of $\tilde{\Omega}_L > n_L$ on the LT-plane is expressed as [22]

$$\frac{\xi^2}{2} + \eta^2 < \frac{1 - n_L}{n_L} - \frac{1}{\lambda_{ci}^2}\left(\frac{2n_L - 1}{4n_L}s_1^2 + \varepsilon\right). \tag{20.20}$$

As shown in Fig. 20.3, when $n_L = 0.5$ assuming $\varepsilon = 0$, the vortex region projected in the LT-plane is equivalent to $\xi^2/2 + \eta^2 < 1$, which is identical to the ellipse defined by $LT_{cri} > 0$. When $n_L > 0.5$, the corresponding projection is still an ellipse, while the size of ellipse becomes smaller with the increase of threshold. The increase of the shearing component s_1 make the ellipse smaller, and such effect of s_1 is equivalently enhanced by a non-zero ε.

20.4.3 Relationships under Non-Zero Threshold Value Condition

The non-zero threshold values for vortex criteria are usually practical utilized to identify intense vortex structures. For these Q series criteria, a non-zero threshold value γ is given with unit (s^{-2}), the corresponding projections of vortex regions extracted by different criteria with the threshold value γ are listed in Table 20.2. Compared with the zero threshold condition, the projection regions under non-zero

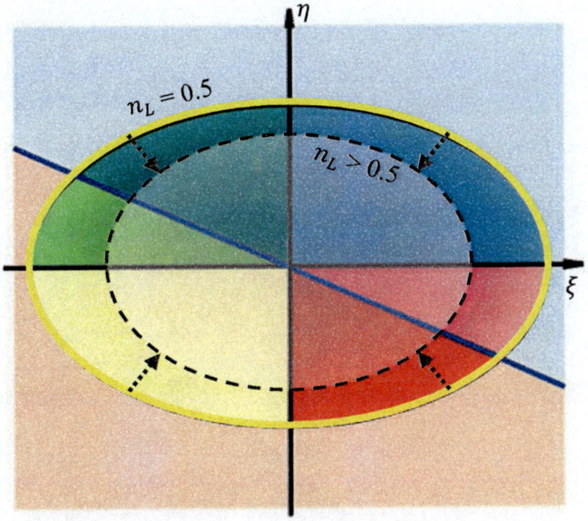

Fig. 20.3 Projections of $\tilde{\Omega}_L > n_L$ criterion in the LT-plane. The yellow ellipse marks the boundary of region defined by $\tilde{\Omega}_L > n_L$ when $n_L = 0.5$ (with $\varepsilon = 0$), and the smaller ellipse bounded by dashed line represents a resulting region when $n_L > 0.5$

threshold condition are all influenced by a multiplier $\left(1 - \gamma/\lambda_{ci}^2\right)$. As shown in Fig. 20.4, the relationships between different criteria with non-zero threshold γ are still similar to those under zero threshold condition that shown in Fig. 20.2. The difference is caused by the size reduction of the ellipse under non-zero threshold condition. The coordinates of points $a'(-2\sqrt{1-\gamma/\lambda_{ci}^2}/\sqrt{3}, \sqrt{1-\gamma/\lambda_{ci}^2}/\sqrt{3})$ and $b'(2\sqrt{1-\gamma/\lambda_{ci}^2}/\sqrt{3}, -\sqrt{1-\gamma/\lambda_{ci}^2}/\sqrt{3})$ are also influenced by $\left(1 - \gamma/\lambda_{ci}^2\right)$, and they act as tangential points on the $P = 0$ line just like point a and b.

Another obstacle is that the choices of non-zero threshold values for different vortex criteria with different units are subjective. The values of different vortex criteria are contaminated by pure shearing and compressing-stretching to varying degree, and

Table 20.2 Projections of Q series criteria with zero and non-zero threshold values in the LT-plane

Vortex criteria	Expressions of projections on the LT-plane	
	Zero threshold condition $\gamma = 0$	Non-zero threshold condition $\gamma > 0$
$Q = \gamma$	$\eta^2 + 2\eta\xi + 1 = 0$	$\eta^2 + 2\eta\xi + \left(1 - \gamma/\lambda_{ci}^2\right) = 0$
$Q_{cri} = \gamma$	$\xi^2/2 + \eta^2 = 1$	$\xi^2/2 + \eta^2 = \left(1 - \gamma/\lambda_{ci}^2\right)$
$Q_D = \gamma$	$\eta - \xi = \pm\sqrt{3}$	$\eta - \xi = \pm\sqrt{3\left(1 - \gamma/\lambda_{ci}^2\right)}$
$Q_M = \gamma$	$\eta - \xi = \pm\sqrt{2 - \frac{s_1^2+s_2^2+s_3^2}{4\lambda_{ci}^2}}$	$\eta - \xi = \pm\sqrt{2\left(1 - \gamma/\lambda_{ci}^2\right) - \frac{s_1^2+s_2^2+s_3^2}{4\lambda_{ci}^2}}$

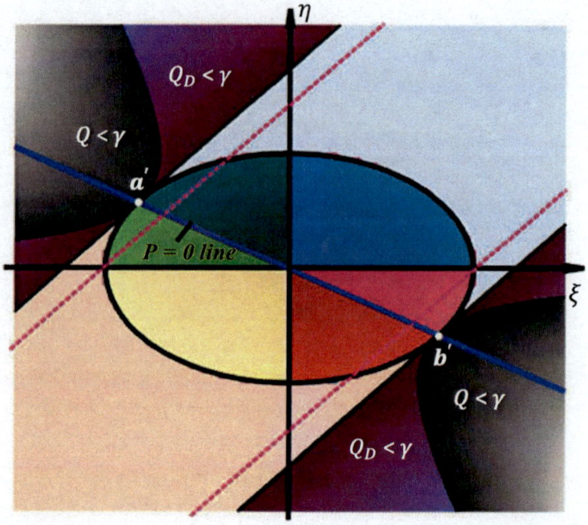

Fig. 20.4 Projections of different vortex criteria in the LT-plane under non-zero threshold value condition

there lacks a principle to determine proper thresholds when comparing their performances. To solve this problem, the concept of equivalent threshold value is proposed to extract similar vortex structures. According to the correlations of different criteria, the equivalent threshold for different vortex identifications is given as [22]

$$(R_L)_{th}^2/2 = (\lambda_{ci})_{th}^2 = (LT_{cri})_{th} = (Q_{cri})_{th} = (Q_D)_{th} = (Q_M)_{th} = \gamma. \quad (20.21)$$

As the term $\lambda_{ci}^2 = \phi(\phi + s_1)$ is explicitly contained in all the Q series criteria, the equivalent threshold values for λ_{ci} and Q series criteria can be simply given only considering the consistency of units. For Liutex-R_L, the content of ϕ is considered. Mathematically, $(R_L)^2/2 = 2\phi^2$ and $(\lambda_{ci})^2 = \phi^2 + \phi s_1$, then the difference between them simply depends on the difference between the strength of rigid-body rotation ϕ and the shearing component s_1.

20.5 Application in Corner Separation Flow of a Compressor

The performances of different vortex criteria are compared in a flow field of a three-dimensional (3D) corner separation flow in a highly loaded axial compressor cascade. The cascade has been experimentally studied [25] with a Mach number of approximately 0.59 and a Reynolds number of up to 7.2×10^5. More information about the blade geometry and the experiments can be found in Ref. [25]. The flow field analyzed

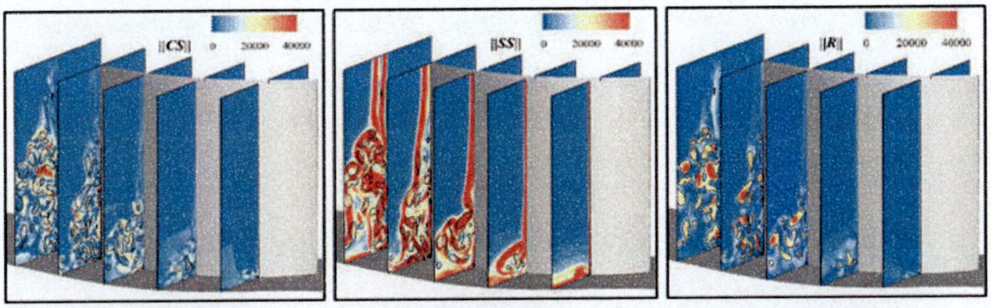

(a) $\|\boldsymbol{CS}\|$ (b) $\|\boldsymbol{SS}\|$ (c) $\|\boldsymbol{R}\|$

Fig. 20.5 Contours of (a) $\|\boldsymbol{CS}\|$, (b) $\|\boldsymbol{SS}\|$, and (c) $\|\boldsymbol{R}\|$ in the streamwise slices of the corner separation flow. The $(R_L)^2/2 = 2 \times 10^8 (\mathrm{s}^{-2})$ iso-lines in black imply the intense vortex regions

is simulated by the DDES method based on the Spalart–Allmaras (SA) model, under an incidence angle of 0°. Ref. [22] provides details about the simulation and its validation.

In Fig. 20.5, the strength of different swirling flow components are represented by the norm of the decomposed tensors in Eq. (20.2). Comparisons are made by streamwise slices of the corner separation flow. The strength of compressing-stretching represented by $\|\boldsymbol{CS}\|$ is entirely weaker than that of pure shearing by $\|\boldsymbol{SS}\|$ and rigid-body rotation by $\|\boldsymbol{R}\|$, and is especially weak in most of the intense vortex regions implied by iso-lines of R_L due to the relatively low compressibility under a subsonic condition with $Ma \approx 0.59$. The pure shearing effect dominates in regions between the rotational separation vortices and main flow, the adjacent regions with two vortex shears, as well as the layers near solid boundaries. The results in Fig. 20.5 indicates that the main contamination of the vortex identification lays on the pure shearing effects, while the contributions by compressing-stretching effect is little.

Comparison of different vortex criteria is done by capturing the three-dimensional vortex structures. As shown in Fig. 20.6, global features of the corner separation vortex in a typical instantaneous flow field are visualized by the iso-surfaces of different vortex criteria, with a proper equivalent threshold value of $2 \times 10^8 \left(\mathrm{s}^{-2}\right)$ The iso-surfaces are colored by the pure shearing component s_1, which is the main contamination in vortex identification.

In Fig. 20.6(a)–(c), vortex structures identified by λ_{ci}, Q_{cri}, and Q_D are nearly identical. As discussed in Sec. 4, the main difference in the values of these three criteria comes from terms containing λ_r and λ_{cr}, which represent the compressing-stretching effect. And as concluded from Fig. 20.5, the global compressing-stretching strength is much weaker than the swirling strength in this flow field, this results in the remarkable similarity in the vortex structures extracted by λ_{ci}, Q_{cri}, and Q_D. None of these three vortex criteria can effectively exclude the boundary layer with high s_1 wrapping the blade suction side, as areas marked by white dashed lines in Fig. 20.6(a)–(c). Moreover, at the onset of the separation marked by red dashed lines, none of λ_{ci}, Q_{cri}, and Q_D can capture the vortex structures composed of multiple vortices, instead, a cloud with large pure shearing s_1 exists here. While

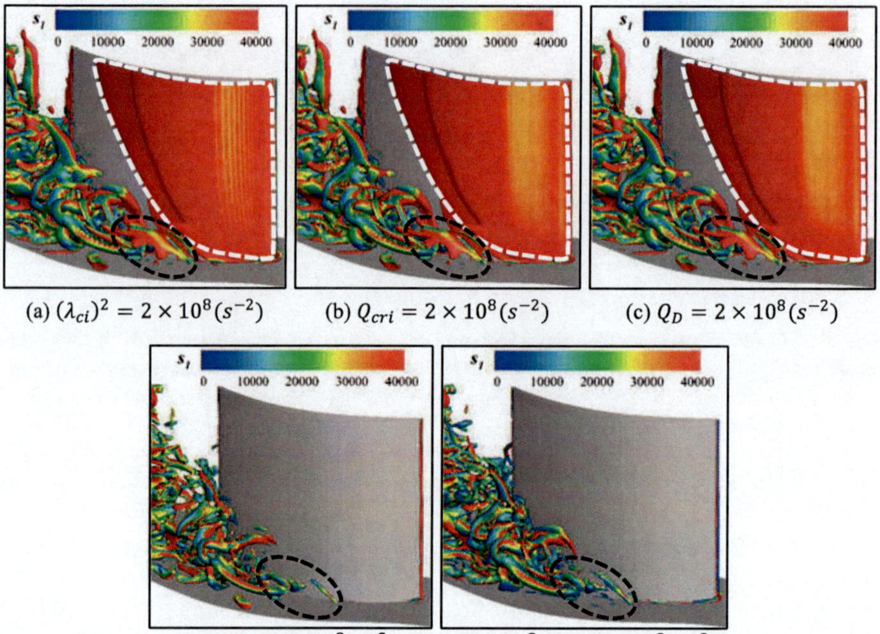

(a) $(\lambda_{ci})^2 = 2 \times 10^8(s^{-2})$ (b) $Q_{cri} = 2 \times 10^8(s^{-2})$ (c) $Q_D = 2 \times 10^8(s^{-2})$

(d) $Q_M = 2 \times 10^8(s^{-2})$ (e) $(R_L)^2/2 = 2 \times 10^8(s^{-2})$

Fig. 20.6 Vortex structures of the corner separation flow extracted by (a) $(\lambda_{ci})^2 = 2 \times 10^8(\mathrm{s}^{-2})$, (b) $Q_{cri} = 2 \times 10^8(\mathrm{s}^{-2})$, (c) $Q_D = 2 \times 10^8(\mathrm{s}^{-2})$, (d) $Q_M = 2 \times 10^8(\mathrm{s}^{-2})$, and (e) $(R_L)^2/2 = 2 \times 10^8(\mathrm{s}^{-2})$. The iso-surfaces are colored by the value of pure shearing component s_1

in Fig. 20.6(d), as the influence of pure shearing is filtered in the value of Q_M, the resulting vortex structures successfully excludes the shear-dominant region wrapping the blade suction side. However, in the region circled by black dashed, although distinguished vortices are shown without noise, some small-scale vortices are omitted, such as the suction side leg of the horseshoe vortex, which is significant in the generation of corner separation.

In Fig. 20.6(e), more small-scale vortices can be clearly captured by the Liutex-R_L. The shear layer wrapping the blade suction side is omitted. Small scale vortices are shown much more clearly at the start of the separation region, and the vortex structures here are relatively more plentiful than that in the results extracted by Q_M. With the equivalent threshold, the vortices extracted by Liutex-R_L are similar to those by λ_{ci}, Q_{cri}, and Q_D without a shear layer, because the Liutex measures the rigid-body rotation strength without the influence of pure shearing.

The normalized Liutex method – $\tilde{\Omega}_L$ criterion is utilized to evaluate the relative strength between rigid-body rotation and deformation. In application, the determination of the small positive parameter ε to avoids division by zero and computational noise is empirical. As shown in Fig. 20.7, the level of ε has a significant influence on capturing the vortex structures under the suggested threshold of $\tilde{\Omega}_L = 0.52$. The iso-surfaces are colored by the Liutex-R_L to show the global rigid-body rotation strength of the extracted vortices. Different levels of ε are obtained by changing the coefficient b_0. From Eq. (20.20), ε equivalently strengthens the local shearing effect and thus minifying the size of the vortex region by $\tilde{\Omega}_L$. In the corner separation flow, a larger ε results in the exclusion of shear layer wrapping the blade suction side, but some significant vortex structures are lost. For example, in Fig. 20.7(a), almost all vortex structures disappear with higher level of ε defined by the suggested $b_0 = 1 \times 10^{-3}$ [24]. As ε becomes smaller, the vortex structures become plentiful, but there appears many clouds (noise) when ε is too small in Fig. 20.7(c). Additionally, the global rigid-body strength of the extracted vortices become weaker as ε becomes smaller, because the pure shearing level enhanced by ε are lower and then the required rigid-body rotation level achieving the balance of $\tilde{\Omega}_L = 0.52$ decreases. Thus, for more accurate identification, the level of ε should be carefully determined in different flow cases.

The performances of these different vortex criteria are further compared in the slice located at 25% span-height of the blade. Shown in Fig. 20.8, comparisons are made between the Q_{cri}, Q_M, Liutex-R_L, and the dimensionless $\tilde{\Omega}_L$ criterion. The maximal vortical regions of each criterion are shown, and the black iso-lines of s_1 are employed to indicate the significantly shearing-contaminated regions. The ε for $\tilde{\Omega}_L$ is determined by an extremely small $b_0 = 1.0 \times 10^{-15}$ to obtain an objective distribution of $\tilde{\Omega}_L$. For the resulting vortical regions, the regions extracted by $Q_{cri} > 0$ and $\tilde{\Omega}_L > 0.52$ are most similar to each other, although regions of $Q_{cri} > 0$ are slightly larger, as the non-zero ε equivalently enhance the shearing effect and then minimizes the vortical region. The $R_L > 0$ which is identical to $\Delta > 0$ determines the largest regions, while much shearing-contaminated regions are excluded by positive Q_M. Some strong vortices identified by Q_{cri}, R_L, and $\tilde{\Omega}_L$, such as those in the wake flow with strong shearing are excluded or identified as weak vortices by Q_M. In Fig. 20.8(d), some vortical regions are measured to be higher than $\tilde{\Omega}_L = 0.9$ although they are identified as weak vortices in blue by the other criteria. The ability of the $\tilde{\Omega}_L$ criterion to capturing vortices with different strength and scales are shown.

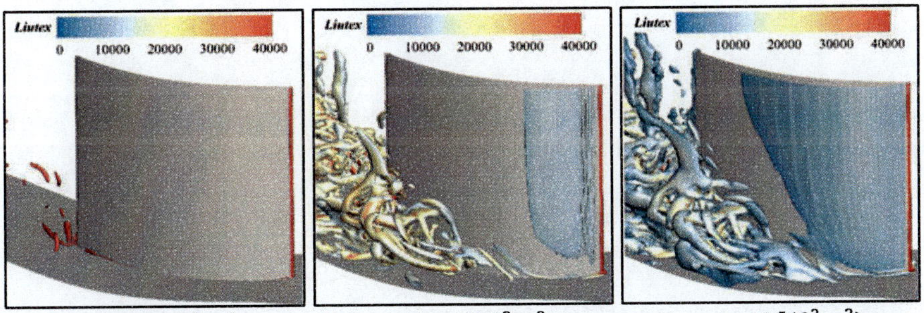

(a) $\varepsilon=1\times10^{-3}(\beta^2-\alpha^2)_{max}$ (b) $\varepsilon=5\times10^{-5}(\beta^2-\alpha^2)_{max}$ (c) $\varepsilon=1\times10^{-5}(\beta^2-\alpha^2)_{max}$

Fig. 20.7 Vortex structures of the corner separation flow identified by $\tilde{\Omega}_L = 0.52$ with (a) $\varepsilon = 1.0 \times 10^{-3}(\beta^2 - \alpha^2)_{max}$, (b) $\varepsilon = 5.0 \times 10^{-5}(\beta^2 - \alpha^2)_{max}$ and (c) $\varepsilon = 1.0 \times 10^{-5}(\beta^2 - \alpha^2)_{max}$. The iso-surfaces are colored by the value of Liutex-R_L

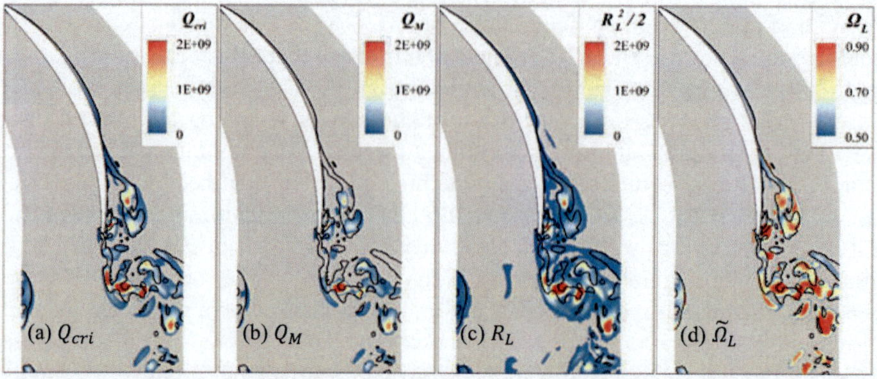

Fig. 20.8 Contours of (a) Q_{cri} in $Q_{cri} > 0$ regions, (b) Q_M in $Q_M > 0$ regions, (c) R_L in $R_L > 0$ regions, and (d) $\tilde{\Omega}_L$ in $\tilde{\Omega}_L > 0.52$ regions in the 25% span-height slices of the corner separation flow. The $s_1 = 2 \times 10^4 (\mathrm{s}^{-1})$ iso-lines in black imply the regions with strong shearing

Vortices always lead to loss in compressor flow. In Fig. 20.9, the correspondences between the vortex regions identified by different criteria and the regions with high pressure loss are analyzed at 25% span-height of the blade. The loss of the flow is represented by the total pressure loss coefficient Y_p, which is defined as

$$Y_p = \frac{p_{t1} - p_t}{p_{t1} - p_1}, \tag{20.22}$$

where p_{t1} is the total pressure at the inlet, p_1 is the static pressure at the inlet, and p_t is the local total pressure. In Fig. 20.9, vortex regions defined by $Q_{cri} > 0$ and $\tilde{\Omega}_L > 0.52$ are similar, and they distribute within the pressure loss regions. There

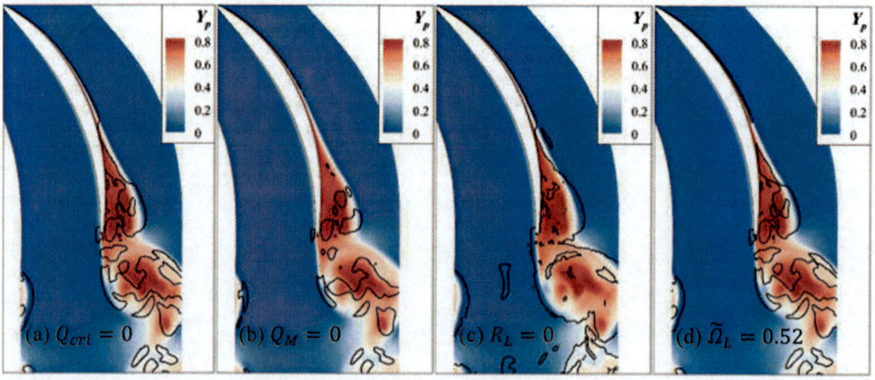

Fig. 20.9 Contours of the total pressure loss coefficient Y_p, vortex regions are implied by black iso-lines of (a) $Q_{cri} = 0$, (b) $Q_M = 0$, (c) $R_L = 0$, and (d) $\tilde{\Omega}_L = 0.52$ in the 25% span-height slices of the corner separation flow

is high consistency between the $Q_M > 0$ regions and the high pressure loss regions with $Y_p \geq 0.7$, but the $Q_M > 0$ regions do not cover the pressure loss areas at the onset of the separation. While in Fig. 20.9(c), the $R_L = 0$ iso-lines outline the pressure loss areas completely and nicely, which means that the Liutex method is an effective vortex criteria that can identify both vortical and loss regions in this compressor flow case.

20.6 Conclusions

In the present study, the Liutex-based velocity gradient decomposition method and the local trace criterion are reviewed and employed to analyze the relationships between different vortex criteria. Applications of these vortex criteria are presented in a practical 3D corner separation flow of a highly loaded compressor cascade. Conclusions are as follows:

(1) The analytical expressions of different vortex criteria can be derived by rigid-body-rotation, pure shearing, and compressing-stretching components of swirling flow based on Liutex decomposition. Relationships between different vortex criteria are graphically revealed in the LT-plane under both the zero threshold and the non-zero threshold conditions. Equivalent threshold values determined based on these relationships help extract similar vortex structures in comparing the performances of vortex criteria with different units.
(2) In the subsonic compressor cascade, the identification of vortex is mainly contaminated by pure shearing effects, while the compressing-stretching affects little. Vortex structures identified by λ_{ci}, Q_{cri}, and Q_D are nearly identical, because these three criteria only differ in compressing-stretching content. Filtering of pure shearing effect by Q_M leads to the clearest vortex structures, while some vortices are omitted. The Liutex captures clear structures of multi-scale vortices by measuring the rigid-body rotation, and the vortical regions with $R_L > 0$ can indicate the total pressure loss in the separation flow.
(3) The normalized Liutex criterion $\tilde{\Omega}_L > 0.5$ is equivalent to $LT_{cri} > 0$ theoretically, but the parameter ε equivalently enhances the shearing strength and minimizes the vortical region in vortex identification. Thus, the level of ε should be carefully determined in different flow cases to achieve proper identification results and to extract vortices with different strength and scales simultaneously.

Acknowledgements This work is supported by the National Natural Science Foundation of China (No. 51976006, No. 52106039, No. 51790513), National Science and Technology Major Project (2017-II-003-0015), and the Aeronautical Science Foundation of China (2018ZB51013).

References

1. A.K.M.F. Hussain, Coherent structures and turbulence. J. Fluid Mech. **173**, 303–356 (1986). https://doi.org/10.1017/S0022112086001192
2. J. Jeong, F. Hussain, On the identification of a vortex. J. Fluid Mech. **285**, 69–94 (1995). https://doi.org/10.1017/S0022112095000462
3. P. Moin, K. Mahesh, Direct numerical simulation: a tool in turbulence research. Annu. Rev. Fluid Mech. **30**, 539–578 (1998). https://doi.org/10.1146/annurev.fluid.30.1.539
4. B. Epps, Review of vortex identification methods. AIAA Paper 2017–0989 (2017). https://doi.org/10.2514/6.2017-0989
5. M.S. Chong, A.E. Perry, B.J. Cantwell, A general classification of three-dimensional flow fields. Phys. Fluids A **2**, 765–777 (1990). https://doi.org/10.1063/1.857730
6. P. Chakraborty, S. Balachandar, R.J. Adrian, On the relationships between local vortex identification schemes. J. Fluid Mech. **535**, 189–214 (2005). https://doi.org/10.1017/S0022112005004726
7. Y.-W. Liu, X.-J. Yu, B.-J. Liu, Turbulence models assessment for large-scale tip vortices in an axial compressor rotor. AIAA J. Propuls. Power **24**(1), 15–25 (2008). https://doi.org/10.2514/1.26134
8. Y.-W. Liu, Y.-M. Tang, A.D. Scillitoe, P.G. Tucker, Modification of shear stress transport turbulence model using helicity for predicting corner separation flow in a linear compressor cascade. ASME J. Turbomach. **142**(2), 021004 (2020). https://doi.org/10.1115/1.4045658
9. Y.-M. Tang, Y.-W. Liu, L.-P. Lu, Solidity effect on corner separation and its control in a high-speed low aspect ratio compressor cascade. Int. J. Mech. Sci. **142**, 304–321 (2018). https://doi.org/10.1016/j.ijmecsci.2018.04.048
10. J.C.R. Hunt, A.A. Wray, P. Moin, Eddies, streams and convergence zones in turbulent flows, in: *Centre for turbulence research proceedings of the summer program* (1988), pp. 193–208
11. V. Kolár, Compressibility effect in vortex identification. AIAA J. **47**(2), 473–475 (2009). https://doi.org/10.2514/1.40131
12. V. Kolář, J. Šístek, Corotational and compressibility aspects leading to a modification of the vortex-identification Q-criterion. AIAA J. **53**(8), 2406–2410 (2015). https://doi.org/10.2514/1.J053697
13. J. Zhou, R.J. Adrian, S. Balachandar, T.M. Kendall, Mechanisms for generating coherent packets of hairpin vortices in channel flow. J. Fluid Mech. **387**, 353–396 (1999). https://doi.org/10.1017/S002211209900467X
14. Y.-S. Gao, C.-Q. Liu, Rortex and comparison with eigenvalue-based vortex identification criteria. Phys. Fluids **30**(8), 085107 (2018). https://doi.org/10.1063/1.5040112
15. C.-Q. Liu, Y.-S. Gao, S.-L. Tian, X.-R. Dong, Rortex-a new vortex vector definition and vorticity tensor and vector decompositions. Phys. Fluids **30**, 035103 (2018). https://doi.org/10.1063/1.5023001
16. S.-L. Tian, Y.-S. Gao, X.-R. Dong, C.-Q. Liu, Definitions of vortex vector and vortex. J. Fluid Mech. **849**, 312–339 (2018). https://doi.org/10.1017/jfm.2018.406
17. Y.-Q. Wang, Y.-S. Gao, C.-Q. Liu, Letter: Galilean invariance of Rortex. Phys. Fluids **30**, 111701 (2018). https://doi.org/10.1063/1.5058939
18. C.-Q. Liu, Y.-S. Gao, X.-R. Dong, Y.-Q. Wang, J.-M. Liu, Y.-N. Zhang, X.-S. Cai, N. Gui, Third generation of vortex identification methods: Omega and Liutex/Rortex based systems. J. Hydrodyn. **31**(2), 205–223 (2019). https://doi.org/10.1007/s42241-019-0022-4
19. Y.-S. Gao, Y.-F. Yu, J.-M. Liu, C.-Q. Liu, Explicit expressions for Rortex tensor and velocity gradient tensor decomposition. Phys. Fluids **31**, 081704 (2019). https://doi.org/10.1063/1.5118948
20. Y.-S. Gao, C.-Q. Liu, Rortex based velocity gradient tensor decomposition. Phys. Fluids **31**, 011704 (2019). https://doi.org/10.1063/1.5084739
21. Y.-W. Liu, Y.-M. Tang, An elliptical region method for identifying a vortex with indications of its compressibility and swirling pattern. Aerosp. Sci. Technol. **95**, 105448 (2019). https://doi.org/10.1016/j.ast.2019.105448

22. Y.-W. Liu, W.-B. Zhong, Y.-M. Tang, On the relationships between different vortex identification methods based on local trace criterion. Phys. Fluids **33**, 105116 (2021). https://doi.org/10.1063/5.0063326
23. V. Kolář, J. Šístek, Vortex and the balance between vorticity and strain rate. Int. J. Aerosp. Eng. **8** (2019), https://doi.org/10.1155/2019/1321480
24. J.-M. Liu, C.-Q. Liu, Modified normalized Rortex/vortex identification method. Phys. Fluids **31**, 061704 (2019). https://doi.org/10.1063/1.5109437
25. Y.-M. Tang, Y.-W. Liu, L.-P. Lu, H.-W. Lu, M. Wang, Passive separation control with blade-end slots in a highly loaded compressor cascade. AIAA J. **58**(1), 85–97 (2020). https://doi.org/10.2514/1.J058488

Chapter 21
Research on Vortex Structure of All-Oxygen MILD Combustion Reheating Furnace Based on Third-Generation Vortex Identification Method

Yongfu Wu, Shuang Zhao, Hongyu Ji, Changpeng Guo, Zhenfeng Wang, and Zhongxing Liu

Abstract The whole oxygen MILD combustion technology mainly relies on the flue gas coiling and reflow effect of MILD combustion technology, so that the whole oxygen combustion can achieve MILD effect. Flue gas coiling backflow is the embodiment of gas eddy motion. There are a lot of eddy structures in coiling mixing process, so the effect of all-oxygen MILD combustion is determined by eddy structure. In this paper, numerical simulation and experimental methods are combined to analyze the vortex structure in heating furnace with the help of Liutex method, and the threshold value 3 of vortex core is selected to draw the vortex core and vortex line. The results show that the vortex structure can directly reflect the vortex motion of gas in the furnace, and the vortex motion in the furnace is mainly concentrated in the initial section of the furnace. The size and quantity of the vortex structure in this area is stronger than that in the middle and rear of the furnace, and it is also the main combustion region of the furnace. Under certain threshold conditions, too much or too little vortex core structure in furnace will affect combustion efficiency. The combination of vortex line and streamline line can better reflect the backflow and mixing of gas in the furnace, and is more conducive to the analysis of vortex motion of gas in the furnace.

Y. Wu (✉) · S. Zhao · H. Ji · C. Guo · Z. Wang · Z. Liu
College of Energy and Environment, Inner Mongolia University of Science and Technology, Baotou, Inner Mongolia, China
e-mail: wyf07@imust.edu.cn

Y. Wu · Z. Wang
School of Metallurgy, Northeastern University, Shenyang, Liaoning, China

Y. Wang et al. (eds.), *Liutex and Third Generation of Vortex Identification*,
Springer Proceedings in Physics 288, https://doi.org/10.1007/978-981-19-8955-1_21

21.1 Introduction

MILD combustion technology can reduce NO_X, CO and other pollutant emissions in the combustion reaction process, and has high combustion efficiency [1–3]. Full oxygen combustion technology is characterized by high combustion efficiency and other aspects. However, the total oxygen combustion technology still has some defects, resulting in a relatively low application rate of oxygen combustion technology in industry [4]. The combination of the two combustion technologies in the combustion process can achieve the dispersion effect of oxygen combustion, and the combustion reaction can achieve high efficiency and low pollution effect. Therefore, the whole oxygen MILD combustion technology has more advantages than MILD combustion technology in the future development.

Vortex structure widely exist in nature and industry, especially in the turbulent there are a large number of different scales, the intensity of vortex structure, in the turbulence plays a key role in the process of generating and maintaining, as a result, accurate identification of vortex structure to understand the flow mechanism, solve problem and turbulence flow control has important significance [5, 6].

According to Liu et al. [7], the development of vortex recognition methods can be divided into three generations. The first-generation method is based on vorticity and considers the antisymmetric part of the velocity gradient tensor (vorticity tensor) derived from cauchy-Stokes decomposition as rotation. However, many researchers have found that the correlation between vorticity and vortex is not very high. In order to overcome the defects of the first generation method, the second generation vortex identification method represented by Q criterion and other methods came into being. However, the second-generation method relies on the threshold values related to specific examples and will be influenced by shearing in practice.

In 2016, Liu et al. [8] proposed the method of Ω vortex identification, and for the first time proposed the concept of further decomposition of vorticity into the rotating part and the non-rotating part. In this method, the vortex was defined as the connected region with vorticity greater than deformation and expressed in relative value. Compared with the second-generation vortex identification method, the method has the advantages of clear physical significance, easy implementation, normalization (value range 0–1), and can capture both strong and weak vortices without significantly adjusting the threshold, which overcomes the problem of artificially adjusting the threshold of the second-generation vortex identification method [7]. Wang et al. [9] applied Ω, Q, Δ and λ_2 vortex recognition methods to Burgers vortices and Sullivan vortices, and found that compared with other vortex recognition methods, Ω could well measure the degree of local fluid tending to rigid body rotation. Gui et al. [10] identified the vortex structure in the flow field by using Q, Ω, and vorticity sizes $\|\omega\|_2$ and λ_2 respectively in the rotating jet, and found that when the main vortex structure in the flow field was clearly displayed, more secondary weak vortex structures could be displayed by using the method for Ω vortex structure identification. Wu yongfu, guo changpeng et al. [11, 12] studied the eddy current movement of mixed gas and adopted three eddy current identification methods to identify the eddy current structure in the oxygen fuel reheating furnace. By comparing the sensitivity of threshold value and the display ability of vortex structure, the optimal vortex recognition method is selected. The results show that the criterion, criterion and criterion can identify the vortex structure in the furnace, but the method can capture both strong and weak vortices, showing more vortex structures and better recognition ability. They also focused on Q-criterion and Liutex vortex structure identification methods, analyzed the formation characteristics of vortex structure in the front half of the furnace and its relationship with MILD combustion, and verified it with experimental data. The results show that the Q-criterion is basically the same as Liutex method in identifying large vortices, but not for small vortices. Therefore, liutex method is more accurate in identifying vortex structures.

The above literatures provide some help to the research of this paper mainly from the aspect of vortex structure identification methods, and also verify that the third generation of vortex identification methods based on the method and Liutex method will become the mainstream method of vortex identification. Liu first systematically proposed how to extract the rotating motion part of rigid body from fluid motion, and named the rotating part of vorticity as Rortex vector, whose direction represents the local rotation axis and size represents rotation angular velocity. In December 2018, Rortex vector was renamed Liutex vector [13, 14].

The Liutex vectors developed in 2017 and 2018 provide a systematized, mathematical definition of local rigid body rotation of fluids, including local rotation axes and rotational intensity. Ω method, Liutex vector and a series of related methods are defined as the third generation of vortex recognition methods [7].

The third generation of vortex identification method successfully answered the six elements of vortex, including absolute intensity, relative intensity, local rotation axis, vortex core location, vortex core size and vortex boundary definition. However, the first generation of vortex identification method could not provide the above elements, and the second generation could only provide the approximate boundary of vortex under the condition of minimal threshold value. Based on the above advantages, the third-generation vortex identification method has been rapidly and widely applied [7].

21.2 Third Generation Vortex Structure Identification Method

Liu et al. [8] proposed an Ω vortex structure recognition method, and the formula is:

$$\boldsymbol{\Omega} = \frac{\|\mathbf{B}\|_F^2}{\|\mathbf{A}\|_F^2 + \|\mathbf{B}\|_F^2} \tag{21.1}$$

In practice, to prevent the problem of dividing by zero, add a small positive number ε to the denominator of Eq. 21.2, so that the expression for Ω becomes

$$\boldsymbol{\Omega} = \frac{\|\mathbf{B}\|_F^2}{\|\mathbf{A}\|_F^2 + \|\mathbf{B}\|_F^2 + \varepsilon} \tag{21.2}$$

Obviously, the value range of Ω is $0 \leq \Omega \leq 1$, which can be understood as the concentration of vorticity. More specifically, it represents the rigidity of fluid movement. When $\Omega=1$, it represents rigid body rotation of fluid. When $\Omega \geq 0.5$ represents the dominance of the antisymmetric tensor B relative to the symmetric tensor A, so Ω slightly greater than 0.5 can be used as the criterion for vortex recognition.

On the basis of Ω vortex structure identification method. Liu et al. [13] also proposed the Liutex vector to accurately define vortices, and Wang et al. [15] derived an explicit expression of the Liutex size in the initial coordinate system:

$$\mathbf{R} = \left(\langle \boldsymbol{\omega}, \mathbf{r} \rangle - \sqrt{\langle \boldsymbol{\omega}, \mathbf{r} \rangle^2 - 4\lambda_{ci}^2} \right) \mathbf{r} \tag{21.3}$$

21.3 Experiment and Simulation

21.3.1 Experiment

In this paper, a 36 KW full oxygen heating furnace composed of two furnaces is adopted for combustion experiment. The nozzle structure is 11-degree asymmetric jet nozzle structure, and the exhaust pipe is at the bottom of the furnace. The furnace is 2 m long, 0.2 m wide and 0.2 m high. The material of the whole oxygen heating furnace is mainly steel plate, thermal insulation cotton and firebrick. Set two observation Windows before and after. There are 24 observation points divided into 8 rows and 3 rows on the furnace wall, and all the monitoring points are symmetrical. The temperature of the furnace is monitored by a thermocouple.

The flow chart of experimental principle is shown in Fig. 21.1 [14]. During the hot state test, propane and pure oxygen were sprayed from the upper nozzle and the lower nozzle respectively. Propane and oxygen are injected into the furnace through high-speed injection flow. The gases meet and mix in the furnace, and are ignited for combustion reaction.

The flame front phenomenon can be observed through the observation port. As can be seen from the MILD combustion process in Fig. 21.2, when the flame is ignited in Fig. 21.2a, there is a clear flame front. As time goes by, the flame front in Fig. 21.2b is obviously dispersed and becomes larger in area, and the whole furnace becomes brighter. Finally, in Fig. 21.2c, the flame front basically disappears and the MILD combustion effect is achieved.

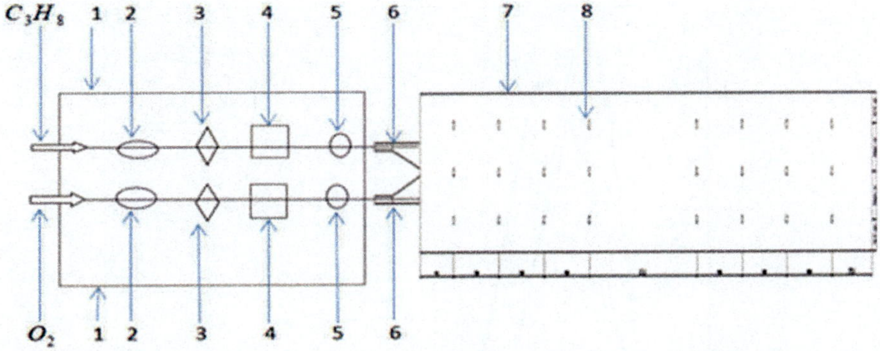

Fig. 21.1 Experimental schematic diagram (1-Multifunctional test bench; 2-Filter voltage regulator; 3-flow regulating valve; 4-Thermal flowmeter; 5-switch; 6-nozzle; 7-Full oxygen heating furnace; 8-temperature sensing point.)

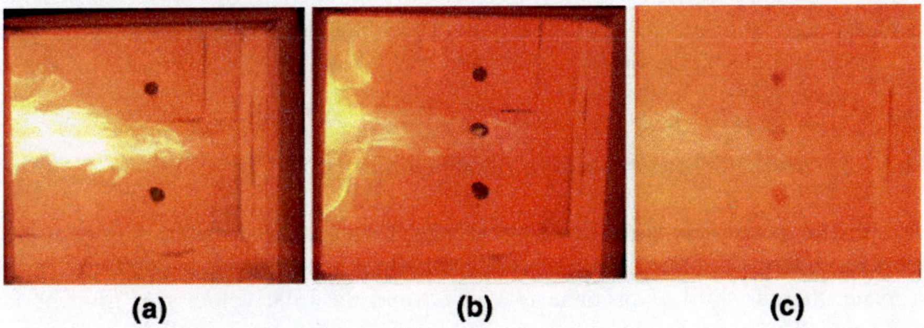

Fig. 21.2 Experimental flame phenomenon diagram

21.3.2 *Numerical Simulation*

21.3.2.1 Modeling and Grid-Free Verification

In the process of numerical simulation, the physical geometry structure of the total oxygen reheating furnace is simplified. The furnace size of the total oxygen reheating furnace is 2000 × 200 × 200 mm. The program with Mesh in Workbench in ANSYS is used to carry out the overall Mesh division of the furnace, and the overall Mesh division of the furnace adopts hexahedral Mesh.

In the process of grid division, grid independence was verified under three different grid numbers. The verification results of grid independence are shown in Fig. 21.3. By comparing the simulation results of different mesh numbers, it can be seen that mesh 73 W is the optimal scheme with both computational efficiency and computational accuracy.

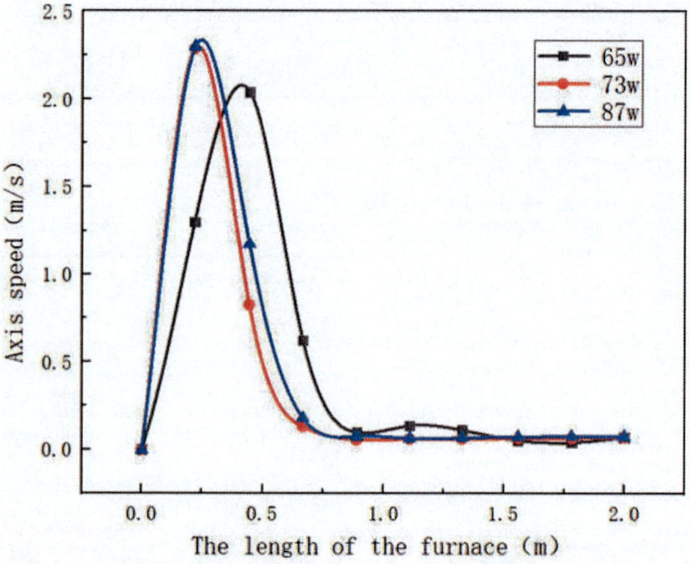

Fig. 21.3 Grid independence verification

21.3.2.2 Model Selection and Boundary Conditions

In terms of model selection, RNG turbulence model and P-1 radiation model are adopted. In terms of boundary condition setting, the upper nozzle is connected with propane, the gas fluid of propane is 1 m^3/h, and the inlet velocity of propane is 39.32 m/s. The gas flow through the lower nozzle is 5 m^3/h, and the oxygen inlet speed is 196.58 m/s. The outlet is a natural outlet. The gas temperature was set at 300 K and the wall temperature was set at 1000 °C. The path ignition was used to conduct the numerical simulation study and calculation of MILD oxygen combustion.

Because it is pure oxygen combustion, too high oxygen concentration will cause temperature rise, and there may be some uncertainty factors, environmental factors and artificial factors error in the experimental process. Therefore, the numerical simulation temperature is generally higher than the experimental temperature.

21.4 Results and Analysis

Figures 21.4 and 21.5 below show the temperature contour cloud diagram and flow diagram of section Z = 0. it can be seen that the high temperature area is mainly at 0.2–1 m and the reflux flow lines in this area are relatively dense, which also verifies that this is the main mixing area of propane and oxygen. Therefore, it can be seen that

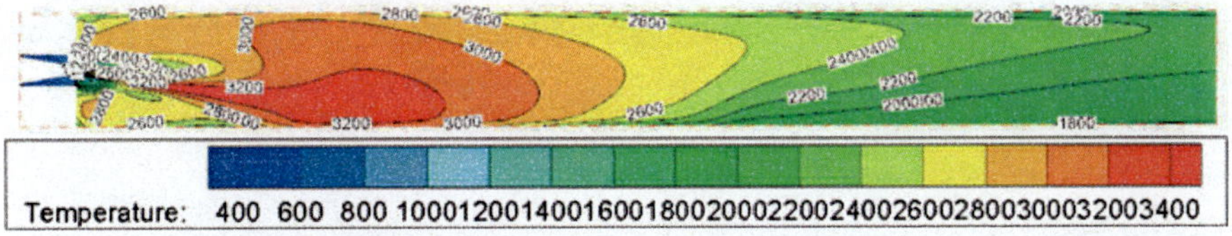

Fig. 21.4 Temperature contour cloud map at Z = 0 cm

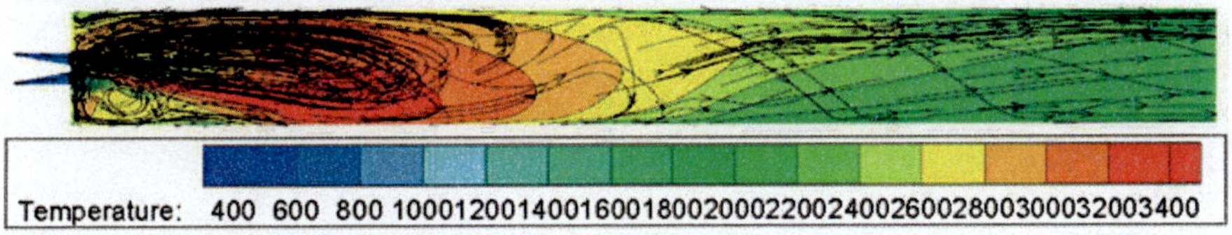

Fig. 21.5 Flow diagram at Z = 0 cm

the high temperature area and the main reaction area of eddy current movement are at the front of the furnace.

According to Liutex vortex structure cloud map Fig. 21.6, there are many vortices with different intensities in the initial section of the furnace, and the initial section of the furnace is a strong mixing zone. The threshold value on the vortex structure identification chart of the middle and rear section of the furnace is basically unchanged, and the intensity of the vortex structure is roughly the same, indicating the existence of weak vortices with large shapes. The vortex structure in the middle and rear of the furnace is weaker than that in the initial section of the furnace.

As can be seen from the vortex structure identification diagram in Fig. 21.6, the vortex structure in the furnace is concentrated in the initial position of the furnace, which is exactly consistent with the previous temperature cloud diagram and flow diagram. It shows that the gas mixture in the furnace can be judged by the eddy current structure diagram. The eddy current structure can directly reflect the eddy current movement of the gas in the furnace.

Liutex vortex structure processing method is used to process the simulated data and import it into Tecplot for analysis. The vortex core and vortex line as shown in

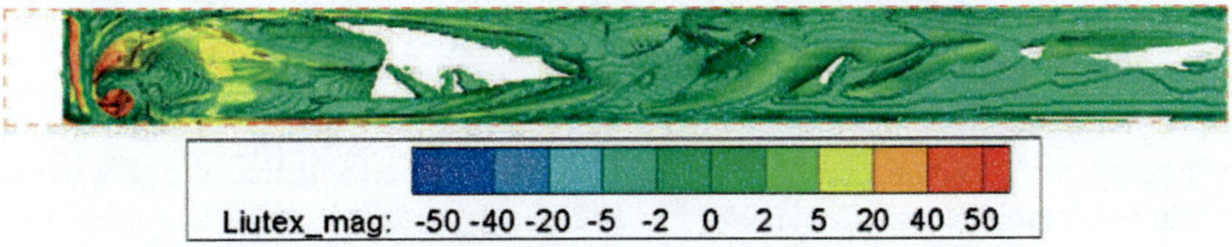

Fig. 21.6 Liutex vortex processing cloud image

Fig. 21.7 Vortex core diagram

Fig. 21.8 Vortex diagram

Figs. 21.7 and 21.8 are displayed with threshold value of 3. The vortex core represents the vortex structure of the intensity at the same threshold, and the vortex line is the streamline passing through the vortex core.

There are basically vortex cores in the X direction of the furnace, but they cannot be spread throughout the whole furnace, because they are only vortex cores under one threshold condition, while vortex cores under other threshold conditions are filled in other positions of the furnace. The vortex lines are similar to the streamline lines, mainly concentrated near the vortex core in the first half of the furnace, indicating that there are vortex structures with different intensities in the furnace, which are conducive to gas reflux and mixing.

When a certain threshold is determined, there should not be too many vortex cores in the furnace, because too many vortex cores with the same threshold intensity can intuitively reflect the vortex structure with the same intensity of vortex movement in the furnace, which is not conducive to the mixing of combustion gas. When the vortex core structure in the furnace is very little or even almost no, it indicates that the vortex structure of gas vortex movement in the furnace basically has no such strength, and there is no flue gas backflow to promote the mixing of combustion gas, and the combustion effect is very poor.

21.5 Conclusion

Through the combination of numerical simulation and experiment, the eddy current structure in the heating furnace is analyzed by Liutex method with the third generation of eddy structure identification method and verified by temperature isoline cloud map and flow chart. The following conclusions are drawn:

(1) Liutex vortex structure can intuitively reflect the vortex motion of gas in the furnace. Vortex motion in the furnace is mainly concentrated in the initial section

of the furnace. The size and quantity of vortex structure in this area is stronger than that in the middle and rear of the furnace, and it is also the main combustion area of the furnace.

(2) Under certain threshold conditions, too much or too little vortex core structure in the furnace will affect the combustion efficiency.

(3) The combination of vortex line and streamline line can better reflect the backflow and mixing of gas in the furnace, and is more conducive to the analysis of vortex motion of gas in the furnace.

References

1. J.A. Wünning, J.G. Wünning, Flameless oxidation to reduce thermal no-formation. Prog. Energy Combust. Sci. **23**(1), 81–94 (1997)
2. P.F. Li, J.C. Mi, B.B. Dally et al., Progress and recent trend in MILD combus-tion. Sci. China Tech. Sci. **54**, 255–269 (2011). ((in Chinese))
3. A. Cavaliere, M. de Joannon, Mild combustion. Prog. Energy Combust. Sci. **30**(4), 329–366 (2004)
4. G.P. Chen, L.I. Hui, D.N. Guo, L.V. Cheng-Zheng, B.J. Tang, The research progress of oxy-fuel combustion technology and on glass melting furnace. Indus. Heat. (2008) (in Chinese)
5. C. Liu, Y. Yan, P. Lu, Physics of turbulence generation and sustenance in a boundary layer [J].Comp. Fluids **102**, 353–384 (2014)
6. J.M. Wallace, Highlights from 50 years of turbulent boundary layer research [J]. J. Turbul. **13**(53), 1–70 (2013)
7. Y.Q. Wang, N. Gui, A review of the third-generation vortex identification method and its applications. Chinese J. Hydrodyn. (2019) (in Chinese)
8. C. Liu, Y. Wang, Y. Yang et al., New omega vortex identi- fication method [J]. Sci. China Phys. Mech. Astron. **59**(8), 684711 (2016)
9. Y. Wang, S. Fu, On the thresholds of vortex identification methods[C], in *Proceedings of the 4th Symposium on Fluid-Structure-Sound Interaction and Control*, (Tokyo, Japan, 2017)
10. N. Gui, L. Ge, P.X. Cheng, et al., Comparative assessment and analysis of rorticity by Rortex in swirling jets [J]. J. Hydrodyn. (2019)
11. Y. Wu, C. Guo, S. Feng, et al., Research on identification of vortex structure in oxy-fuel heating furnace based on vortex identification method[J]. Energy Sourc. Part A Recov. Utiliz. Environ. Effects 1–12 (2020)
12. C. Guo, Y. Wu, Z. Liu, et al., Eddy current research of oxy-fuel heating furnace based on third generation vortex recognition method. (2021)
13. C. Liu, Y. Gao, S. Tian et al., Rortex–a new vortex vector definition and vorticity tensor and vector decompositions [J]. Phys. Fluids **30**, 035103 (2018)
14. Y. Gao, C. Liu, Rortex and comparison with eigen-value-based vortex identification criteria[J]. Phys. Fluids **30**, 085107 (2018)
15. Y.Q. Wang, Y.S. Gao, J.M. Liu, et al., Explicit formula for the Liutex vector and physical meaning of vorticity based on the Liutex-Shear decomposition. J. Hydrodyn. (2019)

Chapter 22
The Effect of Fluidic Pintle Nozzle on the Ablative Features

Zheng Yang, Kan Xie, Xiang Li, Dongfeng Yan, Junwei Li, and Ningfei Wang

Abstract Fluidic Pintle Nozzle (FPN) combines mechanical and fluidic schemes to inject secondary flow at the head of the pintle to change the mainstream gas profile. As a means of mechanical combined with fluidic regulation, this method has not completed systematic research since it was proposed, and the current research is still in the basic theoretical stage, with insufficient research on its basic laws, flow mechanism, influencing factors, ablative protection and so on. This paper focuses on the influence of injecting secondary flow at the head of the pintle on the ablation of the fluidic pintle nozzle of the solid rocket motor, the surface chemical reaction model under fluid–solid coupling was established, and fluent commercial software was used for numerical calculation. The basic law and influencing factors of the ablative rate with a different opening, injection position, and secondary flow temperature were studied on the ablation rate of the nozzle were studied. The results show that the ablation rate of the nozzle is significantly different under the different openings. A low-temperature jet can effectively reduce the temperature of the pintle, reduce the ablation rate of the pintle, and has a significant protective effect on the pintle downstream of the injection hole.

22.1 Introduction

The ability to control thrust, especially random thrust control, including magnitude and direction, is essential to improve vehicle maneuverability. Compared to the thrust-predetermined solid rocket motors, random thrust control can more reasonably distribute the propellant energy according to real-time work needs, which is a development trend for solid rocket motors. Adjusting the thrust by changing the throat area of the nozzle by certain technical means is one of the methods for solid

Z. Yang · K. Xie (✉) · X. Li · J. Li · N. Wang
School of Aeronautics and Astronautics, University of Electronic Science and Technology of China, Chengdu 611731, China
e-mail: xiekan@bit.edu.cn

D. Yan
Civil Aviation Flight, University of China, Guanghan 618307, China

Y. Wang et al. (eds.), *Liutex and Third Generation of Vortex Identification*,
Springer Proceedings in Physics 288, https://doi.org/10.1007/978-981-19-8955-1_22

rocket motors [1]. There are many ways to change the throat area of the nozzle, one of which is using a mechanical valve. This method is represented by the variable thrust motor controlled by a pintle [2–4]. The pintle-controlled method [5, 6] is used to decrease or increase the actual flow area of the nozzle throat by a specific configuration of the pintle by inserting it into or removed from the throat, respectively. The mechanical method of using a moving pintle to change the throat area requires the installation of a drive mechanism system. Owing to the advantages of a wide range and stepless adjustment feature, the pintle-controlled technology is very promising for the thrust control of solid rocket motors (SRM). However, in high-performance engines, the nozzle-throat insert cannot only resist high-speed and high-temperature (up to 3,000 K) gas but also produce extremely high-temperature gradient and thermal stress. For solid rocket motors with high combustion temperature, high flow rate and long working hours, the ablation of the nozzle-throat insert is particularly prominent. The chemical and mechanical ablation has led to the configuration and dimension damage of the nozzle throat, which largely determines the reliability of solid rocket motors. A fluidic pintle nozzle (FPN) is a method that changes the flow area and the shape of the throat of the primary flow as a result of the interaction between the secondary and primary flows (fluidic injection). It is based on the pintle motor and the fluidic nozzle throat (FNT) [7–9] to inject secondary flow at the head of the pintle to change the mainstream gas profile. The secondary flow can provide additional thrust and improve the efficiency of thrust control. The schematic is shown in Fig. 22.1. FPN combines mechanical and fluidic schemes and low-temperature jets can reduce the temperature of the pintle and can reduce the ablation.

Currently, carbon/carbon composites have been developed from being produced with a single ablative material to thermal structural materials and have been applied to a variety of solid rocket motors. This paper considers the ablative characteristics of fluidic pintle nozzle with carbon/carbon composites. The surface chemical reaction

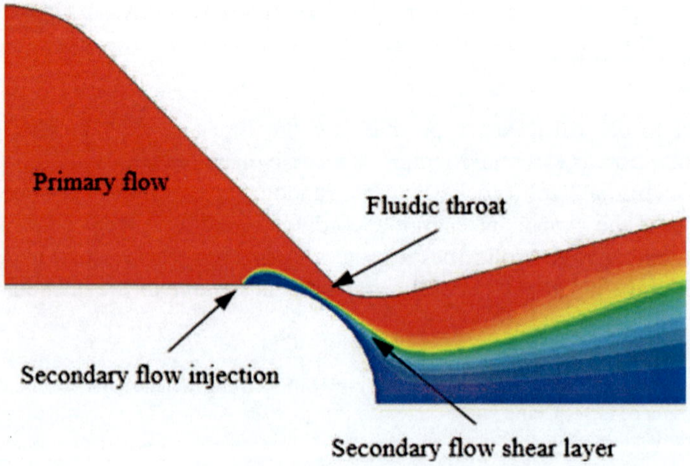

Fig. 22.1 Scheme of secondary injection in pintle motor

model under fluid–solid coupling was established, and the basic law and influencing factors of the jet in pintle nozzle on ablation rate of nozzle and pintle were studied.

22.2 Geometry, Boundary Conditions and Operating Conditions

The established model exploits the axisymmetric geometry of the system. The geometric structure is shown in Fig. 22.2. R_p is the radius of the pintle, L_i is the position of the injection. The relative position ratio of dimensionless parameters L_i/R_p is defined to represent the relative position of the injection. The area between the wall of the pintle and the wall of the nozzle is defined as A_t, and the relative displacement of the pintle is defined as *DE*. The position of the pintle is characterised by the opening O_p, which shows the relationship between the opening and the pintle position. When the area of the airflow channel formed by the position of the pintle and the wall is equal to the area of the throat, the geometrical throat area is the maximum, the O_p is defined as 100%. When the pintle reaches the throat of the nozzle, the geometrical throat area is the minimum, the O_p is defined as 0%. O_p division of three levels based on maximum and minimum [11] (see Table 22.1). The area ratio of the injection port to the throat is 0.2, and the injection angle is 90°.

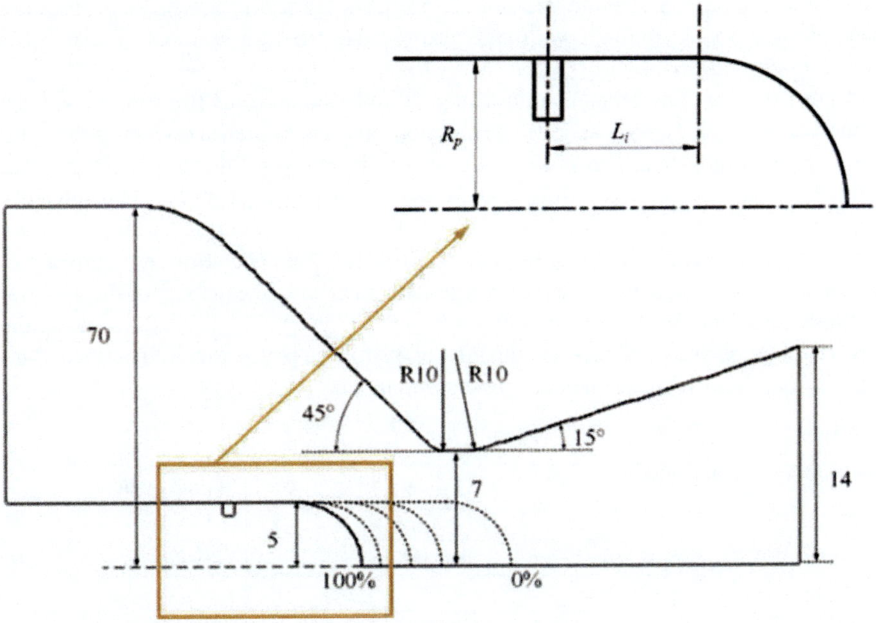

Fig. 22.2 Schematic of the fluidic pintle nozzle

Table 22.1 Relationship between opening and pintle position

O_p	100%	50%	0%
A_t (mm^2)	153.94	114.67	75.40
DE (mm)	8.30	5.84	0.00

In addition, the effective throat area of the FPN is mainly related to the flow ratio of the secondary flow $f_m = \dot{m}_s/\dot{m}_o$ and primary flow and pressure ratio $f_p = p_s/p_o$ (subscript s represents secondary flow). We define the modified parameter $\dot{w} = \dot{m}\sqrt{T/\mu}$ so that the ratio of the secondary flow to the primary modified flow ratio is [10]:

$$f_w = \frac{\dot{w}_s}{\dot{w}_o} = \frac{\dot{m}_s\sqrt{T_s\mu_o}}{\dot{m}_o\sqrt{T_o\mu_s}}$$

where μ_0 and μ_s are the molar mass of the primary and secondary flows, respectively. The definition of a modified flow ratio can eliminate the influence of the temperature and molar mass of secondary flow on the result.

The primary gas is based on the common temperature of the solid propellant gas and published values [11, 12], and the temperature of the jet is also referenced thus: both are at a high temperature and do not participate in the reaction (see Table 22.2). The inlets of the primary and secondary flows are set to be the mass-flow inlet, and their conditions replicate the literature data. The outlet is a pressure outlet with a pressure of 101 kPa and a temperature of 288 K. At the walls, no-slip conditions are used along with an adiabatic wall and zero normal pressure gradient. The mesh is shown in Fig. 22.3.

Based on the flow area, the fluid area is intercepted near the throat, and the carbon/carbon solid area is added. The ablation calculation area is divided into a fluid zone, throat solid zone, and pintle solid zone. The fluid–solid interface is coupled with the heat transfer (including a secondary flow injection port). The interface between gas and throat and pintle is set as a heterogeneous reaction surface as shown in Fig. 22.4. The computational domain is 50 × 26 mm, consisting of a convergent nozzle zone, a throat, a pintle, and a divergent nozzle zone. The density of the material is 1900 kg/m^3 and the thermal conductivity is 98.67 W/(m·K). Table 22.3 shows the gas composition and parameters at the inlet of the fluidic pintle nozzle. The secondary flow is consistent with the primary flow components.

Table 22.2 Gas parameter

	Primary flow	Secondary flow
$\dot{m}$	0.2 kg/s	–
T	3000 K	1789 K
k	1.17	1.17
μ	26.32	26.32

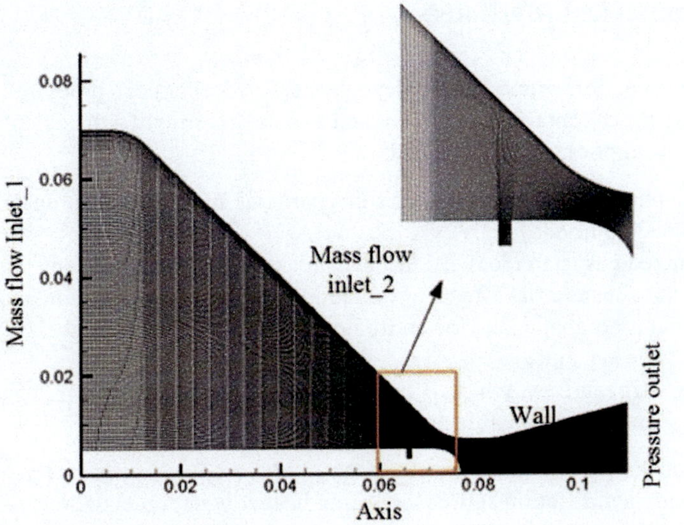

Fig. 22.3 Mesh used to model pintle

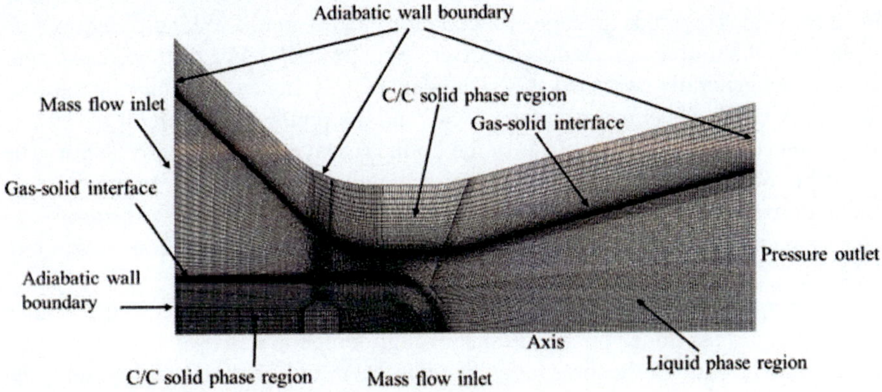

Fig. 22.4 Computational domain and boundary conditions

Table 22.3 Gas composition and parameters at the pressure inlet boundary

Component mass content	Value
Y_{CO_2}	0.27
Y_{H_2O}	0.21
Y_{H_2}	0.01
Y_{CO}	0.11
Y_{HCl}	0.28
Y_{N_2}	0.11
Y_{OH}	0.01

22.3 Numerical Methods

During the injection process, complex physical and chemical processes occur in various parts of the motor combustion and nozzle. To simplify the calculation, the following assumptions are introduced:

(1) Single-phase flow, regardless of solid particles in which the propellant is not completely burned;
(2) The mixed gas is an ideal gas that satisfies the ideal gas equation;
(3) Does not consider the effects of radiation forces such as radiation and gravity;
(4) The wall is adiabatic, and the entire flow field undergoes no heat exchange with the outside system;
(5) Gas is a frozen flow, which is composed of an ideal compressible gas, and its component diffusion conforms to Fick's law.

The thermochemical ablation process of the FPN can be described by the N-S equation with chemical reaction source terms and heat conduction equations of the solid phase region. The N-S equation with the chemical reaction source term, the gas component transport equations, and the heat conduction equation of the solid-phase region can refer to in literature [13], and will not be described here. The convection term in the Reynolds average N-S equation can be set to a second-order upwind scheme and the source term adopts the discrete central difference method. The turbulence model generally uses the k-ε SST model [14, 15]. In heat transfer calculations, the quality of the mesh near the throat wall and the pintle are important. Accuracy is guaranteed only when the mesh in the boundary layer is sufficiently dense. The y + value of the mesh can be used to verify the mesh quality [13]. In the ablation calculation of the nozzle-throat insert, generally $y+ < 2$, which needs to be reassured after each calculation is converged. If $y + > 2$, the further calculation is required, followed by the repetition of the above steps until it is satisfied.

The heterogeneous gas-surface chemical reactions that take place on the surface of C/C composites under the conditions of high temperatures [16, 17]. Table 22.4 lists the kinetic data for the three heterogeneous chemical reactions considered at the surface.

The equation of the chemical reaction rate of carbon is

$$\dot{m}_i = p_i^n \cdot A_i T_w^b exp(-E_i/R_u T_w)$$

Table 22.4 Heterogeneous reaction system at the interface

Surface reaction	A_i(m^2·s·Pa$^{0.5}$)	b	E_i (J·mol^{-1})	n
$C(s) + H_2O \rightarrow CO + H_2$	1.51×10^3	0.0	2.88×10^8	0.5
$C(s) + CO_2 \rightarrow {}_2CO$	28.27	0.0	2.85×10^8	0.5
$C(s) + OH \rightarrow CO + H$	3.61×10^{-3}	−0.5	0.00	1

where p_i represent the partial pressure of species i at the surface, R_u is the general gas constant.

The wall retreat rate caused by the reaction is the ablation rate:

$$\dot{r} = \frac{\dot{m}_{H_2O} + \dot{m}_{CO_2} + \dot{m}_{OH}}{\rho_s}.$$

22.4 Results and Discussion

22.4.1 Influence of Opening on the Ablative of FPN

Due to the different O_p of the pintle, the combustion pressure and throat flow at each opening are also significant differences, and this section provides a comparative analysis for the O_p at 0, 50, and 100%. Through simulation calculations, the temperature nephogram of O_p at 0 and 100%, and the pressure, temperature, and ablation rate distributions of the nozzle wall and pintle of O_p at 0, 50, and 100% are obtained, respectively, which are shown in Figs. 22.5, 22.6 and 22.7.

At different openings, in addition to the different combustion pressure, the injection port also changes with the position of the pintle. In the range of 100–0% opening, the pintle gradually approaches the throat of the nozzle, and the combustion pressure increases. At the same time, the location of the maximum ablation rate on the nozzle moves downstream as the gas sonic surface moves downstream.

We can see from Table 22.5 that when O_p is 0%, the combustion pressure is maximum. At this time, the pintle is closest to the throat, and the nozzle ablation rate is significantly higher than the other two openings, as shown in Fig. 22.7a. When O_p is 50 and 100%, the combustion pressure is 3.5 and 2.74 MPa, by 34 and 48%, while the peak ablation rate of the nozzle decreases from 0.12 to 0.082 mm/s and 0.075 mm/s, by 32 and 38%. However, the peak ablation rate of the nozzle does not change significantly in the range of 50–100%.

The nozzle ablation rate at 100% opening appears as a double-peak phenomenon, as shown in Fig. 22.7a. The nozzle ablation rate at 100% opening appears as a double-peak phenomenon, as shown in Fig. 22.7a. Combined with the analysis in Fig. 22.6a, it can be seen that at 100% opening, a second peak occurs because the pressure rises after the gas passes through the oblique shock wave, causing the ablation rate to rise.

Figure 22.7b shows the ablation rate of the pintle at different openings, the ablation rate upstream of the injection position is significantly higher than that downstream of the injection position, and the ablation rate of the pintle upstream of the injection port increases gradually as the opening decreases. When O_p is 0%, the ablation rate of the pintle reaches its peak due to the acceleration of the gas, the thinning of the boundary, and the increase in the mass flow rate near the pintle wall. In the vicinity of the injection port, the pintle is affected by the low-temperature secondary flow, which

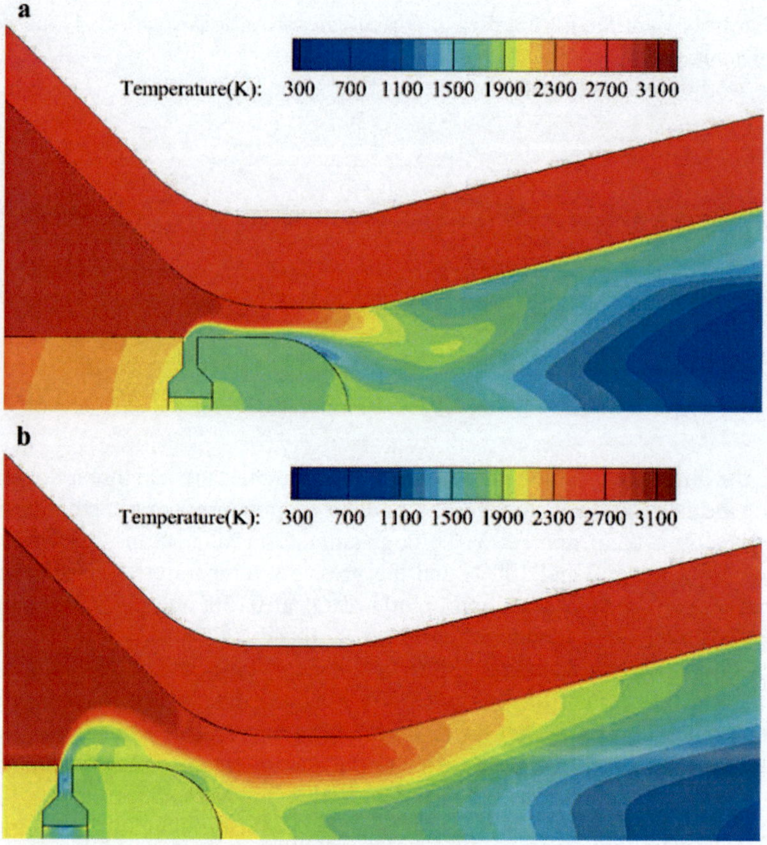

Fig. 22.5 Temperature nephogram at different opening. **a** $O_p = 0\%$, **b** $O_p = 100\%$

reduces the temperature and slows down the increase in mass flow rate, thus causing a rapid decrease in the ablation rate in the area near the injection port. The ablation rate fluctuated downstream of the injection port due to the mass flow rate near the wall, but the ablation rate showed a decreasing trend compared to the upstream, with an average value of 0.015 mm/s.

22.4.2 Influence of the Injection Port Position on the Ablative of FPN

The relative position of the injection port has a significant impact on both the combustion pressure and the flow in the throat. This section provides a comparative analysis of the relative positions L_i/R_p of 0 and 1, respectively, at $T_S = 1789$ K, $f_w = 0.3$, and $O_p = 50\%$ of the operating conditions. Through simulation calculations, the results

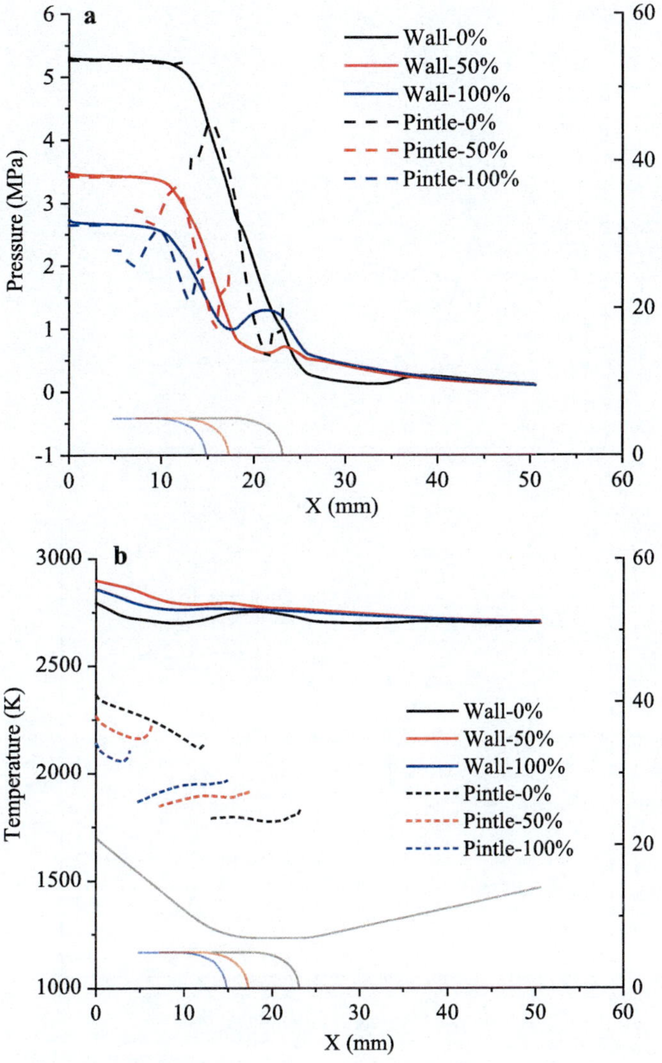

Fig. 22.6 Pressure and temperature distribution at different opening. **a** Pressure, **b** Temperature

with the relative position L_i/R_p of 0 and 1 are obtained, respectively, which are shown in Figs. 22.8, 22.9 and 22.10.

According to Figs. 22.8 and 22.9, there is a significant difference in the pintle temperature at different injection port positions, while the nozzle temperature is not affected by the relative position of the injection port. When the relative position L_i/R_p is 0, the injection port is closer to the throat, which has a better choke state effect on the primary flow and higher pressure in the combustion.

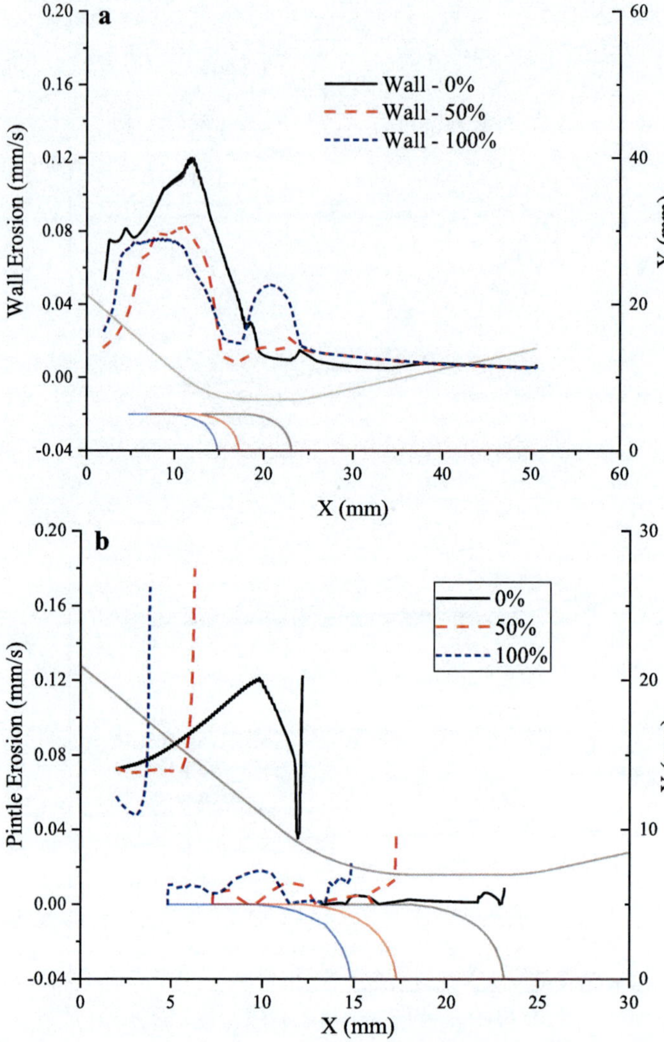

Fig. 22.7 Ablation rate at different opening. **a** Nozzle Wall, **b** Pintle

Table 22.5 Comparison of calculation results with different opening of the pintle

O_p	100%	50%	0%
Combustion pressure (MPa)	2.74	3.5	5.31
Peak ablation rate of the nozzle (mm/s)	0.075	0.082	0.12

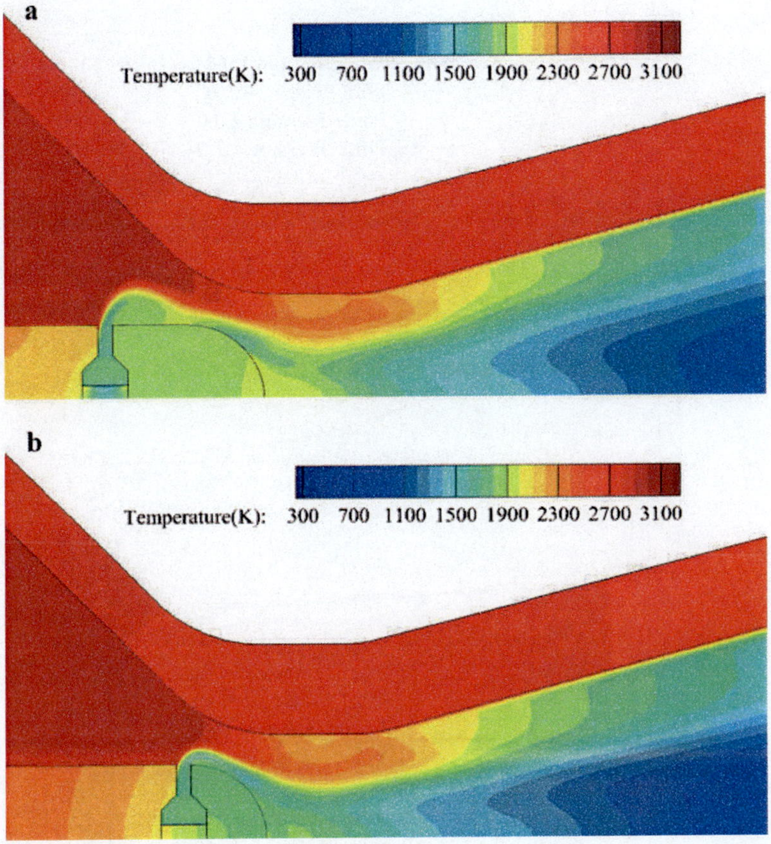

Fig. 22.8 Temperature nephogram at a different injection position. **a** $(L_i/R_p) = 1$, **b** $(L_i/R_p) = 0$

The smaller the relative position L_i/R_p is, the higher the peak ablation rate of the nozzle, as shown in Fig. 22.10. The peak ablation rates for L_i/R_p of 0 and 1 are 0.1 and 0.083 mm/s, respectively, with the peak values at the same locations. For the pintle, the ablation rate downstream of the injection port is lower than that upstream, and the ablation rate peaks at 0.053 mm/s ($L_i/R_p = 0$) and 0.067 mm/s ($L_i/R_p = 1$) in the head recirculation zone near the axisymmetric geometry of the system.

22.4.3 Influence of Secondary Flow Temperature on the Ablative of FPN

The secondary flow temperature has a direct effect on ablation and can be controlled by using propellants with different combustion temperatures or carrying cryogenic gas cylinders to achieve injection temperature control. In this section, for O_p is 50%,

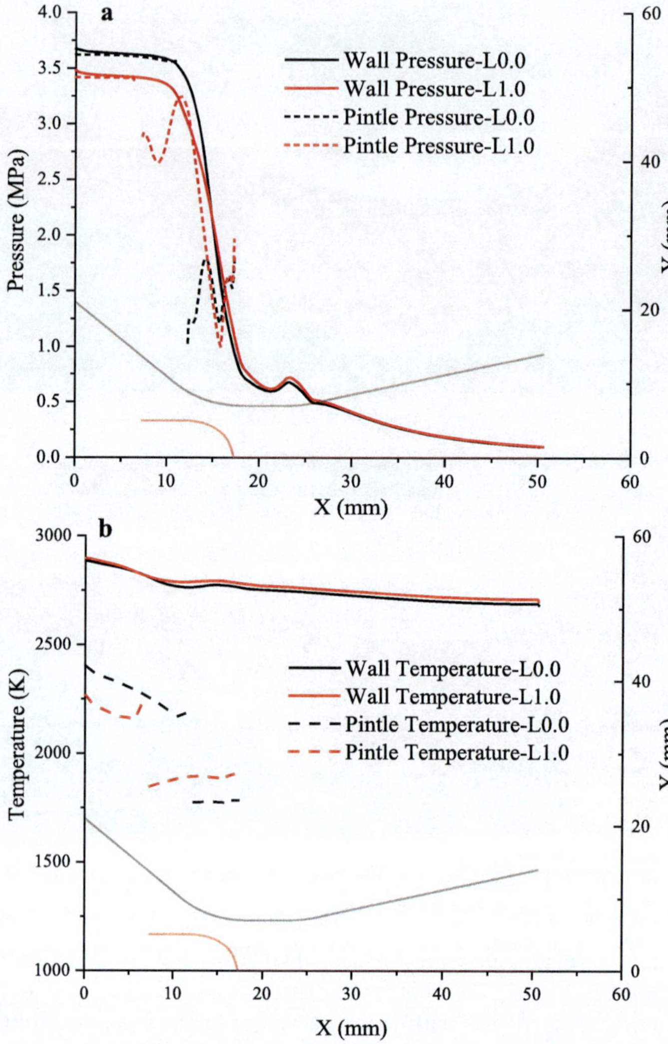

Fig. 22.9 Pressure and temperature distribution at a different injection position. **a** Pressure, **b** Temperature

the secondary flow temperature of 300, 1,000, and 1,789 K are three conditions for comparative analysis. Through simulation calculations, the results with different secondary flow temperatures are obtained, which are shown in Figs. 22.11, 22.12 and 22.13.

Under the action of the secondary flow, the overall temperature of the pintle is significantly reduced, and the temperature of the solid phase area upstream of the

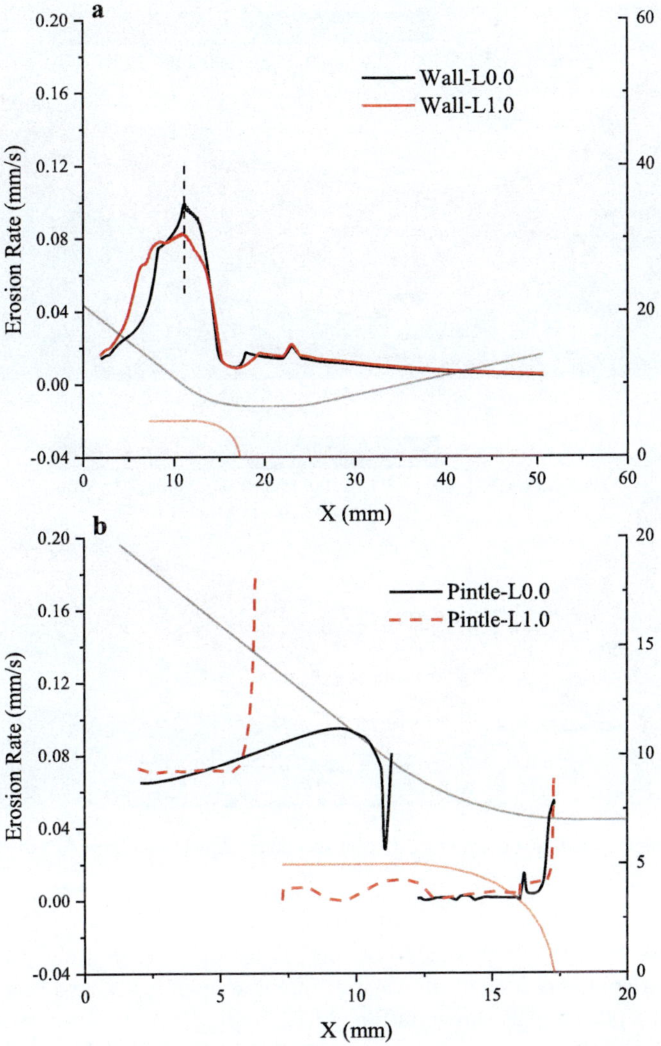

Fig. 22.10 Ablation rate at a different injection position. **a** Nozzle Wall, **b** Pintle

injection port is lower than the primary flow temperature due to the joint action of the primary gas and the heat transfer of the secondary flow, as shown in Fig. 22.11. The overall temperature of the pintle dropped as the secondary flow temperature was reduced. The secondary jet flows downstream along the wall of the pintle to protect the downstream structure, and the temperature of the solid phase region downstream of the injection port is lower than upstream, and temperature is even lower than that of the gas. However, since there is no direct contact between the secondary flow and the throat, the change of the secondary flow temperature has almost no effect on the

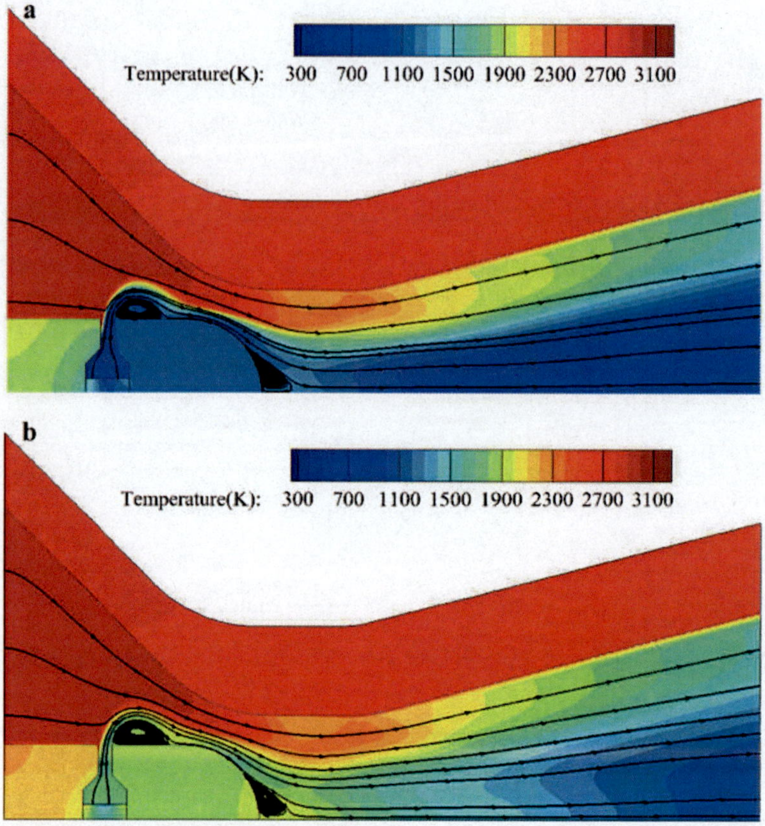

Fig. 22.11 Temperature nephogram at a different secondary flow temperature. **a** $T_s = 1000$ K, **b** $T_s = 1789$ K

throat temperature in the high-speed flow process of the primary flow, as shown in Fig. 22.12b. Also, the secondary flows at different temperatures have less effect on the combustion pressure, as shown in Fig. 22.12a.

Since there is no direct contact between the secondary flow and the nozzle, and the primary flow rate is faster, the nozzle ablation rate amplitude and the change trend are basically the same, not affected by the change in the secondary flow temperature, as shown in Fig. 22.13a.

The ablation rate of the pintle is directly related to secondary flow temperature. The ablation rate of the upstream pintle decreases with the decrease of secondary flow temperature, and the average ablation rates upstream of the injection port are 0.07, 0.02, and 0.0078 mm/s for T_S of 1,789, 1,000, and 300 K. Downstream of the injection port, the pintle temperature is lower than upstream, as shown in Fig. 22.12b, and the ablation rate decrease as the secondary flow temperature decreases. After the

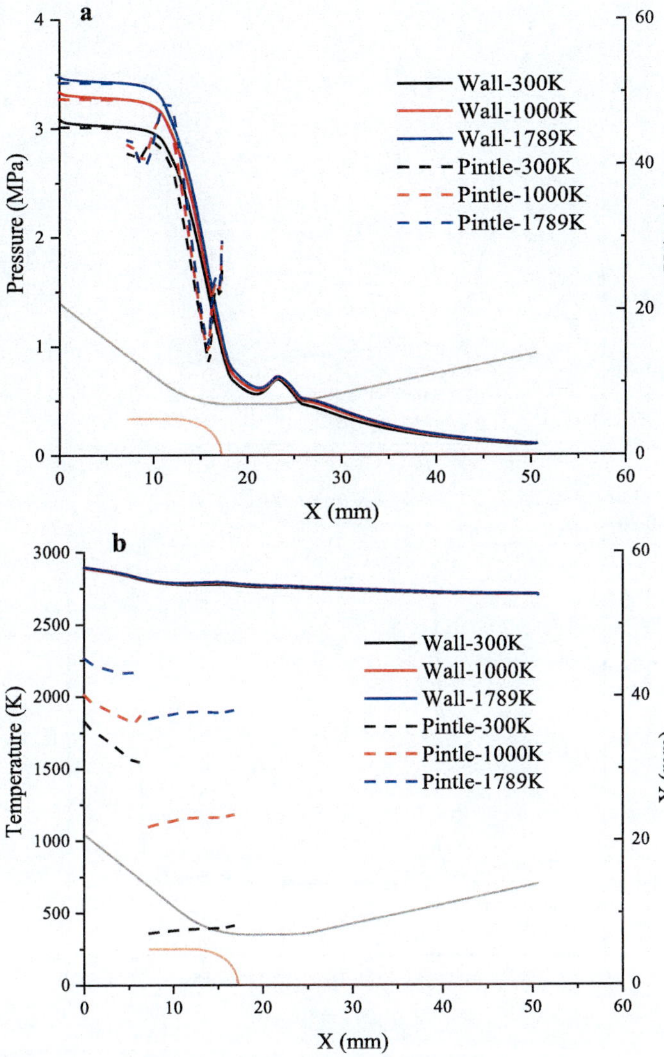

Fig. 22.12 Pressure and temperature distribution at a different secondary flow temperature. **a** Pressure, **b** Temperature

flow separation of the gas in the head of the pintle, the pintle ablation rate starts to increase gradually and reaches a peak near the axisymmetric geometry of the system. The peak pintle ablation rates are 0.067, 0.055, and 0.054 mm/s at T_S of 1,789, 1,000, and 300 K, which generally remained at a low level.

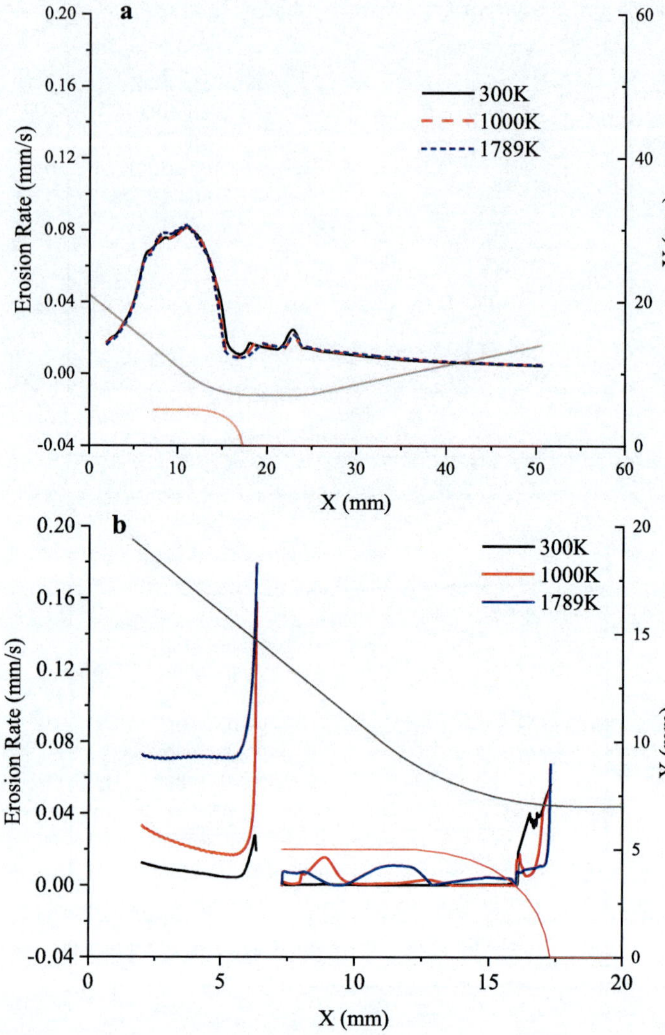

Fig. 22.13 Ablation rate at a different secondary flow temperature. **a** Nozzle Wall, **b** Pintle

22.5 Conclusions

In this paper, by establishing the chemical ablation model of fluid pintle nozzle and using the fluid–solid coupling surface chemical reaction method, the variation law and influencing factors of the ablation rate are studied and analyzed under the condition of changing the injection structure and jet injection parameters. Some conclusions are obtained through principle analysis and the processing of the simulation results.

(1) In a pintle motor, the ablation rate of the nozzle increases gradually with the flow direction of the gas, reaches the peak before the actual throat (throttle surface), and then decreases gradually. The ablation rate of the divergent section is much lower than that of the convergent section.
(2) During the operation of the pintle, the smaller the opening of the pintle, the closer the pintle is to the throat. With the reduction of the opening of the pintle, the ablation rate of the nozzle is slightly increased due to the increase in combustion pressure, and the ablation rate of the nozzle is significantly different under the different openings. Nevertheless, there is little difference in the ablation rate of the pintle wall downstream of the injection port.
(3) The smaller the relative position of the injection is, the higher the combustion pressure is, the higher the ablation rate of nozzle and pintle is, but the position of the peak point of nozzle ablation rate remains unchanged. When the injection port is close to the head of the pintle, the area of the pintle covered by the secondary flow is small and the ablation protection effect on the pintle is poor.
(4) The ablation rate of the pintle is directly related to the secondary flow temperature. The lower the temperature, the lower the ablation rate of the pintle. After the flow separation of the gas at the head of the pintle, the ablation rate of the wall of the pintle began to increase gradually and reached the peak near the axisymmetric geometry of the system.
(5) The change in the secondary flow temperature has almost no effect on the ablation rate of the nozzle wall.

Conflicts of Interest The authors declare that there is no conflict of interest regarding the publication of this paper.

Data Availability The data used to support the findings of this study are available from the corresponding author upon request.

References

1. H.L. Ji, B.H. Park, W. Yoon, Parametric investigation of the pintle-perturbed conical nozzle flows. Aerosp. Sci. Technol. **26**(1), 268–279 (2013). https://doi.org/10.1016/j.ast.2012.05.005
2. M. Ostrander, et al., Pintle motor challenges for tactical missiles, in *36th AIAA/ASME/SAE/ASEE Joint Propulsion Conference and Exhibit* (2000)
3. S. Burroughs, Status of army pintle technology for controllable thrust propulsion, in *37th Joint Propulsion Conference and Exhibit* (2001)
4. J.X. Wang, Z.J. Wei, N.F. Wang, High burning temperature experiment on pintle controlled solid motor. Tuijin Jishu/J. Propuls. Technol. **33**(1), 89–92 (2012)
5. J. Heo, K. Jeong, H.G. Sung, Numerical study of the dynamic characteristics of pintle nozzles for variable thrust. J. Propuls. Power (2015). https://doi.org/10.2514/1.B35257
6. Joung, et al., Investigation of pintle shape effect on the nozzle performance. J. Korean Soc. Aeronaut. Space Sci. **36**(8), 790–796
7. C. Guo et al., Thrust control by fluidic injection in solid rocket motors. J. Propul. Power **33**(4), 1–15 (2017). https://doi.org/10.2514/1.B36264

8. C. Yang, et al., Effect of secondary injection reaction heat on thrust vector control in divergent section, in *2018 Joint Propulsion Conference* (2018)
9. C. Guo et al., One-dimensional theoretical analysis of dilute particle-gas swirling flow in the Laval Nozzle. AIAA J. **56**(3), 1–7 (2018). https://doi.org/10.2514/1.J056214
10. D. Yan, et al., Simulation of thrust control by fluidic injection and pintle in a solid rocket motor. Aerospace Science and Technology 99. https://doi.org/10.1016/j.ast.2020.105711
11. L.I. Juan, et al., Study on performance of pintle controlled thrust solid rocket motor. J. Solid Rock. Technol. (2007)
12. Y.L. Wang, et al., Experiment on non-coaxial variable thrust pintle solid motor. J. Solid Rock. Technol. (2008)
13. P. Thakre, V. Yang, Graphite nozzle material erosion in solid-propellant rocket motors, in *45th AIAA Aerospace Sciences Meeting and Exhibit* (2006)
14. F.R. Menter, Two-equation eddy-viscosity turbulence models for engineering applications. Aiaa J. **32** (1994) https://doi.org/10.2514/3.12149
15. D. Yan et al., Study of the vortex structure of a subsonic jet in an axisymmetric transonic nozzle. Phys. Fluids **32**(7), 076109 (2020). https://doi.org/10.1063/5.0008796
16. D. Bradley, et al., The oxidation of graphite powder in flame reaction zones. Symp. Combust. **20**(1), 931–940. https://doi.org/10.1016/S0082-0784(85)80582-8
17. H.K. Chelliah et al., Modeling of graphite oxidation in a stagnation-point flow field using detailed homogeneous and semiglobal heterogeneous mechanisms with comparisons to experiments. Combust. Flame **104**(4), 469–480 (1996). https://doi.org/10.1016/0010-2180(95)00151-4

Chapter 23
Liutex Investigation of Backflow Leakage in a Shaft-Less Rim-Driven Thruster

Xincheng Wang, Hua Ruan, Xiaorui Bai, Chengzao Han, Huaiyu Cheng, and Bin Ji

Abstract In the current paper, the phenomenon of backflow leakage in a shaft-less rim-driven thruster is numerically investigated. The numerical results simulated with the k-ω SST model are calibrated with the measured parameters. The feasibility of different vortex identification methods (vorticity method, Q-criterion, Ω method, and Liutex method) in the investigation of backflow leakage is discussed. Only the Liutex method can filter the shear layer while retaining the main vortex structure. With the assistance of the Liutex method, the complex components of backflow leakage are identified, which are gap flow, wake vortex, upstream gap vortex, and downstream gap vortex. Moreover, one phenomenon that should be especially concerned is the mixing of wake vortex and downstream gap vortex, because it can aggravate the flux of backflow leakage and raise a significant hydraulic loss.

23.1 Introduction

As a novel propulsion pump, shaft-less rim-driven thruster receives increasing attention in virtue of its significant advantages, such as higher transmission efficiency, better flow regime, and lower noise intensity [1, 2]. For engineering design, the understanding of flow regularity in a shaft-less rim-driven thruster is necessary and urgent. So far, the relevant investigation is still limited, which inspires the authors to conduct a preliminary work to reveal the typic flow structure inside a shaft-less rim-driven thruster.

For the design of shaft-less rim-driven thruster, one of the most concerned regions is the gap between the rim and the shell. As depicted in Fig. 23.1, in the vicinity of the gap outlet, the flow regime gets much more complex as the result of the mixing

X. Wang · X. Bai · C. Han · H. Cheng · B. Ji (✉)
State Key Lab of Water Resources and Hydropower Engineering Science, Wuhan University, Wuhan 430072, China
e-mail: jibin@whu.edu.cn

H. Ruan
Nation Key Laboratory On Ship Vibration and Noise, China Ship Science Research Center, Wuxi 214082, China

Y. Wang et al. (eds.), *Liutex and Third Generation of Vortex Identification*,
Springer Proceedings in Physics 288, https://doi.org/10.1007/978-981-19-8955-1_23

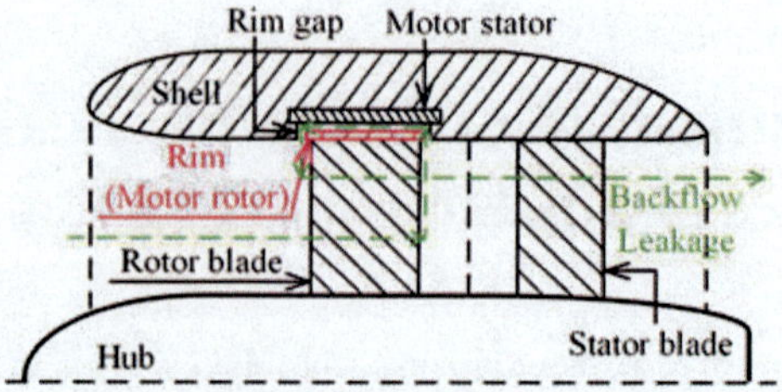

Fig. 23.1 Sketch map of a shaft-less rim-driven thruster

between leakage flow and mainstream, which is followed by a series of detrimental impacts, including operational security and operating efficiency. Meanwhile, challenges also exist in the identification of local flow structure. One factor is that the strong shear motion between the rim and the shell pollutes the identified vortex structure, which further increases the difficulty of flow investigation.

In recent years, aiming at eliminating the drawbacks of traditional vortex identification methods, a series of new methods have been proposed [3, 4]. One remarkable method is Liutex method [5], which possesses a better capacity in distinguishing pure rotation from shearing motion. Under the efforts of many investigators, the potential and superiority of the Liutex method are testified [6–8]. Therefore, the Liutex method is promising in understanding the complex flow problem of engineering machinery.

Considering the advantages of the Liutex method, a preliminary work is conducted to explore the feasibility of the Liutex method in identifying the complex vortical flow inside a shaft-less rim-driven thruster.

23.2 Mathematical and Physical Model

23.2.1 Turbulence Model

Given the huge size of the shaft-less rim-driven thruster, the mesh number demanded to reach a full-resolved large eddy simulation (LES) or direct numerical simulation (DNS) is significantly high, thus a Reynolds-Averaged Navier–Stokes (RANS) simulation is adopted. The governing equations containing continuity and momentum equations can be written as follows:

$$\frac{\partial \rho}{\partial t} + \frac{\partial (\rho u_j)}{\partial x_j} = 0 \tag{23.1}$$

$$\frac{\partial (\rho u_i)}{\partial t} + \frac{\partial (\rho u_i u_j)}{\partial x_j} = -\frac{\partial p'}{\partial x_i} + \frac{\partial}{\partial x_j}\left[\mu_{eff}\left(\frac{\partial u_i}{\partial x_j} + \frac{\partial u_j}{\partial x_i}\right)\right] \tag{23.2}$$

where u_i is the i-component of velocity ($i = 1, 2, 3$), ρ is the density. The effective viscosity μ_{eff} and the modified pressure p' are respectively defined by:

$$\mu_{eff} = \mu + \mu_t \tag{23.3}$$

$$p' = p + \frac{2}{3}\rho k + \frac{2}{3}\mu_{eff}\frac{\partial u_k}{\partial x_k} \tag{23.4}$$

where μ is the laminar viscosity, μ_t is the turbulence viscosity, k is the turbulence kinetic energy. In the current paper, three classic RANS models are tested to decide a suitable model for our investigation, which are standard k-ε [9], RNG k-ε [10], and SST k-ω models [11]. Their main discrepancy is the modular method of the turbulence viscosity μ_t. In the standard k-ε model and the RNG k-ε model, the turbulence viscosity μ_t is defined by

$$\mu_t = C_\mu \rho \frac{k^2}{\varepsilon} \tag{23.5}$$

where C_μ is a constant. In the standard k-ε model, the turbulence kinetic energy k and turbulence dispassion rate ε come directly from two scalar transport equations:

$$\frac{\partial(\rho k)}{\partial t} + \frac{\partial}{\partial x_j}(\rho u_j k) = \frac{\partial}{\partial x_j}\left[\left(\mu + \frac{\mu_t}{\sigma_k}\right)\frac{\partial k}{\partial x_j}\right] + P_k - \rho\varepsilon \tag{23.6}$$

$$\frac{\partial(\rho \varepsilon)}{\partial t} + \frac{\partial}{\partial x_j}(\rho u_j \varepsilon) = \frac{\partial}{\partial x_j}\left[\left(\mu + \frac{\mu_t}{\sigma_\varepsilon}\right)\frac{\partial \varepsilon}{\partial x_j}\right] + \frac{\varepsilon}{k}(C_{\varepsilon 1} P_k - C_{\varepsilon 2}\rho\varepsilon) \tag{23.7}$$

where $C_{\varepsilon 1}$, $C_{\varepsilon 2}$, σ_{k} and σ_{ε} are constants. The turbulence production due to viscous forces is modeled by

$$P_k = \mu_t\left(\frac{\partial u_i}{\partial x_j} + \frac{\partial u_j}{\partial x_i}\right)\frac{\partial u_i}{\partial x_j} - \frac{2}{3}\frac{\partial u_k}{\partial x_k}\left(3\mu_t\frac{\partial u_k}{\partial x_k} + \rho k\right) \tag{23.8}$$

The discrepancy between the standard k-ε model and the RNG k-ε model is the transport equation concerning turbulence dispassion rate ε:

$$\begin{aligned} \varepsilon\frac{\partial(\rho \varepsilon)}{\partial t} + \frac{\partial}{\partial x_j}(\rho u_j \varepsilon) &= \frac{\partial}{\partial x_j}\left[\left(\mu + \frac{\mu_t}{\sigma_{\varepsilon RNG}}\right)\frac{\partial \varepsilon}{\partial x_j}\right] \\ &+ \frac{\varepsilon}{k}(C_{\varepsilon 1RNG} P_k - C_{\varepsilon 2RNG}\rho\varepsilon) \end{aligned} \tag{23.9}$$

where the constant $C_{\varepsilon 1}$ is replaced by a function $C_{\varepsilon 1RNG}$:

$$C_{\varepsilon 1RNG} = 1.42 - f_\eta \tag{23.10}$$

with

$$f_\eta = \frac{\eta\left(1 - \frac{\eta}{4.38}\right)}{\left(1 + \beta_{RNG}\eta^3\right)}, \quad \eta = \sqrt{\frac{P_k}{\rho C_{\mu RNG}\varepsilon}} \tag{23.11}$$

For SST k-ω model, the turbulence viscosity μ_t is defined by the turbulent kinetic energy k and the turbulent frequency ω:

$$\frac{\partial(\rho k)}{\partial t} + \frac{\partial}{\partial x_j}(\rho u_j k) = \frac{\partial}{\partial x_j}\left[\left(\mu + \frac{\mu_t}{\sigma_{k3}}\right)\frac{\partial k}{\partial x_j}\right] + P_k - \beta' \rho k \omega \tag{23.12}$$

$$\begin{aligned} \frac{\partial(\rho\omega)}{\partial t} + \frac{\partial}{\partial x_j}(\rho u_j \omega) &= \frac{\partial}{\partial x_j}\left[\left(\mu + \frac{\mu_t}{\sigma_{\omega 3}}\right)\frac{\partial \omega}{\partial x_j}\right] \\ &+ (1 - F_1)2\rho\frac{1}{\sigma_{\omega 2}\omega}\frac{\partial k}{\partial x_j}\frac{\partial \omega}{\partial x_j} + \alpha_3\frac{\omega}{k}P_k - \beta_3\rho\omega^2 \end{aligned} \tag{23.13}$$

$$\mu_t = \rho\frac{a_1 k}{\max(a_1\omega, SF_2)} \tag{23.14}$$

where the coefficients with subscript 3 denote a linear combination of the corresponding coefficients in k-ε and k-ω models, such as α_3:

$$\alpha_3 = F_1\alpha_1 + (1 - F_1)\alpha_2 \tag{23.15}$$

with the blending functions F_1 and F_2:

$$F_1 = \tanh\left(\arg_1^4\right) \tag{23.16}$$

$$F_2 = \tanh\left(\arg_2^2\right) \tag{23.17}$$

$$\arg_1 = \min\left(\max\left(\frac{\sqrt{k}}{\beta'\omega y}, \frac{500v}{\omega y^2}\right), \frac{4\rho k}{CD_{k\omega}\sigma_{\omega 2}y^2}\right) \tag{23.18}$$

$$\arg_2 = \max\left(\frac{2\sqrt{k}}{\beta'\omega y}, \frac{500v}{\omega y^2}\right) \tag{23.19}$$

23.2.2 Vortex Identification Method

In the current paper, four methods covering three generations of vortex identification methods are employed to visualize the vortical flow inside a shaft-less rim-driven thruster, including vorticity method, Q-criterion [12], Ω method [4], and Liutex method [3]. As the first-generation method, vorticity method or mathematically velocity curl, is the most classical variable to describe the rotation of fluid:

$$\vec{\omega} = \nabla \times \vec{u} \tag{23.20}$$

As a modification method to Cauchy-Stokes decomposition, a second-generation method, namely Q-criterion [12], is proposed to identify vortex, which is represented by the second Galilean invariant of the velocity gradient tensor:

$$Q = \frac{1}{2}\left(\|B\|_F^2 - \|A\|_F^2\right) \tag{23.21}$$

where $\|\cdot\|_F$ stands for the Frobenius norm, A and B are the symmetric part and antisymmetric part of velocity gradient tensor.

To measure the proportion of rotating part to vorticity, Liu et al. [4] proposed a parament Ω, which is defined by a ratio as the relative vortex strength:

$$\Omega = \frac{\|B\|_F^2}{\|A\|_F^2 + \|B\|_F^2 + \varepsilon} \tag{23.22}$$

with

$$\varepsilon = 0.001\left(\|B\|_F^2 - \|A\|_F^2\right)_{\max} \tag{23.23}$$

To define the physical rigid-rotation part of the flow, Gao and Liu [5] utilize a Liutex-shear decomposition instead of the Cauchy-Stokes decomposition and formulate a three-generation method called Liutex vector:

$$\vec{R} = \left(\langle\vec{\omega}, \vec{r}\rangle - \sqrt{\langle\vec{\omega}, \vec{r}\rangle^2 - 4\lambda_{ci}^2}\right)\vec{r} \tag{23.24}$$

where $\vec{r}$ denotes the rotation axis of the local fluid, λ_{ci} represents the pseudo time-average angular velocity.

23.3 Numerical Setup

The bird and clip views of the shaft-less rim-driven thruster investigated herein are depicted in Fig. 23.2 along with special marks of main components (A: shell; B: hub; C: rear-stator; D: runner; E: rim). One observes that the gap between the rim and the shell is so small that the local shear effect is significant. Furthermore, under the effect of high-speed rotation, the laminar-turbulence transition inside the gap can be greatly promoted. Therefore, the identification of the gap vortical flow is a challenge.

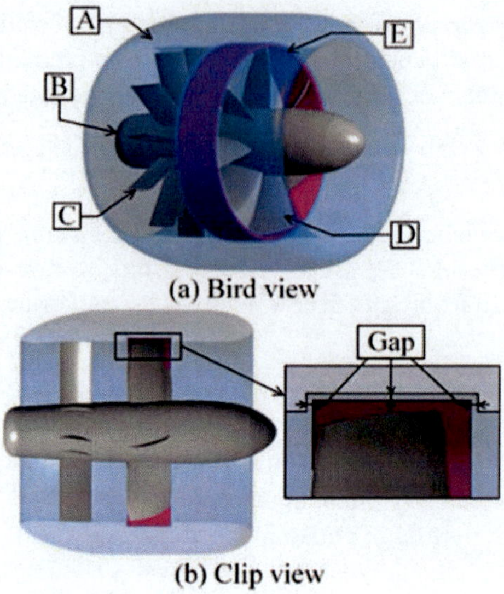

Fig. 23.2 Sketch map of shaft-less rim-driven thruster

Before the numerical investigation, we carried out a series of experiments on the shaft-less rim-driven thruster in China Ship Science Research Center (CSSRC). In the present paper, the experimental condition is strictly obeyed by constructing similar computational domain and boundary conditions shown in Fig. 23.3. Velocity inlet and pressure outlet boundaries are employed to model the inlet and the outlet. Given that the viscosity of the runner wall has little impact on the flow development inside the shaft-less rim-driven thruster, the runner wall is set as a free slip wall. Moreover, all other walls adopt the boundary condition of no-slip wall.

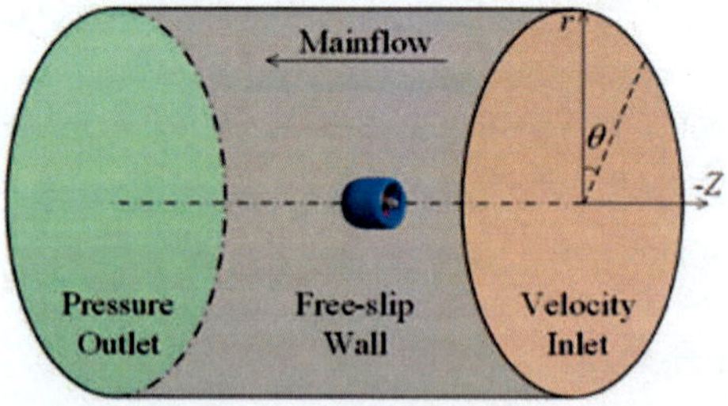

Fig. 23.3 Computational domain and boundary conditions

23.4 Results and Discussion

23.4.1 Validation of Numerical Results

In this section, two aspects are discussed to validate our numerical results, i.e. numerical framework and mesh resolution. A structural mesh with high quality is generated with special refinement in the vicinity of the gap, as shown in Fig. 23.4.

As mentioned above, a series of experiments were conducted to test the performance of this shaft-less rim-driven thruster, where three global parameters are measured, i.e. thrust coefficient, torque coefficient, and efficiency:

$$K_T = \frac{T}{\rho n^2 D^4} \tag{23.25}$$

$$K_Q = \frac{Q}{\rho n^2 D^5} \tag{23.26}$$

$$\eta = \frac{J}{2\pi}\frac{K_T}{K_Q} \tag{23.27}$$

where T denotes the blade trust, n represents rotate speed, D stands for the blade diameter, Q becomes the torque.

We first test the performance of three classic RANS models introduced in Sect. 23.2.1. As depicted in Fig. 23.5, the global parameters predicted with the SST k-ω model present the best agreement with the measured results, and its advantage is significant over the other two models. Therefore, the SST k-ω model is finally employed. Next, the sensitivity of mesh resolution is studied. Three meshes are generated, which share the same mesh topology but with different resolutions. Their total node numbers are 2.01, 3.31, and 5.57 million, respectively. Similarly, the numerical results predicted with different mesh resolutions are compared in Fig. 23.6. It can be observed that the results predicted with Mesh 2 are close to the measured data, and the results predicted with Mesh 3 cannot receive a significant improvement. Therefore, Mesh 2 is the final mesh adopted for our simulation.

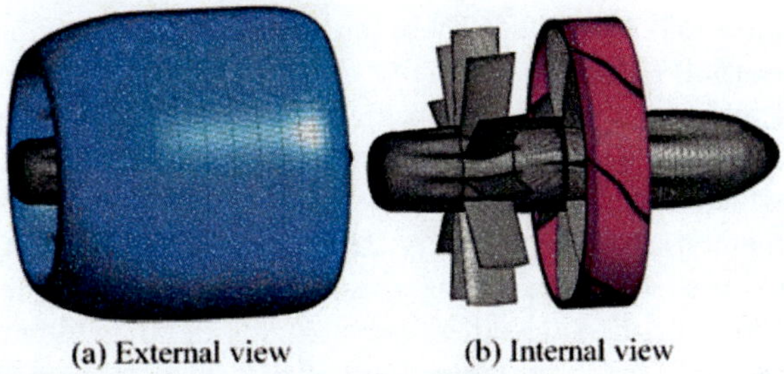

Fig. 23.4 Structural mesh generated around shaft-less rim-driven thruster

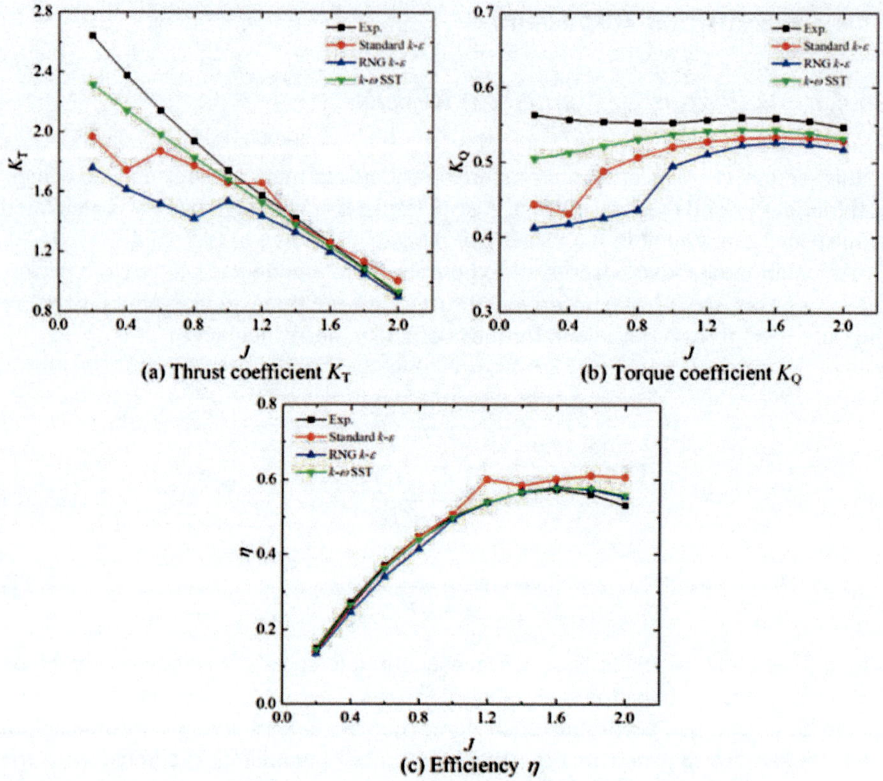

Fig. 23.5 Comparison of different RANS models in the prediction of global coefficients

23.4.2 Comparison of Vortex Identification Methods

As introduced above, the shear motion and the laminar-turbulence transition raise the complex gap flow, thus the identification of the local vortical flow is a challenge. In this section, four vortex identification methods (vorticity method, Q-criterion, Ω method, and Liutex method) are employed to visualize the gap vortical flow. Furthermore, the threshold selection of these methods is very critical, thus four thresholds are tested. These thresholds are linked with an initial threshold V_0, which is 1500 s^{-1} for the vorticity method, 10^4 s^{-2} for Q-criterion, and 100 s^{-1} for the Liutex method. In addition, given that the Ω method is a ratio parament, its thresholds are selected by referring to the vortex structure identified by the vorticity method, which are set to 0.4, 0.48, 0.49, and 0.5. The vortex structures identified by different methods are depicted in Fig. 23.7. It can be observed that the shear structures attached to the shell and the rim significantly pollute the identified vortical flow. Except for the Liutex method, the other three methods all fail to filter the shear layer while retaining the main vortex structure. Particularly, the vortex structure in the vicinity of the gap

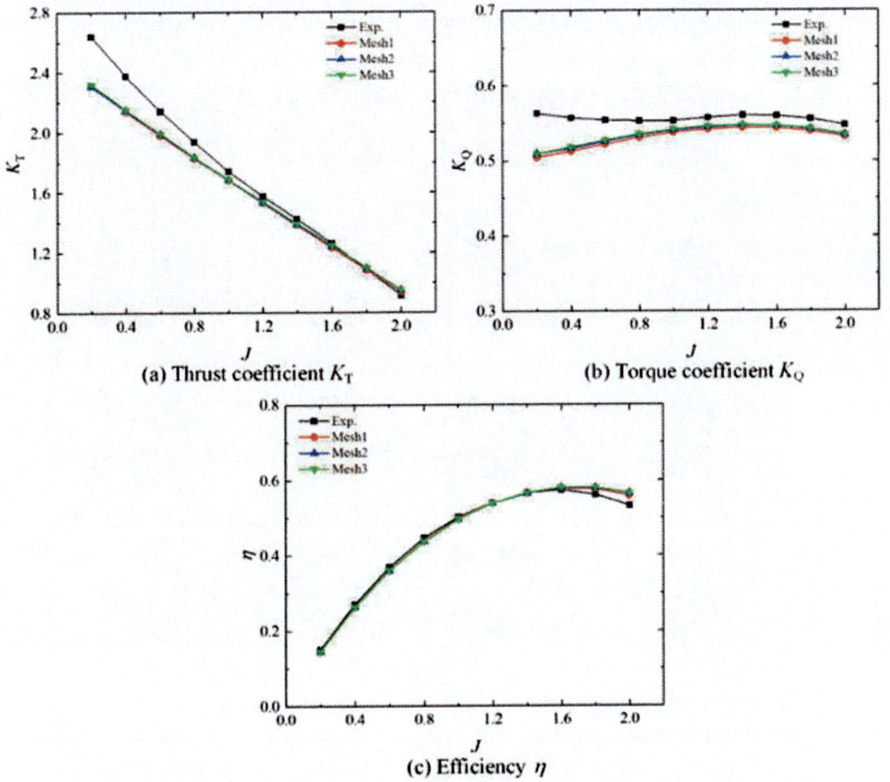

Fig. 23.6 Comparison of different mesh resolutions in the prediction of global coefficients

identified with $R = 2V_0$ can present an excellent effect. The above phenomenon indicates the great advantage of the Liutex method in the investigation of the flow mechanism inside a shaft-less rim-driven thruster.

23.4.3 Correlation Between Vortical Flow and Hydraulic Loss

For engineering design, the mixing loss between gap flow and mainstream is of great concern. This mixing loss is related to not only the leakage at the rim gap but also the velocity mismatching between the mainstream and the gap flow. Therefore, a combined analysis of vortical flow identified by the Liutex method and mixing loss measured with the entropy generation theory is conducted in this section.

The vortex structure near the gap is investigated by discussing the iso-surface of $R = 2V_0$, which is depicted in Fig. 23.8 together with special marks of four typical flow structures. Flow structure 1 is the gap flow between rim and shell, and

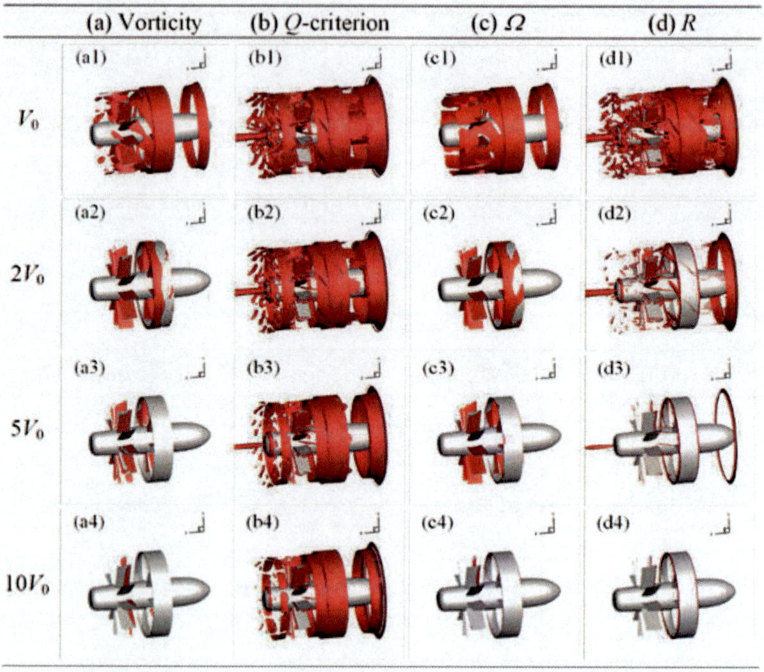

Fig. 23.7 Comparison of different vortex identification methods in identifying vortex flow in a shaft-less rim-driven thruster

flow structure 2 is the wake vortex detached from the rotor. Furthermore, powerful vortex structures, marked as 3 and 4, also exist in the upstream and the downstream gaps. To visualize the evolution process of these flow structures, several locations are especially selected in Fig. 23.9 to depict the contours of Liutex magnitude shown in Fig. 23.10. It can be observed that the upstream and downstream gap vortex 3 and 4 exist at all positions, while the wake vortex 2 only exists near the trailing edge of the blade. Moreover, when the wake vortex 2 is formed (Fig. 23.10f), some changes happen in the downstream gap vortex 3, indicating that the local flow separation can affect the gap flux. The gap flux is of great concern for the design of shaft-less rim-driven thruster. To visualize this effect, the gap streamlines along with the leakage flux distribution are depicted in Fig. 23.11. The direction of streamline is upstream, suggesting the backflow phenomenon inside the gap. One can also observe that the maximum of backflow flux always exists near the region of wake vortex formation. It indicates the flow separation of rotor is an important factor of flow instability, which can not only produce the wake vortex but also increase the leakage flux.

Based on the entropy production theory, further discussion concentrates on the mixing loss raised by the above flow structures. Assuming that the flow inside the shaft-less rim-driven thruster is incompressible and adiabatic, the hydraulic loss due to dispassion mainly consists of direct dissipation and turbulent dissipation [13, 14]:

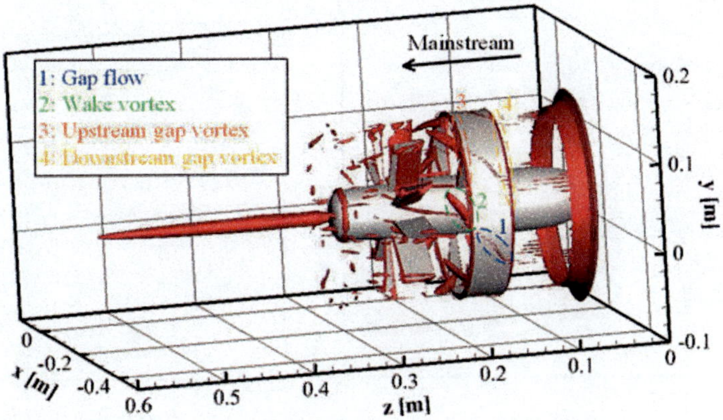

Fig. 23.8 Vortex structure identified by iso-surface of $|R| = 200\ \mathrm{s}^{-1}$

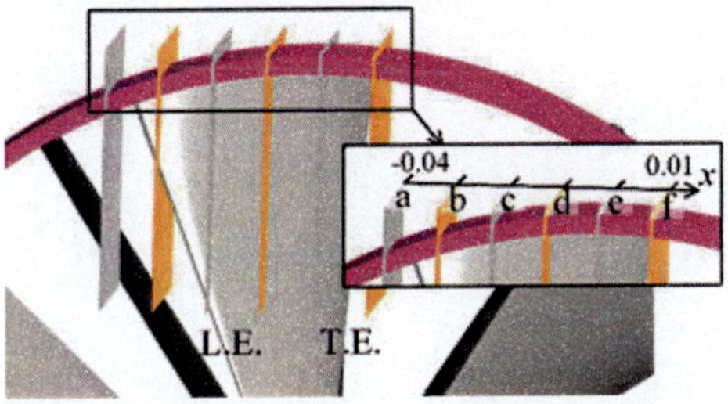

Fig. 23.9 Locations of slice planes

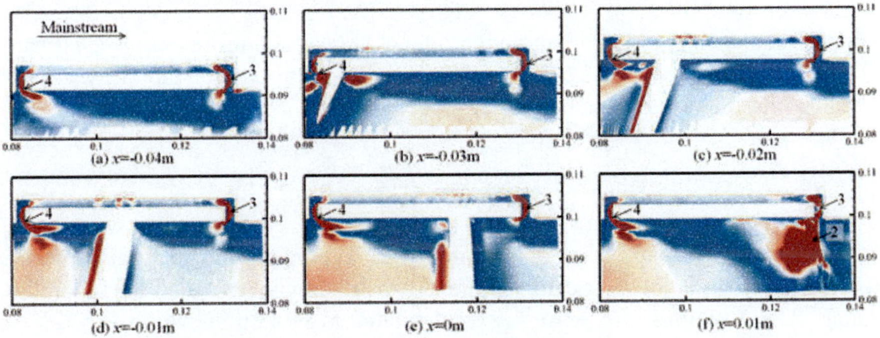

Fig. 23.10 Distribution of Liutex magnitude at different locations

$$\dot{S}_D''' = \dot{S}_{\overline{D}}''' + \dot{S}_{D'}''' \tag{23.28}$$

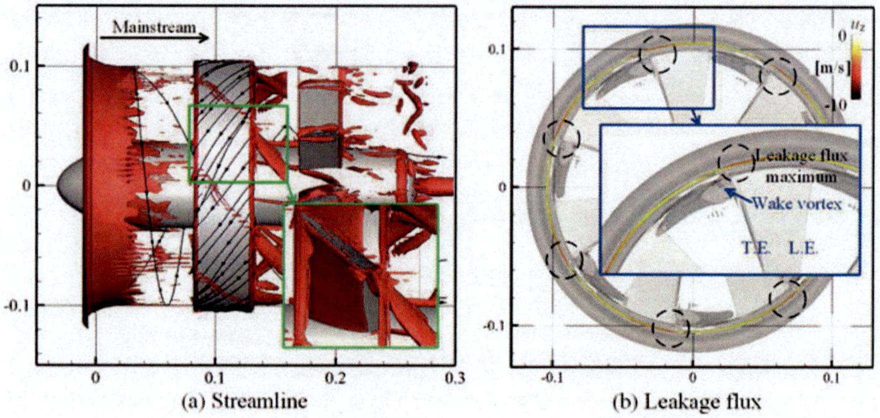

Fig. 23.11 Distribution of gap streamlines along with the leakage flux distribution

$$\dot{S}'''_{\overline{D}} = \frac{\mu_{eff}}{T}\left\{\begin{array}{l} 2\left[\left(\frac{\partial \overline{u}}{\partial x}\right)^2 + \left(\frac{\partial \overline{v}}{\partial y}\right)^2 + \left(\frac{\partial \overline{w}}{\partial z}\right)^2\right] \\ +\left(\frac{\partial \overline{u}}{\partial y} + \frac{\partial \overline{v}}{\partial x}\right)^2 + \left(\frac{\partial \overline{u}}{\partial z} + \frac{\partial \overline{w}}{\partial x}\right)^2 + \left(\frac{\partial \overline{v}}{\partial z} + \frac{\partial \overline{w}}{\partial y}\right)^2 \end{array}\right\} \tag{23.29}$$

$$\dot{S}'''_{D'} = \frac{\mu_{eff}}{T}\left\{\begin{array}{l} 2\left[\left(\frac{\partial u'}{\partial x}\right)^2 + \left(\frac{\partial v'}{\partial y}\right)^2 + \left(\frac{\partial w'}{\partial z}\right)^2\right] \\ +\left(\frac{\partial u'}{\partial y} + \frac{\partial v'}{\partial x}\right)^2 + \left(\frac{\partial u'}{\partial z} + \frac{\partial w'}{\partial x}\right)^2 + \left(\frac{\partial v'}{\partial z} + \frac{\partial w'}{\partial y}\right)^2 \end{array}\right\} \tag{23.30}$$

where the superscript $^{-}$ and $'$ denote the time-averaged and turbulent physical variable, respectively. For the SST k-ω model, the entropy production rate (EPR) due to turbulent dissipation $\dot{S}'''_{D'}$ can be further modeled as:

$$\dot{S}'''_{D'} = \beta\frac{\rho\omega k}{T} \tag{23.31}$$

Seeing from the distribution of EPR in Fig. 23.12, the mixing loss mainly occurs at three regions, i.e. the upstream gap, the downstream gap, and the flow separation region. These regions also exist powerful rotational motion shown in Fig. 23.10, which suggests the uneven distribution of local velocity. Deducing from Eqs. (23.29) and (23.30), this uneven distribution can raise the significant EPR or hydraulic loss. Furthermore, it should be noted that the distribution of EPR in the region where occurs the mixing of wake vortex and downstream gap vortex is especially significant. It indicates that the interaction between wake vortex and downstream gap vortex should be specially paid attention, because it can simultaneously aggravate the flux of backflow leakage and raise a significant hydraulic loss.

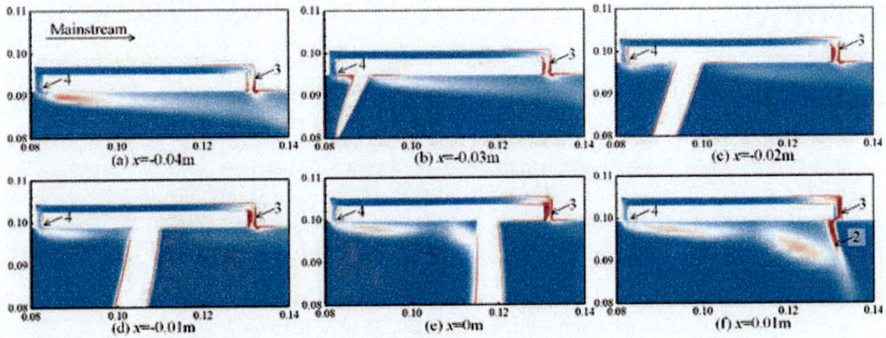

Fig. 23.12 Distribution of entropy production rate at different locations

23.5 Conclusion

In the current investigation, the phenomenon of backflow leakage in a shaft-less rim-driven thruster is numerically investigated. The flow field is simulated with the SST k-ω model and good agreement is received from the comparison between predicted and measured global parameters. Based on the results and discussion, the main conclusions can be written as follows:

1. The feasibility of different vortex identification methods (vorticity method, Q-criterion, Ω method, and the Liutex method) in the investigation of backflow leakage is compared. Only the Liutex method can filter the shear layer while retaining the main vortex structure.
2. With the assistance of the Liutex method, the main flow structures are identified, which are gap flow, wake vortex, upstream gap vortex, and downstream gap vortex. The upstream and the downstream gap vortices exist at all positions, while the wake vortex only exists near the trailing edge of the blade.
3. Hydrodynamic loss is mainly raised by wake vortex, upstream gap vortex, and downstream gap vortex. Great attention should be paid to the mixing of wake vortex and downstream gap vortex, because it can aggravate the flux of backflow leakage and raise a significant hydraulic loss.

Acknowledgements This work was financially supported by the National Natural Science Foundation of China (Project nos. 52176041 and 12102308) and China Postdoctoral Science Foundation (Project no. 2020M682471). The numerical calculations in this paper were done on the supercomputing system in the Supercomputing Center of Wuhan University.

References

1. S.M.A. Sharkh, S.H. Lai, S.R. Turnock, Structurally integrated brushless PM motor for miniature propeller thrusters. IEE Proc. - Electric Power Appl. **151**, 513 (2004)
2. X. Yan, X. Liang, W. Ouyang, Z. Liu, B. Liu, J. Lan, A review of progress and applications of ship shaft-less rim-driven thrusters. Ocean Eng. **144**, 142 (2017)
3. Y. Gao, C. Liu, Rortex and comparison with eigenvalue-based vortex identification criteria. Phys. Fluids **30**, 85107 (2018)
4. C. Liu, Y. Wang, Y. Yang, Z. Duan, New omega vortex identification method. Sci. China-Phys. Mech. Astron. **59**, 684711 (2016)
5. C. Liu, Y. Gao, S. Tian, and X. Dong, Rortex—A new vortex vector definition and vorticity tensor and vector decompositions. Phys. Fluids **30**, 35103 (2018)
6. X. Bai, H. Cheng, B. Ji, X. Long, Z. Qian, X. Peng, Comparative study of different vortex identification methods in a tip-leakage cavitating flow. Ocean Eng. **207**, 107373 (2020)
7. Y. Wang, W. Zhang, X. Cao, H. Yang, The applicability of vortex identification methods for complex vortex structures in axial turbine rotor passages. J. Hydrodyn. **31**, 700 (2019)
8. N. Gui, H. Qi, L. Ge, P. Cheng, H. Wu, X. Yang, J. Tu, S. Jiang, Analysis and correlation of fluid acceleration with vorticity and Liutex (Rortex) in swirling jets. J. Hydrodyn. **31**, 864 (2019)
9. B.E. Launder, D.B. Spalding, *Lectures in Mathematical Models of Turbulence* (Academic, London, 1972)
10. V. Yakhot, and S.A. Orszag, Renormalization group analysis of turbulence I. Basic theory. J. Sci. Comput. **1**, 1 (1986)
11. F.R. Menter, Two-equation eddy-viscosity turbulence models for engineering applications. AIAA J. **32**, 1598 (1994)
12. J.C.R. Hunt, A.A. Wray, P. Moin, *Studying Turbulence Using Numerical Simulation Databases* (1988)
13. F. Kock, and H. Herwig, Local entropy production in turbulent shear flows: a high-Reynolds number model with wall functions. Int. J. Heat Mass Transfer **47**, 2205 (2004)
14. B. Schmandt, H. Herwig, Internal flow losses: a fresh look at old concepts. J. Fluids Eng.-Trans. Asme **133**, 51201 (2011)

Chapter 24
Application of Omega Identification Method in the Ventilated Cavities Around a Surface-piercing Hydrofoil

Yuchang Zhi, Rundi Qiu, Renfang Huang, and Yiwei Wang

Abstract This article summarizes a simulation campaign to study atmospheric ventilation flow patterns, forces, and stability at various speeds. Simulations are based upon cited experiments, and results show good agreement for global forces, cavity shapes, etc. The hydrofoil deceleration process includes three different flow regimes: fully ventilated flow, fully wetted flow, and partially ventilated flow. To obtain the evolution process of the vortex structure, the vortex structure is visualized using the Omega method. The results show that the evolution of the vortex structure mainly includes three types of vortex structures, i.e., a tip vortex, a Karman vortex, and an unstable vortex. The vortex structure is related to the ventilated cavity. Omega can identify weak vortices and strong vortices.

Y. Zhi
School of Aeronautics and Astronautics, Sun Yat-Sen University, Guangzhou 510275, China
e-mail: zhiyuchang0129@163.com

R. Qiu · R. Huang (✉) · Y. Wang
Key Laboratory for Mechanics in Fluid Solid Coupling Systems, Institute of Mechanics, Chinese Academy of Sciences, Beijing, China
e-mail: hrenfang@imech.ac.cn

R. Qiu
e-mail: qiurundi19@mails.ucas.edu.cn

Y. Wang
e-mail: wangyw@imech.ac.cn

R. Qiu · Y. Wang
School of Future Technology, University of Chinese Academy of Sciences, Beijing, China

Y. Wang
School of Engineering Science, University of Chinese Academy of Sciences, Beijing, China

Y. Wang et al. (eds.), *Liutex and Third Generation of Vortex Identification*,
Springer Proceedings in Physics 288, https://doi.org/10.1007/978-981-19-8955-1_24

24.1 Introduction

Hydrodynamic devices operating near the free surface will generate ventilation [1–4] due to the different densities of water and air fluids. Ventilation will affect the hydrodynamic performance of the hydrofoil, resulting in instability of the hull. It is generally believed that flow separation is the cause of ventilation. The formation mechanism of ventilation and the disappearance mechanism of ventilation have attracted the attention of many researchers.

A large number of scholars have carried out drag experiments on ventilation [5–7], they found that the surface-piercing hydrofoil ventilation formation mechanism mainly includes tip-vortex induced formation, stall-induced formation, cavitation-induced formation, perturbation-induced formation, and tail ventilation formation. The formation mechanism of the ventilated cavity promotes the development and stability of the cavity. Young et al. [8, 9] and Harwood et al. [5, 10] found that the elimination mechanism of the ventilated cavity mainly includes washout and re-wetting. The re-entrant jet affects the ventilated cavity stability. Huang et al. [11] carried out the ventilation experiment of a surface-piercing hydrofoil under the high Froude number, and the results showed that the hydrofoil leading edge and the hydrofoil tip had natural cavitation under the high Froude number, and the natural cavitation promoted the formation of ventilation. Zhi et al. [12] used CFD for the first time to simulate the evolution of the ventilated cavity and verified that the re-entrant jet affects the ventilated cavity stability. Numerical results are used to explore details of the flow, including re-entrant jet dynamics and vorticity fields to complement the findings of prior experimental work, the calculated evolution of the ventilated cavity is in good agreement with the experiment. A great deal of research has also been done using CFD to simulate the surface-piercing hydrofoil ventilated cavity [13, 14]. The results show that the CFD can accurately obtain the ventilation cavity and underwater vortex structure of the hydrofoil.

Past studies have shown that the vortex structure has a great impact on cavitation flow. Identifying and analyzing the vortex structure is of great importance to predict its dynamic behavior. The past vortex structure definition methods have certain limitations. For example, the selection of the threshold of the Q criterion [15] in practical applications needs to be adjusted according to different problems, and even the same problem must be adjusted at different times. In the process of using these vortex identification methods, how correctly selecting the threshold becomes the key to vortex structure identification. A new vortex structure identification method has been developed to address the previous problems with vortex structure identification, Liu et al. [16, 17] have proposed the vortex identification method after a series of experiments. Compared with the previous methods, the new vortex structure identification method can decompose the vorticity into two parts: rotating and non-rotating. The new vortex structure identification method can identify strong and weak vortices without artificial adjustment of the threshold.

Liutex method [18] can identify the formation and development process of the U-shaped vortex structure, and at the same time can display the details of the vortex structure shedding during the cavity shedding process. Zhang et al. [19] compared the visualization results of the Omega method with others, and found that the Omega method does not require manual adjustment of the threshold, a fixed threshold of ω = 0.52 can be used, and the Omega vortex identification method can identify strong and weak vortices.

Due to the hydrofoil ventilation, the flow field structure is more complicated, and the vortex structure and the ventilated cavity have a great influence on each other. The main components of the vortex structure around the hydrofoil should be analyzed using a more advanced vortex structure identification method, and to further obtain the relationship between the vortex structure and the ventilated cavity.

24.2 Numerical Model and Computation Setting

24.2.1 Governing Equations

To solve a multiphase flow, continuity and momentum equations be used:

$$\frac{\partial \rho}{\partial t} + \frac{\partial (\rho u_i)}{\partial x_i} = 0 \tag{24.1}$$

$$\frac{\partial (\rho u_i)}{\partial t} + \frac{\partial (\rho u_i u_j)}{\partial x_j} = -\frac{\partial p}{\partial x_i} + \frac{\partial}{\partial x_j}\left(\mu \frac{\partial u_i}{\partial x_j}\right) \tag{24.2}$$

where ρ indicates the density of the mixture, u_i represents velocity in the i direction, μ represents flow viscosity, and p represents pressure.

In this paper, to obtain a more accurate ventilated cavity, the standard $k-\varepsilon$ turbulence [20] model is applied. For kinetic energy and turbulent dissipation rate ε:

$$\frac{\partial}{\partial t}(\rho k) + \nabla \cdot (\rho k \overline{u}) = \nabla \cdot \left[\left(\mu + \frac{\mu_t}{\sigma_k}\right)\nabla k\right] + P_k - \rho(\varepsilon - \varepsilon_0) + S_k \tag{24.3}$$

$$\frac{\partial}{\partial t}(\rho \varepsilon) + \nabla \cdot (\rho \varepsilon \overline{u}) = \nabla \cdot \left[\left(\mu + \frac{\mu_t}{\sigma_\varepsilon}\right)\nabla \varepsilon\right] + \frac{1}{T_e} C_{\varepsilon 1} P_\varepsilon - C_{\varepsilon 2} f_2 \rho\left(\frac{\varepsilon}{T_e} - \frac{\varepsilon_0}{T_0}\right) + S_\varepsilon \tag{24.4}$$

where f_2 denotes a damping function, P_k and P_ε are production terms, σ_k, σ_ε, $C_{\varepsilon 1}$, and $C_{\varepsilon 2}$ denote model coefficients, $T_e = k/\varepsilon$ is the large-eddy time scale, and S_k and S_ε are user-specified source terms. Based on the source terms, T_0 is a specific time scale:

$$T_0 = \max\left(\frac{k_0}{\varepsilon_0}, C_t \sqrt{\frac{\nu}{\varepsilon_0}}\right) \tag{24.5}$$

where ν is the kinematic viscosity, ε_0 is the ambient turbulence value [21], with C_t being a model coefficient.

24.2.2 Vortex Identification Methods

The Omega method [16] can be defined as follow:

$$\Omega = \frac{b}{a + b + \varepsilon} \tag{24.6}$$

To visualize the vortex structure, choose $\Omega = 0.52$ as the parameter. ε is an empirical function about α and β:

$$\varepsilon = c \times (b - a)_{\max} \tag{24.7}$$

where c is a small positive number. The values of a and b are:

$$a = trace(\mathbf{A}^T\mathbf{A}) \tag{24.8}$$

$$b = trace(\mathbf{B}^T\mathbf{B}) \tag{24.9}$$

24.2.3 Numerical Methods

The research object was a symmetrical hydrofoil, as shown in Fig. 24.2. Taking the intersection of the surface-piercing hydrofoil and the free surface as the origin, the range of each coordinate axis of the calculation domain (x along the hydrofoil chord length, y along the hydrofoil width, and z along the hydrofoil span length) is $-4.0c < x < 15.0c$, $-4.0c < y < 4.0c$, and $-5.0c < z < 4.0c$, as shown in Fig. 24.1. In the diagram of Fig. 24.1, the velocity inlet and the pressure outlet are set as the inlet and outlet conditions, respectively, and the other boundaries are set as walls. The quality of the grid directly affects the convergence and divergence of the calculation and the accuracy of the results. Firstly, the hydrofoil model is reconstructed, the hydrofoil surface mesh is processed and surface mesh with triangulation is generated, and a boundary layer is generated based on the surface mesh and trimmed mesh volume mesh, the mesh distribution is shown in Fig. 24.2. The present numerical approach has been described in detail in our previous work [12], and it is proved to be able to accurately capture the unsteady ventilated cavities when compared with the experiments.

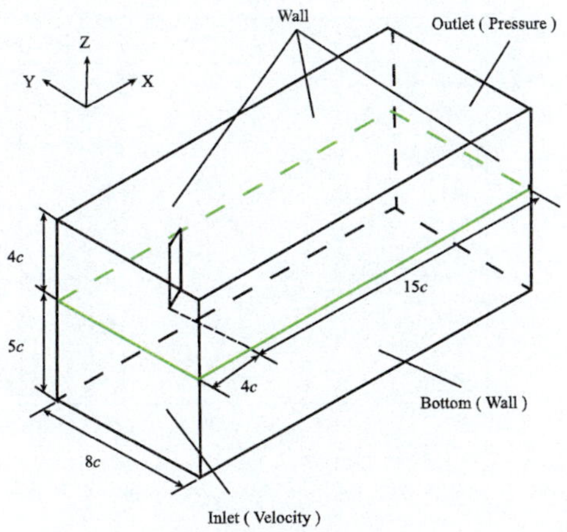

Fig. 24.1 Computational domain and boundary condition of the surface-piercing hydrofoil

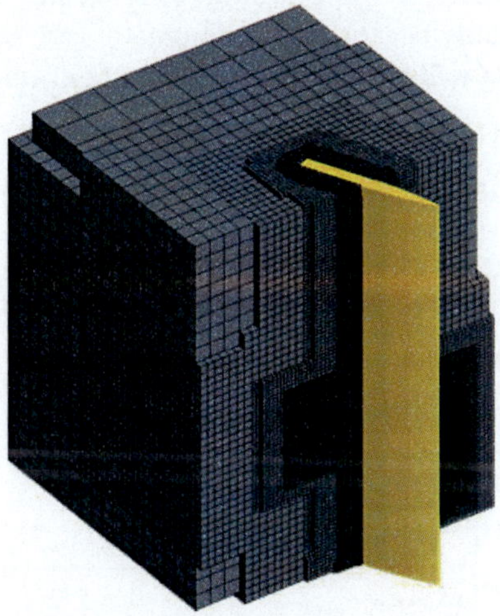

Fig. 24.2 Mesh around the surface-piercing hydrofoil

24.3 Results and Discussion

The calculation conditions are consistent with the experiments. First, the surface-piercing hydrofoil sails at a constant speed, and the hydrofoil forms a fully ventilated (FV) flow. Then the hydrofoil sails at a reduced speed, and the hydrofoil forms a partially ventilated (PV) flow. When the hydrofoil speed reaches $F_{nh} = 0.309$, the hydrofoil forms a fully wetted (FW) flow. The hydrofoil speed and surrounding flow regime are shown in Fig. 24.3.

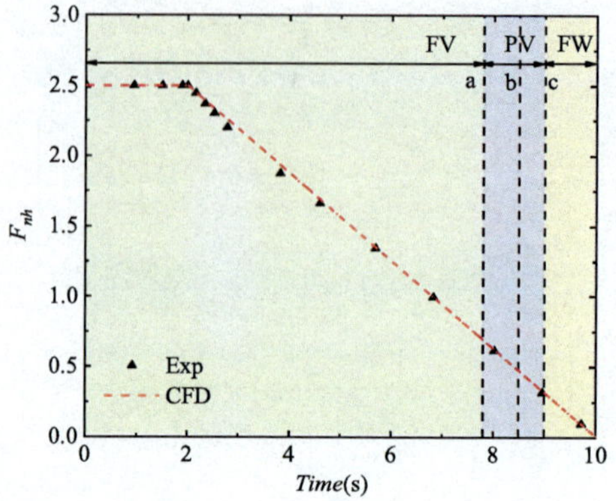

Fig. 24.3 Hydrofoil speed curve

Figure 24.4 shows that the calculated results agree well with the experiment [5]. The hydrodynamic performance of hydrofoils is related to speed. When $F_{nh} < 0.5$, the lift error is large, mainly because the denominators are so small, any influencing factors in the calculation will be amplified, leading to erroneous results for lift and coefficient.

Figure 24.5 shows the evolution of the ventilated cavity. The hydrofoil deceleration process includes three different flow regimes: fully ventilated flow, fully wetted flow, and partially ventilated flow.

Figures 24.5(a-e) shows examples of a fully ventilated cavity. It can be seen from the figure that the ventilated cavity covers the entire leading edge of the hydrofoil. The ventilated cavity is stable, mainly due to the inability of the re-entrant jet to reach the leading edge of the hydrofoil.

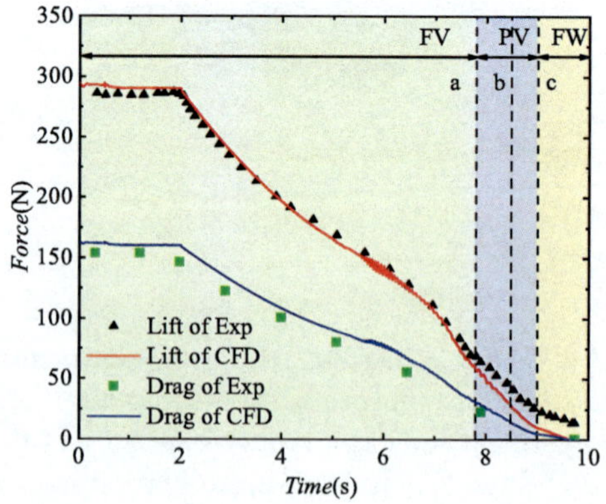

Fig. 24.4 Lift and drag of the hydrofoil

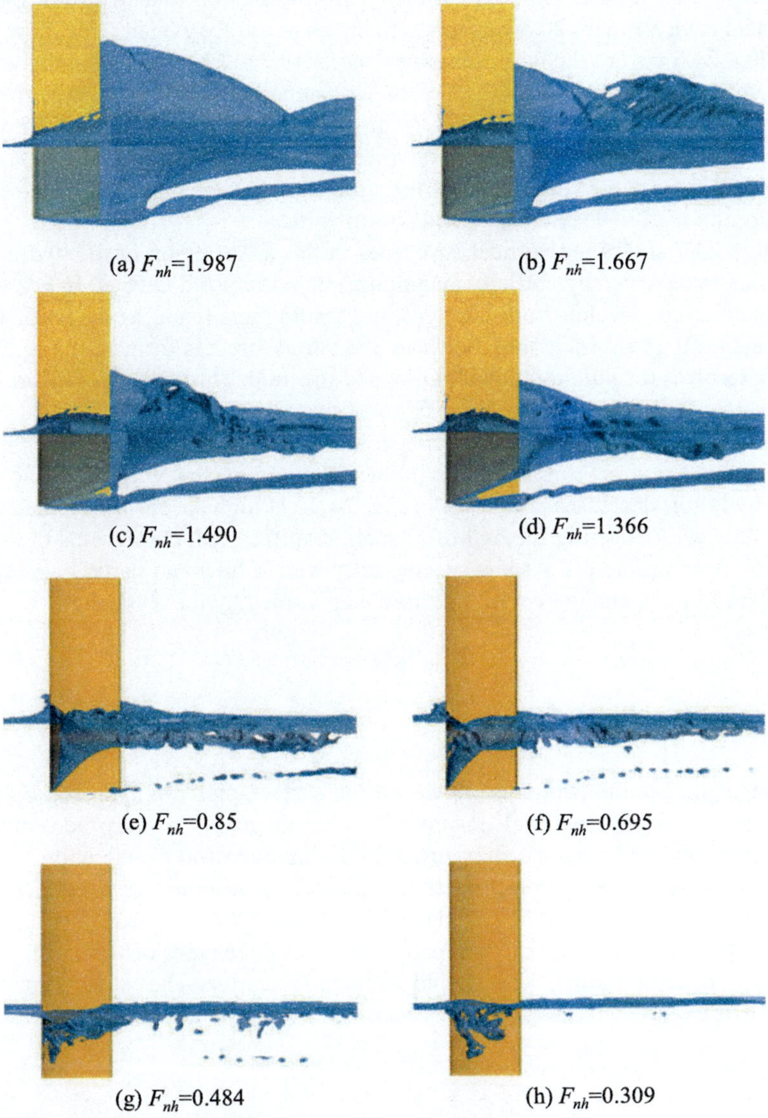

Fig. 24.5 Evolution of the ventilated cavity of the hydrofoil

Figures 24.5(f-g) depicts two examples of partial ventilation. The ventilated cavity cannot cover the entire leading edge of the hydrofoil. At this time, the ventilated cavity is not stable, mainly due to the re-entrant jet acting on the leading edge of the hydrofoil.

Fully wetted flow is denoted by the hydrofoil suction side with no apparent ventilated cavity. Figure 24.5(h) shows examples of a fully wetted (FW) flow.

In Fig. 24.6 we use the isosurfaces of contour $\Omega = 0.52$ to represent the vortical structures around the hydrofoil. The underwater complex vortex structures are obtained by the Omega method. The results show that the evolution of the vortex structure mainly includes three types of vortex structures, i.e., a tip vortex, a Karman vortex, and an unstable vortex. The vortex structure is related to the ventilated cavity. Omega can identify weak vortices and strong vortices.

Figure 24.7 shows the vortical structures in the wake region of the hydrofoil at z/h = 0.5, represented by vorticity magnitude and normalized Omega. In Fig. 24.7a, two shear layers develop from the hydrofoil trailing edge, and as the shear layers alternately roll up and shed into the wake, the vortex street is formed. In Fig. 24.7b, Omega resolves the pure rotational motions of the fluid. Through comparison, it can be found that Omega can identify weak vortices and strong vortices.

Figure 24.8 shows vortical structures contours at z/h = 0.5, colored by Omega. A Kaman vortex street can be seen in the wake region of the hydrofoil as the speed of the hydrofoil decreases, as shown in Fig. 24.9. At high speeds, this is due to the ventilated cavity covering the hydrofoil suction surface, which prevents the vortex structure from falling off. At low speeds, fully wetted flow and partially ventilated flow were formed, and the vortex structure was formed by the shedding.

24.4 Conclusions

This paper studies the two-phase flow past the surface-piercing hydrofoil by using the VOF method and k-e turbulence model. Our main goal is to study the ventilation flow regime and underwater vortex structure during hydrofoil deceleration.

The hydrofoil deceleration process includes three different flow regimes: fully ventilated flow, fully wetted flow, and partially ventilated flow. The results show that the evolution of the vortex structure mainly includes three types of vortex structures, i.e., a tip vortex, a Karman vortex, and an unstable vortex. The vortex structure is related to the ventilated cavity. Omega can identify weak vortices and strong vortices.

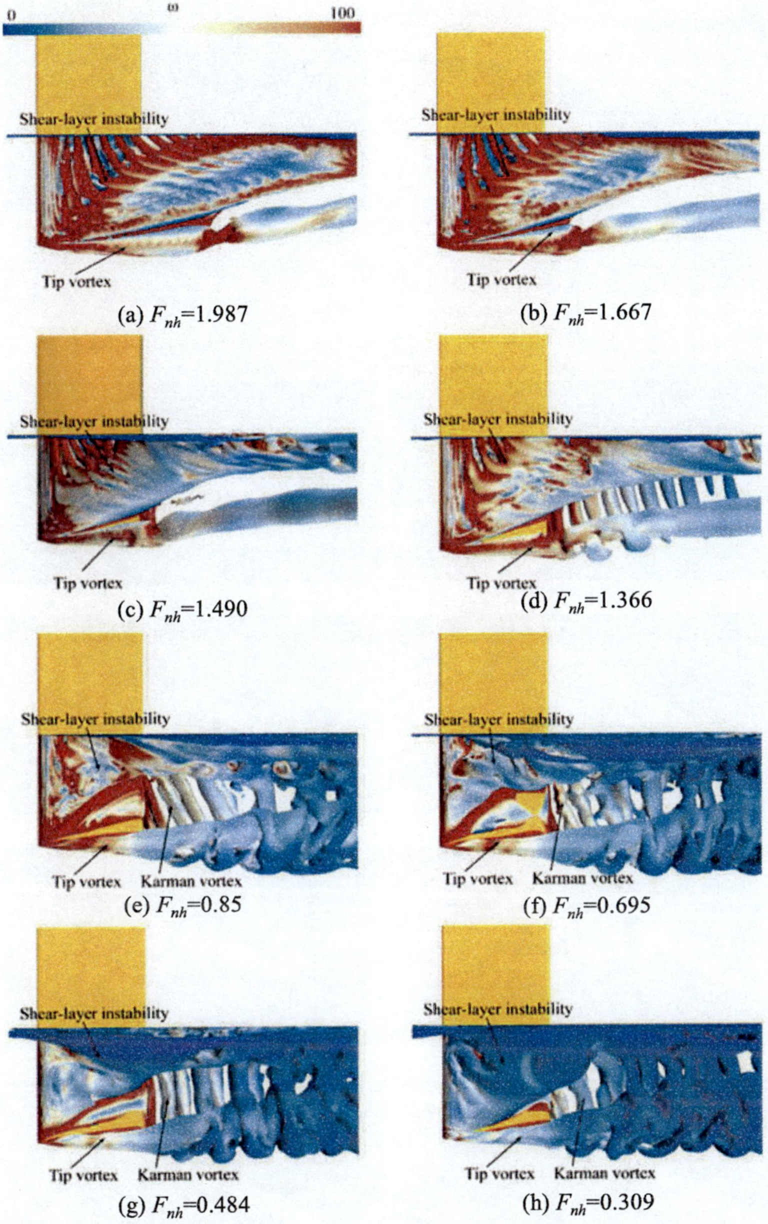

Fig. 24.6 Evolution of vortex structures. Isosurfaces are extracted by contour $\Omega = 0.52$ and colored by vorticity magnitude

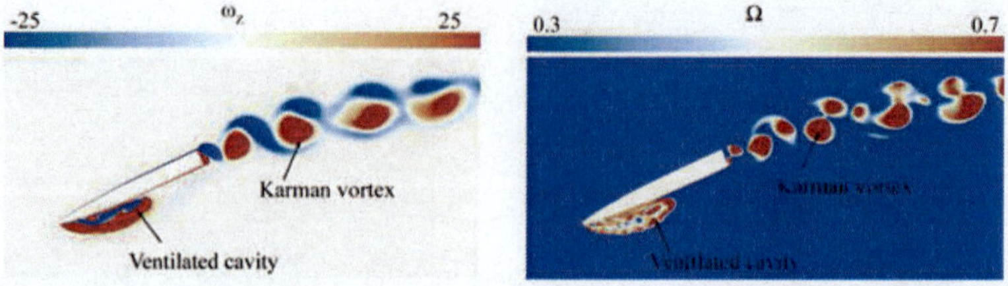

(a) Vorticity (b) Omega

Fig. 24.7 Vortical structures contours at z/h = 0.5

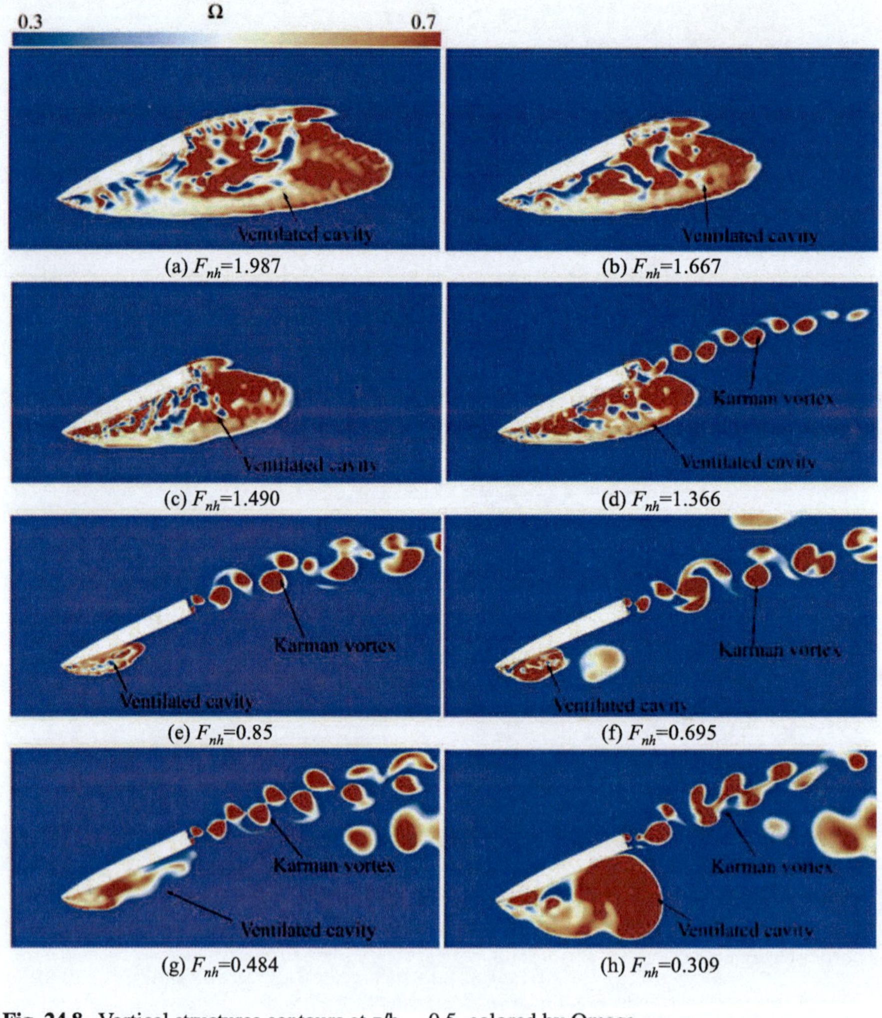

Fig. 24.8 Vortical structures contours at z/h = 0.5, colored by Omega

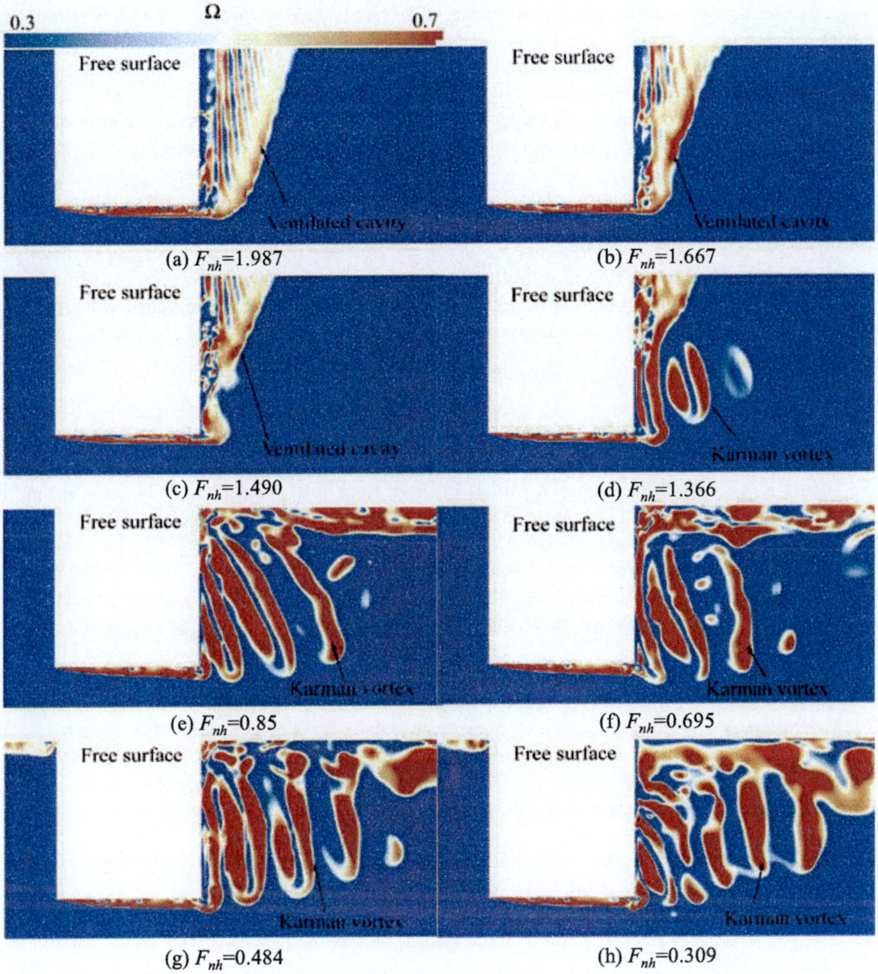

Fig. 24.9 Instantaneous vortical structures contours at y/h = 0, colored by Omega

Acknowledgements This work is accomplished using the code Omega-LiutexUTA, released by Chaoqun Liu at the University of Texas at Arlington. The authors would like to acknowledge gratefully the National Natural Science Foundation of China (No. 52006232, 12122214).

References

1. A. Acosta, Hydrofoils and hydrofoil craft. Annu. Rev. Fluid Mech. **5**, 161–184 (1973)
2. E.P. Rood Jr, Turning maneuver limitations imposed by sudden strut side ventilation on a 200-Ton 80-Knot Hydrofoil Craft Document No. Number (1975)

3. A.M. Kozlowska, S. Steen, K. Koushan, *Classification of Different Types of Propeller Ventilation and Ventilation Inception Mechanism* (Citeseer, 2009)
4. A. Yousefi, R. Shafaghat, Numerical study of the parameters affecting the formation and growth of ventilation in a surface-piercing propeller. Appl. Ocean Res. **104**, 102360 (2020)
5. C.M. Harwood, K.A. Brucker, F. Miguel, Y.L. Young, S.L. Ceccio, *Experimental and numerical investigation of ventilation inception and washout mechanisms of a surface-piercing hydrofoil* (2014)
6. R.S. Rothblum, Investigation of methods of delaying or controlling ventilation on surface piercing struts, University of Leeds (1977)
7. C. Harwood, F. Miguel Montero, Y.L. Young, S. Ceccio, *Experimental Investigation of Ventilation of a Surface Piercing Hydrofoil* (2013)
8. Y.L. Young, C.M. Harwood, F. Miguel Montero, J.C. Ward, S.L. Ceccio, Ventilation of lifting bodies: review of the physics and discussion of scaling effects. Appl. Mech. Rev. **69** (2017)
9. Y.L. Young, S. Brizzolara, J. Binns, R. Brown, N. Bose, *Numerical and physical investigation of a surface-piercing hydrofoil* (2013)
10. C.M. Harwood, Y.L. Young, S.L. Ceccio, Ventilated cavities on a surface-piercing hydrofoil at moderate Froude numbers: cavity formation, elimination, and stability. J. Fluid Mech. **800**, 5–56 (2016)
11. Huang R, Qiu R, Zhi Y, Wang Y (2022) Investigations into the ventilated cavities around a surface-piercing hydrofoil at high Froude numbers. Phys. Fluids **34**, 043304 (2022)
12. Y. Zhi, J. Zhan, R. Huang, R. Qiu, Y. Wang, Numerical investigations into the ventilation elimination mechanism of a surface-piercing hydrofoil. Ocean Eng. **243**, 110225 (2022)
13. Y. Wang, C. Huang, T. Du, R. Huang, Y. Zhi, Y. Wang, Z. Xiao, Z. Bian, Research on ventilation and supercavitation mechanism of high-speed surface-piercing hydrofoil. Phys. Fluids **34**, 023316 (2022)
14. K.I. Matveev, M.P. Wheeler, T. Xing, Numerical simulation of air ventilation and its suppression on inclined surface-piercing hydrofoils. Ocean Eng. **175**, 251–61 (2019)
15. Y.-N. Zhang, X.-Y. Wang, C. Liu, Comparisons and analyses of vortex identification between Omega method and Q criterion. J. Hydrodyn. **31**, 224–30 (2019)
16. C. Liu, Y.-S. Gao, X.-R. Dong, Y.-Q. Wang, J.-M. Liu, Y.-N. Zhang, X.-S. Cai, N. Gui, Third generation of vortex identification methods: Omega and Liutex/Rortex based systems. J. Hydrodyn. **31**, 205–223 (2019)
17. Y. Yu, P. Shrestha, O. Alvarez, C. Nottage, C. Liu, Investigation of correlation between vorticity, Q, λci, $\lambda 2$, Δ and Liutex. Comput. Fluids **225**, 104977 (2021)
18. J. Chen, B. Huang, T. Liu, Y. Wang, G. Wang, Numerical investigation of cavitation-vortex interaction with special emphasis on the multistage shedding process. Appl. Math. Model. **96**, 111–30 (2021)
19. Y.-H. Zhang, X.-J. Hu, W. Lan, Y.-C. Liu, M. Wang, J.-Y. Wang, Application of Omega vortex identification method in cavity buffeting noise. J. Hydrodyn. **33**, 259–70 (2021)
20. Y. Zhi, Z. Zhang, R. Huang, R. Qiu, Y. Wang, Numerical investigations into supercavitating flows and hydrodynamic characteristics of a heaving hydrofoil. Mod. Phys. Lett. B 2150605 (2022)
21. P.R. Spalart, C.L. Rumsey, Effective inflow conditions for turbulence models in aerodynamic calculations. AIAA J. **45**, 2544–53 (2007)

Chapter 25
Application of Third Generation of Vortex Identification for Straight-Bladed Vertical Axis Wind Turbines

Guoqiang Tong, Zhenjun Xiao, Shengbing Yang, Fang Feng, Yan Li, and Kotaro Tagawa

Abstract The Straight-Bladed Vertical Axis Wind Turbine (SB-VAWT) has a high share in the small-scale wind turbine market with the advantages of its simple structure, low noise, smart forms, as well as its suitability for off-grid operation. Currently, the performance of the SB-VAWT is lower than that of Horizontal Axis Wind Turbines (HAWTs). One of the main reasons is the complexity of the SB-VAWT flow field. Accurate identification of vortex in the flow field is critical because an effective means of improving the performance of vertical axis wind turbines is to optimize the vortex. In this study, a four-bladed SB-VAWT with rotor diameter, blade chord length, and blade height of 0.6 m, 0.125 m, and 0.5 m, respectively, is used as the object of study. The Ω method, Liutex, λ_2 criterion, and Q criterion are used to identify the vortex of the SB-VAWT flow field, respectively, when the tip speed ratio is 1.0. The results show that the vortex structure of the SB-VAWT flow field can be obtained by all four methods. Liutex can identify more small-scale vortices, which is not available in other methods. This paper provides a reference for the selection of vortex identification methods for SB-VAWT.

25.1 Introduction

The wind turbine is one of the most important components of the development and utilization of wind energy. The wind turbine could convert natural wind energy into electrical, thermal, and mechanical energy [1]. They are classified as Vertical Axis Wind Turbines (VAWTs) and Horizontal Axis Wind Turbines (HAWTs) based on

G. Tong · Z. Xiao · S. Yang · Y. Li (✉)
Engineering College, Northeast Agricultural University, Harbin 150030, China
e-mail: liyanneau@163.com

F. Feng
College of Arts and Sciences, Northeast Agricultural University, Harbin 150030, China

K. Tagawa
Faculty of Agriculture, Tottori University, Tottori 6808551, Japan

Y. Wang et al. (eds.), *Liutex and Third Generation of Vortex Identification*, Springer Proceedings in Physics 288, https://doi.org/10.1007/978-981-19-8955-1_25

the location of the rotating axis relative to the ground. At present, the HAWT has been a mainstream type of large-scale wind turbine, resulting from years of research. In contrast, the VAWT is mostly employed in small-scale wind energy systems. Currently, the performance of VAWTs is lower than that of HAWTs. Even so, the VAWT has some advantages that the HAWT does not have, such as simple structure, low noise, easy installation and maintenance, smart forms, as well as its suitability for off-grid operation. Straight-bladed Vertical Axis Wind Turbines (SB-VAWTs), a type of lift-type VAWT, have received the attention of most researchers of small-scale wind energy systems around the world.

Despite the obvious advantages of the SB-VAWT, the lower aerodynamic performance is the critical problem that limits its development. Researchers have conducted a lot of studies to investigate this problem. Zhu et al. found that the flow-deflecting-gap blade effectively improves or eliminates the dynamic stall of SB-VAWT during rotation. The stall angle of attack of the blade with flow-deflecting-gap is increased by 2° compared to the normal blade. The lift-to-drag ratio is improved at high angles of attack, and the frequency of vortex shedding is reduced, reducing the lift oscillation amplitude [2]. Tong et al. concluded that deformed blades are one of the most effective ways to increase the lift of a blade, which can increase the maximum lift of a blade by nearly 60% [3]. Hao et al. found that adaptive flap is an effective method to mitigate SB-VAWT flow separation at high solidity. The adaptive flaps will automatically raise to block the return flow during separation return, alleviating the flow separation problem and increasing the blade aerodynamic force. However, in the case of a high tip speed ratio, long flaps have a negative effect since they cannot be retracted in time, while short flaps near the leading edge of the blades will avoid this problem [4]. This conclusion was also confirmed by Syawitri [5]. A study of winglets by Xu et al. found that the installation of blade winglets will reduce the influence of the vortex at the top of the blade on the span direction, thus improving the performance of the blade [6]. The study of auxiliary devices found that the Wind-capture-accelerate device [7] and Wind Gathering Device (WGD) [8] have a significant improvement on the performance of SB-VAWT. The WGD, which can be installed above and below the rotor, improves the aerodynamic performance while also having an improved effect on the vortex at the tip of the blade.

All of the above are better explored by researchers for improving the aerodynamic performance of SB-VAWT, and they have achieved the objective to different degrees. However, to fundamentally improve the aerodynamic characteristics of SB-VAWT, the flow characteristics of SB-VAWT must be considered. The operating mode of the SB-VAWT result in an extremely complex flow field. The vortex is a typical structure of unsteady flow inside the SB-VAWT, resulting in pressure fluctuations, complex operating conditions of the downstream blades, and vibration of the rotor. During the operation of the SB-VAWT, it is always accompanied by vortex generation, shedding, and broken by blade impact, which affects the aerodynamic performance of the SB-VAWT. The optimization of the vortex in the flow field is an effective method to improve the performance of SB-VAWT. The identification of the vortex structure in the flow field is an essential prerequisite.

Currently, there are many methods of vortex identification have been developed. The Q criterion and λ_2 criterion are widely used in vortex analysis. However, the results are highly dependent on the selection of the threshold value. When the threshold is large, a weak vortex may be ignored, while when the threshold is small, some vortexes will not be distinguished [9]. The Ω method, proposed by Liu et al. in 2016, achieves a good balance and avoids one of the main drawbacks of traditional vortex identification methods, which is threshold sensitivity [10]. In 2018, Liu et al. further proposed the third-generation vortex identification method, Liutex method, by using a new decomposition. The objective Liutex vector is defined by forming a spatially averaged spin tensor with the assistance of spatially averaged vorticity, giving a pure velocity gradient tensor, and the defined objective vector does not change its direction and magnitude in the Eulerian rotation coordinate system. The identification results of Liutex avoid contamination by stretching, compression, and shear. Liutex provides a quantified way to study vortices and turbulence [11].

The vortex structure in the SB-VAWT is complex, and therefore a well-defined method is necessary for vortex identification. In this paper, different vortex identification methods, including Liutex, Ω, Q, and λ_2, are used to identify the vortex structure of the SB-VAWT flow field. This paper provides a reference for the selection of vortex identification methods for the SB-VAWT flow field.

25.2 Methods for Vortex Identification

It is well known that vortex identification plays an important role in the study of the vortex. Currently, there are many methods of vortex identification have been developed. In 1988, Hunt et al. proposed the Q criterion, where Q represents the balance between strain rate and vorticity magnitude [12]. It is defined by a positive second invariant, and $Q > 0$ is considered as a vortex. In 1995, Jeong et al. proposed the λ_2 criterion, and $\lambda_2 < 0$ is considered as vortex [13]. In 2016, Liu et al. proposed the Ω method independent of the threshold [10]. Liu et al. divided the vorticity into two parts: one part is the rotation of the flow, and the other part is the deformation. The vortex is defined as the ratio of the vortex vorticity to the whole vortex vorticity, as shown in Eq. 25.1.

$$\Omega = \frac{a}{a + b + \varepsilon} \tag{25.1}$$

where a and b are defined in the reference [12] and ε is a positive infinitesimal quantity to avoid the case where the denominator is zero. Liu et al. verified that the Ω method could identify the vortex structure in different cases within the threshold range of 0.52–0.6.

Liu et al. in 2018, adopted a new decomposition method further proposed the third-generation vortex identification method, Liutex method. Liutex eliminates the contamination by stretching, compression, and shear compared with other methods and thus can accurately quantify the local rotational strength [11], as shown in Eq. 25.2.

$$\Delta \mathbf{v} = A + R + S \tag{25.2}$$

where R is the tensor representing the rigid rotation and $R/2$ is the angular velocity.

In this paper, the above four methods will be used to identify the vortex structure in the SB-VAWT flow field.

25.3 Numerical Simulation

ANSYS FLUENT was used for the numerical simulation. The numerical simulation is based on a small-scale SB-VAWT with a rated power of 100 W. The schematic diagram of the SB-VAWT is shown in Fig. 25.1, including the main components such as blades, beams, flanges, shaft, and generator. During the simulation, the connecting components were ignored, and only the blades were retained. The main geometrical and operational parameters are shown in Table 25.1.

The pressure–velocity coupled and SIMPLE method was used. The Reynolds number for the numerical simulation is 4.05×10^5. In order to obtain a more accurate flow field result for the SB-VAWT, Large Eddy Simulation was used. The turbulent

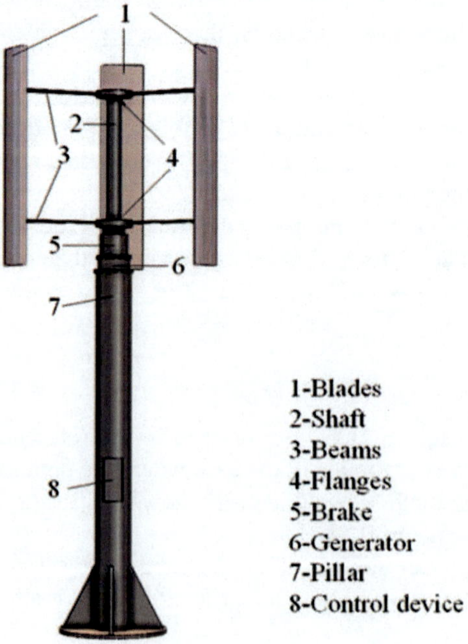

Fig. 25.1 Schematic diagram of the SB-VAWT

Table 25.1 Basic parameters of the SB-VAWT model

Parameters	Value
Blade airfoil	NACA0018
Blade height (*H*)[m]	0.5
Rotor diameter (*D*)[m]	0.6
Blade chord length (*C*)[m]	0.125
Blades number (*N*)	4
Wind speed (*U*)[m/s]	10
Tip speed ratio (λ)	1.0

Table 25.2 Simulation settings

Items	Methods
Turbulence model	LES
Mesh type	Unstructured tetrahedral mesh
Numerical method	Finite volume method
Reynolds number (*Re*)	4.05×10^5
Time step (Δt)[s]	6.54×10^{-4}
Total time (*t*)[s]	3.53

energy and turbulent dissipation rate adopted the second-order upwind style, and the calculated convergence error was set to 10^{-6}. The time step was set to T/360, which is 6.54×10^{-4} s, for 30 iterations was performed in each time step. A total of 15 rotations are performed, and the calculated results are imported into Tecplot for subsequent processing. The simulation settings are listed in Table 25.2.

The computational domain is shown in Fig. 25.2. The figure shows that the computational domain consists of rectangular stationery and cylindrical rotational domains. In order to capture the development of wake and avoid blocking effects, the stationary domain is set to $20\ D \times 20\ D \times 25\ D$, where D is the rotor diameter. The rotational domain diameter and height are $1.5\ D$ and $1.5\ H$, respectively, where H is the blade height. The distances from the center of the rotor to the left and right boundaries are $10\ D$ and $15\ D$, respectively. The distance from the center of the rotor to the surrounding boundaries is $10\ D$. The left boundary of the stationary domain is set to the velocity inlet, and the airflow is input at a constant velocity of 10 m/s. The right boundary and the surrounding boundary of the stationary domain are set as the pressure outlet with the same pressure values as the standard atmospheric pressure better to simulate the operating environment of SB-VAWT under natural conditions. An unstructured tetrahedral mesh was used, with 300 and 500 nodes around the blade section and in the spreading direction, respectively. The growth rate of the mesh is 1.05. The same mesh setting is used at the stationary and rotational domains interface to ensure stable data transmission. A total number of 20 million meshes in the computational domain.

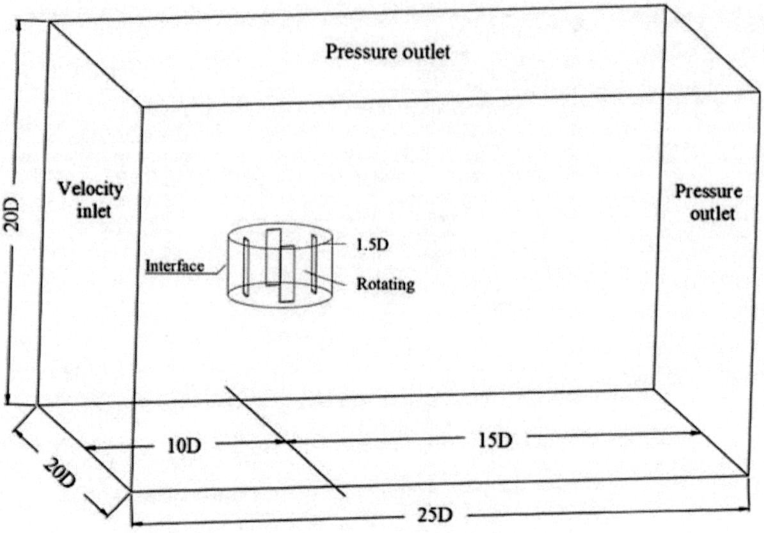

Fig. 25.2 Computational domain

25.4 Results and Discussion

The numerical simulation focused on the 1.0 tip speed ratio of the small-scale SB-VAWT in the rotating state. Four vortex identification methods, λ_2, Q, Ω, and Liutex, were selected to identify the vortex structure in the SB-VAWT flow field.

Figure 25.3 shows the vortex structure of the SB-VAWT. As described in the previous section, the flow field of SB-VAWT is complex. There are a large number of highly unsteady vortexes in the flow field. A large number of small-scale vortices were observed on the blade surface, and they were shed with the rotation of the SB-VAWT. Large-scale vortices were generated at the trailing edge of the blade, with SB-VAWT rotation gradually becoming a horseshoe shape continuing to develop backward. Moreover, the large-scale vortex was also generated at the blade tip. The vortex at the tip of the blade has a significant effect. On the SB-VAWT performance compared to the vortex generated at the blade surface and trailing edge. Optimization of the vortex at the tip of the blade is an effective way to improve the aerodynamic performance of the SB-VAWT. The operating mode of the SB-VAWT caused the downstream blades to impact the vortex generated by the upstream blades during the rotation. This will result in the breakup of the large-scale vortex into smaller-scale vortices, complicating the wake of the SB-VAWT. This also makes the operating conditions of the downstream blades unstable as well as causes possible rotor vibrations.

As shown in Fig. 25.3, the vortex structure of SB-VAWT can be obtained by using different identification methods. There is a consistency in the results obtained using the λ_2, Q, and Ω methods. The results obtained by using Liutex were slightly

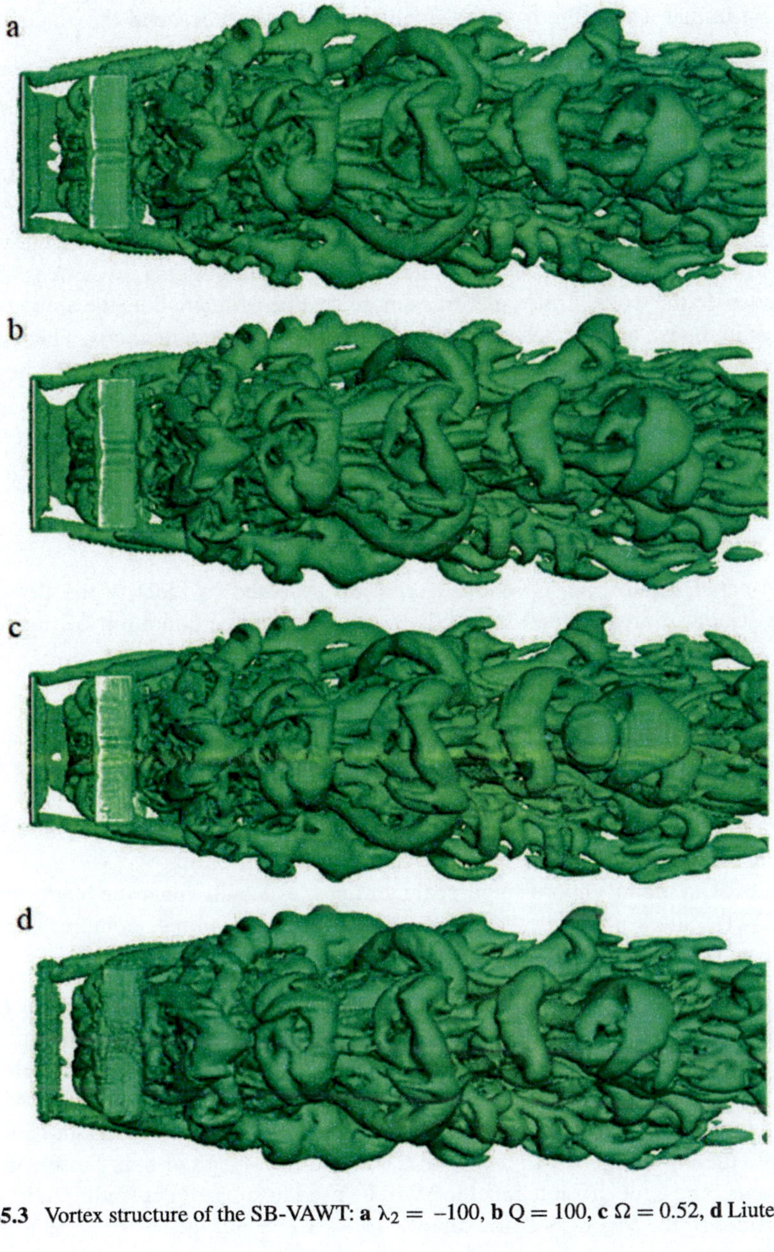

Fig. 25.3 Vortex structure of the SB-VAWT: **a** $\lambda_2 = -100$, **b** $Q = 100$, **c** $\Omega = 0.52$, **d** Liutex = 15

different in that it identified a more small vortex on the surface of the blade, which was not available with other methods.

Figure 25.4 shows the vortex of the SB-VAWT on the central slice. As shown in the figure, there are some differences in the results obtained by the four methods. The vortex identified by the λ_2 criterion and the Q criterion is almost identical. In contrast to the λ_2 and the Q, the Ω method identifies weak vortices in the wake of SB-VAWT that are ignored by these two methods. The results of Liutex are slightly different from the results of the other three methods. The difference is mainly in the vortex around the blades and in the wake. Liutex identified a larger area of vortices around the blades. In the wake, Liutex identifies more vortices. Meanwhile, the area of some vortices in the wake is decreased compared to the other three methods. This may be due to the fact that Liutex avoids contamination from stretching, compression, and shear.

25.5 Summary

In this paper, Liutex, Ω, Q, and λ_2 criterion are used to identify the flow field vortex structure of the small-scale SB-VAWT. Under the condition of this study, the conclusions can be summarized as follows:

1. All methods, including Liutex, Ω, Q, and λ_2, can generally be used to characterize the basic features of the SB-VAWT flow field vortex structure. The Ω method is threshold insensitive, which avoids potential errors caused by choosing inappropriate thresholds.
2. Both Liutex and Ω methods can identify more weak vortices in the flow field. These weak vortices are ignored by the λ_2 and Q criterion.
3. Liutex can identify more small-scale vortices, especially near the blades of SB-VAWT. Liutex eliminates the contamination by stretching, compression, and shear. These are not available in other methods.

Liutex has shown good capability in the identification of vortex structures of the SB-VAWT. At the same time, the identification results of Liutex avoid contamination by stretching, compression, and shear. However, Liutex is sensitive to the selection of the threshold value. In 2019, Dong et al. proposed the Omega-Liutex method. The idea of the Omega method was adopted in Liutex, which allows Omega-Liutex to maintain the advantages of Liutex with the added advantage of being insensitive to thresholds. The Omega-Liutex method will be used in subsequent studies to identify the vortices of the SB-VAWT flow field.

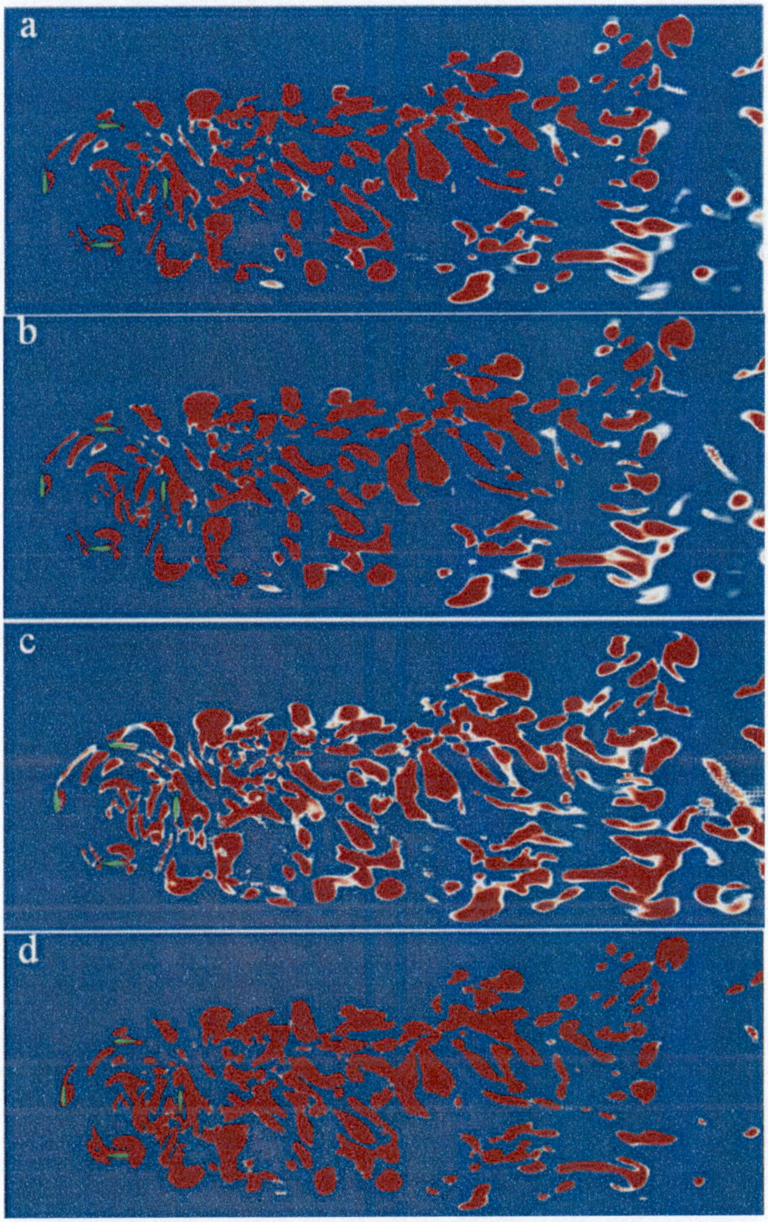

Fig. 25.4 Centralslice extracted from Fig. 25.3: **a** $\lambda_2 = -100$, **b** $Q = 100$, **c** $\Omega = 0.52$, **d** Liutex = 15

References

1. Z. Zhao, D. Wang, T. Wang, W. Shen, H. Liu, M. Chen, Sustainable energy technologies and assessments, 49 (2022)
2. H. Zhu, W. Hao, C. Li, Q. Ding, Renewable energy, 158 (2020)
3. H. Tong, Y. Wang, Renewable energy 173 (2021)
4. W. Hao, M. Bashir, C. Li, C. Sun, Energy conversion and management, 249 (2021)
5. T. Syawitri, Y. Yao, J. Yao, B. Chandra, Sustainable energy technologies and assessments, 49 (2022)
6. W. Xu, G. Li, F. Wang, Y. Li, Energy conversion and management, 205 (2020)
7. L. Kuang, J. Su, Y. Chen, Z. Han, D. Zhou, K. Zhang, Y. Zhao, Y. Bao, Energy, 239 (2022)
8. Y. Li, S. Zhao, C. Qu, G. Tong, F. Feng, B. Zhao, T. Kotaro, Energy conversion and management, 203 (2020)
9. Y. Zhang, K. Liu, J. Li, H. Xian, X. Du, J. Hydrodyn. 30 (2018)
10. C. Liu, Y. Wang, Y. Yang, Z. Duan, Sci. China Phys., Mech. Astron. 59 (2016)
11. C. Liu, Y. Gao, S. Tian, X. Dong, Phys. Fluids 30 (2018)
12. J. Hunt, A. Wary, P. Moin, *Center for Turbulent Research* (1988)
13. J. Jeong, F. Hussain, J. Fluid Mech. 285 (1995)

Chapter 26
Study on Fluid-Borne Noise Around a Cylinder Based on Vortex Sound Theory with Liutex-Shear Decomposition

Zhiyi Yuan and Yongxue Zhang

Abstract The flow-induced noise over a circular cylinder is widely studied as its geometry is typical and the noise is highly mattered on many occasions. The acoustic field for the flow around the cylinder is studied using delayed detached-eddy simulation and Powell vortex sound theory. Combined with Liutex-shear decomposition of vorticity, the effect of the rigid-rotational and shear part of the fluid parcel on the acoustic perturbation is discussed. The results show that the numerical prediction on the acoustic field is in good agreement with the experiment. Shear often makes up a large proportion of vorticity in vortex areas compared to rigid rotational parts, and the vorticity concentration areas with a weaker shearing effect correspond to lower loss. The tonal noise is from vortex shedding, where the shear mainly contributes to noise compared to the rigid rotational part of vorticity. The possible reason is that the sound can be a kind of energy loss and the shearing effect dominates the loss.

26.1 Introduction

Fluid-borne noise control has always been a research focus in many areas, as noise pollution could not only adversely affect the health of humans and wildlife but threaten the secrecy of military equipment. The flow-induced noise over a cylinder is widely studied for its typical geometry on many occasions, like aerodynamic noise in landing gear, automobile convex body, and the hydrodynamic noise in the submarine. Since it is generally believed that the pressure fluctuation caused by vortex shedding leads to noise generation and brings pollution, it is meaningful to investigate the noise generation mechanism from the perspective of the vortex and develop vortex control methods to reduce noise.

Z. Yuan (✉) · Y. Zhang
College of Mechanical and Transportation Engineering, China University of Petroleum, Beijing, China
e-mail: ziyiyuanxy@gmail.com

Key Laboratory of Process Fluid Filtration and Separation, Beijing, China

Y. Wang et al. (eds.), *Liutex and Third Generation of Vortex Identification*,
Springer Proceedings in Physics 288, https://doi.org/10.1007/978-981-19-8955-1_26

The most general method to predict noise is the FW-H method proposed by Ffowcs Williams and Hawkings that the sound source term is divided into monopole, dipole and quadrupole source [1]. Karthik et al. [2] employed FW-H method on the aeroacoustic field of cylinder flow and the ratio of length and diameter was observed to be a key factor to affect vortex shedding and control noise. Cai et al. [3] concluded that the quadrupole noise of flow past cylinder is much smaller than dipole noise except for the primary radiation. Some improvements on FW-H method were implemented to compute the quadrupole source precisely. Choi et al. [4] compared the original FW-H with permeable FW-H in the noise prediction around a submerged cylinder and found that the difference was attributed to a quadrupole source of vortex shedding. Then they employed the formulation Q1As method in the same case and the results were validated by experiments [5]. It is observed that the Q1As method is more accurate for the noise prediction at the peak frequency.

Although these methods based on FW-H are widely used in acoustic calculation, this formula shows less relationship with vortex dynamics to present the sound generation mechanism of vortical flow. Powell [6] indicated that the sound generation is highly related to vortex and proposed vortex sound equation. Takaishi et al. [7] calculated the noise of turbulent flow past a circular cylinder via vortex sound theory. The results show the sound pressure consists of a discrete tone originating from Karman vortices, and the numerical predictions agree well with the experiments at frequency up to 2 kHz. Daryan et al. [8] investigated the sound generation mechanism of vortex reconnection and observed that significant sound occurs due to the coiling and uncoiling of the twisted vortex filaments wrapping. Despite the vortex sound method performing well in analyzing the noise, the vortex identified by vorticity is not precise enough. The vortex can be extremely weak at the turbulent boundary layer where vorticity is concentrated [9]. A rigid vorticity conception (Liutex) initially proposed by Liu et al. [10–12] avoids the shear contamination in vortex identification. Considering the Liutex-shear decomposition of vorticity, it provides a new perspective view to calculate the sound pressure originated from rigid rotational motion and shear.

This paper uses delayed detached-eddy simulation (DDES) method to predict the turbulence flow around a submerged cylinder and vortex sound method for noise calculation. Combined with the Liutex-shear decomposition, three kinds of sound pressure dominated by vorticity, rigid vorticity and shear are investigated. The effect of rigid vorticity and shear on acoustic perturbation is revealed.

26.2 Numerical Simulation Method

26.2.1 Vortex Sound Equation with Liutex-Shear Decomposition

The Powell vortex sound equation can be written as [6]:

$$\frac{1}{c^2}\frac{\partial^2 p_s}{\partial t^2} - \nabla^2 p_s = \nabla \cdot \left[\rho_0(\omega \times u) - u\frac{\partial \rho_0}{\partial t} + \nabla\left(\rho_0 \frac{u^2}{2}\right)\right] \tag{26.1}$$

where c is sound speed, ρ_0 is the density for the incompressible fluid, p_s is pressure perturbation (sound pressure), $\boldsymbol{\omega}$ is vorticity, $\boldsymbol{u}$ is velocity.

For the current study under low Mach number, the sound source term $\nabla \cdot \left(u\frac{\partial \rho_0}{\partial t}\right)$ and $\nabla\left(\rho_0 \frac{\boldsymbol{u}^2}{2}\right)$ can be neglected [13], the vortex sound equation becomes:

$$\frac{1}{c^2}\frac{\partial^2 p_s}{\partial t^2} - \nabla^2 p_s = \nabla \cdot (\rho_0(\omega \times u)) \tag{26.2}$$

As the rigid vortex vector (Liutex) proposed by Liu overcome the shortage of vortex identification contaminated by shearing, Liutex-Shear decomposition of vorticity is [12]:

$$\boldsymbol{\omega} = \boldsymbol{R} + \boldsymbol{S} \tag{26.3}$$

where $\boldsymbol{R}$ is Liutex (rigid vorticity), $\boldsymbol{S}$ denotes shear part of vorticity.

The Liutex magnitude R is twice the minimum angular velocity, which denotes the absolute vortex strength. The explicit formula for the Liutex vector is:

$$\boldsymbol{R} = Rr = \left[\omega \cdot \boldsymbol{r} - \sqrt{(\omega \cdot \boldsymbol{r})^2 - 4\lambda_{ci}^2}\right]\boldsymbol{r} \tag{26.4}$$

where $\boldsymbol{r}$ and λ_{ci} are the real eigenvector and the imaginary part of the complex eigenvalue of velocity gradient tensor respectively.

The relative vortex strength could be valued by Omega [14]:

$$\Omega = \frac{\mathrm{tr}(\boldsymbol{B}^{\mathrm{T}}\boldsymbol{B})}{\mathrm{tr}(\boldsymbol{A}^{\mathrm{T}}\boldsymbol{A}) + \mathrm{tr}(\boldsymbol{B}^{\mathrm{T}}\boldsymbol{B})} \tag{26.5}$$

where $\boldsymbol{A}$ is strain rate tensor, $\boldsymbol{B}$ is vorticity tensor.

According to the Liutex method, the vortex sound equation with different sound sources can be written in three forms:

$$\frac{1}{c^2}\frac{\partial^2 p_{s,\omega}}{\partial t^2} - \nabla^2 p_{s,\omega} = \nabla \cdot (\rho_0(\boldsymbol{\omega} \times \boldsymbol{u})) \tag{26.6}$$

$$\frac{1}{c^2}\frac{\partial^2 p_{s,R}}{\partial t^2} - \nabla^2 p_{s,R} = \nabla \cdot (\rho_0(\boldsymbol{R} \times \boldsymbol{u})) \tag{26.7}$$

$$\frac{1}{c^2}\frac{\partial^2 p_{s,S}}{\partial t^2} - \nabla^2 p_{s,S} = \nabla \cdot (\rho_0(\boldsymbol{S} \times \boldsymbol{u})) \tag{26.8}$$

where $p_{s,\omega}$, $p_{s,R}$ and $p_{s,S}$ are sound pressure generated by vorticity, rigid vortex and shear respectively.

26.2.2 Computational Method

The calculation domain of flow around the cylinder is shown in Fig. 26.1. The diameter of cylinder D is 20 mm. Figure 26.1c shows some monitoring points at the midplane, where points a ~ e locate along with the downstream of the cylinder, point m sets at the bottom side. The monitoring line l_H is from cylinder center O to the outlet boundary, while l_V is from point O to the bottom side. Besides, 16 points are evenly arranged around the cylinder with a radius of r = 3D to investigate the sound direction. The surface mesh with refined in key flow region is shown in Fig. 26.1d.

The commercial software Fluent is applied in this study. For the calculation of fluid flow, the SST k-ω model is used to solve the turbulence. The velocity inlet boundary is 6 m/s of water, and the outlet boundary condition is set as outflow. The steady simulation is initially applied, and then the unsteady simulation of DDES is employed when the steady calculation becomes convergence. The time step is 0.0005 s, and the inner iterations are 20. The convergence criterion is 0.0001.

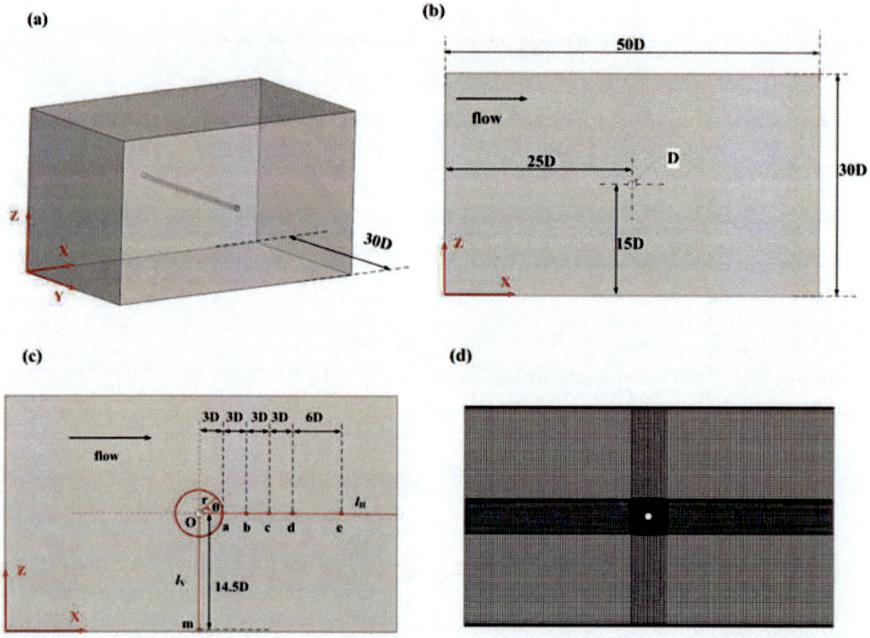

Fig. 26.1 Calculation domain: **a** 3D model, **b** mid-plane, **c** location of monitoring points, **d** structured mesh

The fluid-sound coupling simulation is carried out via user-defined scalars (UDS) function in Fluent. The vortex sound equation can be solved by using three UDS define macros, namely DEFINE_UDS_UNSTEADY, DEFINE_DIFFUSIVITY and DEFINE_SOURCE, which correspond to the unsteady term, diffusive term and source term respectively. The inlet and outlet boundaries of sound simulation are set as absorption conditions satisfying Eq. (26.9), while the wall is the total reflection boundary shown in Eq. (26.10).

$$\frac{1}{c}\frac{\partial p_s}{\partial t} + \nabla p_s \cdot \boldsymbol{n} = 0 \tag{26.9}$$

$$p_s = 0 \tag{26.10}$$

The irreversible loss is evaluated by entropy production rate, which has been validated in many studies [15–18]. It is defined as:

$$S_{pro} = \frac{(\mu + \mu_t)}{T}(\nabla \boldsymbol{u} + \nabla \boldsymbol{u}^{\mathrm{T}})(\nabla \cdot \boldsymbol{u}) \tag{26.11}$$

where μ is dynamic viscosity, μ_t is turbulence viscosity, T is temperature.

The acoustic intensity is measured by sound pressure level (SPL) and total sound pressure level (TSPL):

$$\mathrm{SPL} = 20\,\mathrm{lg}\,\frac{p_{ef}}{p_{ref}} \tag{26.12}$$

$$\mathrm{TSPL} = 10\,\mathrm{lg}\left(\sum_{i=1}^{n} 10^{\mathrm{SPL}_i/10}\right) \tag{26.13}$$

where p_{ef} is effective sound pressure (root-mean-square of p_s), p_{ref} is 1e-6 Pa, subscript i indicates corresponding frequency.

26.2.3 Simulation Validation

The pressure coefficient of cylinder surface is shown in Fig. 26.2, where the number of mesh 1, mesh 2 and mesh 3 is 3.2, 4.2 and 5.4 million, respectively. As the figures stay almost unchanged when the grid number exceeds 4.2 million, mesh 2 was used in this paper. The average y + of the wall is about 0.8.

Figure 26.3 presents the comparison between simulation results and the data from Choi [5]. The spectrum of SPL calculated by vortex sound method agrees well with the experiment and simulation by Choi. Since the maximum errors of main frequcncy and corresponding SPL are less than 8% and 3%, this simulation result is deemed to be reliable.

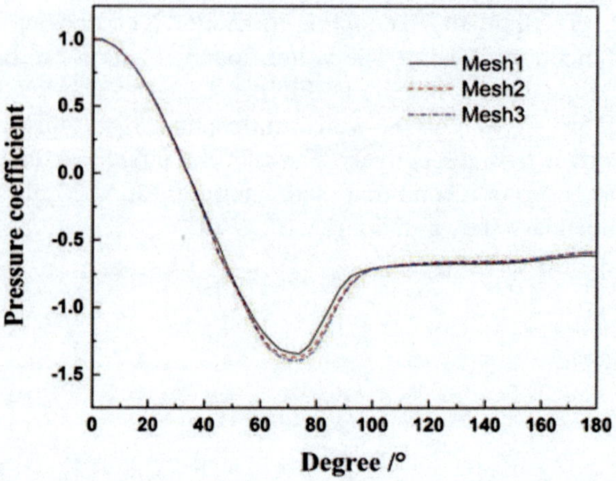

Fig. 26.2 Grid independence validation

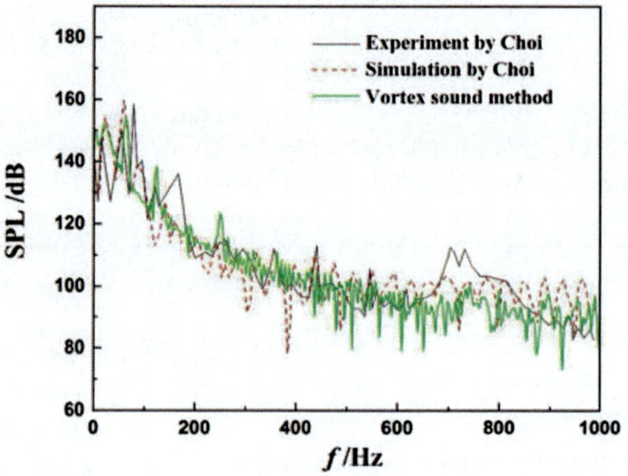

Fig. 26.3 Simulation validation on sound pressure level

26.3 Results and Discussion

26.3.1 Characteristic of Shedding Vortex

Figure 26.4 illustrates the distribution of vorticity and Liutex, and both present the typical phenomenon of Karman vortex street. Since the magnitude of Liutex is much smaller than vorticity in most cases, the contour range of Fig. 26.4b is decreased to display the rigid vortex clearly compared to Fig. 26.4a. It can been that Fig. 26.4b has more bright distinctions on vortex boundaries, and there is no rigid vortex near the wall while vorticity concentration occurs close to the wall, which indicates that Liutex method could avoid shear contamination. Besides, the vortex strength becomes weak along with the downstream flow. In order to analyze the relationship between vortex and irreversible loss, Fig. 26.5a gives the vortex structures identified by iso-surface of Omega = 0.6 and colored by entropy production rate. It is observed that the loss is significant for the vortex near the cylinder wall, while the loss of vortex downstream the cylinder flow decreases. The contour of shear is displayed in Fig. 26.5b, which shows a close correlation between shear and loss. Although a similar distribution of shear and Liutex is shown in shedding vortex areas, the high shear does not equal to strong vortex. Compared with Figs. 26.4b and 26.5b, the shear concentration can be found near the wall instead of Liutex.

Figure 26.6 further presents the detailed relationship among the vortex features of shear, rotation and loss. From Fig. 26.6a, a sharp decrease near cylinder wall and wavy vorticity lines occur downstream, showing the alternate shedding features of vortex street. It is often thought the loss is higher when vorticity is concentrated, while there is a counterexample for the points p_1 and p_2. This is because the loss is dominated by shear instead of vorticity. The vorticity of p_1 is higher than it of p_2, but the loss of p_1 is lower since the shear effect of p_1 is weaker and the rotation effect gets stronger, which also indicates the loss is likely to decrease when the rotation effect

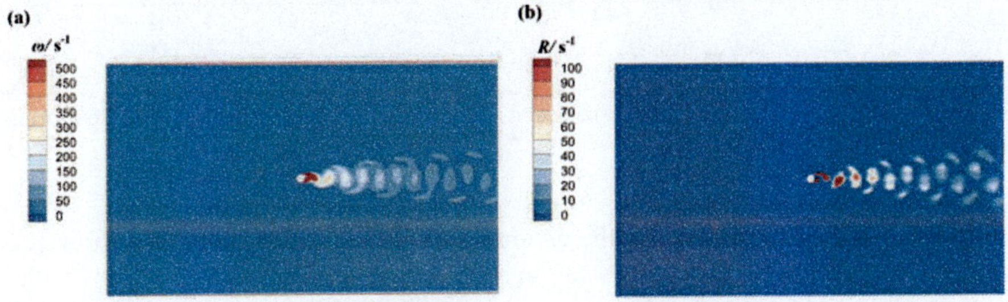

Fig. 26.4 Distribution of vorticity and Liutex: **a** contour of vorticity, **b** contour of Liutex

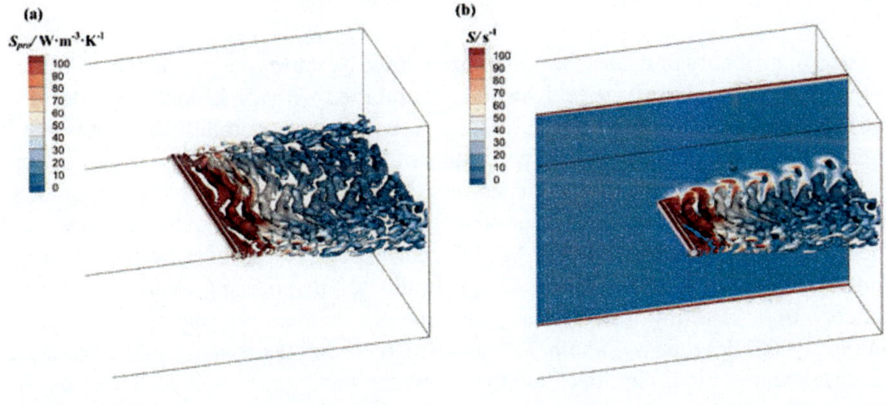

Fig. 26.5 Distribution of entropy production rate and shear: **a** iso-surface of Omega colored by entropy production rate, **b** contour of shear

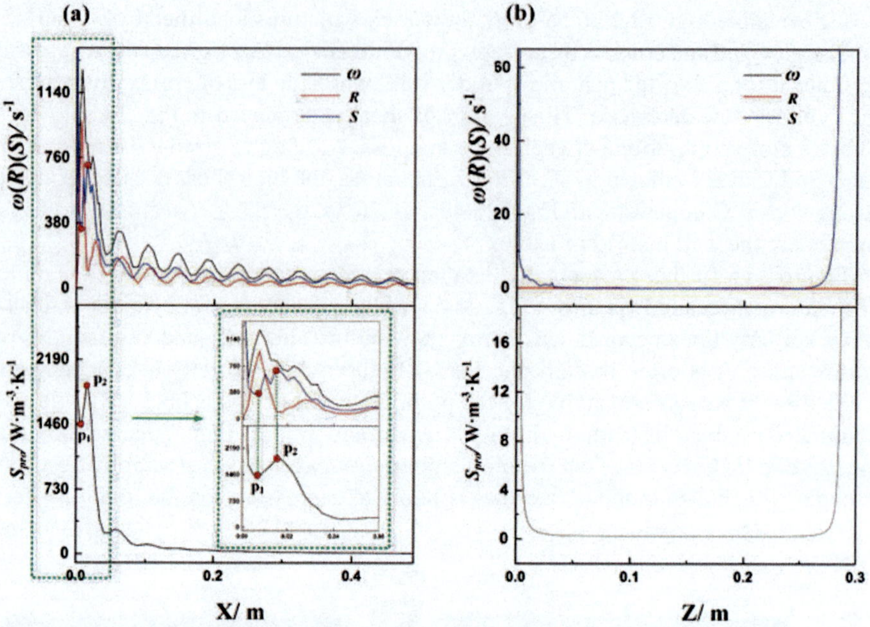

Fig. 26.6 Change of vorticity, Liutex, shear and energy loss: **a** monitoring line l_{H}, **b** monitoring line l_{V}

increases. Figure 26.6b presents the changing trend along line l_{V}, which shows no vortex exists except for the concentration of vorticity, shear and loss near the wall.

26.3.2 Influence of Vortex on the Acoustic Field

Vortex sound theory indicates the noise is generated by vortex and reveals the relationship between sound pressure and vorticity. Combined with the Liutex-shear decomposition method, the sound pressure originating from rigid rotational motion and shear is discussed in this part. The spectrums of sound pressure and pressure fluctuation are shown in Fig. 26.7. For the spectrum diagram of point m, the main frequency of sound pressure and pressure pulsation is f_0, which indicates the frequency of vortex shedding from one side of cylinder, while f_{s} correspond to shedding from the bottom and top sides. The changing trend of $p_{\mathrm{s},S}$ dominated by shearing effect is in accordance with p_s resulting from vorticity, and the $p_{\mathrm{s},R}$ originated from rigid rotation is obviously smaller than $p_{\mathrm{s},S}$ and $p_{s,\omega}$. Figure 26.7b shows the trend of point b located at downstream of cylinder flow, and the characteristic frequency is f_{s}. These trends indicate the shearing effect plays a major role in sound generation. The possible cause is that sound is a kind of energy propagation, and the energy loss of fluid flow is highly related to shear. Thus, the sound pressure change closely relates to the shearing effect.

To further illustrate the influence of rotation and shear effect on sound pressure, Fig. 26.8 gives the spectrums of vorticity, sound pressure and pressure fluctuation, where f_s is the main frequency in all the diagrams. It can be seen from Fig. 26.8a that the vorticity amplitude of the main frequency obeys $p_{s,\omega} > p_{s,S} > p_{s,R}$, and so does the sound pressure. At point e, all these figures get a significant reduction, which indicates vortex strength, sound pressure and pressure become weakened when flowing downstream. These trends also validate the conclusions drawn above.

Figure 26.9 is the total sound pressure change along with the monitoring points, which shows the TSPL decreases gradually downstream and the shearing part of vorticity mainly contributes to noise. The sound direction distribution is shown in Fig. 26.10. By calculating the total sound pressure level of these points around the cylinder, the plot indicates the dipole pattern of the acoustic field around the cylinder, which agrees with the previous studies. It can also be seen that the shearing effect on sound generation is more effective than the rigid rotation.

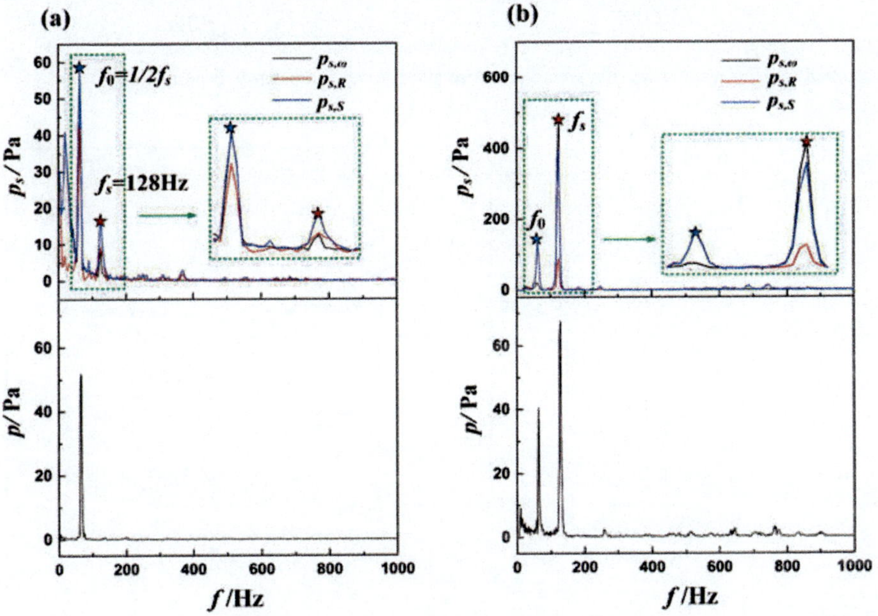

Fig. 26.7 Spectrum of sound pressure and pressure: **a** point m, **b** point d

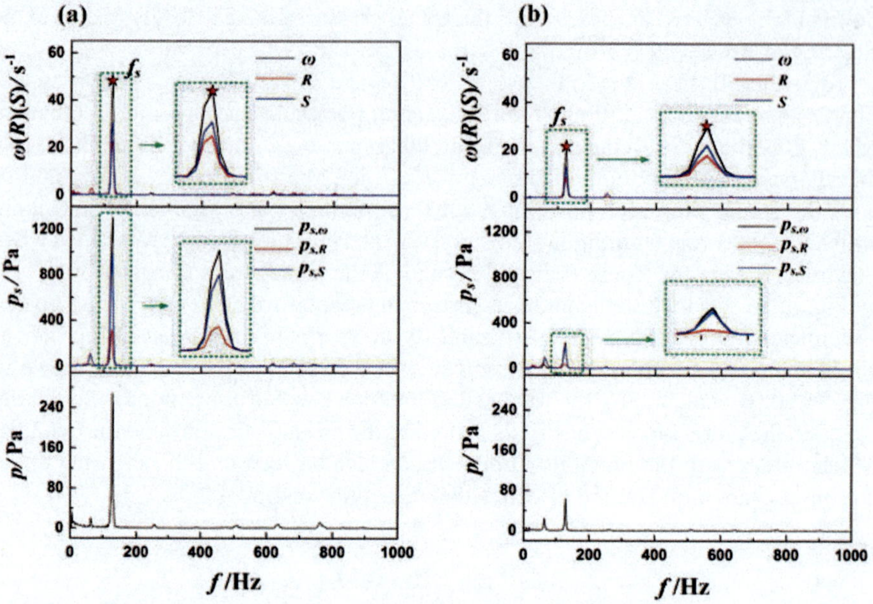

Fig. 26.8 Spectrum of vorticity, sound pressure and pressure: **a** point b, **b** point e

Fig. 26.9 Total sound pressure level

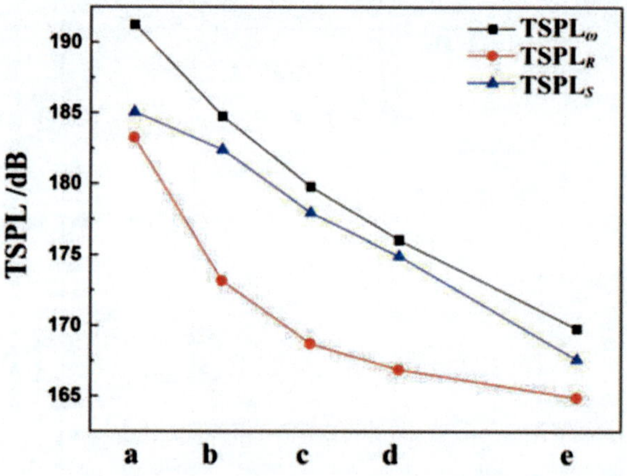

26.4 Conclusions

This paper presents a novel investigation on the effect of rigid vorticity and shear on acoustic perturbation of turbulent flow around a submerged circular cylinder. By combining the vortex sound theory and Liutex-shear decomposition of vorticity, the conclusions can be concluded as follows:

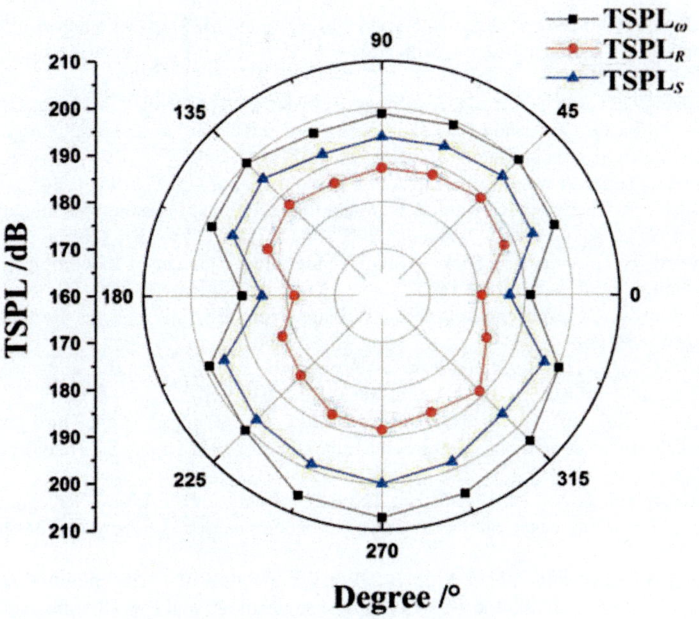

Fig. 26.10 Sound direction pattern at 3D away from the cylinder

(1) Energy loss of shedding vortex drops dramatically near the cylinder wall and decreases slightly along the downstream direction. Shear often makes up a large proportion of vorticity in vortex areas. Vorticity concentration areas with a weaker shear correspond to lower loss, and the loss is also likely to decrease when the rotation effect increases.
(2) Vortex sound method is credible to predict flow-induced noise. The tonal noise is from vortex shedding, where the shearing part of vorticity mainly contributes to noise compared to the rotational part. The possible reason is that the sound can be a kind of energy loss while the loss is dominated by the shearing effect.

Acknowledgements This work was supported by the National Natural Science Foundation of China (Grant No.51876220, No.52179094).

References

1. J.E. Ffowcs Williams, D.L. Hawkings, Sound generation by turbulence and surfaces in arbitrary motion. Philos. Trans. R. Soc. A **264**(1151), 321–342 (1969)
2. K. Karthik, S. Vengadesan, S.K. Bhattacharyya, Prediction of flow induced sound generated by cross flow past finite length circular cylinders. J. Acoust. Soc. Am. **143**(1), 260–270 (2018)
3. J.C. Cai, J. Pan, A. Kryzhanovskyi, A numerical study of transient flow around a cylinder and aerodynamic sound radiation. Thermophys. Aeromech. **25**(3), 331–346 (2018)

4. W.S. Choi, Y. Choi, S.Y. Hong, J.H. Song, H.W. Kwon, C.M. Jung, Turbulence-induced noise of a submerged cylinder using a permeable FW–H method. Int. J. Nav. Archit. Ocean Eng. **8**(3), 235–242 (2016)
5. Y.S. Choi, W.S. Choi, S.Y. Hong, J.H. Song, H.W. Kwon, H.S. Seol, C.M. Jung, Development of formulation Q1As method for quadrupole noise prediction around a submerged cylinder. Int. J. Nav. Archit. Ocean Eng. **9**(5), 484–491 (2017)
6. A. Powell, Theory of vortex sound. J. Acoust. Soc. Am. **36**(1), 177–195 (1964)
7. T. Takaishi, M. Miyazawa, C. Kato, A computational method of evaluating noncompact sound based on vortex sound theory. J. Acoust. Soc. Am. **121**(3), 1353–1361 (2007)
8. H. Daryan, F. Hussain, J.P. Hickey, Sound generation mechanism of compressible vortex reconnection. J. Fluid Mech. 933 (2022)
9. S.K. Robinson, Coherent motions in the turbulent boundary layer. Annu. Rev. Fluid Mech. **23**(1), 601–639 (1991)
10. C. Liu, Y. Gao, S. Tian, X. Dong, Rortex—A new vortex vector definition and vorticity tensor and vector decompositions. Phys. Fluids **30**(3), 035103(2018)
11. C. Liu, Y.S. Gao, X.R. Dong, Y.Q. Wang, J.M. Liu, Y.N. Zhang, N. Gui, Third generation of vortex identification methods: Omega and Liutex/Rortex based systems. J. Hydrodyn. **31**(2), 205–223 (2019)
12. Y.Q. Wang, Y.S. Gao, J.M. Liu, C. Liu, Explicit formula for the Liutex vector and physical meaning of vorticity based on the Liutex-Shear decomposition. J. Hydrodyn. **31**(3), 464–474 (2019)
13. N. Zhang, X.I.E. Hua, W.A.N.G. Xing, B.S. Wu, Computation of vortical flow and flow induced noise by large eddy simulation with FW-H acoustic analogy and Powell vortex sound theory. J. Hydrodyn. B **28**(2), 255–266 (2016)
14. C. Liu, Y. Wang, Y. Yang, Z. Duan, New omega vortex identification method. Sci. China: Phys. Mech. Astron. **59**(8), 1–9 (2016)
15. C. Wang, Y. Zhang, H. Hou, J. Zhang, C. Xu, Entropy production diagnostic analysis of energy consumption for cavitation flow in a two-stage LNG cryogenic submerged pump. Int. J. Heat Mass Transf. **129**, 342–356 (2019)
16. H. Hou, Y. Zhang, Z. Li, A numerically research on energy loss evaluation in a centrifugal pump system based on local entropy production method. Therm. Sci. **21**(3), 1287–1299 (2017)
17. C. Wang, Y. Zhang, Z. Yuan, K. Ji, Development and application of the entropy production diagnostic model to the cavitation flow of a pump-turbine in pump mode. Renew. Energy **154**, 774–785 (2020)
18. Z. Yuan, Y. Zhang, C. Wang, B. Lu, Study on characteristics of vortex structures and irreversible losses in the centrifugal pump. Proc. Inst. Mech. Eng. A: J. Power Energy **235**(5), 1080–1093(2021)

Zhiyi Yuan is a doctoral candidate of Power Engineering and Engineering Thermal Physics at China petroleum university-Beijing. His research focuses on the vortex and fluid-borne noise in turbomachinery.

Chapter 27
Coherent Structures Analysis Across Cavity Interface in Cloud Cavitating Flows Using Different Vortex Identification Methods

Changchang Wang, Guoyu Wang, and Biao Huang

Abstract Cavitating flows are characterized by strong compressibility inside the cavity and weak compressibility outside the cavity. The transition between these two regions with distinct acoustic features is defined as the acoustic shear layer (i.e., from cavity interface to 0.99 sonic seed in pure liquid, about $1 \times 10^{-7} < \alpha_v < 1 \times 10^{-1}$) which has been preliminarily studied in the recent paper (*Ocean Eng.* 209(2020): 107025). The acoustic shear layer is characterized by the large sonic speed gradient which is of great acoustics importance to understand how the disturbance (i.e., shockwave) propagates between these two regions. With varying cavitation numbers and Reynold numbers, cavitating flows present different flow regimes where cloud cavitation is supposed to be the most destructive cavity regime and of great fundamental interest and engineering applications, mainly consisting of two kinds of cavity structures, namely attached sheet cavity and shedding cloud cavity. In this work, to improve the understanding of cavity instabilities associated with wave dynamics, we examine in detail the coherent structures inside the acoustic shear layer of both the attached sheet cavity and the shedding cloud cavity, in particular that across the cavity interface. Numerical simulation of cavitating flows around a NACA66 (mod) hydrofoil was conducted using computational fluid dynamics (CFD) tool. Vortex identification methods including vorticity, Q-criteria, and the Liutex method, are employed to identify the flow structures within the acoustics shear layer. Results show that acoustic shear layer across the attached sheet cavity is thinner than that across the shedding cloud cavity. The acoustic shear layer consists of two regions, i.e., turbulence dominated region and the acoustics dominated region. Specifically, the turbulence dominated region which is identified by the turbulent/non-turbulent interface near the cavity boundary is important for the mass, momentum, and energy transfer characteristics. The acoustics dominated region is significant for the disturbance propagation (i.e., shockwave) between the compressible cavitation region and incompressible pure liquid region. Different vortex identification methods show different performances in terms of identifying the flow structures in these two regions. It is suggested that further work could be done to implement the

C. Wang (✉) · G. Wang · B. Huang
Beijing Institute of Technology, Beijing, China
e-mail: wangchangchang026@vip.163.com

Y. Wang et al. (eds.), *Liutex and Third Generation of Vortex Identification*,
Springer Proceedings in Physics 288, https://doi.org/10.1007/978-981-19-8955-1_27

acoustics characteristics into the vortex identification method to improve the identification performance in acoustics dominated regions, i.e., low void fraction region in gas–liquid flows.

27.1 Introduction

Cavitation generally occurs at local low-pressure regions where the local pressure drops below the vapor pressure in high-speed liquid flows [1]. Consequently, a unique feature in cavitating flows is that the low-pressure gas-filled or gas- and vapor-filled cavities are surrounded by pure liquid with higher pressure. As shown in Fig. 27.1, there exists a shear layer across the cavity interface. Considering the large sonic speed gradient in this region shown in Fig. 27.2, it is defined as the acoustics shear layer as has been studied in [2]. Cavity-liquid exchanges inside acoustic shear layer contribute important surface sources and sinks to the mass, momentum, heat, water vapor transfer in both the cavitation region and pure liquid region, and the interactions between these two regions as well. The physical process in this transport phenomenon is closely associated with the coherent structures inside the acoustic shear layer. However, to the best of our knowledge, no studies have been done to understand the flows structures in this acoustic shear layer across the cavity interface.

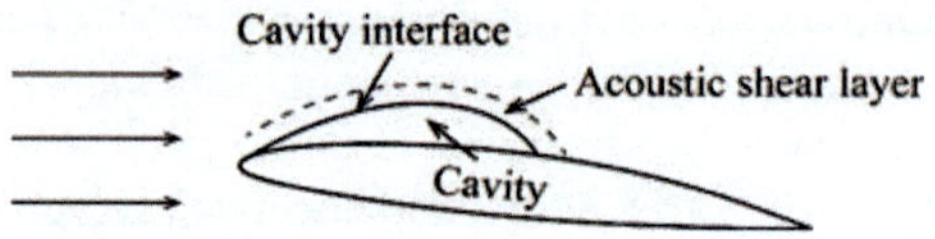

Fig. 27.1 Illustration of acoustic shear layer across the cavity interface

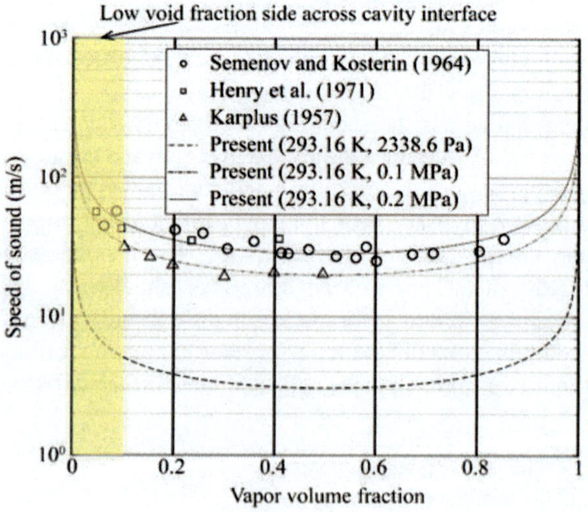

Fig. 27.2 Sonic speed distribution as a function of vapor volume fraction. (Note that the shadow region indicates the quantitative sonic speed distribution across cavity interface)

When the cavity length arrives at certain value, the cavity length oscillates a lot along with periodic attached sheet cavity growth, development, breakup, shedding and collapse process. This cavity regime is called a cloud cavitation. Cloud cavitation is supposed to be the most destructive cavitation stage, especially in the process of cavity breakup, shedding and collapse, and has attracted lots of interests [3–7]. The physics in the cavity break, shedding and collapse process is widely studied, and mainly the two mechanisms are proposed, namely reentrant jet mechanism [3, 4] and shockwave mechanism [5, 8]. For example, Callenaere et al. [4] experimentally investigated the cavitation instabilities and mapped the flow regimes with no shedding, periodic reentrant jet with no shedding and reentrant jet induced periodic shedding. Their work indicates that the adverse pressure gradient is important for the development of reentrant jet that causes periodic shedding. Since Arndt et al. [5] firstly proposed the shockwave dominated shedding mechanism, the role of shockwave in cloud cavitation has long been investigated. Recently, using high speed densitometry, Ganesh et al. [8] visualized the shockwave structures and their propagation inside cavity. Their work shows that when the shockwave propagates to the cavity leading edge, the interactions between shockwave/cavity will cause cavity breakup, followed by the shedding and collapse process, which is called a shockwave mechanism. However, due to the complexity in cloud cavitation, i.e., compressibility effects and shockwave dynamics, the understanding of cloud cavitation mechanism is very limited and requires new perspectives. It will provide new viewpoint to analyze the cloud cavitation mechanism by examining the flow structures in the proposed acoustics shear layer.

In this work, to improve the understanding of cloud cavitating flow mechanisms, the flow structures inside the acoustic shear layer across the attached sheet cavity and shedding cloud cavity, which are the two typical cavity structures in cloud cavitating flows, are numerically studied. Vortex identification methods are used to identify the coherent structures in acoustic shear layer. The objectives of this work are to (1) use CFD tools and vortex identification methods to investigate the flow fields in the acoustic shear layers; (2) quantify the physical mechanism of interaction between the compressible cavitation region and incompressible pure liquid region; and (3) provide suggestions to develop a new coherent structure identification method to improve the identification performance in unique flow regions, i.e., acoustic shear layer.

27.2 Numerical Methodology

27.2.1 Governing Equations

The present study is based on the compressible cavitation solver [2, 9, 10] developed on the open source finite volume computational fluid dynamics code, OpenFOAM-4.0. The two phase gas–liquid cavitating flows are simulated in the homogeneous multiphase flow strategy, considering the compressibility of both liquid and its vapor. The interface between liquid and its vapor is captured by the volume-of-fluid (VOF) approach, and the mass exchanges between liquid and its vapor is modelled by the cavitation model.

The governing equations of the compressible cavitating flows consist of continuity, momentum, and energy equations, along with a transport equation for the vapor volume fraction:

$$\frac{\partial \rho_m}{\partial t} + \nabla \cdot (\rho_m U) = 0 \tag{27.1}$$

$$\frac{\partial \rho_{\mathrm{m}} U}{\partial t} + \nabla \cdot (\rho_{\mathrm{m}} U U) = -\nabla p_{\mathrm{rgh}} - \vec{g} \cdot h \nabla \rho + \nabla \cdot T + \vec{F}_{\mathrm{st}} \tag{27.2}$$

$$\frac{\partial \rho_{\mathrm{m}} e}{\partial t} + \nabla \cdot (\rho_{\mathrm{m}} U e) + \frac{\partial \rho_{\mathrm{m}} K}{\partial t} + \nabla \cdot (\rho_{\mathrm{m}} U K) = \nabla \cdot q - \nabla \cdot (pU) \tag{27.3}$$

$$\frac{\partial \rho_{\mathrm{m}} \alpha_{\mathrm{l}}}{\partial t} + \nabla \cdot (\rho_{\mathrm{m}} \alpha_{\mathrm{l}} U) = \dot{m} \tag{27.4}$$

where α is void fraction, ρ is fluid density, U represents the velocity, $g = 9.81\ \mathrm{m/s^2}$ is the gravity acceleration, p is the pressure, e is the internal energy, and $K \equiv \|u\|^2/2$ is the kinetic energy per unit mass. In OpenFOAM, the pressure gradient and gravity force terms are rearranged numerically as $-\nabla p + \rho g = -\nabla p_{\mathrm{rgh}} - (g \cdot r)\nabla \rho$, where $p_{\mathrm{rgh}} = p - \rho g \cdot r$ and r is the position vector from the wall. $\dot{m}$ represents the mass transfer between liquid and vapor and modelled by the cavitation model. $\vec{F}_{\mathrm{st}}$ is the surface tension force and modelled by the continuum surface force (CSF) proposed by Brackbill et al. [11]. The subscripts m, l and v denote the mixture, liquid, and vapor respectively. In order to consider the compressibility effects of both liquid and vapor, the Tait equations of state [12] is adopted for liquid and ideal gas equation of state for vapor, respectively. More details about this compressible cavitation solver and its performance can be found in references [2, 10].

27.2.2 *Saito Cavitation Model and RANS/LES Turbulence Model*

The Saito cavitation model developed by Saito et al. [13] is used in the present study. This cavitation model is a transport-based model and expressed as

$$\dot{m}^{+} = C_e \alpha^2 (1-\alpha)^2 \frac{\rho_{\mathrm{l}}}{\rho_{\mathrm{v}}} \frac{\max((p_{\mathrm{v}} - p), 0)}{\sqrt{2\pi R_{\mathrm{v}} T}}, \text{if } p < pv \tag{27.5}$$

$$\dot{m}^{-} = C_c \alpha^2 (1-\alpha)^2 \frac{\max((p - p_{\mathrm{v}}), 0)}{\sqrt{2\pi R_{\mathrm{v}} T}}, \text{if } p > pv \tag{27.6}$$

here, $R_v = 461.6$ J/kg K is the gas constant, C_c is the condensation rate coefficient when local pressure exceeds the vapor pressure, and C_e is the evaporation rate coefficient when local pressure below the vapor pressure. In the current work, $C_e = C_c = 0.1$ is used according to the study by Saito et al. [13].

The scale-adaptive simulation SST-SAS turbulence model [14] is employed to simulate the turbulence effects. The SST-SAS turbulence model is a hybrid RANS/LES model, where in the attached flow region, the SST model acts, showing the RANS-like performance, and in unsteady regions, the LES model is active, showing the LES-like performance. The transition between the RANS and LES is determined by the von Karman length-scale, L_{vk}, which is different from that based on local grid size. The SAS term is implemented into the ω equation as a sensor for detecting the unsteadiness, and expressed as follows

$$Q_{SAS} = \max\left[\rho_m \xi_2 \kappa S^2 \left(\frac{L_{k\omega}}{L_{vk}}\right) - C \cdot \frac{2\rho_m k}{\sigma_\Phi} \max\left(\frac{|\nabla\omega|^2}{\omega^2}, \frac{|\nabla k|^2}{k^2}\right), 0\right] \quad (27.7)$$

here, $\xi_2 = 3.51, \sigma_\Phi = 2/3, C = 2, P_k = \mu_T S^2, L_{k\omega}$ is the local length scale modelled by RANS method and L_{vk} is the von Karman length-scale. More details about this turbulence model can be found in [2, 10].

27.2.3 *Data Set, Test Case, Computational Domain, and Boundary Conditions*

The data set used is the 3D simulation results in Wang et al. [2, 10] calculated using the compressible cavitation solver on OpenFOAM platform [15]. Here the simulation details are only briefly presented. The cloud cavitating flow studied here is characterized by alternative reentrant jet and shockwave dynamics which has been experimentally studied by Leroux et al. [6, 7]. The geometry model used is NACA66 (mod) which is from the NACA66 series, and characterized by a chord length of $c = 0.15$ m, the relative maximum thickness is $\tau = 12\%$ at 45% from the foil leading edge, and the relative maximum chamber is 2% at 50% from the foil leading edge. According to the experiment data, the attack of angle is fixed at 6°, the inlet flow velocity is $U = 5.33$ m/s, and the cavitation number $\sigma = 1.25$. The water temperature is at 293.15 K. The computational domain has an extent of about 2 c upstream and 6 c downstream of the foil, and 0.3 c along the foil spanwise direction as has been used by Sagaut [16] and Ji et al. [17]. To avoid the pressure wave reflection at the inlet and outlet boundaries, disturbing the flow fields, the non-reflect boundary conditions are used for both inlet and outlet boundaries for pressure fields.

27.3 Typical Vortex Identification Methods

27.3.1 Vorticity (ω) Method

The vorticity field is defined as the anti-symmetry tensor of the velocity gradient tensor which describes the average deformation and rotation of a fluid element. The vorticity threshold criterion shows that a vortex structure exists in the region where the magnitude of vorticity is larger than the certain threshold, $\omega_{\text{threshold}}$,

$$|\omega| > \omega_{\text{threshold}} \tag{27.8}$$

27.3.2 Q-criterion

The Q-criterion is one of the most popular vortex identification methods proposed by Hunt et al. [18]. Q is defined as the second invariant of the velocity gradient tensor, which can be expressed as

$$\nabla \mathrm{U} = \frac{1}{2}\left(\mathrm{G} + \mathrm{G}^{\mathrm{T}}\right) + \frac{1}{2}\left(\mathrm{G} - \mathrm{G}^{\mathrm{T}}\right) = \boldsymbol{A} + \boldsymbol{B} \tag{27.9}$$

$$Q = \frac{1}{2}\left[trace(G)^2 + \|\boldsymbol{B}\|_F^2 - \|\boldsymbol{A}\|_F^2\right] \tag{27.10}$$

here, $\boldsymbol{A}$ and $\boldsymbol{B}$ are the symmetric and antisymmetric parts of the velocity gradient tensor, respectively, and $\| \|_F$ represents the Frobenius norm. The vortex structure is defined by $Q > 0$, and the vortex structure strength is indicated by the magnitude of Q. Usually, a threshold $Q_{\text{threshold}}$ must be used to define the region with $Q > Q_{\text{threshold}}$ as a vortex in practice.

27.3.3 Liutex-Criterion

The Liutex citerion is proposed by Liu et al. [19], and has been shown advantages in identifying the vortex structures in turbulence flows. The Liutex vector is defined as the rotational part of the vorticity, which can clearly represent both the direction and magnitude of the rotational motion. The Liutex vector can be described as

$$\boldsymbol{R} = \boldsymbol{\beta} - \boldsymbol{\alpha} \text{ if } \alpha^2 - \beta^2 < 0, \beta > 0 \tag{27.11}$$

$$\boldsymbol{R} = \boldsymbol{\beta} + \boldsymbol{\alpha}, \text{ if } \alpha^2 - \beta^2 < 0, \beta < 0 \tag{27.12}$$

$$R = 0 \text{ if } \alpha^2 - \beta^2 \geq 0 \tag{27.13}$$

here,

$$\alpha = \frac{1}{2}\sqrt{\left(\frac{\partial V}{\partial y} - \frac{\partial U}{\partial x}\right)^2 + \left(\frac{\partial V}{\partial y} + \frac{\partial U}{\partial x}\right)^2} \tag{27.14}$$

$$\beta = \frac{1}{2}\left(\frac{\partial V}{\partial x} - \frac{\partial U}{\partial y}\right) \tag{27.15}$$

where, $\boldsymbol{U}$, $\boldsymbol{V}$ are in the XYZ-frame, and there exists a corresponding transformation between ∇V and ∇v,

$$\nabla V = \tilde{Q} \nabla y \tilde{Q}^{-1} \tag{27.16}$$

$$\tilde{Q}^{-1} = \tilde{Q}^T \tag{27.17}$$

here, $\tilde{Q}$ is the rotation matrix.

27.4 Coherent Structures Across the Attached Sheet Cavity Interface

In this section, the flow structures in attached sheet cavity are examined, as shown in Fig. 27.3. Vortex structures are mainly located inside the cavity, and different vortex identification methods including vorticity, *Q*-criterion, and Liutex can predict the vortex structures inside cavitation region. However, different vortex identification methods show different performances in terms of local flow structures, i.e., near the cavity interface and at cavity wakes. Generally, vorticity overpredicts the vortex structures near foil surface while underpredicts the vortex structures near cavity interface as presented in Fig. 27.3b. *Q*-criterion is better than vorticity at the cavity interface, while at cavity wakes it overpredicts the vortex structures as well as shown in Fig. 27.3c. Compared with vorticity method and *Q*-criterion, the Liutex method can predict the vortex structures near cavity interface and at cavity wakes well as seen in Fig. 27.3d.

To further illustrate the flow structures in the acoustic shear layer near cavity interface, contours of flow structures on foil mid-plane are presented in Fig. 27.4 corresponding to those in Fig. 27.3. We find that the acoustic shear layer consists of two regions, namely turbulence dominated region near cavity and acoustics dominated region near liquid as illustrated in Fig. 27.4. It is worth noting that all the vortex identification methods (vorticity, *Q*-criterion, and Liutex) can predict the turbulence dominated region near cavity interface, while all fails to predict the flow structures

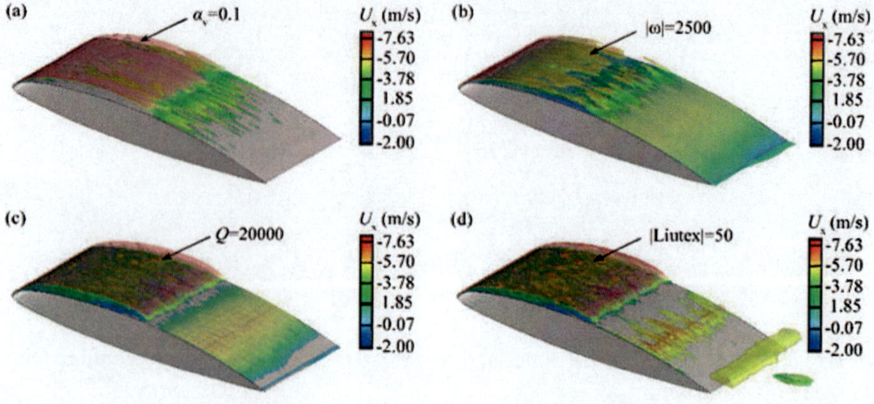

Fig. 27.3 Typical 3D flow structures in the process of cavity growth identified by **a** vapor fraction, **b** vorticity, **c** Q-criterion, and **d** Liutex. The isofurfaces are colored by streamwise velocity u_x

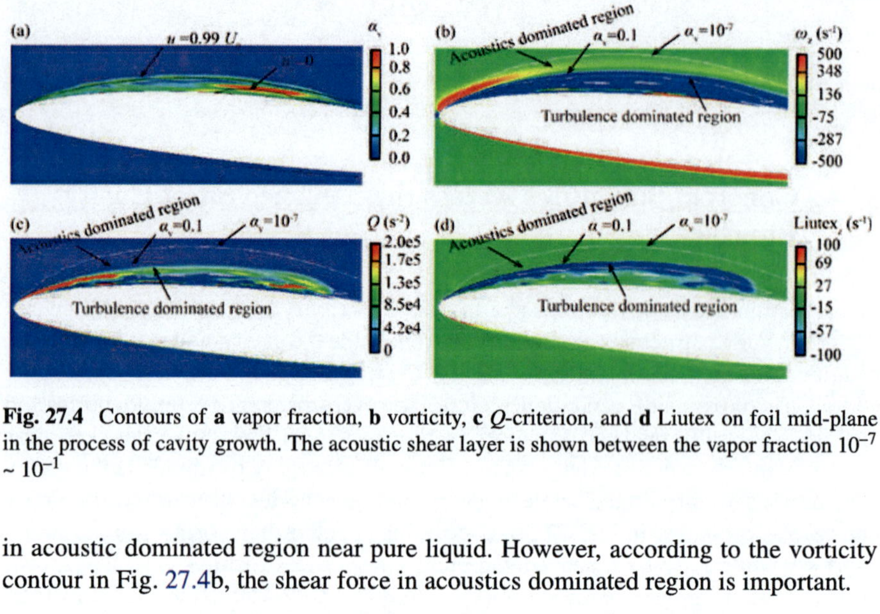

Fig. 27.4 Contours of **a** vapor fraction, **b** vorticity, **c** Q-criterion, and **d** Liutex on foil mid-plane in the process of cavity growth. The acoustic shear layer is shown between the vapor fraction 10^{-7} ~ 10^{-1}

in acoustic dominated region near pure liquid. However, according to the vorticity contour in Fig. 27.4b, the shear force in acoustics dominated region is important.

27.5 Coherent Structures Across the Shedding Cloud Cavity Interface

Figure 27.5 shows the flow structures in the process of cavity cloud shedding identified by vapor fraction, vorticity, Q-criterion, and Liutex, respectively. Different from that in the process of cavity growth controlled by the near wall flow structures, flow

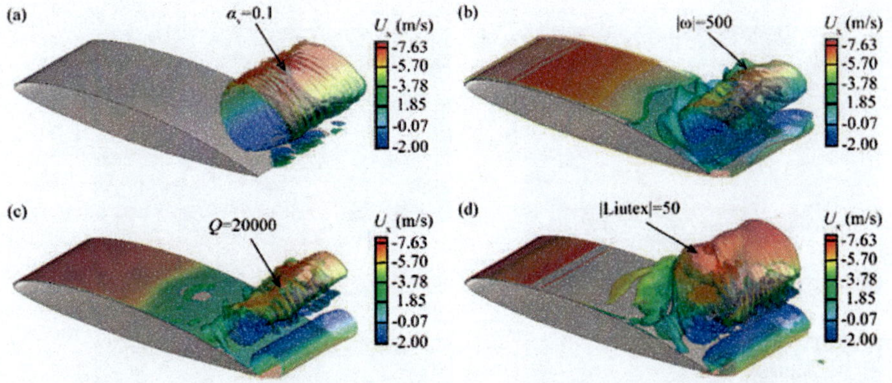

Fig. 27.5 Typical flow structures in the process of cavity cloud shedding identified by **a** vapor fraction, **b** vorticity, **c** *Q*-criterion, and **d** Liutex. The isofurfaces are colored by streamwise velocity u_x

structures in the shedding process are mainly controlled by free shear turbulence. Vortex structures are mainly located near the foil leading edge, and around the shedding cavity cloud. Different vortex identification methods show different prediction performances in these two regions. Both the vorticity and *Q*-criterion overpredicts the vortex structures near foil leading edge as also seen during the cavity growth stage in Fig. 27.3. However, compared with the vorticity and *Q*-criterion, Liutex has the advantages of predicting the vortex structures around shedding cloud as seen in Fig. 27.5d.

Figure 27.6 presents the contours of flow structures on foil mid-plane corresponding to those in Fig. 27.5. The vortex structures inside cavity cloud can be predicted by vorticity, *Q*-criterion and Liutex well. Vorticity overpredicts the vortex structures near foil trailing edge. Compared with vorticity, *Q*-criterion and Liutex predict the flow structures in acoustic shear layer better especially in the region between shedding cloud and foil surface. Moreover, Liutex can predict vortex structures outside the cavitation region, showing advantages in identifying the coherent structures in free shear flows.

27.6 Discussion and Conclusions

Coherent structures across the cavity interface (i.e., acoustics shear layer) are studied using different vortex identification methods, including vorticity, *Q*-criterion, and Liutex. The data set used is based on the compressible numerical simulation results of cloud cavitation around a NACA66 (mod) hydrofoil. The main findings are as follows:

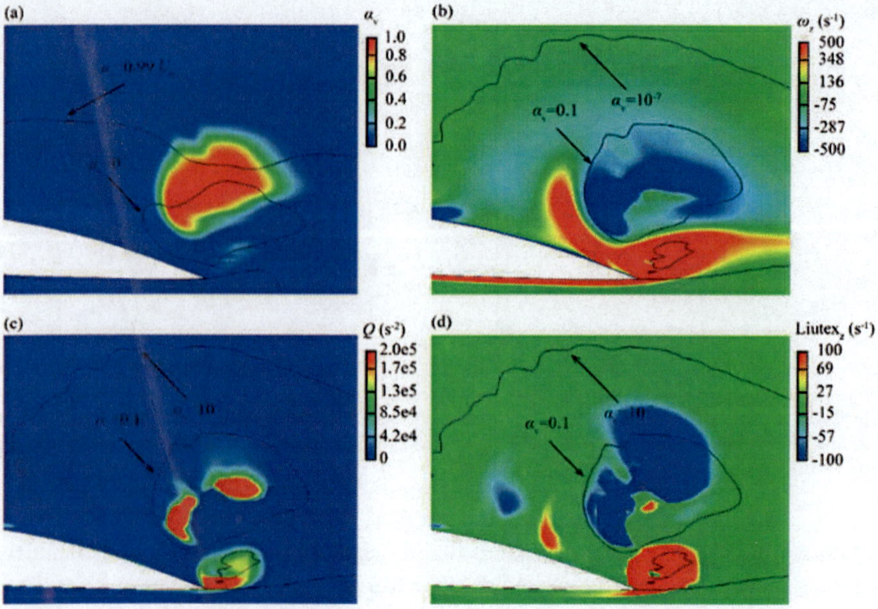

Fig. 27.6 Contours of **a** vapor fraction, **b** vorticity, **c** Q-criterion, and **d** Liutex on foil mid-plane in the process of cavity cloud shedding. The acoustic shear layer is shown between the vapor fraction $10^{-7} \sim 10^{-1}$

(1) Coherent structures across the cavity interface, i.e., acoustics shear layer, mainly consist of turbulence dominated region and acoustics dominated region. The current vortex identification methods can predict the flow structures inside the turbulence dominated region, while fail inside the acoustics dominated region.
(2) Compared with vorticity and Q-criterion, Liutex method has the advantages to predict the flow structures inside cavitation region of both attached sheet cavity and shedding cloud cavity. Vorticity usually overpredicts the vortex structures near foil surface and at cavity wakes.
(3) Coherent structures in acoustic shear layer especially in the acoustics dominated region requires developing new vortex identification method to implement the influence of acoustics characteristics. Different from the current vortex identification methods which are mainly based on velocity fields, the new method should take into account the acoustics characteristics (i.e., sonic speed, Mach number).

Acknowledgements Financial support of the China Scholarship Council (CSC, Grant No.: 201906030144) is appreciated.

References

1. C.E. Brennen, *Cavitation and Bubble Dynamics* (Oxford University Press, Oxford, 1995)
2. C. Wang, G. Wang, B. Huang, Dynamics of unsteady compressible cavitating flows associated with the cavity shedding. Ocean Eng. **209**, 107025 (2020)
3. Y. Kawanami, H. Kato, H. Yamaguchi et al., Mechanism and control of cloud cavitation. J. Fluids Eng. **119**, 788–794 (1997)
4. M. Callenaere, J.P. Franc, J.M. Michel et al., The cavitation induced by the development of a re-entrant jet. J. Fluid Mech. **444**, 223–256 (2001)
5. R.E.A. Arndt, C.C.S. Song, M. Kjeldsen, et al., Instability of partial cavitation: a numerical/experimental approach, in *Symposium on Naval Hydrodynamics, France* (2001)
6. J.B. Leroux, O. Coutier-Delgosha, J.A. Astolfi, A joint experimental and numerical study of mechanisms associated to instability of partial cavitation on two-dimensional hydrofoil. Phys. Fluids **17**, 052101 (2005)
7. J.B. Leroux, J.A. Astolfi, J.Y. Billard, An experimental study of unsteady partial cavitation. J. Fluids Eng. **126**(1), 94–101 (2004)
8. H. Ganesh, S.A. Mäkiharju, S.L. Ceccio, Bubbly shock propagation as a mechanism for sheet-to-cloud transition of partial cavities. J. Fluid Mech. **802**, 37–78 (2016)
9. C. Wang, G. Wang, B. Huang, Characteristics and dynamics of compressible cavitating flows with special emphasis on compressibility effects. Int. J. Multiph. Flow **130**, 103357 (2020)
10. C. Wang, Q. Wu, B. Huang et al., Numerical investigation of cavitation vortex dynamics in unsteady cavitating flow with shock wave propagation. Ocean Eng. **156**, 424–434 (2018)
11. J.U. Brackbill, D.B. Kothe, C. Zemach, A continuum method for modeling surface tension. J. Comput. Phys. **100**(2), 335–354 (1992)
12. R.H. Core, *Underwater Explosion* (Princeton University Press, Princeton, 1948)
13. Y. Saito, R. Takami, I. Nakamori et al., Numerical analysis of unsteady behavior of cloud cavitation around a NACA0015 foil. Comput. Mech. **40**, 85–96 (2007)
14. Y. Egorov, F.R. Menter, Development and application of SST-SAS Turbulence model in the DESIBER project, in *Second Symposium on Hybrid RANS-LES Methods* (2007)
15. H.G. Weller, A new approach to VOF-based interface capturing methods for incompressible and compressible flow. OpenCFD: 2008, Technoque report TR/HGW/04
16. P. Sagaut, *Large Eddy Simulation for Incompressible Flows* (Springer, 2002)
17. B. Ji, X. Luo, R.E.A. Arndt, X. Peng, Y. Wu, Large eddy simulation and theoretical investigations of the transient cavitating vertical flow structure around a NACA66 hydrofoil. Int. J. Multiphas. Flow **68**, 121–134 (2015)
18. J. Hunt, A. Wray, P. Moin, Eddies, streams and convergence zones in turbulent flows. Report CTR-S88, Center for Turbulence Research (1988)
19. C. Liu, Y. Gao, X. Dong et al., Third generation of vortex identification methods: Omega and Liutex/Rortex based systems. J. Hydrodyn. **31**(2), 205–223 (2019)